terminal eocene events

FURTHER TITLES IN THIS SERIES

1. A.J. Boucot
EVOLUTION AND EXTINCTION RATE CONTROLS

2. W.A. Berggren and J.A. van Couvering
THE LATE NEOGENE – BIOSTRATIGRAPHY, GEOCHRONOLOGY AND PALEOCLIMATOLOGY OF THE LAST 15 MILLION YEARS IN MARINE AND CONTINENTAL SEQUENCES

3. L.J. Salop
PRECAMBRIAN OF THE NORTHERN HEMISPHERE

4. J.L. Wray
CALCAREOUS ALGAE

5. A. Hallam (Editor)
PATTERNS OF EVOLUTION, AS ILLUSTRATED BY THE FOSSIL RECORD

6. F.M. Swain (Editor)
STRATIGRAPHIC MICROPALEONTOLOGY OF ATLANTIC BASIN AND BORDERLANDS

7. W.C. Mahaney (Editor)
QUATERNARY DATING METHODS

8. D. Janóssy
PLEISTOCENE VERTEBRATE FAUNAS OF HUNGARY

Developments in Palaeontology and Stratigraphy, 9

terminal eocene events

Edited by

Ch. Pomerol

University of Paris, Paris, France

and

I. Premoli-Silva

University of Milan, Milan, Italy

ELSEVIER
Amsterdam – Oxford – New York – Tokyo 1986

ELSEVIER SCIENCE PUBLISHERS B.V.
Sara Burgerhartstraat 25
P.O. Box 211, 1000 AE Amsterdam, The Netherlands

Distribution for the United States and Canada:

ELSEVIER SCIENCE PUBLISHING COMPANY INC.
52, Vanderbilt Avenue
New York, N.Y. 10017, U.S.A.

Library of Congress Cataloging-in-Publication Data

Terminal Eocene events.

(Developments in palaeontology and stratigraphy ; 9)
"The main results achieved by Project no. 174 on 'Geological Events at the Eocene-Oligocene Boundary' of the International Geological Correlation Program (IGCP), sponsored by UNESCO during its five year duration are presented"--P.
1. Geology, Stratigraphic--Eocene. 2. Geology, Stratigraphic--Oligocene. 3. Paleontology--Eocene. 4. Paleontology--Oligocene. I. Pomerol, Charles. II. Premoli-Silva, I. III. International Geological Correlation Programme. Project no. 174 on "Geological Events at the Eocene-Oligocene Boundary." IV. Series.
QE692.2.T47 1986 551.7'84 86-4497
ISBN 0-444-42623-X (U.S.)

ISBN 0-444-42623-X (Vol. 9)
ISBN 0-444-41142-9 (Series)

Printed in The Netherlands.

PREFACE

In the present book entitled "Terminal Eocene Events" the main results achieved by Project n° 174 on "Geological Events at the Eocene-Oligocene boundary" of the International Geological Correlation Program (IGCP), sponsored by UNESCO during its five year duration are presented.

The aim of the Project was on one side an exhaustive inventory and description of the most complete sedimentary sequences across the Eocene-Oligocene boundary from all over the world, and on the other side a synthesis of the biotic and chemico-physical events detected at the Eocene-Oligocene transition. These two aspects of the research are treated in the second and third parts of the volume, respectively.

The principal motivations of this research may be summarized as follows : the Eocene-Oligocene boundary appeared as one of the most important breaks within the Cenozoic, even more important than the Paleogene-Neogene boundary. Moreover, continuous sequences at the Eocene-Oligocene transition were rarely described in the literature from terrestrial and from deep-sea deposits. Particularly, the deep-sea sediments, which were expected to contain a complete record, appeared to be frequently affected by strong dissolution and/or reworking in this interval.

Despite such a bias, apparently continuous sequences through the Eocene-Oligocene boundary were recovered at a few places,mainly from the open marine realm but also,to a minor extent, in neritic facies. A synthesis of the successions investigated from the various locations,with their sedimentary, biological, geochemical, and physical characters is shown in Table 1. Among these, the sequences which appear to be useful for reconstructing the succession of events at the Eocene-Oligocene boundary are from central Italy (Gubbio, Visso), southern Spain (Fuente Caldera), Barbados (Bath Cliff), Armenia (Landzhar), and New Zealand (Cape Foulwind) in the pelagic realm; Alabama (St Stephen Quarry), Australia (Port Willunga), and Romania (Brebi) in neritic facies; and finally,northern Italy (Priabona) and Borneo (Melinau) in shallow water carbonate platform facies. No complete sequences straddling the Eocene-Oligocene boundary are recorded from South America, while the African sequences exposed in Tanzania (Lindi area), Libya, Tunisia, and Morocco need still to be investigated in detail.

Among the deep-sea sites (see Table 2) the most important for the purpose of this Project are Sites 516, 522 and 523 in the South Atlantic, Site 549 A and Eureka 67-128 in the North Atlantic, Sites 277, 292 and 592 in the South and Equatorial Pacific.

The sections from Alabama, Barbados, Umbria (Italy), and southern Spain have been visited by several members participating to this Project during three field trips in 1982 (USA and Barbados) and in 1984 (Europe). The reports of those trips were included in eleven Newsletters (TEE News) available from the Project leader along with reports on other meetings and general activities of the various working groups participating in the Project.

"The Eocene-Oligocene transition : events and boundary" co-authored by Ch. Pomerol and I. Premoli Silva and treated in the first part of this volume represents a synthesis of all the data collected by the numerous participants in the Project implemented with data derived from the recent literature. After five years of studies, taking into account all the investigated aspects (sedimentology, biostratigraphy, biogeography, chemiostratigraphy, magnetostratigraphy, paleooceanography and tectonics) , it results that : 1) the Eocene-Oligocene boundary considered as a major evolutionary break of the Cenozoic is characterized by a gradual, and not abrupt, environmental change which started in the Middle Eocene and continued into the earliest Oligocene for some 300,000 years.2) There is not a single terminal Eocene event, but a series of events concentrated close to the Eocene-Oligocene boundary as testified by an acceleration of the rate of overturn among most of the taxonomic groups, by an intensification of the climatic deterioration and by a major reorganization of the oceanic water masses.

The final meeting of the Project was held in Paris on March 28-29, 1985 and was attended by 50 participants from ten countries. Our warm thanks are extended to all the scientists who contributed with their work and/or support to the success of the project, to the IGCP Board and to the French IGCP committee for their financial contribution.

Ch. POMEROL and I. PREMOLI-SILVA

Manuscript delivered to the publisher on December 16, 1985.

Addresses of authors missing on their paper are given at the end of the volume.

TABLE 1

Selected land sections at the Eocene-Oligocene boundary in pelagic (P) and neritic (N) environment.

	ENVIRONMENT	LARGER FORAMINIFERA	BENTHIC FORAMINIFERA	PLANKTONIC FORAMINIFERA	CALCAREOUS NANNOFOSSILS	OSTRACODA	POLLEN AND DINOFLAGELLATES	RADIOLARIA	OXYGEN 18	CARBON 13	TRACE ELEMENTS	VOLCANICS	PALEOMAGNETISM	TEKTITES	IRIDIUM	RADIOMETRIC DATING
Gubbio Contessa Quarry (Italy)	P		X	X	X				X	X	X	X	X			X
Gubbio Contessa Highway (Italy)	P		X	X	X							X	X			
North Visso (Italy)	P	X	X	X	X	X	X		X			X	X			
Fuente Caldera (Spain)	P	X	X	X	X				X	X	X					
Bath Cliff (Barbados)	P		X	X	X			X	X	X		X		X	X	X
Cape Foulwind (New Zealand)	P		X	X	X				X							
Landzhar (S. Armenia, USSR)	P		X	X	X		X									
Nicapo (Mexico)	P			X												
Lindi (Tanzanie)	P	X		X												
Brebi (Romania)	N/P	X	X	X	X	X			X							
Kuban River (N. Caucasus, USSR)	N/P		X	X	X	X										
Priabona (Italy)	N	X	X		X	X						X				
Port Willunga (Australia)	N		X	X			X									
St Stephen Quarry (Alabama, USA)	N		X	X	X	X										
Thrace (Turkey)	N	X	X									X				
Momijiyama (Japan)	N		X	X	X											
Melinau (Borneo)	N	X	X													

TABLE 2
DSDP SITES
AND PISTON CORES

	INDIAN OCEAN			PACIFIC OCEAN				ATLANTIC OCEAN						
	219	216	242	277	592	167	292	529	516	548 A	549 A	Eureka 67/128	522/523	366
PLANKTONIC FORAMINIFERA		X	X	X	X	X	X	X	X	X	X	X	X	X
BENTHIC FORAMINIFERA				X	X	X	X	X	X	X	X	X	X	X
CALCAREOUS NANNOFOSSILS		X	X	X	X	X	X	X	X	X	X	X	X	X
OSTRACODA									X					
RADIOLARIA						X	X							
DIATOMS	X					X	X							
$CaCO_3$ DISSOLUTION		X						X	X				X	X
POOR RECOVERY		X	X	X		X	X						X	X
STABLE ISOTOPES	X				X		X	X			X		X	
MICROTEKTITES						X	X							
IRIDIUM							X							
VOLCANICS					X		X							
PALEOMAGNETISM											X		X	

CONTENTS

PART I

THE EOCENE–OLIGOCENE TRANSITION: EVENTS AND BOUNDARY

THE EOCENE-OLIGOCENE TRANSITION : EVENTS AND BOUNDARY

Charles POMEROL[1] and Isabella PREMOLI SILVA[2]

1) Lab. de Géologie des Bassins Sédimentaires et UA/CNRS n° 319, Univ. Paris VI, 4, place Jussieu, 75252 - Paris Cedex 05 - (France)

2) Univ. degli Studi di Milano, Dipartimento di Scienze della Terra, Via Mangiagalli 34, 20133 - MILANO - (Italy).

BIOSTRATIGRAPHIC EVENTS

Planktonic Foraminifera

Several extinctions and originations among the planktonic Foraminifera occurred during the Late Eocene, with earlier Eocene species and/or genera gradually replaced by new forms which would continue into the Oligocene. Some of those events have proven to be stratigraphically useful and can be used for world-wide correlation. The most important and widespread events are (from older to younger):

1) the LO of Morozovella, Acarinina (including Truncorotaloides), and Planorotalites;
2) the FO of Globigerinatheka semiinvoluta;
3) the FO of Turborotalia cocoaensis;
4) the LO of G. semiinvoluta and Turborotalia pomeroli;
5) the LO of Globigerinatheka index associated with a marked decrease in abundance of all Globigerinatheka;
6) the LO of the Turborotalia cerroazulensis group (including T. cunialensis);
7) the LO of Hantkenina, Cribrohantkenina, large-sized Pseudohastigerina micra, and P. danvillensis;
8) the FO of "Globigerina" tapuriensis, associated with an increase in abundance of Turborotalia ampliapertura, followed by the FO of "Globigerina" presaepis.

Events 1) and 7) have been used to define the lower and upper boundaries of the Late Eocene, respectively. Event 1) also defines the base of Zone P15, whereas Event 7) is suggested by Nocchi et al. (this volume) to define the top of Zone P17 and is corroborated by the co-occurrence of primitive "Globigerina" tapuriensis with the last Eocene taxa. It is worth mentioning that in his original definition Blow (1969 and 1979) dated the entirety of Zone P17 as latest Eocene and not as Late Eocene to Early Oligocene as later reported by several authors (see also Berggren et al., 1985). This misinterpretation probably occurred because of the disappearance of the hantkeninids in the middle of

Zone P17 in the Lindi sections observed by Blow (1969). It has now been demonstrated that the T. cerroazulensis group(≐Globorotalia centralis of Blow 1969) disappear prior to the LO of the hantkeninids, which latter taxa, however, may be missing because of their rarity and/or dissolution removal.

The identification of Zone P16 presents several problems. This zone is defined by the total range of Cribrohantkenina inflata. This taxon, however, is missing in several continuous sections, or when present, is rare particularly in the lower portions of its range. As mentioned by Blow (1969) the LO of G. semiinvoluta occurs slightly after the FO of C. inflata. So, in the absence of the nominate taxon, the LO of G.semiinvoluta would allow placement of the base of Zone P16 just below it and, according to the data from Spanish sections (Molina et al., this volume), after the FO of Isthmolithus recurvus (calcareous nannofossil).

Concerning the upper boundary, the FO of Subbotina gortanii cannot be used as a substitute for the LO of C. inflata because S. gortanii is strongly latitudinally controlled (Boersma and Premoli Silva, this volume) and at middle latitudes evolves from Subbotina praeturritilina much earlier within Zone P16 (see also Keller, 1985). Data from Spanish sections suggest that the LO of G.index and the strong decrease in abundance of Globigerinatheka (Event 5) approximate the Zone P16 upper boundary better than the FO of S. gortanii.

Based on planktonic foraminiferal evolution, the Eocene-Oligocene boundary should be placed after the extinction level of the last Eocene taxa(=Event 7). Event 6 (=LO of T. cerroazulensis gr.) then occurs prior to the Eocene-Oligocene boundary as defined above. The age-difference between Events 6 and 7 ranges from about 0.3 m.y. in Barbados, to 0.13 m.y. in Umbria and to less than 0.1 m.y. in southern Spain (see Saunders et al., 1985; Molina et al., this volume; Nocchi et al., this volume). The older age, estimated for Event 6 in Barbados, is probably the result of poor sediment preservation. Consequently, the choice of the LO of T.cerroazulensis gr. to identify the Eocene-Oligocene boundary, particularly in the presence of even slightly dissolved sediments, is incorrect.

Event 5 deserves some comments. It appears that Globigerinatheka index became extinct at high latitudes in the South Pacific much later than in the middle latitudes or than in the Atlantic Ocean. Jenkins and Hoskins and Morgans (both this volume) equated the Eocene-Oligocene boundary to the LO of G. index. On the other hand, Lindsay and Mc Gowran (this volume) stated that G.index in southern Australia disappears much earlier than the Eocene-Oligocene boundary, which according to them is better and more easily identified by the

LO of Globigerina linaperta. The latter taxon was also used to define the Eocene-Oligocene boundary in the Tasman Sea by Jenkins and Srinivasan (in Kennett et al., 1985) (see also Keller, this volume). Discrepancies in the vertical range of G. index can be attributed to the paleooceanographic conditions of this critical area close to Antarctica (see Murphy and Kennett, in press). Unfortunately, no independent way exists to check how these events in the austral region correlate with the record from lower latitudes of either the Pacific or the Atlantic. Currently, no paleomagnetic stratigraphy is available at any of the locations studied from that region.

Benthic Foraminifera

No major change in small benthic Foraminifera is associated with the planktonic foraminiferal extinctions at the Eocene-Oligocene boundary. Small deep-water benthic faunas undergo gradual turnover from the Middle Eocene up to the Early Oligocene. Corliss et al. (1985) stated that the major faunal change in deep benthic Foraminifera during the Middle to Late Eocene is associated with a migration to greater depths of the earlier Eocene abyssal taxa, whose available niches were colonized by new abyssal, long-ranging, cosmopolitan forms through the Late Eocene.

Several extinctions and originations occurred from Late Eocene through Early Oligocene. Nuttallides truempyi, Alabamina dissonata, Clinapertina ssp., Abyssammina ssp., and Aragonia ssp. are known to disappear before the Eocene/Oligocene boundary, balanced by the later appearances of new species of the genera Uvigerina, Pullenia, Laticarinina, and Spaeroidina. Boersma (this volume) pointed out that most of these evolutionary events occur close to or at the oxygen isotope excursion which, via paleomagnetic calibration, is earliest Oligocene in age.

By contrast the larger Foraminifera, which inhabit shallow water carbonate platforms, underwent mass extinctions at the Eocene-Oligocene boundary, which seems to coincide with that based on planktonic Foraminifera. As shown by Adams (this volume), very few Late Eocene species continued into the Oligocene and Early Oligocene species originated predominantly above the boundary. So, Late Eocene faunas have little in common with those of the Early Oligocene from either the American or Indo-Pacific (including the Mediterranean) bioprovince.

Adams (this volume) also reported that in carbonate facies disconformities, unconformities and marked lithologic changes which terminate Eocene limestones are common features at the Eocene-Oligocene boundary. In the few places where apparently continuous sequences were deposited across the boundary, such as Melinau (Borneo) and Priabona (Italy), Late Eocene limestones containing larger

Foraminifera pass upwards into algal-rich limestones lacking age-diagnostic Foraminifera at the base, but within which the typical Oligocene faunas gradually appear (see also Barbin and Bignot, this volume). Although the incompleteness of the sedimentary sequences across the Eocene-Oligocene boundary may emphasize the drastic change in larger foraminiferal faunas, the apparently complete records from Melinau and Priabona demonstrate that no Oligocene species originates during the Late Eocene. Because a similar change occurred at the Middle to Late Eocene boundary, it appears that shallow water larger Foraminifera behave differently than the other groups and undergo true mass extinctions at these times. For example, during the Late Eocene planktonic foraminiferal faunas are progressively enriched in new Oligocene species which replace the older Eocene representatives (see Boersma and Premoli Silva, this volume); such changes are much less rapid than among the larger Foraminifera.

Correlation between carbonate sequences and open marine successions remains uncertain because of the lack of age-diagnostic planktonic organisms in most carbonate facies. Paleomagnetic measurements of three sections from the Priabona area failed to provide reliable data. All magnetic signals obtained were very vitiated attributable to strong leaching of sediments in that area (Roggenthen, personal communication, 1984). Using the most accepted correlations from the literature (Cavelier and Pomerol, 1986), it seems that the major extinction events among larger Foraminifera correlate with the extinction levels of the last warm water-preferring and/or specific oxygen-nutrient related water-preferring species in any given planktonic foraminiferal fauna. The Middle to Late Eocene overturn in larger foraminiferal faunas can be correlated with the extinction level of the last warm Morozovella and Acarinina. Larger foraminiferal mass extinctions at the Eocene-Oligocene boundary seem to coincide with the extinction of hantkeninids, cribrohantkeninids, large pseudohastigerininids, and the very last Globigerinatheka.

Radiolarians

A high degree of biotic overturn among radiolarian faunas punctuate the entire period from the latest Middle Eocene into the earliest Oligocene resulting in a high stratigraphic resolution. As well documented in Barbados (Mueller-Merz and Saunders, this volume), in the older part of this interval up to the middle of the Late Eocene Calocyclas bandyca Zone originations largely exceeded extinctions (by 7 to 2). Later in the Late Eocene close to the Calocyclas bandyca/Cryptoprora ornata zonal boundary radiolarians underwent somewhat mass extinctions. There in a very short interval ten species became extinct and no originations of new species were recorded. Finally, originations and extinctions were equal in the interval from just after the LO of hantkeninids to the positive oxygen isotope shift in the earliest Oligocene. As clearly shown in Barbados (Mueller-Merz and Saunders, this volume), few biostratigraphic events among radiolarian faunas occurred at the same level as those either among planktonic Foraminifera or calcareous nannofossils. In most of the cases, radiolarian events alternated with calcareous plankton events throughout the measured sequence. If none of the extinctions among radiolarians appear to occur within the microtektite-bearing layer, five radiolarian species simultaneously became extinct within an iridium-rich level in the earliest C. ornata Zone of the Late Eocene (Sanfilippo et al., 1985).

Ostracodes

Ostracode faunas exhibit only minor phylogenetic changes at the Eocene-Oligocene transition and the majority of them do not coincide with the LO of hantkeninids. This statement applies to either shallow or deep marine environments. However, according to van den Bold (this volume), Haplocytheridea montgomerensis became extinct at the Eocene-Oligocene boundary in shelf areas of the US Gulf Coast associated with the almost total disappearance of the genus Haplocytheridea. In other shallow areas, such as the Aquitaine basin, ostracode faunal renewal in the Early Oligocene is thought to be related to local environmental changes due to a decrease in temperature and an increase in humidity (lower salinity) (Ducasse and Peypouquet, this volume).

In the deep-sea environment, a moderate overturn starting in the late Middle Eocene accelerated its rate close to the end of the Eocene when at lower bathyal depths "psychrospheric" ostracodes replaced the older, deeper thermospheric faunas (Benson and Peypouquet, 1983). The arrival of the new forms is expected to be diachronous in different basins when cold waters began to form because of the presence of thresholds in Southern Atlantic. It appears that the break-down of the Rio Grande Rise through the Vema Channel allowed the cold bottom water from Antarctic region to flow into the northern basins (Benson and Peypouquet, 1983). Then, by the end of the Eocene, "psychrospheric" faunas were widespread in the entire Atlantic and entered also the Mediterranean Tethys (see Benson in Nocchi et al., this volume).

Calcareous Nannofossils

There are few events, either extinctions or originations, among calcareous nannofossils from Middle Eocene through Early Oligocene and none that corresponds at the Eocene-Oligocene boundary based on planktonic Foraminifera (Perch-Nielsen, this volume). Some changes, however, occur within the calcareous nannofossil assemblages, which through the Late Eocene loose the warm-water rosette-shaped discoasters and became enriched in cool-water indicators, such as Isthmolithus recurvus, even at low latitudes. By the end of the Eocene the nannoflora exhibits a much cooler aspect than at the beginning of the Late Eocene. Because the forms mentioned above are latitudinally controlled, their appearances and/or disappearances are time-transgressive through latitude (Cavelier, 1979; Corliss et al., 1985).

The FO of I. recurvus and the LO of the rosette-shaped Discoaster barbadiensis and D. saipanensis define biostratigraphically the beginning of Zone CP 15b (= Zone NP 19 of Martini, 1971) and of Zone CP 16a (= Zone NP 21), respectively. Because of their stratigraphic importance, it is important to try to estimate the difference in age of a single event from higher to lower latitudes relative to the Late Eocene planktonic foraminiferal events and to the magnetic sequence.

In the Contessa Highway section Discoaster barbadiensis and D. saipanensis disappear contemporaneously, about 0.78 m.y. before the hantkeninid extinction. In Barbados both rosette-shaped discoasters disappear about 0.46 m.y. before the last Eocene planktonic Foraminifera. At Site 522 in the South Atlantic, only D. barbadiensis disappears at a level coeval with that in the Contessa Highway section, whereas the LO of D. saipanensis occurs later, about 0.5 m.y. prior to the Eocene-Oligocene boundary (Backman, pers.comm., 1985). The LO of rosette-shaped discoasters in the Contessa Highway section lies within magnetic Chron 13r and corresponds to the III normal event of Nocchi et al. (this volume). No correlation of this event to the magnetic sequence is available at any of the other locations to test the estimated ages reported above. Whatever the estimated age of the LO of D. barbadiensis and D. saipanensis one uses, it is clear that both taxa disappear well before the extinction of the hantkeninids; therefore, this or these event(s) cannot be utilized for identifying the Eocene-Oligocene boundary.

To estimate the age-difference through latitude of the FO of I. recurvus is somewhat more difficult because of the lack of precise tie-points to either the planktonic foraminiferal succession or to the magnetic sequence. I. recurvus seems to occur in the Contessa Highway section at a level about 2.0 m.y. prior to the hantkeninid extinction level and at the top of Chron 16n, whereas recent investigations by Backman (pers.comm., 1985) suggest that at Site 523 in the South Atlantic I. recurvus is recorded since Chron 16r, dating the appearance of this taxon about 2.8 m.y. older than the Eocene-Oligocene boundary. Through its range I. recurvus also exhibits useful fluctuations in abundance. Specifically, a shift in abundance from rare to more common seems to occur just after the LO of rosette-shaped discoasters in the

Contessa Highway section. Monechi (in press) demonstrated that the abundance increase of I. recurvus occurs at several sites in the Atlantic as well as in southern Spain and appears to be coeval with that in the Contessa section. The duration of this so-called acme is approximately 1 m.y., and it straddles the Eocene-Oligocene boundary.

Diatoms

Evolutionary speaking, planktonic diatoms changed very little at the Eocene-Oligocene transition. Few are, in fact, the species that originated or became extinct either in the Late Eocene or in the Early Oligocene. However, diatom floras markedly changed across the Eocene-Oligocene boundary. Firstly, the abundance of diatoms per g of sediment decreased about one order of magnitude from the latest Eocene to the Early Oligocene. Secondly, species characteristic of the Late Eocene decreased strongly in abundance close to the Eocene-Oligocene boundary and they became a minor component of the Early Oligocene assemblages. Finally, specific diversity among diatom floras increased substantially through the latest Eocene and reached its maximum in correspondence with the positive $\delta 18_O$ shift in the earliest Oligocene. The increase in diversity, however, is largely related to an increased abundance of neritic species. At DSDP Site 366 on Sierra Leone Rise associated with the latter event, there was an increase either in occurrence or in abundance of fresh water diatoms and phytoliths, both of terrestrial origin. According to Fenner (this volume) the occurrence of terrestrial forms must be related to an increase in wind intensity and possibly aridity on the nearby African continent by that time.

Because the turnover among diatoms occurred world-wide including high and low latitudinal belts (Corliss et al., 1984), changes in specific diversity, abundance and assemblage composition of diatom floras through the latest Eocene-Early Oligocene interval appear to be controlled by the intensity of the surface water currents and by the availability of nutrients, particularly at lower latitudes, rather than by surface water temperature.

Dinoflagellates

Biostratigraphic resolution based on dynocysts is very poor close to the Eocene-Oligocene boundary being biased by absence in several localities (i.e. Paris Basin, southern England, most of the Italian sections), when present by the occurrence of atypical, commonly monospecific, assemblages (southern England), and, finally, by several sedimentary gaps in the sequences. According to Chateauneuf (this volume) the most reliable biostratigraphic data reside in the fact that 1) several species which appeared earlier in the Eocene disappeared during the Late Eocene, 2) Wetzeliella gochtii seems to appear only in the Early Oligocene, and finally, 3) Early Oligocene assemblages are poorly diversified in comparison with the high diversity of the Eocene dynocyst assemblages.

Moderately rich dinoflagellate assemblages have been recovered from the topmost part of the Visso North section (Umbria, Italy). According to Biffi (in Nocchi et al., this volume), four dinoflagellate species characteristic of the Eocene disappeared at the same level as the hantkeninids. W. gochtii was not found higher in the sequence.

Charophytes

Detailed investigations carried out on charophytes from northwestern Europe (see Riveline, this volume) demonstrated that this group underwent a gradual change through the Late Eocene-Oligocene transition. Forty-seven charophyte species are recorded in the Late Eocene, but only sixteen of them still occur in the Early Oligocene. The species which became extinct are those which appeared in late Middle or in early Late Eocene. Several new species originated gradually, replacing the extinct forms, and which would continue into the Oligocene. Although the degree of overturn among the charophytes could not be exactly quantified, it appears that the stratigraphic resolution which can be reached with this group is one order higher than any other in the continental realm, being very close to that obtained with marin plankton. According to Riveline (this volume), the Eocene-Oligocene boundary should be placed between the Stephanochara vectensis (latest Eocene) and the Stephanochara pinguis (earliest Oligocene) Zones.

Vegetation

Generally speaking, biostratigraphic events among land vegetations have a significance, if not local, at most on a regional scale. Since the distribution of vegetation is strongly latitudinally and longitudinally controlled even during times of equable climate, the evolutionary patterns from one area cannot be applied to another area even if the latter belongs to a similar latitudinal belt , but to a different continent. The specific composition of the floras at the two localities may be different. Moreover, the floral record through geologic time is rarely continuous and, if present, was rarely exhaustively studied.

Floral changes at the Eocene-Oligocene transition are known from few areas, i.e. the North American continent (Wolfe, 1978 cum lit.), northwestern Europe (Chateauneuf, this volume), and to a lesser degree from Asian USSR (Krasheninnikov et al., Akhmetiev et al., both this volume).

A common feature in the floral evolution at the Eocene-Oligocene boundary from so distant and latitudinally widely ranging regions is that the vegetation markedly changed worldwide as a consequence of a climatic cooling event, one of the most important of the entire Tertiary. At the end of the Eocene, within a geologically short period of time, mid and high latitude areas in North America that had been occupied by broad-leaved evergreen forest became occupied by temperature broad-leaved deciduous forest. A decline in mean annual temperature of about 12-13°C at 60° latitude to about 10-11°C at 45° latitude was estimated for northwestern America associated with a strong increase in the temperature annual gradient (mean annual range of from 3-5°C in the Middle Eocene to at least 21°C in the Early Oligocene) (Wolfe, 1978).

In western Europe, the Eocene-Oligocene boundary is characterized by the disappearance of a large number of thermophile taxa through the Late Eocene replaced at the end of the Eocene by a coniferous and herbaceous flora, indicating a drier and cooler climate. In western Europe, the FO of Boehlensipollis hohli seems to be a reliable biostratigraphic event. This taxon appeared just after the notable increase in abundance of conifers in the Paris basin and in the southern Rhine graben as well as after the Nummulitic Limestones in southern France.

In southeastern Kazakhstan the apparently continuous sequence from the Lake Zaisan Basin comprises several levels which yielded rich floras associated and/or interbedded with vertebrate-bearing layers. There, close to the Eocene termination the subtropical "Poltava" floral elements were gradually replaced by the temperate "Turgai" floral elements. This change is interpreted as a signal of a generalized climatic cooling associated with a decrease to almost a deficit of moisture (Akhmetiev et al., this volume).

The gradual climatic deterioration registered by the vegetation at the Eocene-Oligocene transition could be calibrated to the calcareous plankton evolutionary sequence in south Armenia (Krasheninnikov et al., this volume). There, in the apparently continuous Landzhar section, rich spore and pollen assemblages were recovered from the same layers as the calcareous plankton. Three continental microfloral assemblages close to the Eocene-Oligocene boundary were identified : 1) the older assemblage, from the Late Eocene P15 to middle P 16 zonal interval (planktonic Foraminifera), characterized by 8.5 to 16 % of subtropical spores, by less than 50 % of Pinus, 20 to 50 % of Cedrus, up to 20 % of Ephedra, and up to 12 % of Taxodiaceae among the gymnosperms, and by 5 to 24 % of angiosperm pollens. Subtropical forms present throughout this interval are Quercus, Castanopsis, Palmae, Nissa, Rhus, Magnoliaceae and Myrica. 2) a second assemblage, from the upper part of the Late Eocene Zone P 16 to the lower part of the Early Oligocene Zone P18, characterized by very few angiosperm pollens (2%) and a large amount of Pinus (70 to

75 %) whereas the other gymnosperm species decreased in abundance. By the end of the latest Eocene Zone P17 Pinus percentages markedly decreased to 55 %.
3) later in the Early Oligocene (late Zone P18) palyno-assemblages contained a well diversified and common angiosperm pollens with Fagus, oak, and chestnut (up to 24 %) and a high total gymnosperm content (70-90%) in which, however, Pinus had lost its previous importance (down to 30-55%).

Based on this floral succession, it appears that the cooling episode started in the middle of the Late Eocene Zone P16 and reached its maximum at the end of Zone P17 beginning of Zone P18 in the very late Eocene-earliest Oligocene.

VERTEBRATES

Russell and Tobien in their review for this volume stated that North American mammal faunas underwent high rates of change over a period of at least 10 m.y. which included the Eocene-Oligocene boundary. Among the European mammal faunas, originations exceeded extinctions from the Middle to Late Eocene through the Early Oligocene interval except during the Late Eocene when the trend was reversed. The rate of originations, however, markedly increased in the Early Oligocene (almost twice that in earlier times), so Stehlin's "Grande Coupure" appears to concern more the appearance of new taxa than the disappearance of those already in place ; similar evolutionary patterns at the Eocene-Oligocene transition are recorded among the European amphibians and reptiles (see Rage, this volume), except for the turtles, which seem to cross the Eocene-Oligocene boundary without any apparent change. Asian mammal faunas exhibited major changes in the Middle Eocene and to a less extent in the Late Eocene, whereas the Eocene-Oligocene boundary is marked by a much lower turnover in which originations and extinctions were almost equal in number. Nevertheless, Dashzeveg and Devyatkin (this volume) suggested that the appearance of the genera Entelodon, Bothriodon and Brachyodus characterized the beginning of the Oligocene in central Asia.

GEOCHEMICAL EVENTS

This chapter deals with stable isotopic events ($\delta 18_O$ and $\delta 13_C$) and changes in trace elements (Sr or Mg).

Stable isotope events

The isotopic analyses are mostly based on monospecific samples of benthic or planktonic Foraminifera. In a few cases (site 522, Contessa Quarry of Gubbio) magnetostratigraphy provides a precise control for the isotopic events.

The $\delta 18_O$ isotopic event

Since the late Early Eocene $\delta 18_O$ values became gradually more positive to about 1,5 ‰ culminating in a sharp positive shift of $\delta 18_O$ close to the

E/O boundary ; this was observed at several DSDP Sites in all the oceans and at Barbados (Saunders et al., 1984). Specifically this sharp positive shift is calibrated to magnetostratigraphy at Site 522 (South Atlantic) where it straddles the Chron 13 r - Chron 13 n boundary. Biostratigraphically the sharp positive oxygen shift occurs after the extinction level of the hantkeninids and other Eocene planktonic Foraminifera. The same succession of events was observed at Barbados and the Fuente Caldera Section (South Spain) (Berthenet 1985), whereas the oxygen isotope shift was not evidenced at Gubbio (Contessa section) (Renard et al. this volume). At Barbados (Bath Cliff section, Saunders et al., 1984) the $\delta 18_O$ shift occurs 3-4 m above the last occurrence of <u>Hantkenina</u>, which means some 200,000 to 300,000 years later than the foraminiferal event, according to the estimated accumulation rate at this locality above the E/O boundary. The same order of duration was also calculated at the other mentioned localities. The radiometric dating is approximately 35.6 Ma (Montanari et al., this volume) and the duration of the shift is approximatively 100,000 years.

The amplitude of the $\delta 18_O$ shift recorded by benthic Foraminifera is about 1 ‰ . Based on planktonic Foraminifera the $\delta 18_O$ shift decreases when passing from high latitudes (0,8 ‰) to middle latitudes (0,6 %) and to low latitudes (0.4 to 0 ‰).

The deep bottom water drop in temperature was of 2° to 5° C, worldwide. Surface water cooling occurs in high and mid latitudes, but we can not exclude that low latitudes experienced a temperature increase. As a consequence, the latitudinal temperature gradient may have increased considerably from Eocene to Oligocene time (Vergnaud - Grazzini and Oberhänsli, this volume).

The most widely accepted interpretation of this major cooling event is the isolation of Antarctica and the formation of cold bottom water from the Antarctic continent (Shackleton and Kennett 1975). The drop in temperature is related to sea ice formation or perhaps to continental glaciation. The restriction of the east-west Tethyan sea-way in the Middle East, hypothezed by Ricou et al. (this volume), may also have pushed the system towards a cooler climate.

<u>The $\delta 13_C$ isotopic event</u>

The Eocene-Oligocene transition corresponds to a positive culmination of the trend of $\delta 13_C$ curve (Renard et al.; Vergnaud - Grazzini et al., both this volume). $\delta 13_C$ increases at 1 ‰ from Early Eocene to Late Eocene and it decreases at 0·75 ‰ during the Oligocene. Close to the Eocene-Oligocene boundary we observe a decrease of $\delta 13_C$ planktonic values of 1,5 ‰ and the $\delta 13_C$ gradient among the planktonic Foraminifera strongly decreased to 2 about one fourth of the gradient previously registered up to Zone P 16. Coeval with the $\delta 18_O$ shift the carbon isotope values of benthic Foraminifera approached the carbon isotope planktonic values (Boersma and Premoli-Silva, this volume; Saunders et al., 1984). The very low inter-specific gradient among planktonic Foraminifera at the termination of the Eocene and continuing through the Oligocene may represent a decrease in oceanic productivity associated with a major reorganization of the water masses (Boersma, this volume).

Trace Elements events

Chemical analysis of bulk pelagic carbonates from oceanic sites and land sections allowed the reconstruction of the fluctuations in abundance through time of some trace elements. Strontium and magnesium content have been measured at the Eocene-Oligocene boundary.

The strontium event

According to Renard et al. (this volume) long-term fluctuations of Sr are related to variations in oceanic ridge hydrothermal activity. During high activity periods Ca input in the ocean increases and Sr/Ca ratio of sea water decreases. Short term variations related to transgression-regression cycles are superimposed. During transgression Ca and Sr are trapped in platform carbonates, but because oceanic residence of Sr is 6 times greater than that of Ca, the Sr/Ca ratio increases. Short term variations were used to establish a chemiostratigraphy (see correlation with biostratigraphy in Cavelier and Pomerol 1986).

The low Sr values of the Early-Middle Eocene (400 to 900 ppm) correspond to a period of high hydrothermal activity of the oceanic ridges. To explain the high values (1000-1300 ppm) close the Eocene-Oligocene boundary Renard thinks that "hydrothermal activity and sea-floor spreading was considerably slakened at the end of Middle Eocene and increases again at the end of the Late Eocene". In the Contessa Quarry section a rapid excursion (negative shift followed by a positive one) occurs at the top of Chron 13 n. The four small negative shifts observed at the Eocene-Oligocene transition during a period of general trend of increasing Sr/Ca ratios are due to small regressive episodes.

The magnesium event

In the Contessa section an important negative shift (700 ppm) occurs in the early Late Eocene. It is followed by a gradual increase of Mg up to 1750 ppm close to the top of Chron 13 n. Such an increase in Mg content coïncides with the Sr excursion considered above. It is worth mentioning that, since the

Cenomanian, Sr and Mg concentrations fluctuate in an opposite way. By Late Eocene time, however, Sr and Mg contents begin to fluctuate in a similar way. According to Renard (1984) most of these fluctuations are controlled by the variations in the hydrothermal activity of the ocean ridges.

GEOCHRONOLOGY

During the last decade numerous dates have been proposed for the Eocene-Oligocene boundary, in a range from 32 to 38 Ma. The more recent investigations slightly reduced the range from 33 to 37 Ma. According to Odin (1982) "an age of 33^{+2}_{-1} Ma seems appropriate. An age older than 35 Ma can definitely be eliminated." After Montanari (this volume) the unaltered volcanic biotite from the Contessa Section near Gubbio provides an interpolated age of the Eocene-Oligocene boundary of 35.7 $\pm$ 0.4 My calibrated to the biostratigraphy and magnetostratigraphy.

The age of the microtektite layer of Bath Cliff Section (Barbados) was calculated as 35.5 $\pm$ 0.4 Ma (Glass, this volume). This level is situated 26 m below the LO of hantkeninids. Based on a sediment accumulation rate of 27 m/My estimated by Saunders et al., 1984, for the Bath Cliff section the age of the E/O boundary and of the Early Oligocene positive shift of $\delta^{18}O$ would be 34.5 $\pm$ 0.4 Ma and 34.2 Ma respectively.

At the end of his review Berggren (this volume) points out that values between 35-37 Ma would appear to more acceptable for the E/O boundary. He notes that the potential sources of error are multiple and suggests an integrated approach, i.e. magneto-bio-radiochronology, like in the Umbria area (Italy).

Another result of the investigations of Montanari et al. is an unexpectedly young age of 36.4 $\pm$ 0.3 Ma for the early Late Eocene Zone 15 from the Contessa Quarry (top of the Chron 17 n in the lower part of the Priabonian stage). This date makes the duration of the Priabonian about 1 Ma instead of 3,5 Ma as predicted by other time scales. This considerable short duration of the Late Eocene would imply a strong acceleration of the sea floor spreading rate as suggested by Renard and challenged by Berggren. If one accepts such a younger age for the Zone P 15 the sediments at the Contessa Quarry section during the Late Eocene would have been accumulated at a rate three times higher than in the other portions of the Scablia Formations. Scaglia accumulation rate throughout the whole Tertiary is about 6m My.

In conclusion, our feeling is that the age of the Eocene-Oligocene boundary is probably bracketed between 34 and 36 Ma (see also Cavelier and Pomerol 1977). The dates provided by fission-track and K-Ar in microtektites (34.5 Ma) and K-Ar and Rb-Sr on biotites (35.7 Ma) are within the suggested range with a time lag of 1.2 Ma between the two measurements.

MAGNETOSTRATIGRAPHIC EVENTS

Three important results have recently been obtained in this field. Firstly, precise correlation has been established between magnetic lineations and biostratigraphy at the Eocene-Oligocene transition. There are three localities where this correlation could be demonstrated : at DSDP Sites 522 and 523 (pars) in the South Atlantic (Tauxe et al., 1984), in the Valle della Contessa (Gubbio) in the Contessa Quarry and Contessa Highway sections (Lowrie et al., 1982; Nocchi et al., this volume), and in the Visso North section (Cascia area) (Nocchi et al., this volume). In the latter locality the magnetostratigraphic record is limited to a short interval and is somewhat poor (Nocchi et al., this volume).

Secondly, as pointed out by Lowrie (this volume), at Site 522 and in the Contessa Highway section the well identified Chron 13r contains three discrete almost equally spaced normal events. It also contains the Eocene-Oligocene boundary as defined by planktonic Foraminifera, thus confirming the observations of Lowrie et al. (1982), which boundary falls just above the youngest normal event in the upper part of Chron 13r (=Event 1 of Nocchi et al., op. cit.) in both the Contessa Highway section and Site 522. And finally, the oxygen isotope shift, related by several authors to a major cooling episode (Matthews and Poore, 1980; Shackleton et al., 1984; Oberhänsli et al., 1984), occurs within the earliest part of Chron 13n . Therefore, it must be dated as earliest Oligocene in age and does not coincide with the Eocene-Oligocene boundary based on planktonic Foraminifera (Oberhänsly and Toumarkine, 1985).

SEDIMENTOLOGIC EVENTS

The main sedimentologic events at the Eocene-Oligocene boundary are :

1) the increase in calcium carbonate accumulation,
2) the deepening of the CCD,
3) the scarcity of siliceous sediments,
4) a progressive increase of illite, chlorite, and in some place kaolinite, and a concomitant decrease of smectites and fibrous clay,
5) the development of hiatuses and sediment reworking.

It appears they are the consequence of :

- an acceleration of deep sea cold, oxygenated water currents,
- a global cooling,
- modifications in surface water productivity,
- a regression well documented in Europe, Australia, Africa, Gulf Coast ..., but of a minor magnitude.

In a general way the Late Eocene was a time characterized by a low accumulation rate of carbonates, whereas the Early Oligocene was one of higher accumulation. Close to the Eocene-Oligocene boundary, the continuous sequences exhibit a marked carbonate dissolution facies. According to Thunnel and Corliss (this volume) these changes are related to fluctuation in productivity, which would also apply to the different locations. The trend towards an increasing of the amount of carbonate sediments at depth continued through the Oligocene.

The dramatic deepening of the CCD at the Eocene-Oligocene transition has been demonstrated by van Andel (1975) (see Thunnell and Corliss, fig. 6, this volume). The decline is abrupt in the Pacific (4200 m) (mainly equatorial Pacific 5000 m) and more gradual, towards 4200 m in the Atlantic and Indian Oceans. An explanation of the CCD fluctuations has been developed by Berger and Winterer (1974) and Sclater et al., 1979. According to their model, during transgressions the CCD is relatively shallow because carbonate deposition is concentrated on the shelves. Alternately, during regressions the CCD is depressed and the deep oceans become the major site of carbonate sedimentation. On the other hand the development of sea ice around Antarctica increases the turnover of bottom waters and provides more oxygenated water, with relatively lower concentration of CO_2 and therefore lower acidity which favoured the preservation of carbonates in the deeper basins (Berger 1973).

The siliceous sediments are less represented in the Late Eocene than in the Middle Eocene when chert layers (= horizon A) were widespread also in the North Atlantic. Poverty of siliceous sediments was related to a decrease of primary productivity.

According to Chamley (1985) the clay mineral trend at the Eocene-Oligocene transition is characterized by a progressive increase of illite, chlorite, irregular mixed-layers and sometimes kaolinite and a concomitant decrease of smectite, fibrous clay and often kaolinite. The cause would be the Cenozoic worldwide cooling favoring on land the physical alteration process instead of the chemical ones.

The number of sedimentary hiatuses increase around the Eocene-Oligocene boundary both on shelf and in ocean basins. Continuous sequences across the Eocene-Oligocene boundary are rarely recorded either in a neritic environment (except in some places of Australia, Alabama, Borneo, Italy) or in the pelagic environment (some DSDP sites and in Italy, Spain, Barbados, New-Zealand).

Hiatuses on the shelves may be related to low stand of sea level and/or tectonics. In the oceanic basin we have seen that deep water currents became more vigorous at the termination of the Eocene. The acceleration of the deep sea circulation caused submarine erosion and therefore widespread hiatuses. The number of hiatuses and their duration depended upon their geographic location. For example, in the South Atlantic Ocean they lasted longer in the western part because of the deviation towards the west of the Antarctic bottom current (fig. 1).

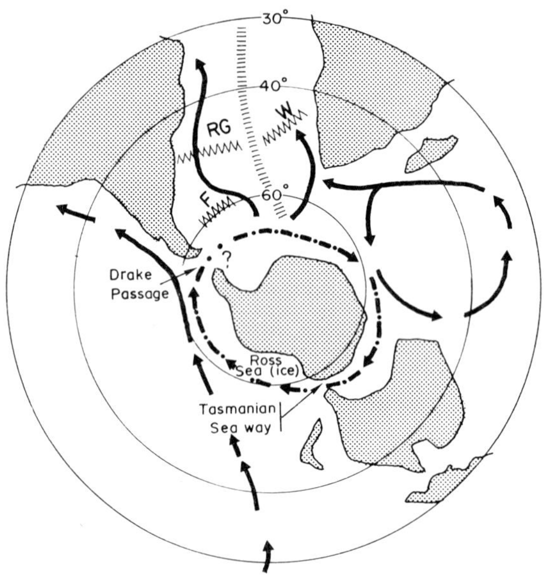

Fig. 1. Tentative reconstruction of surface (dashed line) and bottom (solid line) circulation of the Southern Ocean during the Eocene-Oligocene transition (after Haq, 1980 and Kennett, 1982, modified) F. Falkland Plateau; R.G. Rio Grande Rise; W. Walvis Ridge.

Thiede et al., 1981, recorded the occurrence of reworked and displaced fossils in the Mesozoic and Cenozoic of the Central Pacific Ocean. The maximum age difference between oldest reworked fossil components and the host sediment increased from the Maastrichtian to the Eocene-Oligocene boundary. Suddenly, at that time, the age difference reached its maximum of 70 Ma. Such a high age difference appears to be coeval with the maximum distribution of the erosional hiatuses and resulted from the acceleration of the oceanic bottom circulation.

TECTONIC AND VOLCANIC EVENTS

Three major tectonic events occurred at the Eocene-Oligocene transition: the opening of the Tasmanian sea-way, a strong compressive Alpine phase, and the restriction of the Tethys. During the Mesozoic through the Middle Eocene Australia and Antarctica were connected; relatively warm waters were then reaching high latitudes in the southern hemisphere. By Late Eocene, the Tasmanian seaway opened and at the same time a possible surface connection was created between Atlantic and Pacific Ocean through the Drake passage (fig. 1). It resulted in a thermal isolation of Antartica which lead to the development of a circum-Antarctic current. Consequently cool surface waters from high latitudes of the Southern Indian Ocean would have been transported around Antarctica and may have triggered sea-ice formation and production of cold bottom waters in the world Ocean (Kennett 1982). The sea-ice and perhaps the ice cap on Antarctica increased the albedo and contributed to accelerate the drop of temperatures.

The Triassic break up of the Pangea had created an East-West seaway which encircled the earth at low latitudes (Tethyan Ocean). During the Late Eocene, the collision between Africa and Eurasia in the Middle East cut off the deep latitudinal circulation at least in the depths between the Indian and the Atlantic Ocean. This fact contributed to the climatic deterioration of the earth.

Moreover, it was at that time that the barrier erected by the Rio Grande Rise and Walvis Ridge system broke down through the Vema Channel and the Faroë-Shetland Channel in the Northern Atlantic. Consequently, the deep cold waters that originated at high latitudes could flow and reached the deep basins at lower latitudes. They are also thought to be responsible for the frequent hiatuses and increased reworking of the pelagic deposits.

Close to the Eocene-Oligocene boundary also occurs one of the paroxysmal orogenic compressive alpine phase (Pyrenean or Illyrian orogenic phase) which have resulted from collision between Africa-India and Eurasian plates. This orogeny explains the frequent and widespread hiatuses and disconformities in neritic environment between Late Eocene and Oligocene formations. As a possible consequence of this orogenic phase, one would expect an increase of the emerged areas to contribute to the acceleration of the climatic deterioration (increase of altitudes and albedo).

At the same time a pulse of volcanic activity is observed in New Zealand, Australia, Japan, Philippines, Peru, ... which resulted from the development of an Australian/Pacific plate boundary and from the greatest Late Phanerozoic change in Pacific plate motion, associated with widespread tectonism throughout the Pacific region.

A number of pale green volcanic laminae are recorded at site 592 and 593 (Tasmian Sea region) during the Late Eocene-Early Oligocene. They indicate intensified volcanism in the New Zealand region (Kennett et al., 1985).

In the Umbrian sections and particularly in the Gubbio area several layers of marly limestone contain bentonite layers which provided abundant euhedral biotite flakes used by Montanari et al. (this volume) for radiometric dating. According to these authors the biotite and the volcanic quartz and feldspar associated with it have been transported by wind from the Alps into the Umbrian basin.

COSMIC EVENTS

Several cosmic events have been suggested for explaining the Eocene-Oligocene transition such as a giant meteoritic impact followed by a darkening due to atmospheric dust (HSÜ, 1980) or to the formation of a ring comparable to those of Saturn encircling the earth and resulting in the screening of sunlight (O'Keefe 1980).

Although the cause of the Eocene-Oligocene cosmic events is unknown the occurrence of a large strewn field of microtektites in North America (microtektite strewn field) (Gulf of Mexico, Caribbean Sea, Barbados and in the equatorial western Atlantic) speaks in favour of an extraterrestrial input (Glass, this volume). The stratigraphic position of the microtektite bearing layer is difficult to precise. It is close to the top of the *Globigerinatheka semiinvoluta* Zone (P 15) or in the *Turborotalia cerroazulensis* Zone (P 16). According to Keller (this volume) three microtektite horizons are present in Late Eocene sediments - one in the Upper *G. semiinvoluta* Zone, and two closely spaced layers in the lower part of *T. cerroazulensis* Zone. This assumption is challenged by Glass (this volume). For the latter author there is only one microtektite layer. "Later occurences or horizons, appear to be due to reworking from the North American microtektite layer". As mentioned above the Bath Cliff microtektite layer, situated 26 m below the LO of hantkenines is probably 1 My older than the Eocene-Oligocene boundary.

Clinopyroxene-bearing spherules and iridium anomaly

Clinopyroxene-bearing spherules (cpx spherules) were found associated with the North American microtektites or just below the microtektite layer (25 cm) in core RC9-58 from the Caribbean Sea. Cpx spherules are also found across the equatorial Pacific and in the eastern Indian Ocean (total mass estimated at 10 g metric tons)(fig. 2). An iridium anomaly is associated with the cpx spherule layer which provides further support for the impact origin of these spherules.

In Barbados the cpx spherule layer has not been found, but there, an iridium anomaly occurs 27 cm below the microtektite layer and it is about 10,000 years older. The extinction of several species of radiolarians occurs in the

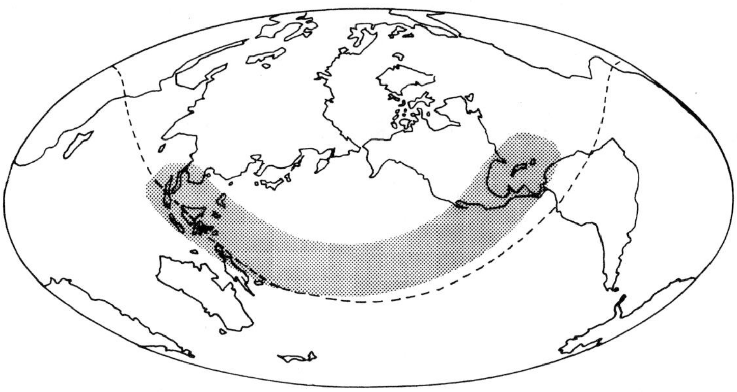

Fig. 2. The microtektites and the clinopyroxene-bearing spherule strewn fields (the latter about 10000 years older) occur 1 My before the Eocene-Oligocene boundary, from the Caribbean Sea (including to Barbados and Georgia to the eastern Indian Ocean (after Glass, 1979, modified).

cpx spherule layer and/or in the associated iridium anomaly layer (Glass and Zwart, 1977).

CONCLUSIONS

The Events

On the basis of the review presented in the previous pages and on the papers included in the book implemented by recently published works, one can make the following conclusions :

1) the Eocene-Oligocene boundary is not characterized by a single event either catastrophic or other in origin;

2) most of the numerous and important changes that characterized the Eocene-Oligocene transition have their origin as far back as the late Middle Eocene or the early Late Eocene;

3) such a change occurred gradually, but its rate accelerated approaching the termination of the Late Eocene;

4) a generalized event was a marked cooling registered in all realms which culminated in a drop of temperature on the order of 5°C some 300,000 y. later than the extinction level among the Eocene planktonic Foraminifera. It is testified by a positive shift of $\delta^{18}O$ in the oceans (fig. 3).

5) the vertical and latitudinal thermal gradients gradually decreased during the Late Eocene-Early Oligocene interval and resulted in a reduced upwelling rate in the oceans (Lipps and Mitchell, 1976);

6) a more rapid decrease of the $\delta^{13}C$ vertical gradient started just prior to the hantkeninid extinction level and culminated in an overall gradient reduced to about one-half that previously known in the late Eocene, in correspondence

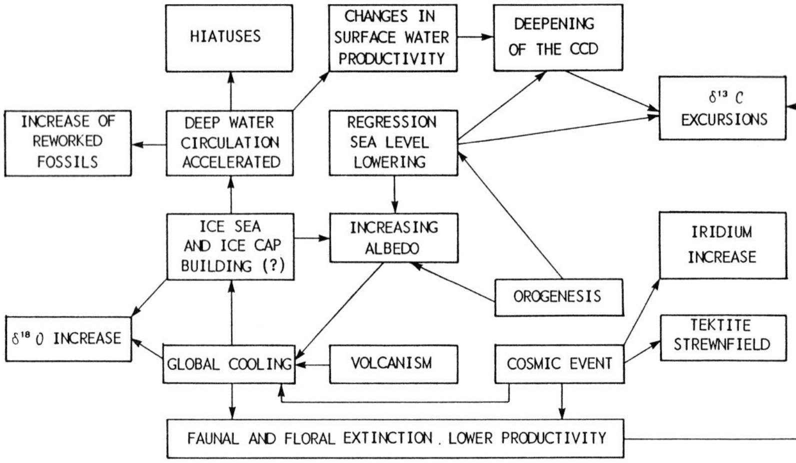

Fig. 3. Proposed relationship between cosmic, tectonic, volcanic, sedimentological, geochemical, faunal, floral, oceanographic and climatic events at the Eocene-Oligocene transition (after Cavelier et al., 1980, completed and modified). It is evident that all these events are not coeval. For example the global cooling is initiated at the Late Middle Eocene and the $\delta 18_O$ increase occurs in the earliest Oligocene (see text and fig. 4).

to the $\delta 18_O$ shift of the earliest Oligocene.

7) a gradual decrease of the fertility in the oceans (including the carbonate platform environment) affected by the end of the Eocene all the trophic levels (Lipps and Mitchell, 1976; Lipps, 1986; Boersma and Premoli Silva, this volume);

8) In relation to the changes previously mentioned, a major reorganization of the watermasses occurred. Ocean waters cooled from the bottom to the surface; near-surface water homogenized as testified by the low oxygen isotope gradient between near-surface and deeper dwelling planktonic Foraminifera.The near-surface at the Eocene-Oligocene transition contained a low productivity and only moderately oxygenated zone overlying an essentially thermally uniform, well mixed, low nutrient and low oxygen subsurface. The deeper oxygen minimum and subsurface oxygen maximum were eliminated.

9) Major tectonic events contributed to create and/or emphasize some of the environmental changes close to the end of the Eocene.

- The separation of Antarctica from Australia following the opening of the Tasmian Sea associated with a possible surface current through the Drake Passage caused cooler conditions in the Antarctic region.
- The break-down of the Rio Grande Rise through the Vema Channel allowed the

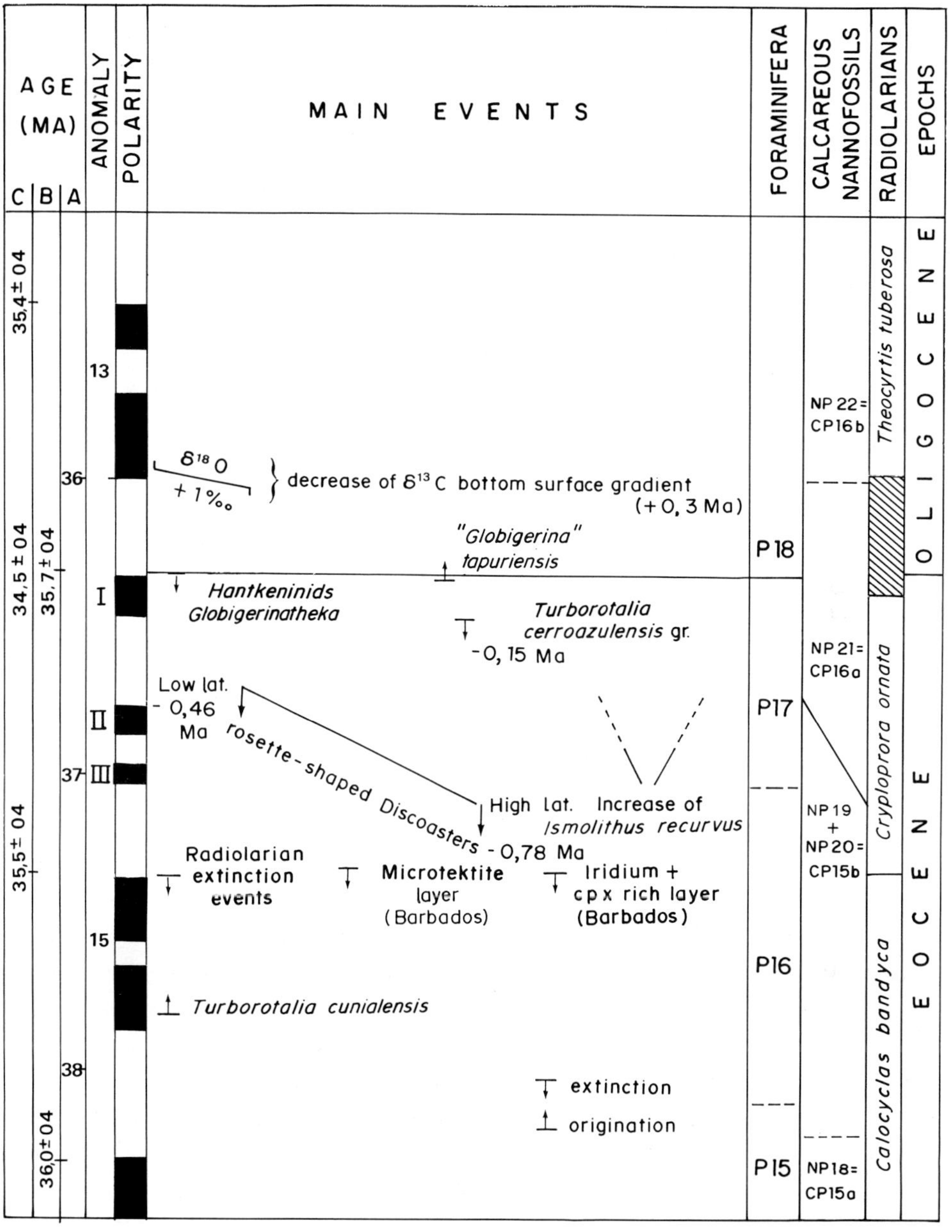

Fig. 4. Significant stratigraphic events at the Eocene-Oligocene boundary plotted against the magnetic sequence. Time scale after La Brecque et al., 1977 (A). Radiometric ages after Montanari et al. (B) and Glass (C), both this volume.

deep cool waters from the Antarctic region to flow north into the North Atlantic basin, increasing erosion at the bottom and thus hiatuses and/or reworking.

- The restriction of the Tethys sea way in the Middle East as a consequence of the African-Asian collision interrupted at least at depth the important current system flowing circumtropically since the Mesozoic, causing a major reorganization of the oceanic circulation.
- Finally, the major Alpine tectonic phase which occurred prior to the Eocene-Oligocene boundary possibly increased the land/ocean ratio, then the increased albedo resulted in increasing cooling.

10) Extraterrestrial events testified by the occurrence of microtektite-rich layers and/or iridium anomaly are not associated with any of the major changes, either biologic or chemicophysical, except in radiolarians among which several species disappear at the level of the iridium anomaly in Barbados. They may have helped in accelerating some processes, but they do not appear to be so essential.

11) The major cooling ($\delta 18_O$ shift) in the earliest Oligocene, possibly associated with a glaciation in Antarctica, seems to partially reverse the trend as testified by a deepening of the Carbonate Compensation Depth and higher carbonate accumulation in the deep sea basin, whereas the carbonate platform community started once again to diversify.

The Eocene-Oligocene boundary

The succession of the major biostratigraphic events used in the literature to place the Eocene-Oligocene boundary, plotted against the magnetostratigraphy and other chemico-physical events, is summarized in figure 4. It is worth noticing that the hantkeninids and associated forms are the last Eocene characteristic taxa to become extinct. The LO of the Turborotalia cerroazulensis group is from almost 0.1 My to 0.3 My older than the LO of hantkeninids, while the extinction level of the rosette-shaped discoasters precedes that of hantkeninids from 0.46 My at low latitudes to 0.78 My at higher latitudes. Moreover, the positive $\delta 18_O$ shift, corresponding to the major cooling event, is some 0.3 My younger than the hantkeninid extinction. Whatever biostratigraphic event among those mentioned above will be chosen for use in the future to define the Eocene-Oligocene boundary, they all occur within the magnetic Chron 13r.

According to the data presented in this volume, in our opinion, it appears that the LO of hantkeninids, of the last Globigerinatheka and Pseudohastigerina danvillensis (Event 7 among planktonic Foraminifera) is apparently the most reliable biostratigraphic event upon which to define the Eocene-Oligocene boundary. It falls just above the younger normal event in late Chron 13r (Nocchi et al., this volume), dated, according to Montanari et al. this volume, at 35.7 $\pm$ 0.4 Ma.

In terms of historical stages the Eocene-Oligocene boundary can be placed between the top of the Priabonian and the base of the Stampian, or of the Rupelian s.l. including the late Tongrian at its base.

It will be a duty of the International Subcommission on Paleogene Stratigraphy to decide which one of these events could be the most useful to define the Eocene-Oligocene boundary and to choose the stratotype section.

REFERENCES

Benson, R.H. and Peypouquet, J.P., 1983. The upper and mid-bathyal Cenozoic Ostracode faunas of the Rio Grande Rise found on leg 72 Deep Sea Drilling Project. In : Barker, P.F., Carlson, R.L., Johnson, D.A. et al., Init. Repts. DSDP, 72 : 805-818.

Berger, W.H., 1973. Cenozoic sedimentation in the eastern tropical Pacific. Geol. Soc. Am. Bull 84 : 1941-1954.

Berger, W.H., and Winterer (1974). Plate stratigraphy and the fluctuating carbonate line. In K.J. Hsü and H.C. Jenkins edit. Pelagic sediments on land and under the sea. Int. Assoc. Sedimentol., Spec. Publ., 1 : 11-48.

Berthenet, F., 1985. Evénements géochimiques à la limite Eocene-Oligocene. Application aux corps de la région de Grenade et de Biarritz. DEA offset, Univ. Paris VI : 30.

Cavelier, C., and Pomerol, Ch. ,1986 . Proposition d'une échelle standard pour le Paléogène. Newsl. Stratigr. 6 (1) : 56-65.

Cavelier, C., and Pomerol, Ch. ,1986 . Stages and stratigraphy of the Paleogene. Bull. Soc. Geol. France, VIII, 2, n° 2, in press.

Corliss, B.H., Aubry, M.P., Berggren, W.A., Fenner, J.M., Keigwin, L.D., jr , and Keller, G., 1984. The Eocene-Oligocene boundary event in the Deep Sea. Science, 226 : 806-810.

Glass, B.P., Swincki, M.B., and Zwart, P.A., 1979. Australian, Ivory-Coast and North America tektites strewnfields : size, mass and correlation with geomagnetic reversals and other Earth events in proceed tenth Lunar and Planet. Sci., Pergamon press, pp. 2535-2545.

Haq, B.U., 1984. Paleoceanography : a synoptic overview of 200 My. of Ocean History in Marine Géol. and oceanog. of Arabian sea and Coastal Pakistan. Haq and Milliman edit. Van Nostrand Reinhold comp., New-York, pp. 201-231.

Hsü , K.J., 1980. Terrestrial catastrophe caused by cometary impact at the end of Cretaceous. Nature, 285 (5762) : 201-203.

Kennett, J.P., 1982. Marine Geology, 1 vol. : 814. Prentice-Hall publisher.

Kennett, J.P., et al., 1985. Palaeotectonic implications of increased Late Eocene-Oligocene volcanism from South Pacific DSDP sites. Nature, 316 : 507-511.

Lipps, J.H., 1986. Extinction Dynamics in pelagic ecosystems. In :D.K.Elliott, Dynamics of Extinction. John Wiley and Sons, Inc., New-York, in press.

Lipps, J.H., and Mitchell, E., 1976. Trophic model for the adaptive radiations and extinctions of pelagic marine mammals. Paleobiology, 2 : 147-155.

O'Keefe, J.A., 1980. The terminal Eocene event : formation of a ring system around the Earth. Nature 285 (5763) : 310-311.

Saunders, J.B., Bernouilli, D., Müller-Merz, E., Oberhänsli, H., Perch-Nielsen , K., Riedel, W.R., Sanfilippo, A, and Torrini Jr, R., 1984. Stratigraphy of the Late Middle Eocene to Early Oligocene in the Bath Cliff Section, Barbados, West Indies. Micropaleontology, 30, 4 : 390-425.

Shackleton, N.J., and Kennett, J.P., 1975. Paleotemperature history of the Cenozoic and the initiation of Antarctic glaciation : oxygen and carbon isotope analyses in DSDP sites 277, 279 and 281. Init. Rep. DSDP, 29 : 743-755.

Sclater, J.G., Boyle, E., and Edmond, J.M., 1979. A quantitative analysis of some factors affecting carbonate sedimentation in the Ocean in Deep Sea drilling results in the Atl. Ocean : Continental margins and paleoenvironment. Am. Geophys. Union, pp. 235-248.

Setiawan, J.R., 1983. Foraminifera and microfacies of the type Priabonian. Utrecht Micropal. Bull., 29 : 1-161.

Thiede, J., Boersma A., Schmidt, D.R., and Vincent, E., 1981. Reworked fossils in Mesozoic and Cenozoic pelagic Central Pacific Ocean sediments, DSDP sites 463, 464, 465 and 466, leg 62. Init. Rep. of DSDP, vol. LXII : 495-512.

Thierstein, H., and Berger, W.H., 1978. Injection events in Earth history, Nature 276 : 461-464.

Van Andel, T., 1975. Mesozoic/Cenozoic calcite compensation depth and the global distribution of Calcareous sediments. Earth Planet. Sc. Lett., 26 : 187-194.

Verhallen, P.J.J.M., and Romein, A.J.T., 1983. Calcareous nannofossils from the Priabonian stratotype and correlations with the parastratotypes. Utrech Micropal. Bull., 29 : 163-173.

Wolfe, J.A., 1978. A paleobotanical interpretation of Tertiary climates in the Northern Hemisphere. American Scientist, 66 : 695-703.

PART II

INVESTIGATIONS ON EOCENE–OLIGOCENE SEQUENCES: EUROPE, ASIA, NEW-ZEALAND, AUSTRALIA, AFRICA, NORTH AND SOUTH AMERICA AND DEEP SEA SITES

THE EOCENE-OLIGOCENE BOUNDARY IN THE UMBRIAN PELAGIC SEQUENCES, ITALY

NOCCHI M., PARISI G., MONACO P.
Dipartimento di Scienze della Terra, Università di Perugia

MONECHI S., MADILE M., NAPOLEONE G., RIPEPE M.
Dipartimento di Scienze della Terra, Università di Firenze

ORLANDO M., PREMOLI SILVA I.
Dipartimento di Scienze della Terra, Università di Milano

BICE D.M.
Department of Geology and Geophysics, University of California, Berkeley

INTRODUCTION

Detailed biostratigraphic and paleomagnetic studies on Eocene-Oligocene pelagic sequences from Umbria (Central Italy) revealed that deposition across the Eocene-Oligocene boundary was continuous and undisturbed in that area. The aim of the present paper is to synthetize the data from 8 stratigraphic sequences spanning the interval from the Early to Middle Eocene boundary through the Early Oligocene with specific emphasis on the events at the Eocene-Oligocene boundary. The reader can refer to Nocchi et al. and to Premoli Silva et al. (both, in press) for a more complete description of each section studied.

Pelagic biomicritic sediments are widespread in the Umbrian region. They consist primarily of calcareous marls and marly limestones, formerly nanno-foraminiferal oozes, of the Scaglia Formation, which changes color from dominantly pink-red (in the Eocene) to gray (in the Oligocene). In the southern part of Umbria (Visso and Cascia area), calcareous turbidites rich in shallow-water skeletal debris are interbedded within the pelagic sequence; this characterizes the so-called "sud-umbrisches facies" of Renz (1936). The source area of the skeletal debris was identified as the Abruzzi carbonate platform located southeast of the studied sections (Baldanza et al., 1982).

Since Renz (1936), several authors studied in detail the Umbrian Scaglia Formation (i.e. Luterbacher and Premoli Silva, 1962, 1964; Luterbacher, 1964). Baumann and Roth (1969) and Baumann (1970) were the first to carry out detailed studies of the calcareous nannofossils and planktonic Foraminifera on the Eocene-Oligocene boundary. Their investigations revealed that sedimentation was continuous across this boundary in several sections from both northern and southern Umbria. More recently (starting in 1974), biostratigraphic studies at 1 cm intervals were made on typical Scaglia sequences in order to calibrate the Late Cretaceous through Paleogene magnetic reversal sequence near Gubbio (Alvarez et al., 1977; Lowrie et al., 1982; Napoleone et al., 1983). Among the Gubbio sections already investigated, only the Contessa Quarry section extended from the Eocene into the Oligocene, but several meters of section are missing at the Eocene-Oligocene boundary because of a minor fault.

In order to have a complete biostratigraphic and paleomagnetic record, the Contessa Highway section was sampled upwards into the north-western quarry - which mirrors the Contessa Quarry of Lowrie et al. (1982) - where the Eocene-

Oligocene boundary is undisturbed. Because of its completeness, the Contessa Highway section is here chosen as the reference section to which all the other sections are correlated (see Fig. 1).

Lithologically, the Scaglia Formation is very monotonous except for the major color changes mentioned above. Visual correlations between sections are limited to minor changes which are not easily detected. Some of these changes, however, appear to be of larger (not local) significance. They are, from bottom to top:

1) The presence of chert in the Early and early Middle Eocene interval, identified as the "cherty member" by Lowrie et al. (1982);
2) The biotite-bearing layers which bracket the Eocene-Oligocene boundary. Maximum concentrations of biotite flakes occur some 10 below and about 6 m and 9 m above the boundary (see also Montanari et al., this volume); and
3) The last red colored layer which occurs at 6 m below the Eocene-Oligocene boundary and includes one of the biotite-bearing layer.

The last two lithologic features have been useful for locating the Eocene-Oligocene boundary on the field.

MAGNETOSTRATIGRAPHY

1. Contessa Highway section

Previous study of this section (Lowrie et al., 1982) included the 161 m above the Cretaceous-Tertiary boundary which has been correlated to equivalent sections else where in the Contessa Valley and to the Bottaccione sequence at Gubbio (Napoleone et al., 1983). The upward continuation has been sampled in the nearby quarry that contains the complete Scaglia above meter 161. Therefore the magnetic polarity sequence and its associated lithologic log shown in Figure 1 are composite from two adjacent sections along the Contessa Highway. The connection of the two sections was accomplished by matching both the base of Zone P16 (= *T. cerroazulensis*) and the top of magnetic Chron 16n between the two sections. Thus, the composite Contessa Highway sequence contains no gaps or overlaps in the biostratigraphy or magnetostratigraphy.

The average sampling interval is spaced about 0.7 m, ranging between 0.1 to 2.0 m, in a uniform lithology unaffected by tectonic disturbance or weathering (see Premoli Silva et al., in press.). Magnetostratigraphically, all the samples were quite viable, being well clustered on the stereoplots and very consistent on the Zijderveld diagrams. Therefore, the polarity intervals marked by the VGP sequence were well defined.

The entire Contessa Highway section reaches meter 199.5 above the Cretaceous-Tertiary boundary where it is suddendly interrupted by a significant fault, that cuts the facing Contessa Quarry section at lower level (see above). By correlating all the magnetozones identified in the Scaglia Cinerea (gray Scaglia) to those of the facing Contessa Quarry section not affected by the fault at this level, it appears that the Highway section is missing 36 m of the Oligocene.

As it has been deviced since the first magnetostratigraphy at Gubbio, the polarity sequence at the Contessa Highway has been linked to the sea-floor anomaly sequence by using biostratigraphic tie-points. We use the same biostratigraphic tie-points in the 8 sections reported in this paper to construct the magnetostratigraphic sequence for our sections. Moreover, the geochronological implications already established for Gubbio (Napoleone et al., 1983) are assumed for our sequences in order to identify the sedimentation rates for each biozone and quantify the sedimentary events interrupting normal pelagic deposition.

In addition, two major events were found in polarity zone 13 r, a) the presence of 3 well-defined, short, normal polarity events; and b) the biostratigraphic Eocene-Oligocene boundary.

2. Visso North section

The magnetostratigraphy of the Visso North section, shown in Fig. 1, is based on measurements of 41 samples from a 30 m portion of the section. Due to the weathered deformational fabric in the Scaglia Cinerea at this locality,most samples yielded only one measurable specimen, and the sampling interval was constrained to more competent levels. The presence of magnetite as the carrier of the magnetization is indicated by studies of the acquisition of an isothermal remanent magnetization at room temperature as well as the behavior upon thermal demagnetization. Hematite is present in a few samples, based on demagnetization curves, but the directions recorded by the hematite are both normal and reversed, and so the hematite component is not a recent addition. Thermal demagnetization above 350°C shows no sign of the expected dehydration of geothite to hematite as discussed by Lowrie and Heller (1982), so geothite does not appear to be present in these limestones.

The samples were demagnetized in either alternating fields up to 100 mT in 5 mT and 10 mT steps, or thermally up to 650°C in 50°C steps. The characteristic sample directions were then used to calculate virtual geomagnetic pole latitudes, from which the sequence of reversals was determined. The data as a group are roughly antipodal, and are thus taken to be original directions of magnetization. However, two samples record anomalously shallow VGP latitudes. These samples may reflect acquisition of magnetization during or near the time of a reversal, or, alternatively, they may result from errors in the orientation of the cores.

BIOSTRATIGRAPHY

Calcareous nannofossils and planktonic Foraminifera are the major fossil components of the Scaglia Formation. Benthic Foraminifera are present throughout, but they, as well as the ostracodes,occur in very low percentages. Autochthonous megafossils are absent, whereas displaced and/or reworked large Foraminifera associated with shallow-water skeletal debris occur in the interbedded calcarenitic layers in sections belonging to the southern Umbrian facies.

Biostratigraphic studies were carried out on calcareous plankton which produced a very continuous record. Radiolarians, present only in the short Early to early Middle Eocene interval (= cherty member) could not be used for zoning most of the sequences.

Thirty-six biostratigraphic events based on calcareous nannofossils and planktonic Foraminifera were recognized in the 8 Umbrian sequences in the interval spanning the Early - Middle Eocene boundary to the Early Oligocene.

1. Planktonic Foraminifera

The succession of events recognized in tropical areas are well represented in the Umbrian sequences, with a few minor differences. Blow's 1969 planktonic foraminiferal zonation (see also Berggren et al., 1985) could be applied to the studied interval in most of the cases (see Fig. 1).

Uncertainities primarily involve Zone P13 and the Late Eocene zonal boundaries.

They have been solved in the following ways:

1) In the absence of *Orbulinoides beckmanni*, the index species of Zone P13, this short zone may be recognized indirectly because of a) the appearance of *Turborotalia cerroazulensis* s. str. in the middle of zone; b) the extinction of *Acarinina bullbrooki* gr. close to the base of the zone; and in addition to the latter event, c) marked decrease in abundance of *Morozovella spinulosa*. Moreover, high-spired subbotinids of the *S. corpulenta* gr. become frequent just after the extinction of *O. beckmanni*.
2) The P14/P15 zonal boundary, here equated to the Middle to Late Eocene boundary, is placed at the extinction level of the last spinose acarininids and of the smooth walled *Planorotalites* (i.e. *P. subscitulus* gr.). *Globigerinatheka semiinvoluta* appears just after the latter event.
3) The identification of boundaries of Zone P16 is somewhat indirect because of the absence or rarity of the index species *Cribrohantkenina inflata* in the Umbrian sequences. As reported in Molina and Molina et al. (both, this volume), Zone P16 was identified in the Spanish sections based on the total range of *C. inflata*. There, the appearance of the latter taxon occurs slightly before the extinction level of *Turborotalia pomeroli* and *Globigerinatheka semiinvoluta*, and after the appearance of *Isthmolithus recurvus* (calcareous nannofossil). The extinction of *C. inflata* slightly postdates the extinction of *Globigerinatheka index*.
4) According to the Umbrian data, Zone P17 of Blow (1969) must be confined to the latest Eocene with its lower boundary as mentioned above, and the upper boundary as defined by the extinction of *Cribrohantkenina*, *Hantkenina* and the very last *Globigerinatheka tropicalis*. Such a new definition is supported by the overlap of primitive *"Globigerina" tapuriensis*, the index species of Zone P18, with the latter Eocene taxa.

   The Eocene-Oligocene boundary is here placed to coincide with the P17/P18 zonal boundary above the extinction level of the last Eocene taxa mentioned above.
5) In the Umbrian sequences Zones P18 and P19 could not be separated, whereas the extinction of the last representatives of *Pseudohastigerina (P. naguewichiensis* and *P. barbadoensis)* marked the P19/P20 zonal boundary.

Several other events allowed us to refine the biostratigraphic resolution and, thus, to increase the number of correlation points. They are as follows, from bottom to top:

1 - FO (first occurrence) of *Turborotalia cerroazulensis* s. str. within Zone P13.

2 - FO of *Globigerinatheka index* towards the top of Zone P14.

3 - FO of *Turborotalia cocoaensis* in middle Zone P15 (as defined here).

4 - FO of *Turborotalia cunialensis* in middle Zone P16 (as defined here).

5 - LO (last occurrence) of *Globigerinatheka index* in the upper part of Zone P16.

6 - FO of *Turborotalia ampliapertura* within the upper half of Zone P16.

7 - FO of *Subbotina gortanii* in Zone P17 prior to the LO of the *T. cerroazulensis* gr.

8 - LO of *Turborotalia cerroazulensis* gr. slightly prior to the extinction level of hantkeninids and of the last *Globigerinatheka*, as already mentioned by Baumann (1970).

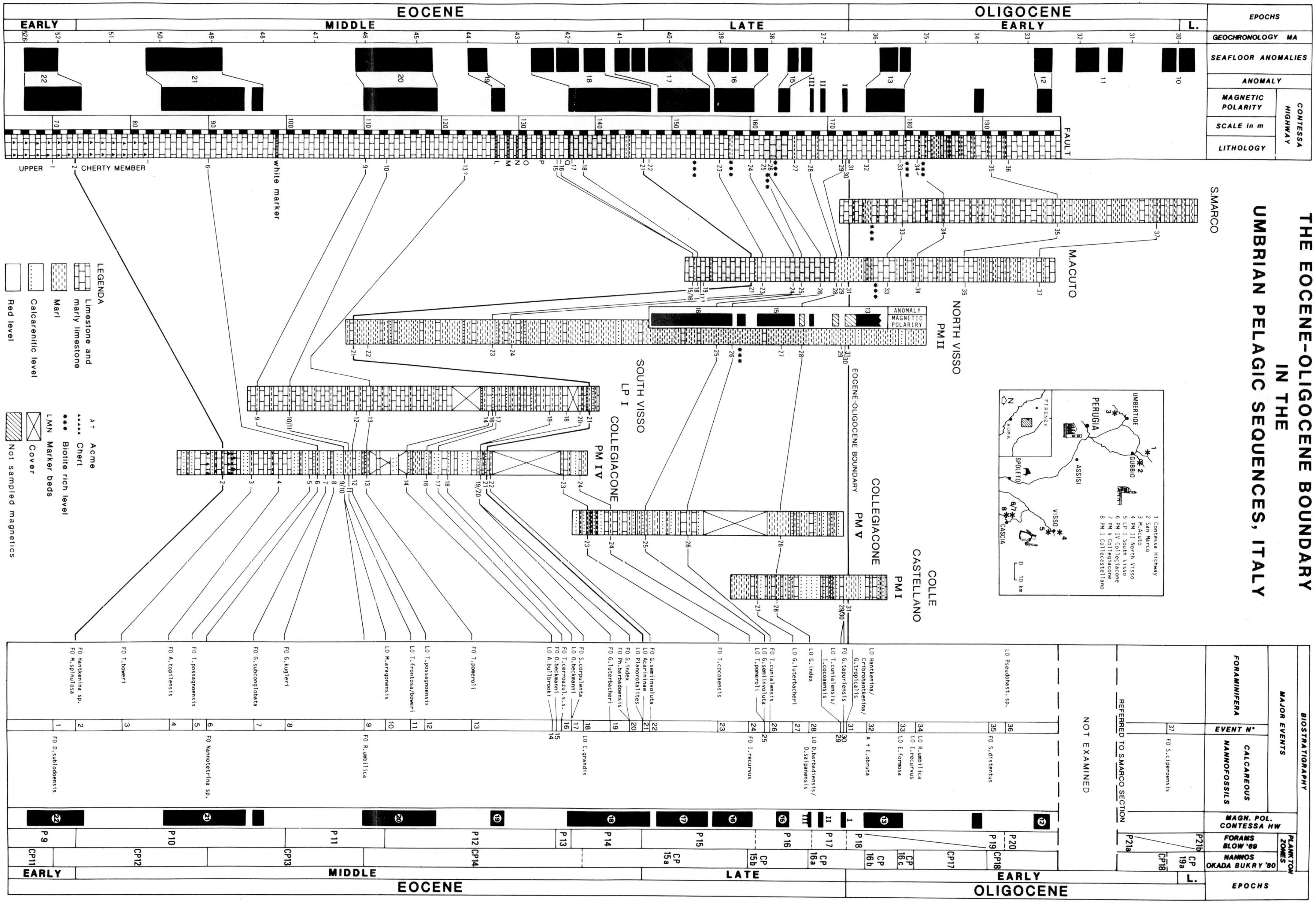
THE EOCENE-OLIGOCENE BOUNDARY
IN THE
UMBRIAN PELAGIC SEQUENCES, ITALY
EOCENE
OLIGOCENE
EARLY
MIDDLE
LATE
EARLY
L.
EPOCHS
GEOCHRONOLOGY MA
SEAFLOOR ANOMALIES
ANOMALY
MAGNETIC POLARITY
SCALE in m
LITHOLOGY
CONTESSA HIGHWAY
FAULT
UPPER
CHERTY MEMBER
white marker
S.MARCO
M.ACUTO
NORTH VISSO
PM II
SOUTH VISSO
LP I
COLLEGIACONE
PM IV
COLLEGIACONE
PM V
COLLE CASTELLANO
PM I
EOCENE-OLIGOCENE BOUNDARY
ANOMALY
MAGNETIC POLARITY
LEGENDA
Limestone and marly limestone
Marl
Calcarenitic level
Red level
Acme
Chert
Biotite rich level
LMN Marker beds
Cover
Not sampled magnetics
UMBERTIDE
PERUGIA
GUBBIO
ASSISI
SPOLETO
VISSO
CASCIA
FIRENZE
ROMA
1 Contessa Highway
2 San Marco
3 M.Acuto
4 PM II North Visso
5 LP I South Visso
6 PM IV Collegiacone
7 PM V Collegiacone
8 PM I Collecastellano
0 10 km
BIOSTRATIGRAPHY
MAJOR EVENTS
FORAMINIFERA
EVENT N°
CALCAREOUS NANNOFOSSILS
NOT EXAMINED
REFERRED TO S.MARCO SECTION
MAGN. POL. CONTESSA HW
PLANKTON ZONES
FORAMS BLOW '69
NANNOS OKADA BUKRY '80
P9
P10
P11
P12
P13
P14
P15
P16
P17
P18
P19
P20
P21a
P21b
CP11
CP12
CP13
CP14
CP 15a
CP 15b
CP 16a
CP 16b
CP 16c
CP17
CP18
CP 19a

9 - LO of *Pseudohastigerina danvillensis* and of large sized (>125 μm) *P. micra* at the same level as the hantkeninids.

It is worth mentioning that the ranges of *G. index* and *T. cunialensis* overlap for some time late in Zone P16. Nocchi et al. (in press) favor the formalization of a subzone on the basis of the concurrent range of these two taxa.

2. Calcareous Nannofossils

Calcareous nannofossils in all investigated sections are abundant and assemblages are well diversified, although sometimes very poorly preserved due to recrystallization and/or dissolution. Reworking is a minor problem involving only a few Late Cretaceous to Early Eocene specimens.

The succession of events, known from the tropical areas, has been recognized in all the Umbrian sequences with a few minor differences. The calcareous nannofossil zonation by Okada and Bukry (1980) rather that by Martini (1971) was applied in these sequences. In a few cases the zonal boundaries were identified as follows:

1 - the lower boundary of Zone CP13 was defined by the appearance of representatives of the genus *Nannotetrina*. Poor preservation prevented identification at the specific level, so the index species *N. fulgens* could not be determined.

2 - The first occurrence of *Reticulofenestra umbilica*, marking the lower boundary of Zone CP14 (=NP16/NP17), occurs in the correct order within the upper part of planktonic foraminiferal Zone P11 (=Anomaly 20) in all the sequences except the Collegiacone section (PMIV) where a hiatus is present. Berggren et al. (1985) reported the FO of *R. umbilica* much higher within the upper part of Zone P12 and Chron 18r. Such a discrepancy is related to the different interpretation of this taxon by various authors.

3 - In the Umbrian sequences because of the heavy dissolution, the delicate rhabdospheres, including *R. gladius*, could not be identified.

4 - All *Chiasmolithus* are very rare in all sections. The extinction of *C. grandis*, associated with the first occurrence of *C. oamaruensis* defines the base of Zone CP15a (=NP18), and occurs at the top of planktonic foraminiferal Zone P14 in the South Visso section. In several other sections, however, *C. grandis* disappears at the base of Zone P14.

Several events proved important for correlating within the Middle Eocene-Early Oligocene interval. They are:

1) the coeval last occurrences of *Discoaster barbadiensis* and *D. saipanensis* which coincide with the last occurrence of *Globigerinatheka index*. Moreover, the latter events occur at an older level than the last occurrence of the *T. cerroazulensis* gr. and the hantkeninids. Therefore, the discoaster extinctions occur prior to the Eocene-Oligocene boundary based on planktonic Foraminifera.

2) *Isthmolithus recurvus*, a cold water indicator, increase in abundance after the last occurrence of *Discoaster barbadiensis* and *D. saipanensis*. This event was also noted in low latitude sites, where, however, it directly precedes the extinction of rosette-shaped discoasters. Therefore, the abundance peak of *I. recurvus* allows us to confirm that the last occurrence of the rosette-shaped discoasters is delayed at low latitudes; furthermore, it appears to be a potentially reliable stratigraphy event (see Monechi, in press).

3) *Ericsonia formosa* slightly decreases in abundance after the last occurrence of rosette-shaped discoasters.

4) *Lanternithus minutus* slightly increases in abundance after the abundance peak of *I. recurvus*, but prior to the Eocene-Oligocene boundary.
5) Finally, an acme of *Ericsonia obruta* was observed after the Eocene-Oligocene boundary and extending up to Zone CP16c (=NP22).

CALIBRATION OF MAGNETIC REVERSAL SEQUENCE

As shown in Figure 1, the correlation between biostratigraphy and magnetostratigraphy proposed by Lowrie et al. (1982) and by Napoleone et al. (1983) is here confirmed and complemented by the additional data. The more important results are, from bottom to top:

1) The base of Zone CP14 correlated with the base of Chron 20n.
2) The short foraminiferal Zone P13 straddles the Chron 18r/18n.
3) The Middle to Late Eocene boundary, equated to the P14/P15 foraminiferal zonal boundary, occurs almost at the top of Chron 18n.
4) The base of the nannofossil Zone CP15b correlates with the top of Chron 16n.
5) The LO of *Discoaster barbadiensis* and *D. saipanensis* (=base of Zone CP16a) and of *Globigerinatheka index* correlate with the 3° normal event within Chron 13r.
6) The LO of *Turborotalia cerroazulensis* gr. occurs just prior to the 1° normal event within Chron 13r.
7) The LO of the hantkeninids, *Globigerinatheka*, and *Pseudohastigerina danvillensis*, equated to the Eocene-Oligocene boundary, occurs just above the 1° normal event cited above within a reversed interval of Chron 13r.
8) The acme of *Ericsonia obruta* (=base of Zone CP16b) correlates with the base of Chron 13n.
9) The LO of *Ericsonia formosa* occurs at the top of Chron 13n.

From the Umbrian data, it appears that the correlations proposed by Berggren et al. (1985) need to be reconsidered, specifically around the Eocene-Oligocene boundary.

RADIOMETRIC DATING

Radiometric dating with K-Ar and Rb-Sr methods was attempted on the biotites isolated from several layers in the studied sections. The most reliable ages were obtained by Montanari et al. (this volume) on material from the Contessa Quarry section (Lowrie et al., 1982). An age of 35.7 $\pm$ 0.4 m.y. was estimated for the Eocene-Oligocene boundary.

BENTHIC ORGANISMS

1. Deep water benthic assemblages

This section focuses on the small benthic foraminiferal assemblages from the Visso e Cascia area and with ostacode faunas from the upper part of the North Visso and Collecastellano sections.

1.1. Benthic Foraminifera

The diversity of benthic foraminiferal assemblages was low through Zone P9 and most of P10, but by Zone P11 increases due to the addition of a number of large *Heterolepa* and *Cibicidoides*. This is an event of worldwide significance. From Zone P13, at a level corresponding to the appearance of *Turborotalia cerroazulensis* s. str. up to the base of Zone P15, benthic foraminiferal assemblages were changed, containing frequent *Pleurostomella*, *Heterolepa grimsdalei*, *H.*

tuxpamensis, *Karreriella chapapotensis*, *Globocassidulina globosa*, and *Anomalinoides capitata*. *Uvigerina (U. multistriata* and *U. rippensis)* is more abundant than *Bolivina* associated with *Bulimina*. In Zones P15 and P16 *Bolivina (B.antegressa* and *B. nobilis)* becomes more abundant whereas *Uvigerina* decreases conversely.

Close to the Eocene-Oligocene boundary at the extinction level of *Turborotalia cunialensis*, the average size of the benthic Foraminifera decreases sharply, although diversity remains at the rather higher values attained in the underlying levels.

Two other events associated with the decrease in size, are the occurrence of *Cyclammina* and other agglutinated forms (*Haplophragmoides*) and a bloom of small-sized *Bolivina* (<125 μm) associated with the disappearances of *Nuttallides truempyi* and *Alabamina dissonata* and with the appearance of *Anomalinoides pompilioides*.

After the Eocene-Oligocene boundary, coeval with the recovery of planktonic species to their normal sizes in Zone P18, benthic Foraminifera regained the normal sizes and once more *Uvigerina* became frequent and dominated over the much rarer *Bolivina*.

Benthic assemblages consist of cosmopolitan species and are indivative of a lower to middle-lower bathyal environment.

1.2. Ostracodes (R. Benson, personal communication)

Ostracode faunas were preliminarily examined in some samples from the Visso North and Collecastellano sections. They are never abundant, but they give some indications of the paleoenvironment.

The assemblages primarily contain forms belonging to the genera *Krithe*, *Abyssocythere*, *Agrenocythere*, *Cytherella*, and *Oxycythereis*. This fauna is characteristic of the lower to middle bathyal, open oceanic environment. It is similar to the faunas described from the Upper Eocene of Moravia by Pokorny (1975, 1977), from Trinidad and in many DSDP cores, and was found also in northern Italy (Benson, not publ.). It belongs to the essentially "modern" psychrospheric fauna in its formative stage.

The estimated paleodepth at the two locations considered is from 1500 to 2000 m. The absence of *Agenocythere*, a common form of the shallower Priabonian sections such as Possagno, also suggests al lower bathyal more than a mid or upper bathyal environment for the Scaglia basin. Finally, the presence of *Abyssocythere*, which is common in the Atlantic, would indicate an oceanic connection at depth without restrictive thresholds.

Displacement from shallower depths is indicated by one allochthonous specimen each of *Hornibroockella* and *Actinocythereis* in the Collecastellano section.

2. Shallow-water to outer shelf displaced skeletal debris

As shown in Figure 1, the several calcareous turbidites scattered throughout the southern Umbrian sequences first occur in Zone P10. Those resedimented layers are largely composed of bioclasts derived from a shallow water carbonate platform, the Abruzzi-Lazio platform located to the south and se (Baldanza et al., 1982). Coarser fractions of the turbidite beds yielded abundant, age-diagnostic larger Foraminifera associated with red algae, some corals, and bryozoans.

The oldest assemblage, occuring in the P10 through P 12 zonal interval, consists largely of reworked forms ranging in age from Maastrichtian through Late Paleocene. The associated discocyclinids, *Nummulites* ex *N. burdigalensis* gr.

and common, poorly cut *Fasciolites* may have an age closer to that of the accompanying planktonic faunas.

A second (younger) assemblage was identified from the top of Zone P12 through Zone P14. In addition to abundant discocyclinids, large *Nummulites* of the *N. millecaput* gr. and *N. perforatus*, and *Fabiania cassis* characterize this assemblage. Alveolinids tend to decrease in abundance upwards. In Zone P14 a single specimen of *Heterostegina* (*Grzybowskia*-type) is recorded at Visso North. Reworking from older levels apparently is less important in this interval.

By Zone P15 to the end of the Eocene (according to the planktonic Foraminifera), larger foraminiferal assemblages change markedly. Associated with discocyclinids, specifically with abundant *Asterocyclina*, and with a large amount of red algae, there are several new genera among which are abundant *Chapmanina gassinensis*, *Heterostegina* s. str., *Spiroclypeus (S. granulosus)*, *Pellatispira (P. mandaraszi)*, abundant *Gypsina linearis*, and *Baculogypsinoides*. The few *Nummulites* are mainly of small size and are identified as *N. incrassatus*, *N. chavannesi*. Representatives of the *N. fabianii* lineage are very rare and mainly belong to the *N. retiatus* stage. The latter taxon occurs from Zone P16 upwards. Reworked species are common in some layers; they belong to the genera *Fasciolites*, *Orbitolites*, *Assilina*, *Omphalocyclus* and rare large *Nummulites (N. millecaput* gr.*)*.

The same assemblage is found also above the Eocene-Oligocene boundary, but it includes much older elements reworked from Paleocene-Early Eocene and Maastrichtian platforms. The only new taxon recorded just after the boundary at Visso North is a single specimen of *Borelis*, possibil, *B. vonderschmitti* (Schweighauser). Species indicative of an Early Oligocene age, such as the reticulate *Nummulites intermedius* or *Praerhapydionina*, were not found. Therefore, it seems that at the end of the Eocene the carbonate platform stopped growing or its lateral extent was reduced. During the Early Oligocene, therefore, only elements of the older platform were eroded and redeposited in the south Umbrian basin.

Although reworking strongly masks the stratigraphic signal, it appears that, at least in some cases, the ages inferred from the larger Foraminifera are consistent with those based on calcareous plankton, thus confirming most of the correlations proposed in literature.

PALYNOMORPHS (U. Biffi, personal communication)

Forthy-seven samples from the Visso North section were processed for palynological analyses using the standard preparation method. Most of them resulted barren of palynomorphs. Only six samples collected close to the Eocene-Oligocene transition yielded assemblages moderately rich in dinoflagellates. The three samples corresponding to the uppermost two meters of the sequence below the extinction level of hantkeninids contain the following age-diagnostic taxa: *Areospheridium multicornutum* Eaton, *A. arcuatum* Eaton, *Cordosphaeridium funiculatum* Morgenroth, and *Heteraulacacysta ? leptalea* Eaton, of a Late Eocene age. Because all the latter forms are absent in the overlying samples, it appears that the Eocene-Oligocene boundary might be placed above their extinction level. Such a location is consistent with that inferred from the planktonic Foraminifera.

Other dinoflagellate species are recorded in the recovered assemblages. The most common are *Systematophora placantha* (Deflandre and Cookson), *Spiniferites* spp., *Nematosphaeropsis* sp., *Impagidium pallidum* Bujak, *Thalassiphora pelagica*

(Eisenack), and *Deflandrea phosphoritica* Eisenack.

The continental elements are represented by rare disaccate pollens associated with scattered *Classipollis* spp. and *Callialasporites dampieri* reworked from Early Cretaceous.

EVOLUTION OF CALCAREOUS PLANKTON AND PALEOENVIRONMENTAL INTERPRETATION

Planktonic foraminiferal assemblages in the early Middle Eocene are dominated by acarininids; *A. bullbrooki* is very frequent and is associated with common morozovellids and a few subbotinids. Soon after the appearance of *Hantkenina*, low-spired subbotinids increase in number, morozovellids decrease, whereas acarininids remain abundant. In Zone P11, after the appearance of medium-sized *Globigerinatheka (G. kugleri)*, *Morozovella aragonensis* becomes very rare before getting extinct. In Zone P12 *Morozovella spinulosa*, *Hantkenina*, and *Clavigerinella* become more frequent and occasionally common. Acarininids and *Globigerinatheka* occur in the same abundance as in Zone P11, but are associated with rare to few turborotaliids and rare *Catapsydrax*.

By the P12/P13 zonal boundary large *Globigerinatheka* associated with *Orbulinoides beckmanni*, large subbotinids with abortive chambers *(S. corpulenta)*, and *Cataspydrax* become important components of the planktonic faunas. From the middle of Zone P13 up to the lower part of Zone P14, *Turborotalia pomeroli* and *T. cerroazulensis* s. str. are common, whereas the morozovellids become rare. *Acarinina bullbrooki- A. spinuloinflata* almost disappear replaced by small-sized, round acarininids and by *Truncorotaloides rohri*. *Planorotalites* is abundant in the small fraction.

Planktonic foraminiferal faunas change markedly through Zone P14 when high-spired subbotinids *(S.praeturritilina)*, large *Globigerinatheka (G. luterbacheri)*, and turborotaliids gradually increase in importance, while acarininids decrease. *Planorotalites* is still common. Both spinose acarininids and *Planorotalites* disappear at the base of Zone P15.

From Zone P15 up to the base of Zone P16 planktonic assemblages contain frequent turborotaliids of the *T. cerroazulensis* lineage, common *Catapsydrax*, and chiloguembelinids. *Pseudohastigerina barbadoensis* and *Turborotalia increbescens* appear. Hantkeninids are always rare and *Globigerinatheka semiinvoluta* is never abundant.

Close to the base of Zone P16, however, *Isthmolithus recurvus*, a cold water indicator among the calcareous nannofossils, appears. Through Zone P16 the *Turborotalia cerroazulensis* lineage, always frequent, reaches the stage of *T. cocoaensis*, and finally of *T. cunialensis;* this latter species is always less abundant than the former. Towards the top of the zone several taxa successively become extinct starting with *T. pomeroli* and *G. semiinvoluta*, then *G. luterbacheri*, and *G. index*.

This last event coincides with a strong decrease of *Globigerinatheka;* only rare specimens of *G. tropicalis* continue till the end of the Eocene. Moreover, after the extinction of *G. index*, hantkeninids become more frequent, *Pseudohastigerina micra* and *P. danvillensis* increase their size to > 125 μm, whereas high-spired subbotinids continue to increase in abundance associated with *Subbotina yeguaensis* and *"Globoquadrina" galavisi*. In this interval calcareous nannofossil assemblages loose the warm-water, rosette-shaped discoasters, whereas the cold-water *I. recurvus* undergoes an increase in abundance.

Close to the top of Zone P17 turborotaliids of the *T. cerroazulensis* lineage abruptly disappear, just after having decrease markedly in size. Conversely,

Turborotalia pseudoampliapertura, *T. increbescens* and *Globigerina* sp. 1 become common. *T. ampliapertura* is still rare. In corrispondence with the extinction of *T. cerroazulensis* gr., there is a large bloom of *Subbotina gortanii* associated with a large variety of abortive-chambered high-spired subbotinids. At the end of Zone P17, equated to the end of the Eocene, at the same level as the extinction of hantkeninids, of the very last *G. tropicalis*, and of *P. danvillensis*, *Pseudohastiregina micra* decreases in size to less than 125 µm.

A small increase in abundance of *I. recurvus* occurs just prior to the Eocene-Oligocene boundary. Planktonic foraminiferal assemblage at the boundary consists of very large-sized, high-spired subbotinids associated with forms of a generalized smaller size than below or above. *Pseudohastigerina naguewichiensis* and *P. barbadoensis* replaced *P. micra* which becomes rare. Chiloguembelinids are very abundant; tenuitellids and *Globorotaloides wilsoni* are common.

Above the base of Zone P18 planktonic species regain normal size, *"G." tapuriensis* display typical morphology, and high-spired subbotininds decrease in importance; whereas *T. pseudoampliapertura*, *T. increbescens*, and *T. ampliapertura* become frequent associated with common *Catapsydrax* and a few *"Globigerina" presaepis*. In the small fraction abundant chiloguembelinids, tenuitellids and pseudohastigerinids are associated with rare *Cassigerinella*.

These planktonic faunas continue to the top of Zone P19, where pseudoastigerinids become rare and then disappear.

The major changes in the planktonic Foraminifera correlate with the major overturns in the bathyal benthic assemblages as well as the shallow-water communities, although in the latter the signals are strongly biased by reworking. The evolutionary trends described from Umbrian sequences exactly compare with evolution in areas outside the Mediterranean area and are of global significance. This becomes particularly evident when comparing the evolutionary trends of planktonic Foraminifera from the Umbrian sequences with those from the Atlantic Ocean (see Boersma and Premoli Silva, 1985, in press). According to the paleoclimatic framework inferred for the Atlantic, the Umbrian sequences were located within the warm to temperate latitudinal band. This assumption is supported by the fact that 1) diversity was high all way through; and 2) after Zone P12, that means after the first important cooling, species characteristic of mid latitude assemblages such as the subbotinids become more abundant earlier than in the equatorial/tropical region. The same patterns are recorded by the calcareous nannofossils, as suggested by the earlier disappearance of the rosette-shaped discoasters or the larger number of *Isthmolithus recurvus* in the Umbrian sections with respect to tropical sites.

SELECTED REFERENCES

Alvarez, W., Arthur, M.A., Fischer, A.G., Lowrie, W., Napoleone, G., Premoli Silva, I. and Roggenthen, W.M., 1977. Upper Cretaceous-Paleocene magnetic stratigraphy at Gubbio, Italy.- V. Type section for the Late Cretaceous-Paleocene geomagnetic reversal scale. Geol.Soc.Amer.Bull., 88:383-389.

Baldanza, A., Colacicchi, R. and Parisi, G., 1982. Controllo tettonico sulla deposizione dei livelli detritici nella Scaglia cretaceo-paleocenica (Umbria orientale). Rend. Soc.Geol.It., 5:11-14.

Baumann, P., 1970. Mikropalaeontologische und stratigraphische Untersuchungen der obereozaenen-oligozaenen Scaglia im zentralen Apennin (Italien). Eclogae geol. Helv., 63: 1133-1211.

Baumann, P. and Roth, P.H., 1969. Zonierung des Obereozaens und Oligozaens des Monte Cagnero (Zentralapennin) mit planktonischen Foraminiferen und Nanno= plankton. Eclogae geol. Helv., 62: 303-323.

Berggren, W.A., Kent, D.V. and Flynn, J.J., in press. Paleogene geochronology and chronostratigraphy. In: N.J. Snelling (Editor), Geochronology and the Geological Record. Geol.Soc. London, Spec. Paper.

Blow, W.H., 1969. Late middle Eocene to Recent planktonic foraminiferal Bio= stratigraphy. In: Proc. 1st Plankt. Conference, Geneva 1967, Ed.Brill, Leiden, 1: 199-422.

Boersma, A. and Premoli Silva, I., 1985. Terminal Eocene events: Planktonic Foraminifera and isotopic evidence. (This volume).

Kirschvink, J.L., 1980. The least-squares line and plane and the analysis of paleomagnetic data. Geophys. Journ., 62: 699-718.

Lowrie, W., Alvarez, W., Napoleone, G., Perch-Nielsen, K., Premoli Silva, I. and Toumarkine, M., 1982. Paleogene magnetic stratigraphy in Umbrian pela= gic carbonate rocks: the Contessa sections, Gubbio. Geol.Soc.Amer.Bull., 93: 414-432.

Lowrie, W. and Heller, F., 1982. Magnetic properties of marine limestones. Rev. Geophys. Space Phys., 20: 171-192.

Luterbacher, H.P., 1964. Studies in some Globorotalia from Paleocene and lower Eocene of the Central Apennines. Eclogae geol. Helv., 57: 631-730.

Luterbacher, H.P. and Premoli Silva, I., 1962. Note preliminaire sur une révi= sion du profil de Gubbio, Italie. Riv.Ital. Paleont.Strat., 68: 253-288.

Luterbacher, H.P. and Premoli Silva, I., 1964. Biostratigrafia del limite Cretaceo-Terziario nell'Appennino centrale. Riv.Ital.Paleont.Strat.,70:67-129.

Martini, E., 1971. Standard Tertiary and Quaternary calcareous nannoplankton zonation. In: A. Farinacci (Editor), Proc. 2nd Plankt. Conference, Roma 1970, 1:739-785.

Molina, E., 1985. Description and biostratigraphy of the main reference section of the Eocene/Oligocene boundary in Spain: Fuente Caldera section. (This volume).

Molina, E., Monaco, P., Nocchi, M. and Parisi, G., 1985. Biostratigraphic cor= relations between Subbetic central sector (Spain) and Umbro-Marchean (Italy) pelagic sequences at the Eocene-Oligocene boundary. (This volume).

Monechi, S., in press. Calcareous nannofossil events around the Eocene-Oligocene boundary in the Umbrian sections (Italy). Submitted to Paleo.Paleo.Paleo.

Montanari, A., Drake, R., Bice, D.M., Alvarez, W., Curtiss, G.H., Turrin, B.D. and DePaolo, D.J., 1985. Radiometric dating of the Eocene-Oligocene boundary at Gubbio, Italy. (This volume).

Napoleone, G., Premoli Silva, I., Heller, F., Cheli, P., Corezzi, S. and Fi= scher, A.G., 1983. Eocene magnetic stratigraphy at Gubbio, Italy, and its implications for Paleogene geochronology. Geol.Soc.Amer.Bull., 94: 181-191.

Nocchi, M., Parisi, G., Monaco, P., Madile, M. and Monechi, S., in press. Bio= stratigraphy and paleoenvironmental changes of the Foraminifera and calcar= eous nannofossil assemblages from middle Eocene to early Oligocene south eastern Umbria Scaglia sequences, Italy. Submitted to Paleo. Paleo. Paleo.

Okada, H. and Bukry, D., 1980. Supplementary modification and introduction of code numbers to the low latitude coccolith biostratigraphic zonation. Marine. Micropaleont., 5: 321-325.

Pokorny, V., 1975. Discovery of deep-sea ostracodes in the Paleogene of the Czechoslovak Carpathians and its significance. Věst. ustred. Ust. geol. Českosl., 50: 37-39.

Pokorny, V., 1977. The genus Agrenocythere (Ostracoda, Crust.) in the Paleo= gene of Moravia, Czechoslovakia. Čas.Mineral.Geol., 22: 383-393.

Premoli Silva, I., Monechi, S., Madile, M., Napoleone, G., Orlando, M. and Ripepe, M., in press. Calcareous plankton Biostratigraphy and magnetostra= tigraphy at the Eocene-Oligocene transition from the Gubbio area, Italy. Submitted to Paleo. Paleo. Paleo.

Renz, O., 1936. Stratigraphische und mikropalaeontologische Untersuchungen der Scaglia (Obere Kreide-Tertiaer) im zentralen Apennin. Eclogae geol. Helv., 29: 1-149.

RADIOMETRIC DATING OF THE EOCENE-OLIGOCENE BOUNDARY AT GUBBIO, ITALY

ALESSANDRO MONTANARI, ROBERT DRAKE, DAVID M. BICE, WALTER ALVAREZ, GARNISS H. CURTIS
Department of Geology and Geophysics, University of California, Berkeley, CA 94720

BRENT D. TURRIN
United States Geological Survey, 345 Middlefield Rd., Menlo Park, CA 94025

DONALD J. DEPAOLO
Department of Earth and Space Sciences, University of California, Los Angeles, CA 90024

ABSTRACT

Unaltered volcanic biotite has been extracted from five bentonite levels in the upper Eocene-Oligocene pelagic carbonate sequence of the Contessa quarry, near Gubbio. The K-Ar and Rb-Sr age determination of the biotite separates are used here to calibrate, with closely spaced points, the late Paleogene biostratigraphic and magnetostratigraphic time scales. Two samples from the upper Eocene have yielded unexpectedly young ages making the duration of the Priabonian age 0.8 $\pm$ 0.4 m.y., instead of 3.5 m.y. as predicted by other time scales. The interpolated age of the Eocene-Oligocene boundary is 35.7 m.y. with a 2-standard-deviation uncertainty of 0.4 m.y.

INTRODUCTION

The late Paleogene sequence of fossiliferous pelagic carbonates continuously exposed in various sections in the Contessa quarry near the city of Gubbio have been studied in detail by Lowrie and others (1982), who provided a tight correlation between magnetostratigraphic zones and foraminiferal and nannoplankton zones. The biostratigraphy of the same sequence has been defined in numerous other exposures in the Umbria and Marche Apennines by Baumann and Roth (1969), and Baumann (1970). In more recent years Perch-Nielsen and others (this volume), and Nocchi and others (this volume) have further refined the biostratigraphy and the magnetostratigraphy of the Contessa sections and of other sections in southern Umbria. Because of all these studies, the Contessa sections are considered the best currently available magnetostratigraphic and biostratigraphic marine record for the late Paleogene. While studying the section in the southeastern face of the quarry, Lowrie and co-workers discovered several layers of marly limestone containing abundant euhedral biotite flakes. Through further field work, we have found biotite-rich layers in other stratigraphic levels in the same section, and on different correlative sections exposed in the Contessa quarry. The biotite is actually concentrated in thin bentonite layers and is spread into the pelagic marly limestones immediately above and below by bioturbation mostly represented by Zoophycos and Planolites ichnofossils. The sporadic occurrence of euhedral biotite flakes in deep water sediments and the relative scarcity of equant silicate grains of volcanic origin associated with them (mostly quartz grains, microcrystalline volcanic rock fragments, and minor amounts of plagioclase and K-feldspar grains) suggest that the biotite and the bentonite layers are distal air-fall deposits of pyroclastic material selectively transported by wind to the Apennine pelagic basin. The late Eocene and Oligocene volcanism associated with the orogenesis in the Alps is by far the best candidate for the source of this material. However, further geochemical investigation is needed for a precise identification of the volcano(es) that produced the Contessa biotites. Nevertheless, this volcanic material offers the unique opportunity to radiometrically date well determined biostratigraphic levels and geomagnetic chrons from the base of the Priabonian to the lower Chattian. In the present paper we will present and discuss the analytical data and the K-Ar and Rb-Sr age determinations of unaltered biotite separates from the Contessa quarry and the Massignano section. As will be shown in a future paper, biotites partially altered to vermiculite yielded ages inconsistent with those of the unaltered biotites extracted from the same layer at different sites in the Contessa quarry. This experience has clearly shown that altered biotite can not be used as a reliable geochronometer.

POTASSIUM-ARGON AND RUBIDIUM-STRONTIUM DATING

The analytical data in Table 1 were obtained from black, shiny, pure biotite separates carefully extracted from the rock and concentrated by magnetic and wet-sieving techniques. No hydrochloric acid was used to break down the carbonate matrix of the biotite-rich rock samples. The rock was crushed and pulverized mechanically to prevent possible leaching of potassium and loss of argon from the biotite. Loss of argon was also prevented by drying the mineral separates at temperatures less than 60°C. In the last step of the separation procedure, the biotite concentrates were bathed in dilute 10% HCl for 30 seconds or less, just enough to eliminate unwanted residual calcite. The potassium content of each sample was obtained by duplicate flame photometry analysis using a lithium internal standard, following the procedures described by Carmichael and others (1968). Argon extractions and isotopic analyses were carried out using standard isotopic dilution methods and a 10 cm Reynolds-type gas-source mass spectrometer according to the procedures described by Dalrymple and Lamphere (1969). The argon extraction line has been calibrated by measuring air-Ar ratios, and cross-calibrated with an internal biotite standard and the international standard muscovite P2-07.

With modern high-precision mass spectometric measurement of $^{87}Sr/^{86}Sr$ ratios it is possible to obtain very precise ages on Tertiary biotites if the initial $^{87}Sr/^{86}Sr$ ratios of the biotites can be established. In an attempt to do this we have measured the $^{87}Rb/^{86}Sr$ ratio and the $^{87}Sr/^{86}Sr$ ratio in HCl-insoluble residues from each of the layers from which biotite has been separated and measured (Table 2).

TABLE 1. Analytical data for ^{40}K-^{40}Ar dates of biotite separates from volcanic bentonites in the late Paleogene pelagic sequences of the Contessa quarry and Massignano section.

Sample Number	Sample Name	Sample Weight (grams)	K (%)	$^{40}Ar^{a}$ (x 10^{-11} mole/gram)	$^{40}Ar^{a}$ (%)	Age ± 2σ (m.y.)	Biostratigraphic and Magnetostratigraphic Location after Lowie and others (1982)
CONTESSA QUARRY							
KA 4480	CQ/83-GAR	0.45691	6.718	32.99	81.9	28.1 ± 0.3	upper *Globorotalia opima opima*, upper NP23 (NP24?), upper CP18, lower anomaly 9N
KA 4480R		0.28625		33.05	66.0	28.2 ± 0.6	
KA 4484	CQ/83-BOB	0.22007	7.108	39.02	75.4	31.4 ± 0.6	mid *Globigerina ampliapertura* and *Cassigerinella chipolensis*, lower NP23, lower CP18, upper anomaly 12R
KA 4484R		0.18653		40.56	67.5	32.6 ± 0.9	
KA 4078	Cinerea-C	0.19050	6.058	41.81	78.2	34.8 ± 0.3	*Pseudohastigerina micra*, upper NP21, upper CP16b, top anomaly 13N
KA 4078R	CQ/82-232	0.10375		42.50	63.5	35.4 ± 0.5	
KA 4485	CQ/83-ODI	0.17995	7.237	45.19	77.5	35.7 ± 0.7	
KA 4279R	CB/82-178	0.24316	6.940	43.60	70.0	35.9 ± 0.3	
KA 4491	CQ/83-ETT	0.03009	6.462	41.76	39.0	36.9 ± 1.3	upper *Globigerinatheka semiinvoluta*, upper NP18, upper CP15a, top anomaly 16N
KA 4728	CQ/84-210	0.11805	6.841	43.22	76.2	36.1 ± 0.5	lower *Globigerinatheka semiinvoluta*, between NP18 and NP17, mid CP15a, upper anomaly 17N
KA 4728R		0.15470		43.84	68.8	36.6 ± 0.5	
KA 4883	CQ/84-210B	0.13128	6.617	42.25	76.8	36.5 ± 0.7	
MASSIGNANO SECTION							Biostratigraphic Location after Baumann (1970)
KA 4727	MAS/84-2	0.16443	7.128	42.01	87.0	33.7 ± 0.4	upper *Globorotalia cerroazulensis*
KA 4727R		0.15026		42.42	67.4	34.0 ± 0.8	
KA 4492	MAS/83-1	0.27743	6.966	40.53	68.3	33.2 ± 0.5	lower *Globorotalia cerroazulensis*
KA 4492R		0.35987		44.00	80.0	36.1 ± 1.0	
KA 4726	MAS/84-1	0.21448	6.818	41.51	53.4	35.1 ± 0.8	
KA 4726R		0.20788		41.38	75.8	34.7 ± 0.9	

Calculations are based on the radio-active decay for $^{40}K\lambda\beta = 4.962 \times 10^{-11} yr^{-1}$ and $\lambda_c + \lambda_{e'} = 0.581 \times 10^{-10} yr^{-1}$ and on the isotopic abundance $^{40}K = 0.01167\%$ of total K.

Note - The radiometric ages of the Massignano biotites were not used to date the Eocene-Oligocene boundary because of stratigraphic uncertainty (see text for explanation).

The biotite-residue pairs then effectively make a "two-point isochron" that gives the age of the volcanic material. The insoluble residues coarser than 64um are mostly made of volcanic rock and mineral fragments, but also contain some minor amounts of Textulariina shell fragments (these are benthonic foraminifers which built their shells by agglutinating silt-sized particles of hard silicate minerals such as quartz and feldspar). In order to check the common origin of the residues and the biotites, we have determined the initial $^{143}Nd/^{144}Nd$ ratios of both (given in terms of parameter ЄNd in Table 2; De Paolo and Wasserburg, 1978). Identity of the initial ЄNd values is good evidence that the HCl-insoluble residue coarser than 64um and the biotite in the same bentonite layer are derived from the same magma. This is the case for samples CQ/83-GAR and CQ/83-ETT which show excellent agreement between the biotite and the residue ЄNd values. Somewhat poorer agreement was obtained for sample MAS/83-2. The low Nd concentrations obtained for the residues (except MAS/83-2) are as expected for materials dominated by feldspar or a feldspar-quartz mixture.

TABLE 2. Rb-Sr and Sm-Nd isotopic data on biotite separates and associated insoluble residues from bentonites of the Contessa CQ and Massignano sections.

Sample Name	Nd (ppm)	$^{147}Sm/^{144}Nd$	εNd* (± 2σ)	Sr (ppm)	$^{87}Rb/^{86}Sr$	$^{87}Sr/^{86}Sr$ (±2σ)	Age ± 2σ** (m.y.)
CQ/83-GAR (B)	1.121	0.1117	-6.4 ± 0.5	33.30	29.390	0.717256 ± 34	27.8 ± 0.2
CQ/83-GAR (I)	4.301	0.1053	-6.5 ± 0.4	11,200	0.0082	0.708029 ± 31	
CQ/83-ETT (B)	1.914	0.1196	-7.3 ± 0.4	53.9	18.37	0.717358 ± 28	35.4 ± 0.2
CQ/83-ETT (I)	1.419	0.0950	-7.6 ± 0.4	1,413	0.0371	0.708142 ± 28	
MAS/83-1 (B)	2.295	0.0964	-8.4 ± 0.4	14.60	74.73	0.746447 ± 44	36.0 ± 0.5
MAS/83-1R (B)	1.493	0.0973	-7.7 ± 0.5	14.39	74.76	0.746860 ± 17	36.5 ± 0.6
MAS/83-2 (B)	2.556	0.1039	-7.9 ± 0.4	16.96	65.39	0.741716 ± 25	34.4 ± 0.2
MAS/83-2 (I)	30.41	0.1123	-9.2 ± 0.4	127.8	2.685	0.711050 ± 31	

For the stratigraphic location of these samples, see Table 1.
(B) Biotite; (I) Insoluble residue;
* Initial value calculated using determined age and $^{147}Sm/^{144}Nd$
** Uncertainty is the maximum allowed by the analytical uncertainties on the measurements of $^{87}Sr/^{86}Sr$ and $^{87}Rb/^{86}Sr$. The uncertainty on the latter is about 0.5%. The uncertainty given does not include any error in the ^{87}Rb decay constant ($1.42 \times 10^{-11} yr^{-1}$). For sample MAS/83-1 the initial $^{87}Sr/^{86}Sr$ ratio is assumed to be 0.7082±0.0003.

One problematic aspect of the data is the anomalously high Sr concentration measured in the residue from sample CQ/83-GAR. This could be an artifact caused by poor mixing between the Sr in the sample and that added in the monoisotopic tracer. If the concentration given is incorrect, and the true concentration were to be 10 to 20 times lower, it would yieldan increase of the age by 0.1 to 0.2 m.y. A similar size adjustment could also apply to sample CQ/83-ETT.

DISCUSSION AND CONCLUSIONS

The radiometric ages of unaltered volcanic biotites recovered from the late Paleogene pelagic sequence of the northern Apennines (Tables 1 and 2) pose two major problems in the attempt to date the Eocene-Oligocene boundary.

The first problem comes from the significant difference between the dates obtained from the Contessa samples and those obtained from the Massignano samples. The interpolated age for the boundary from the Contessa biotites is 35.7 ± 0.4 m.y., while at Massignano the radiometric age for the latest Eocene is 34.0 ± 0.2 m.y.

The second problem deals with the unexpectedly young ages of the late Eocene samples from the Contessa quarry. These dates would reduce the duration of the Priabonian age from 3.5 m.y. -- as predicted by widely accepted time scales (i.e. Ness and others, 1980) -- to 0.8 ± 0.4 m.y.

In order to find a reasonable solution to these problems, it is necessary to evaluate the uncertainties in the stratigraphic analysis of the Massignano and Contessa sections. The stratigraphy of the Contessa quarry is based on combined lithologic, nannoplanktonic, foraminiferal, and paleomagnetic analyses of closely spaced samples from three correlative sections. These are the CH section along the highway, the CQ section on the southeastern face of the quarry, and the CB section on the northwestern face of the quarry. On the other hand, the stratigraphy at Massignano is based only on planktonic foraminiferal identification in samples collected at intervals of about 2.5 m (Baumann, 1970). For this reason, an accurate correlation with the Contessa sections is questionable. Although we can not say a priori that Baumann's biostratigraphic determination is erroneous, it definitely contains more uncertainty than the detailed multi-disciplinary stratigraphical analysis of the Contessa sections carried out by Lowrie and others (1982). Therefore, the radiometric ages of the biotites from the Massignano section can not be used to calibrate the Eocene-Oligocene boundary until magnetostratigraphic and further micropaleontologic analyses are carried out with the same methodology used for the stratigraphic study of the Contessa sections.

A resolution to the second problem, which arises from the unexpectedly young radiometric age determination of the lower upper Eocene biotite samples from the Contessa quarry, can be provided by invoking argon loss during diagenesis. It must be pointed out, however, that the two samples at the bottom of the sequence (see Table 1) have been recovered from different types of rock at the same biotite-rich level: sample CQ/84 210.5 comes from a soft bentonite layer, while sample CQ/84 210.5B has been extracted from the indurated marly limestone immediately above the bentonite layer. Because of this lithologic difference, it seems unlikely that diagenesis would have had exactly the same alteration effect on these two biotite samples, leading to an identical loss of argon. The absence of mineral alteration in the biotite, as confirmed by X-ray analysis, and the high content of potassium further support the inference that no argon loss can be attributed to diagenetic processes. Uncertainty may still exist in the correlation between the upper Eocene magnetostratigraphic zones of the Contessa sequence and the marine magnetic anomalies recorded on the bottom of the oceans. An independent test to evaluate this uncertainty is, however, difficult since the Contessa sequence itself has been used to calibrate the marine magnetic anomalies, and not the other way around.

A synthesis of the present work is shown in Figure 1. Here the K-Ar and Rb-Sr ages of unaltered biotites from the Contessa quarry are compared with magnetostratigraphic time scales published in recent years, and with radiometrically determined age calibration points for the late Paleogene proposed by Curry and Odin (1982). Within the 2-standard-deviation uncertainty in the analytical error, the ages of Oligocene stratigraphic levels in the Contessa sequence are consistent with most of the time scales published in the last five years. However, the ages for the late Eocene obtained from the Contessa samples are significantly younger than those included in the envelope in Figure 1. This age envelope represents time scales constructed on rather uncertain bases, such as the assumption of constant sea-floor spreading rate through periods of time in the order of 10 m.y, (Ness and others, 1980; Palmer, 1983; Hsü and others, 1984), attribution of absolute ages to biozonal boundaries (Alvarez and Lowrie, 1981), and interpretation of the paleomagnetic record in terrestrial deposits (Prothero and others, 1982) which are typically characterized by frequent sedimentary hiatuses. The radiometric ages for the late Eocene proposed by Curry and Odin (1982) are consistent with our dates even though they bear larger uncertainties in their analytical and stratigraphic determinations.

In conclusion, the radiometric ages of the volcanic biotites from the Contessa quarry presented in this paper constitute the most accurate and precise chronostratigraphic record for the late Paleogene available today. The interpolated age for the biostratigraphically and magnetostratigraphically defined Eocene-Oligocene boundary is 35.7 m.y. with a 2-standard deviation uncertainty of 0.4 m.y. The unexpectedly young ages for the upper Eocene biotite samples attribute a duration of only 0.8 $\pm$ 0.4 m.y. to the Priabonian age. This considerable shortening of the late Eocene seems to imply an anomalously high sea-floor spreading rate of the oceans in the period of time between anomaly 13 and anomaly 17. A shorter duration of the late Eocene would also put a constraint on the attempt to understand the possible cause-and-effect relationship among the so-called terminal Eocene events, which include marine and terrestrial biological extinctions, a world-wide cooling event, one or more extraterrestrial impacts, and other geochemical anomalies recorded in marine sediments.

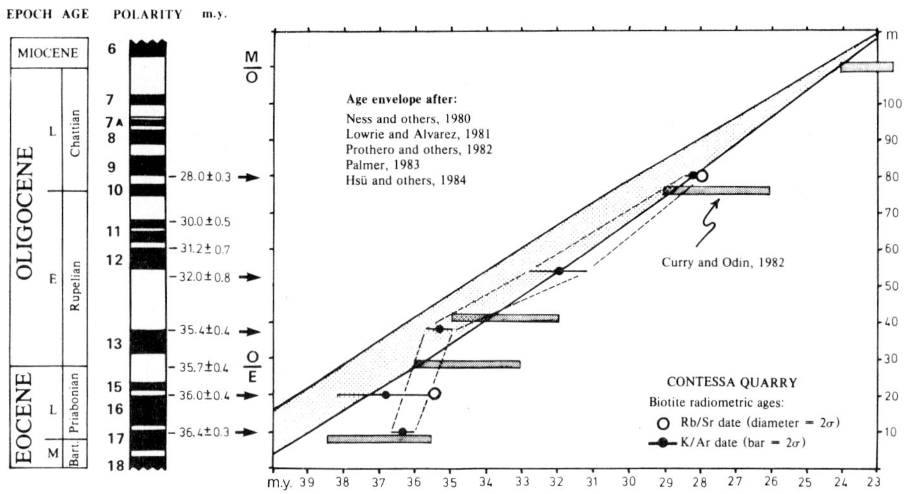

Fig. 1. Radiometric-age calibration of five stratigraphic levels in the composite magnetostratigraphic sequence of the Contessa quarry (sections CQ, CH, and CB). The ages of magnetic reversals 11, 12 and 16, and of the biostratigraphically defined Eocene-Oligocene boundary have been interpolated between the radiometric dates shown in the plot (dashed lines). The K-Ar and Rb-Sr ages of the biotites are indicated by arrows, and compared with recently published numerical time scales and radiometric calibration points for the late Paleogene.

REFERENCES

Baumann, P., 1970. Micropaläontologische und stratigraphische Untersuchung der obereozänen-oligozänen Scaglia im Zentralen Apennin (Italien). Eclogae Geologicae Helvetiae. 63: 1133-1211.

Baumann, P., and Roth, P. H., 1969. Zonierung des Obereozäns und Oligozäns des Monte Cagnero (Zentralapennin) mit planktonischen Foraminiferen und Nannoplankton. Eclogae Geologicae Helvetiae. 62: 303-323.

Carmicheal, I. S. E., Hample, J., and Jack, R. N., 1968. Analytical data on the U.S. Geological Survey Standard Rocks. Chemical Geology, 3: 59-64.

Curry, D., and Odin, G. S., 1982. Dating of the Paleogene. In: G. S. Odin, (Editor) Numerical Dating in Stratigraphy, John Wiley & Son, 607-630.

Dalrymple, G. B., and Lanphere, M. A., 1969. Potassium-argon dating. W. H. Freeman Company, San Francisco, 258 pp.

De Paolo, D. J., and Wasserburg, G. J., 1978. The sources of island arcs as indicated by Nd and Sr isotopic studies. Geophysical Research Letters, 4: 465-468.

Hsü, K. J., La Breque, J., Percival, S., Wright, R. C., Gombos, A. M., Pisciotto, K., Tucker, P., Peterson, N., McKenzie, J. A., Weissert, H., Karpoff, A. M., Carman, M. F. Jr., and Schreiber, E., 1984. Numerical ages of Cenozoic biostratigraphic datum levels: results of South Atlantic Leg 73 drilling. Geological Society of America Bullettin, 95: 863-876.

Lowrie, W., Alvarez, W., Napoleone, G., Perch-Nielsen, K., Premoli Silva, I., and Toumarkine, M., 1982. Paleogene magnetic stratigraphy in Umbrian pelagic carbonate rocks: The Contessa sections, Gubbio. Geological Society of America Bullettin, 93: 414-132.

Ness, G., Levi, S., and Couch, R., 1980. Marne magnetic anomaly time-scales for the Cenozoic and late Cretaceous: A précis, critique and synthesis. Reviews of Geophysics and Space Physics, 18: 753-770.

NEW PROPOSAL FOR AN EOCENE-OLIGOCENE BOUNDARY ACCORDING TO MICROFACIES FROM THE PRIABONIAN-TYPE SECTION

V. BARBIN and G. BIGNOT

Université P. & M. Curie - Département de Géologie Sédimentaire. Laboratoire de Stratigraphie (UA 319) du C.N.R.S. - 4, place Jussieu - 75230 Paris cédex 05 (France).

ABSTRACT

The position of eocene-oligocene boundary in Mesogea goes through stratotypical Priabonian knowledge. The analysis of Priabona "lectostratotype" microfacies led us to put this boundary above Bryozoan marls and limestones, and beds with small Nummulites (*N. fabianii, N. rosai*) and below recif limestones with *Praerhapydionina delicata* and *Peneroplis evolutus*. This boundary does not show any significant paleogeographical event.

The Priabonian stratotype and the main references sections are situated in Northern Italy, Vicentin, more precisely in the Lessini (Granella, Bressana, Buco della Rana...) and in the Colli Berici (Brendola...) (1).

The observed succession at the top of the Priabona section, above Bryozoan Marls, shows five carbonated formations which each present, when examined, through thin sections, well characterized microfacies. From top to bottom :

1) Bryozoan limestones (3m) : biosparite (packstone) with bioclasts (more than 90% can be attributed to Bryozoan) broken, rolled, classed up and in successive laminas disposed.

2) Massive limestones with small *Nummulites* and Bryozoan : biosparite (packstone) beginning with a sudden apparition of plenty of small *Nummulites* from *N. rosai* Tellini, 1888 group and *N. fabianii* (Prever, 1905). The association has still small-sized *Chapmanina cf. gassinensis* (Silvestri, 1931) often broken, as a lot of Bryozoan and Corallinacean fragments. Bioclasts are sorted out and classed up like preceding microfacies.

3) Thin calcareous bed (0,04 to 0,10m). Biopelmicrite (wackestone) only present in the Priabona section, offering a particular microfacies with pellets, Corallinacean thalli and some Miliolidae. Bryozoans and *Nummulites* have disappeared.

4) Massive limestones with rhodoliths and corals (5m). Beside a lot of rhodoliths and corals, we found, in this biolithite (floatstone), rare broken and undetermined *Nummulites*, accompanied by gasteropods and prisms of Pinnidae.

5) Massive limestones with porcelaneous Foraminifera (5m). Biosparite (grainstone-packstone) giving plenty of porcelaneous Foraminifera, Miliolidae, *Spirolina cylin-*

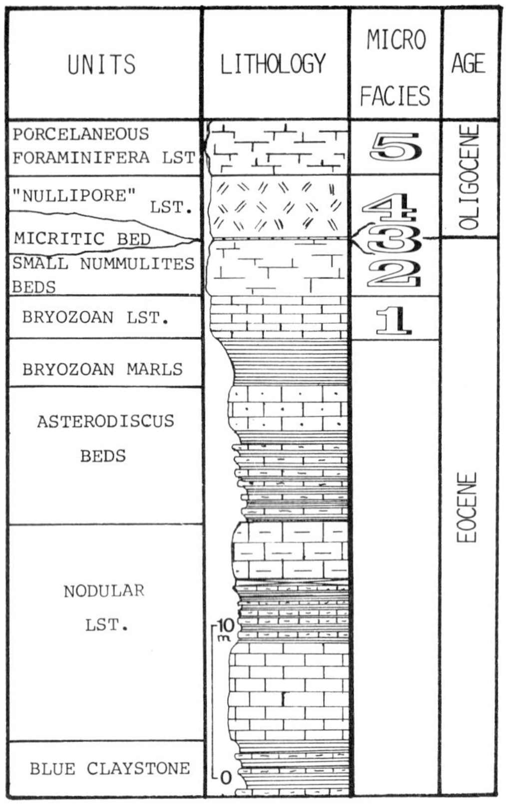

Fig. 1 : Eocene-oligocene boundary in the Priabonian-type section.

dracea Lamarck, 1804, *Peneroplis ssp.* of which *P. evolutus* Henson, 1950, *Praesorites* sensu Henson, 1950, *Praerhapydionina delicata* Henson, 1950, *Austrotrillina (?) sp.*

The section is going on until top of Monte Cassaron through an alternance of units 4 and 5. Such a succession is widespread in all the country, in particular in Buco della Rana. In Bressana (also called Ghenderle) and Brendola, the passage is occuring within marls truncated in their upper part by a channels system and topped with a reef calcareous mass. The carbonate interbeded in marls show microfacies which are not fundamentally different from those described in Priabona.

Considered as the last Eocene stage, the eocene-oligocene boundary has by definition to be placed at the top of the Priabonian stratotype.

Its interest justified plenty of works :

The eocene-oligocene boundary was first fixed by Bayan (1870) between Bryozoan Marls and Rhodoliths Limestones which he called then "Nullipores". These last over being for him partly comparable with Latdorf formations. In 1893, Munier-Chalmas and de Lapparent define Priabonian stage in using two sections, chosen in the Lessini : Granella for the base and Priabona for the median part. Thanks to a third section, the Brendola one, situated at a distance of 20 kilometers, the authors fix their top of their new stage and eocene-oligocene boundary within the Brendola Marls.

Correlating these Brendola Marls to Priabona Bryozoan Marls without evidence, Fabiani (1915) put the Eocene limit at the top of those marls. Schweighauser (1953) doesn't agree with him and places Bryozoan Marls in Oligocene.

Determining *Nummulites intermedius* in the last levels of Granella section,

Roveda (1961) thought he could so attribute them to Oligocene. Since then, this species was never found again and it was established that last Granella levels can be correlated Priabona nodular limestones sections, situated far under Bryozoan Marls and *Asterodiscus* beds. Roveda's ideas who attributed stratotypical Priabonian upper part to Oligocene must therefore be abandoned.

Shortly after Cita and Piccoli (1964) present some conclusions not far from Fabiani in placing eocene-oligocene boundary in Bryozoan Marls.

Finally proposing Priabona section as "type section of Priabonian stage" during Paris Colloquium on Eocene (1968), Hardenbol permitted to bring out a general agreement in order to put the boundary at the top of Bryozoan Marls. This opinion was adopted by Setiawan (1983) who, by another way, maintains that small *Nummulites* and Bryozoan limestones give a small *Nummulites* association which "has an Oligocene aspect". We dont think so for the association has, at least, one species (*N. fabianii*) which doesn't exceed Eocene, and none typical of Oligocene. Therefore, small *Nummulites* and Bryozoan limestones have to be attributed to Priabonian.

The thin level n° 3 does not contain any significant information. Considering its thinness and feeble extension, it seems to us it was overlooked.

Praerhapydionina delicata and *Peneroplis evolutus* being in unit n° 5 limestones involve an Oligocene age (2). These species and those which accompany them indicate that the sediment containing them has settled in a perirecifal environment of lagoon type. Although situated underneath, without any stratigraphically significative fossils, formation n° 4, because of corals and rhodoliths it contains, seems to represent a lateral equivalent to formation n° 5. Consequently we propose to place eocene-oligocene boundary between small *Nummulites* and Bryozoan limestones (formation n° 2) on one hand and recifal limestones with calcareous red algae and with porcelaneous Foraminifera (formations n° 4 and 5) on the other hand.

The Priabonian stratotype contains a succession of facies which fits in quite well with the Arni's model (1965), though the difficulty there is to recognize a real "*Nummulites* bank". The paleogeographical evolution during Priabonian times is divided into two phases : a transgression which favours the installation of a "*Nummulites* bank" followed by a slight regression with a reduction of the detritic deposits which allows the installation of a "lens-shaped to dome-shaped coral reef type" in the Colli Berici (Geister and Ungaro, 1977). Real reefs are missing in Priabona and the observed deposits seem to have taken place in a patch-reef lagoon (Frost, 1981). The facies change at the eocene-oligocene boundary could testify to a simple biotope shift in a shallow marginal basin.

---

(1) The precise geographical location and detailed sections are given in a memoir under publication (Barbin, 1985).

(2) These two species are known in the Tunisian off-shore, on Lampione islot, inside limestones that Bonnefous and Bismuth (1982) attribute, we don't think with any decisive evidence, to Priabonian.

ACKNOWLEDGMENTS : The research was supported by D.G.R.S.T. (Contrat 83071).

REFERENCES

Arni, P., 1965. L'évolution des *Nummulitinae* en tant que facteur de modification des dépôts littoraux. Mém. B.R.G.M., 32 (Coll. intern. Micropal. 1963) : 7-20.

Barbin, V., 1985. Le Priabonien dans sa région-type (Vicentin, Italie du Nord). Stratigraphie. Micropaléontologie. Essai d'intégration dans l'échelle chronostratigraphique. Mém. Sc. Terre Univ. Curie, Paris, sous presse.

Bayan, M.F., 1870. Sur les terrains tertiaires de la Vénétie. Bull. Soc. Géol. Fr., (2), 27 : 444-487.

Bonnefous, J. et Bismuth, H., 1982. Les faciès carbonatés de plate-formes de l'Eocène moyen et supérieur dans l'offshore Tunisien Nord Oriental et en mer Pélagienne : implication paléogéographique et analyse micropaléontologique. Bull. Centres Rech. Explor. Prod. Elf-Aquitaine, 6 : 337-403.

Cita, M.B. et Piccoli, G., 1964. Les stratotypes du Paleogene d'Italie. Mém. B.R.G.M., 28 (Coll. Paris Paléogène, 1962) : 672-684.

Fabiani, R., 1915. Il Paleogene Veneto. Mem. Ist. Geol. Univ. Padova, 3, 1 : 1-336.

Frost, S.H., 1981. Oligocene reef coral biofacies of the Vicentin. Northern Italy. SEPM Special Publ., 30 : 483-539.

Geister, J. et Ungaro, S., 1977. The Oligocene coral formation of the Colli Berici (Vicenza, Northern Italy). Ecl. geol. Helv., 70 : 811-823.

Hardenbol, J., 1969. The "Priabonian" type section (a preliminary note). Mém. B.R.G.M., 58 (Coll. Paris Eocène, 1968) : 629-634.

Munier-Chalmas, E. et de Lapparent, A., 1893. Note sur la nomenclature des terrains sédimentaires. Bull. Soc. Géol. Fr., (3), 21 : 438-488.

Roveda, V., 1961. Contributo allo studio di alcuni macroforaminiferi di Priabona. Riv. Ital. Pal. Strat., 67 : 153-224.

Schweigauser, J., 1953. Mikropaleontologische und stratigraphische Untersuchungen im Paleocaen und Eocaen des Vicentin (Norditalien). Schweiz. Paläont. Abh., 70 : 1-97.

Setiawan, J.R., 1983. Foraminifera and microfacies of the type Priabonian. Utrecht Micropal. Bull., 29 : 1-161.

DESCRIPTION AND BIOSTRATIGRAPHY OF THE MAIN REFERENCE SECTION OF THE EOCENE / OLIGOCENE BOUNDARY IN SPAIN: FUENTE CALDERA SECTION

Eustoquio MOLINA*

1 INTRODUCTION

The Eocene-Oligocene boundary has been acknowledged in different Spanish regions, being located as much in continental as in continuous marine series.

In the North of Spain the limit is situated in continental facies. In the Pyrenees region of Aragón and Navarra thick sections crop out that show clearly that Eocene marine facies last as long as the Late Eocene in certain cases, with the Eocene/Oligocene limit in continental facies. In most of Catalonia and Navarra continental basins there are thick evaporitic formations that reached the top of the Eocene (Reguant, 1984). Therefore, these deposits do not allow a good biostratigrafic study of such a boundary.

In the South of Spain, geologically corresponding to the Betic Cordillera, many continuous successsions at the Eocene-Oligocene boundary can be found in marine facies. The island of Mallorca, however, is an exception; the boundary shows carbonaceous deposits with shallow marine intervals interbedding. In the region of Levante, more precisely in the area around Alicante, a continuous marine section has been quoted (Cremades, 1981) but this section does not seem to offer good possibilities due to its reduced thickness and inappropiate lithology.

The best continuous marine sections can be found in the central sector of the Betic Cordillera, more precisely in the North of the province of Granada. Most of these sections have been described by Martinez-Gallego (1977) and Molina (1979) in their doctoral theses respectively. The upper Eocene and lower Oligocene have been recognized in the sections of the Navazuelo, the Pinarejas and others. Among these, the most outstanding due to their important thickness and good exposure are: the Torre Cardela section, published by Martinez-Gallego and Molina (1975), the Molino de Cobo section, and the Fuente Caldera section, the best found so far.

*Department of Paleontology
University of Zaragoza (Spain)

2 LOCATION

The Fuente Caldera section is geographically located in the Gavilan ravine, in the township of Pedro Martinez (province of Granada), 6.5 km NE from the village itself and 1 km NE from the Fuente Caldera farmhouse. It can be found in the sheet of Huelma 20-39 (970) of the Spanish military map 1:50,000. The U.T.M. coordinates of the points delimiting the section are; base: 30SVG836571 and top: 30SVG835575. The section is accessible by a path from the road between Pedro Martinez and Villanueva de las Torres, which leads to the farmhouse of Fuente Caldera.

Geologically, the Fuente Caldera section is located in the Betic Cordillera that is the most western mountain system of the European

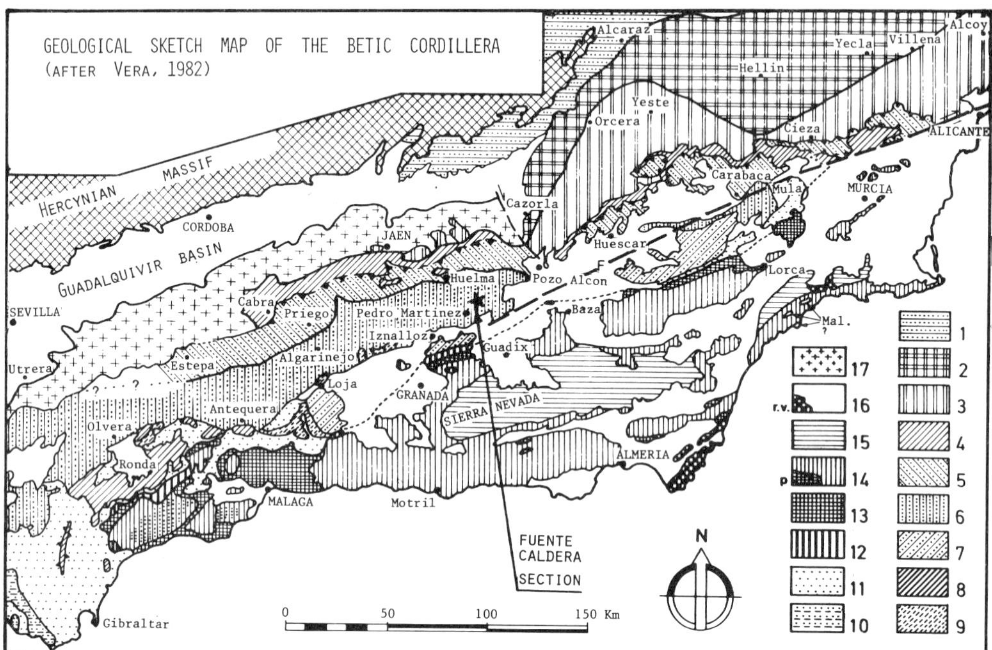

Fig. 1. Geological sketch map showing the situation of Fuente Caldera section: 1) Tabular cover of the Hercynian massif of the Meseta (Triassic and Jurassic); 2) External Prebetic; 3) Internal Prebetic; 4) Intermediate Units (or intermediate realm); 5) External Subbetic; 6) Median Subbetic; 7) Internal Subbetic; 8) Penibetic; 9) Ultrainternal Subbetic and units of dorsalian affinities and flysch substratum; 10) Tectonically underlying Campo de Gibraltar Units; 11) Campo de Gibraltar Units; 12) Rondaides or Betic dorsal; 13) Malaguide; 14) Alpujarride (p:peridotites); 15) Nevado-Filabride; 16) Upper Miocene-Pliocene-Quaternary (r.v.: volcanic rocks); 17) Guadalquivir allochtonous units (Olisthostromes of Subbetic origin inside Miocene materials).

alpine chains. It ranges all along the S and SE region of the Iberian Peninsula, with a general WSW-ENE direction. They extend along 600 Km with a variable width of about some 200 Km and it is bounded northwards, from W to E, by the Guadalquivir Basin, the tabular cover of the Hercynian massif in the Meseta and the Iberic Cordillera. Generally, the importance of folding decreases from S to N, so that in the septentrional realm (Prebetic) overthrusts cannot be individualized. Two main units can be identified from S to N: Internal Zones (or Betic Zone s. str.) and External Zones (Subbetic and Prebetic Zones).

The Fuente Caldera section is situated in the External Zones of the Betic Cordillera, more precisely within the Median Subbetic realm, which appears to be a subsident trough during the Eocene. The Palaeogene - lower Miocene marine materials are laid down in that realm and may reach up to more than 1,000 m in thickness.

The sediments of the Fuente Caldera section correspond to the Cañada Formation (Eocene-Aquitanian) of the Cardela Group established by Comas (1978) as formal lithostratigraphical units for Median Subbetic sequences of the Montes Orientales. The Cañada Formation, which may range up to a thickness greater than 500 m, is composed in general by detritic limestones of turbiditic origin interbedding rhythmically with hemipelagic marly levels.

3 LITHOLOGY AND STRUCTURE

On the flanks of the Gavilan ravine there is a thick series of materials comprised between the uppermost part of the lower Eocene and the Aquitanian. The exposure of the strata is excellent, especially in the interval corresponding to the upper Eocene and lower Oligocene, which allows a very good sampling of the boundary.

The materials of the Eocene-Oligocene boundary as a whole are generally made up of a rhythmic succession of bioclastic calcarenites and calcirudites alternating with marls, which are the predominant lithology. This light grey-greenish marly sediment shows a concoidal fracture in balls and they are easily disintegrated by the laboratory washing. The last appearance of the typical Eocene planktonic foraminifera do not correspond to any sudden lithological change.

Although less frequently there are also certain levels of marly limestones and thin calcareous conglomerate, as we go further on in to the Oligocene, where an olisthostrome is located. It involves materials of different lithologies, including volcanic rocks as well.

The series this paper is concerned with, offers in its top a mainly calcarenitic interval, which is not totally included here because it already belongs to the middle Oligocene.

The section here considered for studying the Eocene/Oligocene boundary is 200 m thick and shows an almost vertical dip, forming the North flank of a wide syncline of a general ENE-WSW direction, with its nucleus located 1 km south and delimited by two hills:Cerro Caldera and Maquina. This syncline leans on pink and white marly limestones dated as Paleocene and Late Cretaceous.

4 SEDIMENTOLOGY

The Fuente Caldera section is of great interest from the sedimentological point of view (Comas, Martinez-Gallego and Molina, 1981), due to the presence of different facies of turbidites and pelagic sediments. In this sequences the autochthonous marly interval is largely developed and they show very little reworking.

The major facies in the allochthonous interval is made up of calcareous turbidites where Bouma a interval can be recognized and sometimes a bioclastic-calciclastic coarse-grained layer appears under it, corresponding to a grain-flow episode. The bioclastic calcarenites generally range from coarse to fine types: wakestone, packstones or pseudograinstone. They show micrite intraclasts in a proportion between 3 and 20% in respect to the fragments of calcareous algae and macroforaminifera. The medium size of the grains in the basal interval is usually larger than 2 mm., and that is why they should be considered as fine calcirudites.

The recognized sequences are of the Tabade, Tab/e, Ta/c/e, types and some of them of the Tbc/e type. They show basal flat surfaces with a development of only a few sole marks: some grooves, load and tool marks. The calciruditic-calcarenitic beds are usually between 30 and 180 cm thick. For some of very incomplete sequences their thickness decreases to 7 cm. Every turbiditic bed is interbedded between marl intervals whose thickness ranges between 50 cm and 5 m, therefore autochthonous pelagic sediments are dominant. If the large quantity of interturbiditic materials alone is considered here as a indicative parameter of distality, they should consequently be considered as distal turbidites.

The origin and direction of the paleocurrent is difficult to establish due to the scarcity of structures they display, which seem to have come from the W. The place of deposit for these associated facies would be a submarine talus or slope, probably in relation to a large submarine fan.

In the upper part of the series an eight-metre thick olisthostroma is located, which is included in the biozone of *G. tapuriensis*. It is formed by a chaotic conglomerate mass, blocks and pebbles, that is the result of a gravity flow of a considerable volume, involving materials of various lithologies and origins. Taking into account the regional data, it may be assumed that the olisthostrome must have fallen from the NE.

5 GEOCHEMISTRY, MAGNETOSTRATIGRAPHY AND RADIOMETRIC DATING.

The geochemical study of the pelagic carbonates of this series have been carried out by Berthenet et al., this volume . They have analyzed the fluctuation of the chemical composition of the oceanic water, with the aim of determining the main geochemical events during the Eocene/Oligocene boundary, concluding that there is a good concordance between the chemiostratigraphy and the bioestratigraphy.

In the Fuente Caldera section five main geochemical events can be recognized. The event G 1 well marked by the $\delta^{13}C$ and $\delta^{18}O$ and moderately marked by the Sr correspond to the boundary of the foraminiferal zones. *P. semiinvoluta/C. inflata*. The event G 2, very neat for the Sr and moderate for the $\delta^{18}O$ corresponds to the limit of the nanno-zones NP 20/NP 21. The event G 3 (very strong for the $\delta^{13}C$ and the Sr) are situated at the boundary of the foraminiferal zones *C. inflata/C. lazzarii*. The event G 4 ($\delta^{13}C$ and Sr) corresponds to the boundary E/O defined by the extinction of the typical Eocene planktonic foraminifera. The event G 5 ($\delta^{13}C$ and $\delta^{18}O$, Sr) is situated in NP 21 and could eventually correspond to the boundary CP 16a/CP 16 b.

Concerning the magnetostratigraphic study, a first sampling followed by a study of the magnetization intensity in roder to determine the real possibilities of this section, was carried out in 1982 by Rasplus who observed a remanent mangnetization. Consequently, the Fuente Caldera section allows the establishment of a magnetostratigraphy.

No studies directed towards radiometric datings have been undertaken so far. Therefore, the real possibilities of the Fuente Caldera section in this sense are still unknown.

6 FLORAL AND FAUNAL CONTENT

The Fuente Caldera section is remarkable for its fossil content, mainly microfossils. However, the macrofossils such as Equinoderms Bivalves and others are very difficult to identify due to the massive fragmentation they have undergone. They appear in small proportion in the calcarenitic strata, associated with Coralline algae, Bryozoa and macroforaminifera.

The macroforaminifera forming part of this association in the turbiditic levels are mainly the following: *Discocyclina sp.*, *Asterocyclina sp.*, *Aktinocyclina sp.*, *Nummulites sp.*, *Operculina sp.*, *Heterostegina sp.*, *Spiroclypeus sp.*, *Amphistegina sp.*, and *Gypsina sp.* It must necessarily be emphasized that the percentage of the typical Eocene Discocyclinidae decreases gradually towards the upper part of the series, but nevertheless, they are still well represented in the biozone of *G. tapuriensis*.

The autochthonous marly intervals contain an extraordinarily varied quantity of calcareous nannoplankton, planktonic foraminifera, small benthonic foraminifera and some ostracoda as well. These sediments are very rich in calcareous nannoplankton and, although their preservation is not ideal, it is good enough to allow a detailed study and to establish an accurate biozonation.

The rich calcareous nannofossil content allows Monechi to recognize several events from Late Eocene to Early Oligocene. The zonation of Bukry and Okada (1980) with the designation CP were used. The zonation of Martini (1971) with the designation NP was also reported for correlation (Monechi, this volume).

To demonstrate the variation in the abundance of the most important species a semiquantitative analysis was carried out and it has been possible to identify a succession of events. The last occurrence of *C. protoannula* and *C. reticulatum* takes place before the extinction of the rosette shaped discoasters. The synchronous extinction of *D. barbadiensis* and *D. saipanensis* at the same level of the LO of *G. index*. A small increase in the abundance of the *I. recurvus*, cold water indicator, was observed between the LO of rosette shaped discoasters and the E/O boundary defined by planktonic foraminifera. The species *E. formosa* decreases around the LO of rosette shaped discoasters.

The planktonic foraminifera are especially abundant reaching 85% in almost all the samples of the marly intervals. They are quite well preserved although the chambers are filled in with sediments. They usually show no deformation whatsoever.

Taking into account the fossil content and sedimentation, it may be assumed that the paleoenvironment would be an open marine basin that would correspond to an almost 2.000 m deep bathyal zone, situated in a subtropical or temperate area.

7 BIOSTRATIGRAPHY BY MEANS OF PLANKTONIC FORAMINIFERA.

The biozonation established for the Oligocene (Molina, 1979) is still mantained in its general outline. But, nevertheless, a more detailed study is attempted in this paper, in order to determine more accurately the range of the different species and to extend the study to include the upper Eocene as well.

The vertical range of the main species, from the biostratigraphical point of view, has been shown in Figure 2. Some other species have also been identified and some of them are very abundant, but most of them appear along the whole Eocene-Oligocene transition. The following ones have been recognized among them: *Globigerina tripartita* Koch, *Globigerina venezuelana* Hedberg, *Globigerina angiporoides* Hornibrook, *Globigerina galavisi* Bermudez, *Globigerina corpulenta* Subbotina, *Globigerina officinalis* Subbotina *Globigerina praebulloides* Blow, *Globorotalia (T.) opima nana* Bolli, *Globorotaloides suteri* Bolli, *Catapsydrax unicavus* Bolli, Loeblich and Tappan, *Chiloguembelina cubensis* (Palmer), *Chiloguembelina victoriana* Beckmann, *Globigerina ouachitaensis* Howe and Wallace, and *Globorotalia (T.) increbescens* (Bandy).

The *Globorotalia cerroazulensis* group is very common throughout the upper Eocene. The author would rather consider these forms at the species level because of the different morphology between the more primitive and the evolutionary ones. In the interval studied, *G. (T.) pomeroli* became extinct at the same time as *P. semiinvoluta;* later *G.(T.) cerroazulensis* became extinct after the disappearance of *G. index*. Finally, *G.(T.) cocoaensis* and *G.(G.) cunialensis* became extinct simultaneously. *G.(G.) cunialensis* is the worst represented since it is very difficult to find good keeled specimens and it

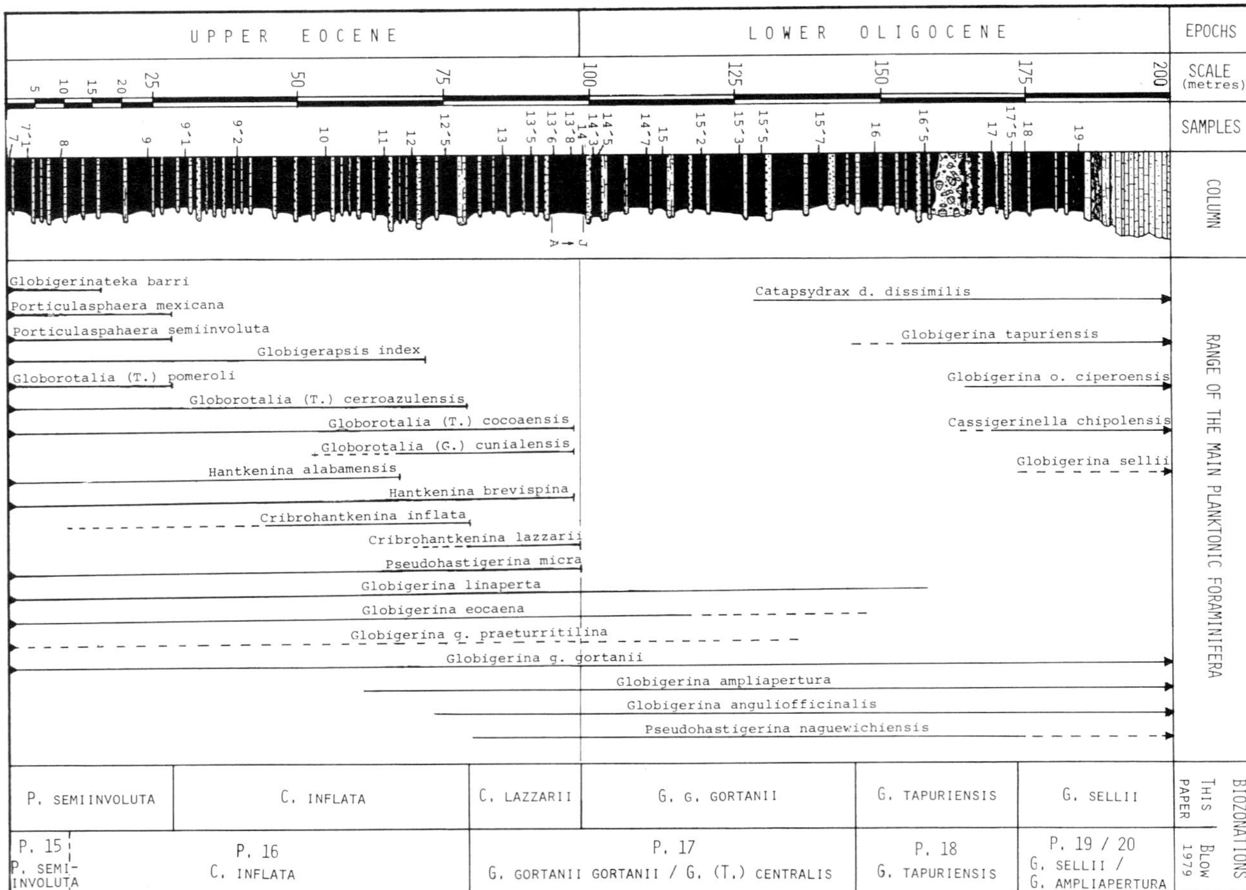

Fig. 2. Column showing the upper Eocene and lower Oligocene in Fuente Caldera section.

appears only in the uppermost Eocene.

The *Hantkenina* and *Cribrohantkenina* genera are frequent in most of the samples. The second one is well represented by the species *C. inflata*, *C. lazzarii* and probably a third one, as was pointed out by Dieni and Proto Decima (1964) in Italy. *C. inflata* has more inflated chambers than *C. lazzarii* which have almost triangular chambers with a rough surface and a few more chambers in the last whorl.

In the Spanish sections the *Pseudohastigerina micra* (Cole) extinction is a very clear datum plane. However, the range of this species does not overlap wiht that of *Cassigerinella chipolensis* Cushman and Ponton) as was pointed out in other regions of the world (Bolli, 1957,66). The explanation for such an event could be:

(1) *P. micra* extinction in low latitudes could be prior to that in mid-latitudes.

(2) *C. chipolensis* appearance in low latitudes could be prior to that in mid-latitudes.

(3) *P. micra* could be found reworked after its last extinction or could be confused with *P. n.naguewichiensis* and *P. n.barbadoensis*; these two subspecies are less compressed and smaller than *P. micra* a species typically larger than 150 microns. The small ones are very frequent from the uppermost Eocene to the lower - Oligocene.Recently, these forms larger than 150 microns have been regarded by Blow (1979), as belonging to *P.danvillensis* Howe and Wallace, 1932) but I would rather keep considering it as junior synonyms of *P. micra* (Cole, 1927) like Cordey, Berggren and Olsson (1970).

The species *Catapsydrax dissimilis* (Cushman and Bermudez) has been found only in the Oligocene, contrary to that which some specialists indicated for other regions of the world. In Spain *C.d. dissimilis* appears in the upper part of the *G.g.gortanii* zone and *C.d. ciperoensis* evolves from the former in the top of the *G. tapuriensis* zone.

The following biohorizons have been considered as the most important and used to establish the biozonation: extinction of *Porticulasphaera semiinvoluta* (Keijer), extinction of *Cribrohantkenina inflata* (Howe), extinction of *Cribrohantkenina lazzarii* (Pericoli), first appearance of *Globigerina tapuriensis* Blow and Banner, first appearance of *Globigerina sellii* (Borsetti).

Similarly, the following biohorizons can be considered clear on the scale of this particular central sector of the Betic Cordillera: extinction of *Globigerinatheka barri* Bronnimann, extinction of *Globigerapsis index* (Finlay), simultaneous extinction of *Hantkenina brevispina* Cushman, *Globorotalia* (*T.*) *cocoaensis* Cushman and *Globorotalia* (*G.*) *cunialensis* Toumarkine and Bolli, being followed inmediately by the extinction of *Pseudohastigerina micra* (Cole) at the same time as *Cribohantkenina lazzarii* (Pericoli).

The Eocene/Oligocene boundary has been sampled in detail (see Fig. 3) and therefore it has been possible to conclude that the extinction of the typical Eocene species is not simultaneous. The same conclusion has also been reached in the Torre Cardela section (Martinez-Gallego & Molina, 1975). As the *Pseudohastigerina micra* and *Cribohantkenina lazzarii* are the last typical Eocene species to became extinct, they have been chosen as the datum plane to mark out such a boundary.

The biozonation established in this paper has been correlated to that which offers more similarities (Blow, 1979), but there are some important differences; so, the top of the *G. semiinvoluta* zone, is delineated by the last appearance of the nominal species instead of by the first appearance of *C. inflata*. The zone P 17 of Blow

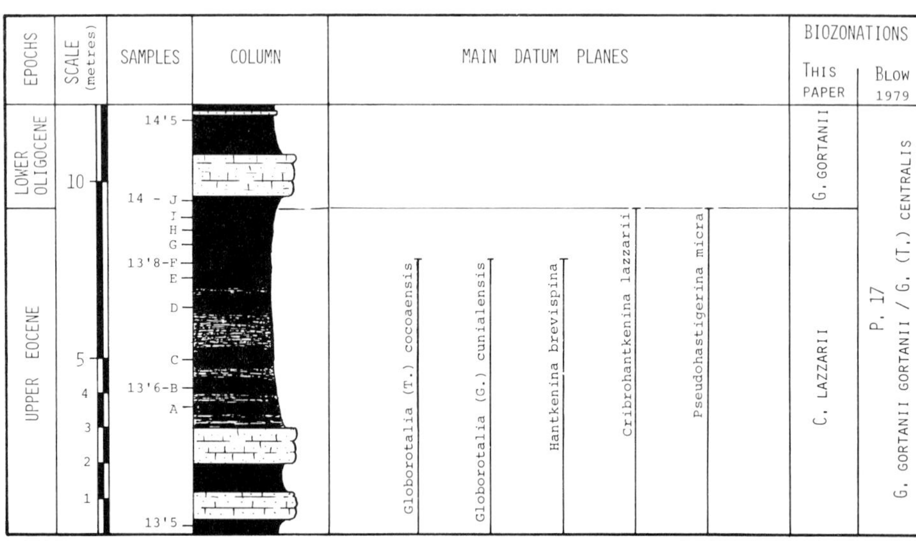

Fig. 3. Column showing a detailed sampling of the Eocene/Oligocene boundary in Fuente Caldera section.

has been divided into two zones: *C. lazzarii* and *G. g. gortanii*. The latter is an interval zone situated between the extinction of *C. lazzarii* and the first appearance of *G. tapuriensis*. Here the planktonic foraminifera bear affinity to those of the Oligocene; and the species *G.g.gortanii* has been found throughout the upper Eocene and Oligocene, contrary to what Blow pointed out.

Finally it should be noted that some biohorizons, such as *G. tapuriensis* and *G. sellii* that have been used in the biozonation, show a very gradual appearance. On the other hand, many extinction biohorizons seem very instantaneous. The typical Eocene species stand out clearly in this sense. Nevertheless, when the Eocene/Oligocene boundary is being sampled in detail, it can be observed that the extinction of the typical Eocene forms are not all simultaneous, as it could appear. Besides that, no lithological change is observed. Consequently, it can be said that there is a transition of biological and sedimentological events at the Eocene-Oligocene in the Fuente Caldera section.

REFERENCES

Berthenet, F. 1985. Evénements géochimiques á la limite E/O. Mém. DEA, option strat. Univ. P.& M. Curie, Paris, Roneo,30 p.
Blow, W.H. 1979. The Cainozoic Globigerinida. Leiden, Brill, 3 vol.
Comas, M.C. 1978. Sobre la geología de los Montes Orientales: sedimentación y evolución paleogeográfica desde el Jurásico al Mioceno inf.(Zona Subbetica, Andalucia).Tesis, Univ.Bilbao, 323 pp.
Comas, M.C., Martinez-Gallego, J., and Molina, E. 1981. Facies y sucesión estratigráfica del Eoceno y Oligoceno al N del Mencal (Zona Subbetica, Prov. de Granada). Cuad. Geol., 12, (in press).
Cremades, J., 1981. Contribución al conocimiento de los foraminiferos planctónicos y al estudio bioestratigráfico del Eoceno y Oligoceno del sector oriental de las Cordilleras Béticas. Tesis Doc.
Martinez-Gallego, J., 1977. Estudio micropaleontologico del Nummulitico de un sector comprendido entre Moreda-Piñar-Pedro Martinez (Zona Subbetica). Tesis Doct. Univ. Granada. 175.
Martinez-Gallego, J. and Molina, E. 1975. Estudio del transito Eoceno-Oligoceno con foraminíferos planctónicos al Sur de Torre Cardela (Prov. de Granada, Zona Subbetica). Cuad.Geol. 6: 177-195.
Molina, E. 1979. Oligoceno-Mioceno inferior por medio de foraminíferos planctónicos en el sector central de las Cordilleras Béticas (España). Tesis Doct. Public. Univ. Granada y Zaragoza. 342 pp.
Molina et al.(in press). Biostratigraphic correlation between the central Subbetic (Spain) and Umbro-Marchean (Italy) pelagic sequences at the E/O boundary using foraminifera. This volume.
Perch-Nielsen, K. et al. (in press). Calcareous nannofossils at the Eocene/Oligocene boundary. This volume.
Renard, M. 1985. Géochimie des carbonates pélagiques. Thése Univ. P.& M. Curie, n°84-16 et Mém. BRGM n°85, (1985), 660 p.
Reguant, S. 1984. The Geological events at the Eocene-Oligocene boundary in Spain. TEE News, 8.

BIOSTRATIGRAPHY OF FUENTE CALDERA SECTION BY MEANS OF CALCAREOUS NANNOFOSSILS

S. MONECHI

Dipartimento di Scienze della Terra, Via La Pira 4, 50121 Firenze (Italy)

The calcareous nannofossil assemblage is very abundant and well diversified; preservation varies from poor to moderate. Reworking is an important phenomenon present throughout the section and is characterized by Late Cretaceous and Early Eocene specimens. The rich calcareous nannofossil content allows to recognize several events from Late Eocene to Early Oligocene.

The zonation of Bukry and Okada (1980) with the designation CP was used. The zonation of Martini (1971) with the designation NP was also reperted for correlation.

In Fig. 1 the vertical distribution of the most important species and the recognized biozones are reported. The base of the I. recurvus Subzone (CP 15a) is not present. The Late Eocene assemblage is characterized by the presence of C. pelagicus (Wallich), C. floridanus (Roth and Hay), D. bisectus (Hay, Mohler and Wade), E. formosa (Kamptner), L. minutus Stradner, Z. bijugatus (Deflandre and Fert); by common rosette-shaped discoasters; by common Helicosphaera such as H. euphratis Haq, H. compacta Bramlette and Wilcoxon. The genus Sphenolithus is common and characterized by S. moriformis (Bronnimann and Stradner), S. predistentus Bramlette and Wilcoxon and sporadic forms of S. pseudoradians Bramlette and Wilcoxon. Few species of C. oamaruensis (Deflandre and Fert) are present throughout the section.

In order to evidentiate the variations in the abundances of the most important species around the E/O boundary a semiquantitative analysis was carried out. The methodology used is that described by Backman and Shackleton (1983). The abundance is expressed as number of specimens per square millimeter. The counts were done on about 100 view-fields at magnification 1250.

From these data it has been possible to identify a succession of events:

- the last occurrence (LO) of C. protoannula Gartner and C. reticulatum Gartner and Smith occurs before the extinction of the rosette shaped discoasters.

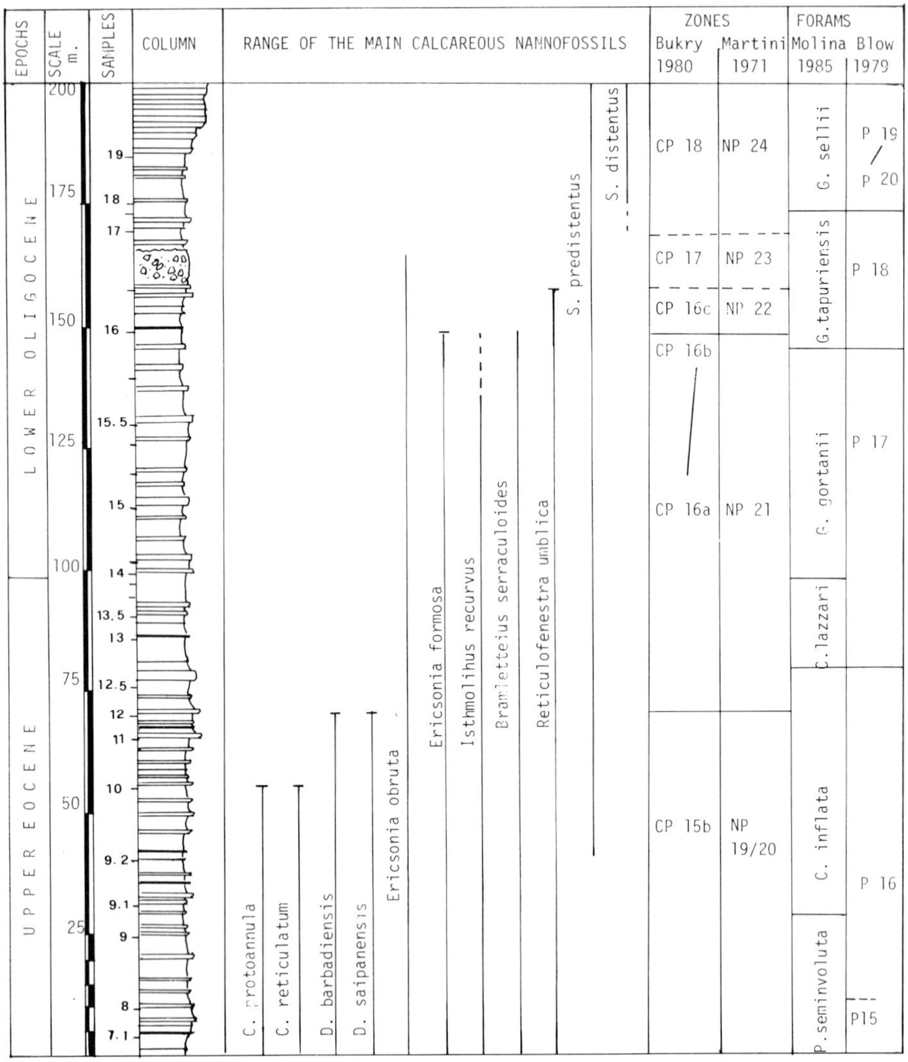

Fig. 1 Distribution of the main calcareous nannofossils along the Fuente Caldera section. Correlation between calcareous nannofossil and planktonic foraminifera zonations. (Compiled from Monechi and Perch-Nielsen, unpublished data)

- the synchronous extinction of D. barbadiensis Tan Sin Hok and D. saipanensis Bramlette and Riedel at the same level as the LO of G. index. This event has been also recognized in several italian sections and in many other E/O boundary sections.
- a small increase in the abundance of I. recurvus Deflandre and Fert, cold water indicator, is observed between the LO of the rosette-shaped discoasters and the E/O boundary defined by planktonic foraminifera through the extinction of P. micra and C. lazzarii.
- E. formosa decreases around the LO of rosette-shaped discoasters.

E. obruta (i.e. E. subdisticha or E. fenestrata according to other authors) is rare in the Late Eocene and becomes common in the upper part of the NP 21 Zone. In this section it has not been possible to identify the Subzone CP 16b defined by the end of the acme of E. subdisticha (E. obruta) because the end of the acme of E. obruta overlaps the LO of E. formosa.

In the Oligocene the assemblage is still rich and does not differentiate from the one already described for the Late Eocene. The LO of I. recurvus, event found in the italian sections coincident with the LO of R. umbilica (Levin), occurs in this section before the LO of E. formosa. The LO of R. umbilica, marker of the Zone CP 17, and B. serraculoides Gartner occurs at sample 16.5 . Few species of I. recurvus, R. umbilica and B. serraculoides were found in sample 17 and are considered reworked.

The first occurrence (FO) of S. distentus (Martini), marker of the base of the Zone CP 18, is found in the upper part of the section. L. minutus and Z. bijugatus, usually common throughout the section, decrease abruptly in abundance in the latter zone.

As a conclusion, we can summarize the previous results stating that the E/O boundary is not marked by a drastic change in the calcareous nannofossil assemblage, but is rather characterized by sequential extinctions of C. protoannula, C. reticulatum, D. barbadiensis, D. saipanensis and variations in the abundances of I. recurvus and E. formosa (Nocchi et al. (1985), Perch-Nielsen et al. (1985)).

REFERENCES

Backman, J. and Shackleton, N.J., 1983. Quantitative biochronology of Plio-

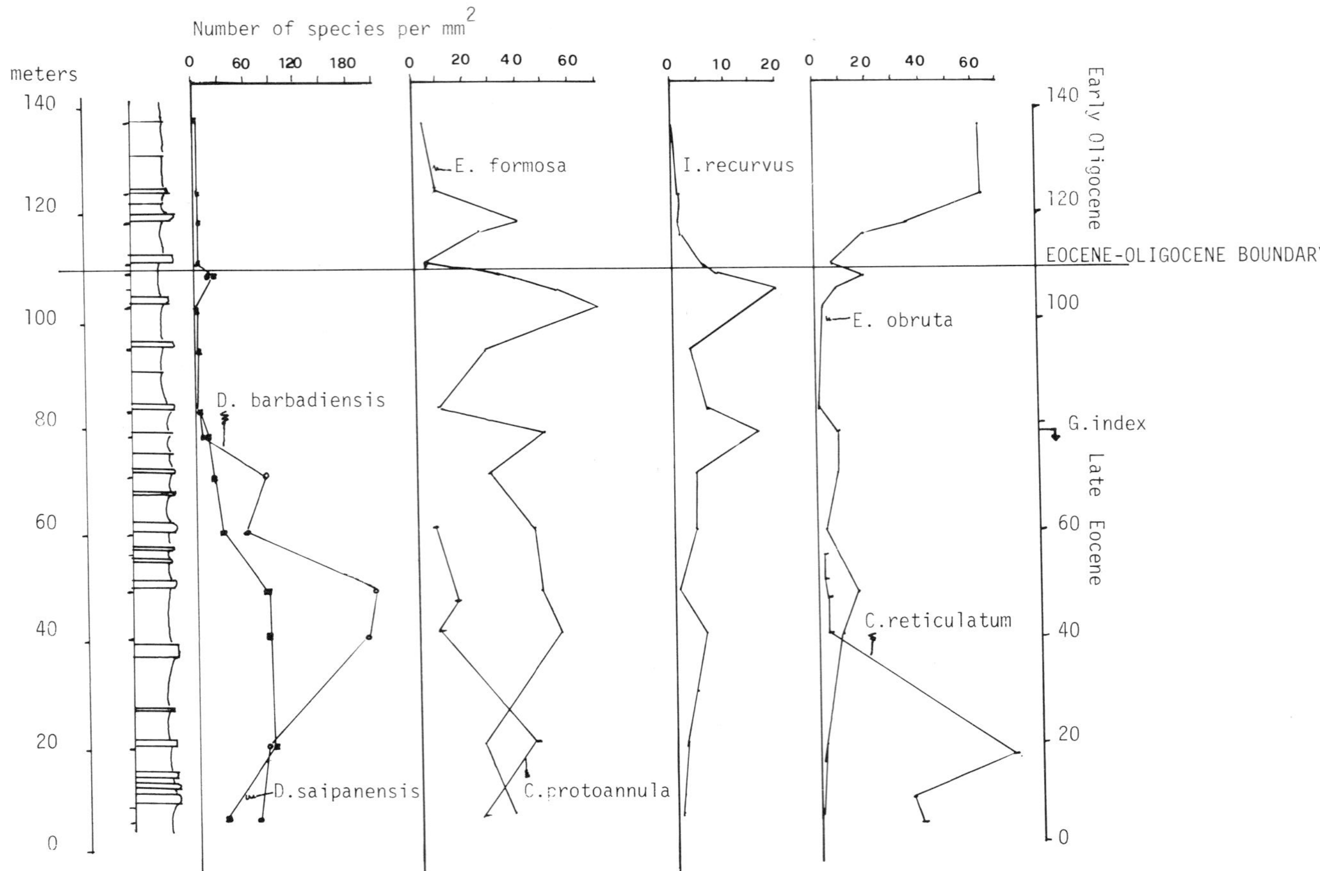

Fig. 2 Semiquantitative data of distribution of seven species of Fuente Caldera section (Data from Monechi and Perch-Nielsen, unpublished data)

cene and Early Pleistocene calcareous nannofossils from Atlantic, Indian and Pacific oceans. Mar. Micropaleont., 8: 141-170.

Okada, H. and Bukry, D., 1980. Supplementary modification and introduction of code numbers to the low latitudes coccolith biostratigraphic zonation. Mar. Micropaleont., 5: 321-325.

Martini, E., 1971. Standard Tertiary and Quaternary calcareous nannoplankton zonation. In Farinacci, A., (Editor) Proceed. Second Planktonic Conference. Roma 1970. Tecnoscienza, 2: 739-785.

Nocchi, M., Parisi, G., Monechi, S., Madile, M., Monaco, P., Napoleone, G., Orlando, M., Premoli Silva, I., Ripepe, M., Benson, R., and Bice, D., 1985. The Eocene-Oligocene boundary in the Umbrian pelagic sequences, Italy. Special Issue IGCP 174. Elsevier (in press).

Perch-Nielsen, K. et al., 1985. Calcareous nannofossils at the Eocene-Oligocene boundary. Special Issue IGCP 174. Elsevier (in press).

GEOCHEMISTRY OF THE FUENTE CALDERA SECTION (Spain).

F. BERTHENET [1], S. CLAUSER [1-2], and M. RENARD[1]

[1]Dept. Geol. Sed. and LA CNRS 319, Univ. P. & M. Curie, 4, Place Jussieu, 75252 - PARIS Cedex 05 - (France).

[2]BRGM, BP 6009, 45060 ORLEANS Cedex (France).

The geochemical investigation was performed on the same samples studied by Molina (this volume) for the determination of the planktonic Foraminifera and by Monechi (this volume) for the Nannofossils.

The characteristics and details of this section can be found in Molina (this volume).

I - STABLE ISOTOPES (fig. 1)

A) Carbon 13

Following a certain stability ($\delta 13_C$ + 1,50 °/oo) during the P.semiinvoluta Zone, the $\delta 13_C$ strongly decreases in the course of the C. inflata Zone and drops to a minimum ($\delta 13_C$ = 0 °/oo) at the base of the C. lazzarii Zone. The signal remains very low throughout the major part of this zone. An abrupt positive shift appears at the top of the C. lazzarii Zone (92 m), just below the Eocene-Oligocene boundary. The $\delta 13_C$ slightly and progressively increases within the base of the G. gortanii Zone (with a maximum of + 1,50 °/oo) and reaches a stable high level at the top of the section.

At first approximation, the fluctuations of the $\delta 13_C$ are linked to the variations of the primary productivity, which in turn correlate according to the Broecker model (1982) with transgressive/regressive cycles. The correlation between regression and low levels of $\delta 13_C$ in pelagic carbonates has been demonstrated by Renard (1984) for the main regressive phases since the Late Jurassic. Assuming the model to be correct two major events could be inferred in the vicinity of the Eocene-Oligocene boundary in this section.

1) The maximum of regression would be located at 85 m (below the Eocene-Oligocene boundary). Taking into account the rather bad sampling, this event might coincide with the limit between the C. inflata and C. lazarii Zones.

2) A sudden transgressive pulse might take place around 92 m just below the Eocene-Oligocene boundary.

Hence there is good agreement between $\delta 13_C$ and planktonic Foraminifera zoning, since the boundaries of the zones P. semiinvoluta/ C. inflata,

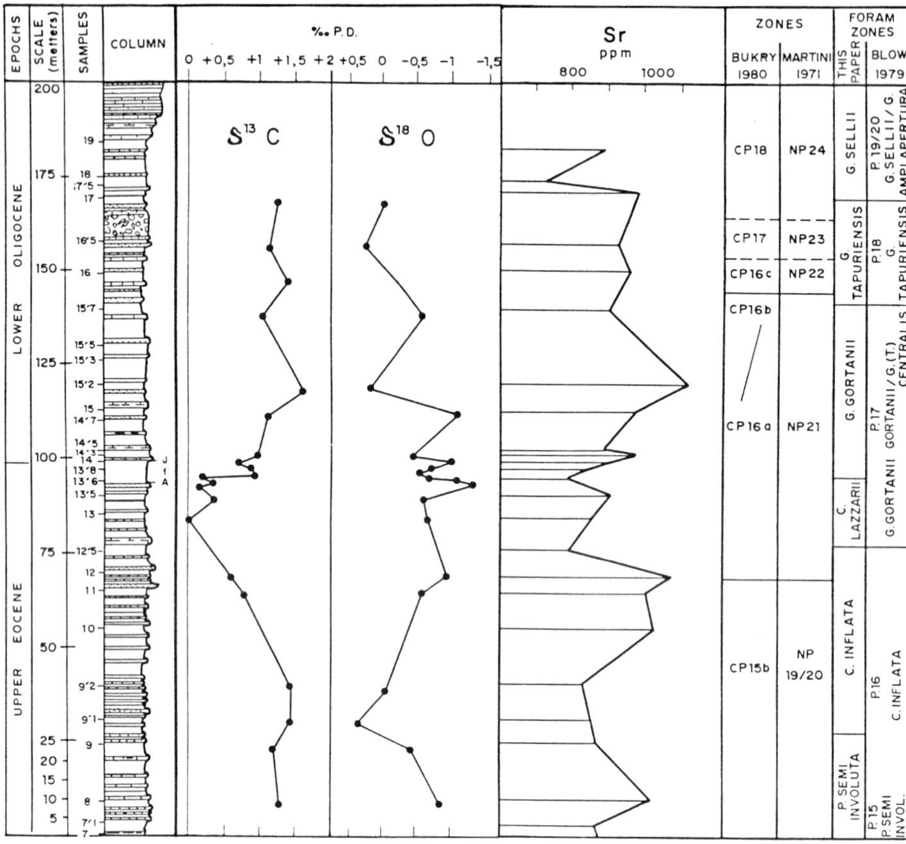

Fig. 1. Stable isotopes and strontium contents of bulk carbonates from the Fuente Caldera section.

C. inflata/C. lazzarii and C. lazzarii/G. gortanii are well obvious in geochemistry. However, the major event concerning $\delta 13_C$ seems located at 85 m (transition between transgressive and regressive overlap) at the limit of C. inflata/C. lazzarii zones rather than at the Eocene-Oligocene boundary. Conversely, nannofossil zoning appears to be rather independent of geochemistry.

B) Oxygen 18

The $\delta 18_O$ first increases from negative values (- 0,80 ‰) to reach a maximum (+0,25 ‰) at the transition from the P. semiinvoluta to the C. inflata Zones. The isotopic ratio then decreases greatly as far as 110 m. In fact, the decrease mainly takes place within the first part of the C. inflata Zone, whence the $\delta 18_O$ fluctuates around - 0,75 ‰. Quenching of the values occurs around the limit NP 20/NP 21.

A sudden positive excursion operates between 110 m and 120 m and results in a change of 0,8 from - 1 °/₀₀ to 0,30 °/₀₀.

The present study thus confirms the lag of the oxygen shift with respect to the Eocene-Oligocene boundary (Vergnaud-Grazzini et al., this volume).

During the Oligocene the values remain high and fluctuate around 0 °/₀₀. After an initial phase of cooling during the P. semiinvoluta Zone, warming during the C. inflata and C. lazzarii Zones and the base of the G. gortanii Zone is observed. A rather sudden drop of temperature starts next, leading to the low values of the Oligocene. It is however not easy to account specifically for the influences of climatic fluctuations and for modifications of the oceanic circulation on the evolution of the $\delta 18_O$ (Renard this volume).

II - STRONTIUM (fig. 1)

The short-term evolution of this signal is more or less directly linked with transgression and regression cycles (Renard 1984).

Following high values (1050 ppm around 10 m), the Sr concentrations decrease within the P. semiinvoluta and C. inflata Zone (down to 900 ppm) and reincrease up to a maximum at 65 m. They then drop suddenly (Sr = 880 ppm) at the vicinity of the C. inflata/ C. lazzarii limit. They remain low within the C. lazzarii zone and reincrease progressively to reach a maximum (1100 ppm) at 115 m. The limit C. lazzarii - G. gortanii is characterized by depression of the Sr content. The values fluctuate later on around 1000 ppm. during the Lower Oligocene.

CONCLUSION :

Figure 2 summarizes the evolution of the 3 geochemical markers studied. The main geochemical events are indicated by stars, the number of which underlines the importance of the event. Five geochemical breaks are observed :

- Boundary G_1, strongly indicated by $\delta 13_C$ and $\delta 18_O$ and to a lesser extent by Sr, corresponds to the limit of foraminiferan P. semiinvoluta/C. inflata Zones.
- Boundary G_2 is prominent for Sr and medium for $\delta 18_O$; it corresponds to the limit of NP 20/NP 21 Nannozones.
- Boundary G_3 (very strong for $\delta 13_C$ and Sr) includes the limit of the foraminiferan C. inflata and C. lazarii Zones.
- Boundary G_4 ($\delta 13_C$ and Sr) is the Eocene-Oligocene boundary.
- Boundary G_5 ($\delta 13_C$, $\delta 18_O$ and Sr) is within NP 21 and might correspond to CP 16A/CP 16B nannozone transition.

The present study confirms the satisfactory agreement between chimiostratigraphy and biostratigraphy.

The Eocene-Oligocene boundary is located in the middle of a sequence of geochemical events which affect the ocean all over the world and it is not easy to stress one of them.

However, at first approximation and although final conclusions are presently impossible because of the small number of observations, it appears that the limits of Nannozones correlate preferentially with geochemical events which occur around the maximum of transgressive phases, whereas the limits of planktonic Foraminifera zones correlate preferentially with geochemical events linked to regressions.

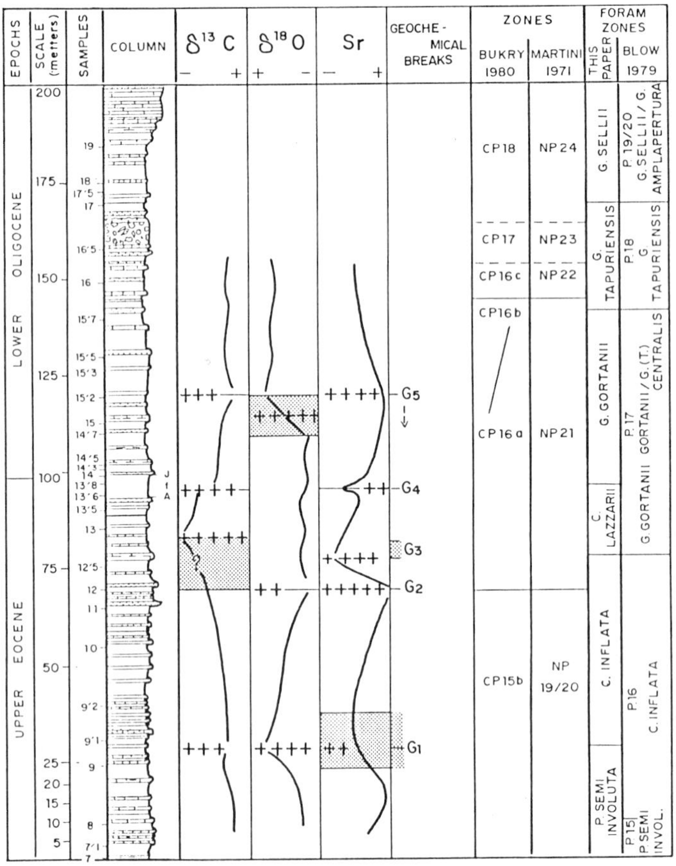

Fig.2. Correlation between geochemical breaks and biostratigraphic data

REFERENCES

BERTHENET F., 1985. Evénements géochimiques à la limite Eocene-Oligocene. Application aux coupes de la région de Grenade (Espagne) et de Biarritz (Pyrénées Atlantique). Mém., DEA, option stratigraphie. Université P. & M. Curie, Paris, Roneo. , 30 pp.

BROECKER W.S., 1982. Ocean Chemistry during glacial time. Geoch. Cosmoch. Acta, vol. 46, pp. 1689-1705.

RENARD M., 1984. Géochimie des carbonates pélagiques. Thèse Univ. P. & M. Curie, n° 84-16 et Mém. BRGM n° 85, 1985, 660 pp.

BIOSTRATIGRAPHIC CORRELATION BETWEEN THE CENTRAL SUBBETIC (SPAIN) AND UMBRO-MARCHEAN (ITALY) PELAGIC SEQUENCES AT THE EOCENE/OLIGOCENE BOUNDARY USING FORAMINIFERA.

EUSTOQUIO MOLINA*, PAOLO MONACO**, MARISA NOCCHI** and GUIDO PARISI**

ABSTRACT

The most significant events, which occurred during the time interval spanning from the Late Eocene to the Early Oligocene, are preserved in the continous pelagic sequences of the Subbetic basin (Spain) and Umbro-Marchean basin (Italy). A comparative analysis has been carried out between events common to both areas, based on planktonic and benthic foraminifera.

The relationship between Hantkeninidae and other selected taxa were used to identify and correlate the biozones identified in the two areas. We show that below the Eocene/Oligocene boundary several successive extinctions occur in the same order both in Spain and in Italy. Particularly, the Turborotalia cerroazulensis group extinction is slightly prior to the Hantkeninidae and large Pseudohastigerina extinctions, which latter mark the Eocene/Oligocene boundary.

INTRODUCTION

This paper deals with the comparison of the biostratigraphic data concerning the Eocene/Oligocene transition from southern Spain (Molina et al., this volume) and Umbro-Marchean region (Italy) (Nocchi et al., this volume). Such a comparison is one of the results of the strict collaboration among scientists from different countries, which occurred under sponsorship of IGCP Project 174 on "Terminal Eocene Events".

The sequences here considered, from both Spain and Italy, are well exposed, continous, pelagic and offer a good biostratigraphic resolution. Sections of this type and age are seldom found in the Mediterranean area because of the general regressive trend related to the Alpine orogenesis.

The stratigraphic interval studied extends from the Middle Eocene/Upper Eocene boundary to the Lower Oligocene/Middle Oligocene boundary, from P14/P15 to P20/P21 of Blow's (1969) biozonation. This comparative study based mainly on planktonic foraminifera, allowed us to recognize many events common to both sequences and to demonstrate a close isochrony between Spanish and Italian foraminiferal events.

* Department of Paleontology, University of Zaragoza, 50009 Spain.
**Department of Earth Sciences, University of Perugia, 06100 Italy.

This comparison permits to resolve some biostratigraphic problems concerning the correlations between Mediterranean sections and the standard biozonation of Blow (1969).

GEOLOGICAL SETTING AND LITHOSTRATIGRAPHY

The Umbro-Marchean and Subbetic sequences (see fig.1) consist of deep water marls and calcareous marls. These pelagic and hemipelagic sediments contain intercalations of calcareous turbidites derived from adjacent carbonate platforms in both areas.

The Spanish sequences of Fuente Caldera, Molino de Cobo and Torre Cardela belong to the Cañada Formation of the Cardela group defined by Comas (1978). The sediments of this formation were deposited in the Median Subbetic trough, which was subsiding during the Eocene. This trough belongs to the external Zone of the Betic Cordillera, which represents the westernmost mountain system of the European Alpine chain.

The coeval Umbro-Marchean sequences of Collecastellano, Visso and M. Cagnero sections, belong to the informally named "Scaglia variegata" and "Scaglia cinerea" units. The M. Cagnero section was previously studied by Baumann and Roth (1969) and was revisited to carried out the present comparison.

During the Eocene and Oligocene, the Umbro-Marchean area was characterized by a system of basin, seamounts and platforms connected to the Adriatic promontory, which was part of the African plate (Channel et al., 1979).

Pelagic sediments, rich in calcareous plankton, accumulated in the basin at the edge of which, in southern part, the calcareous turbidites intercalated testify the vicinity of the Latium-Abruzzi carbonate platform.

The dominant litotypes of the "Scaglia" and Cañada Formation are pelagites and hemipelagites. In the Umbro-Marchean sequences the pelagic limestones and marls are reddish and gray in the Upper Eocene, and gray at the top of the Eocene and in the Lower Oligocene (see Monaco et al., in press), while Spanish sequences consist entirely of gray marls. In addition, the sediments of the Spanish sequences contain more clay and less carbonate then the Italian sections. The Subbetic marly intervals have varying amounts of microfossils as well as rare glassy spherules. The planktonic foraminifera are well preserved and abundant, costituting 85% of almost all the washed samples.

BIOSTRATIGRAPHIC ANALYSIS

The stratigraphic interval in this study extends from the Middle Eocene/Upper

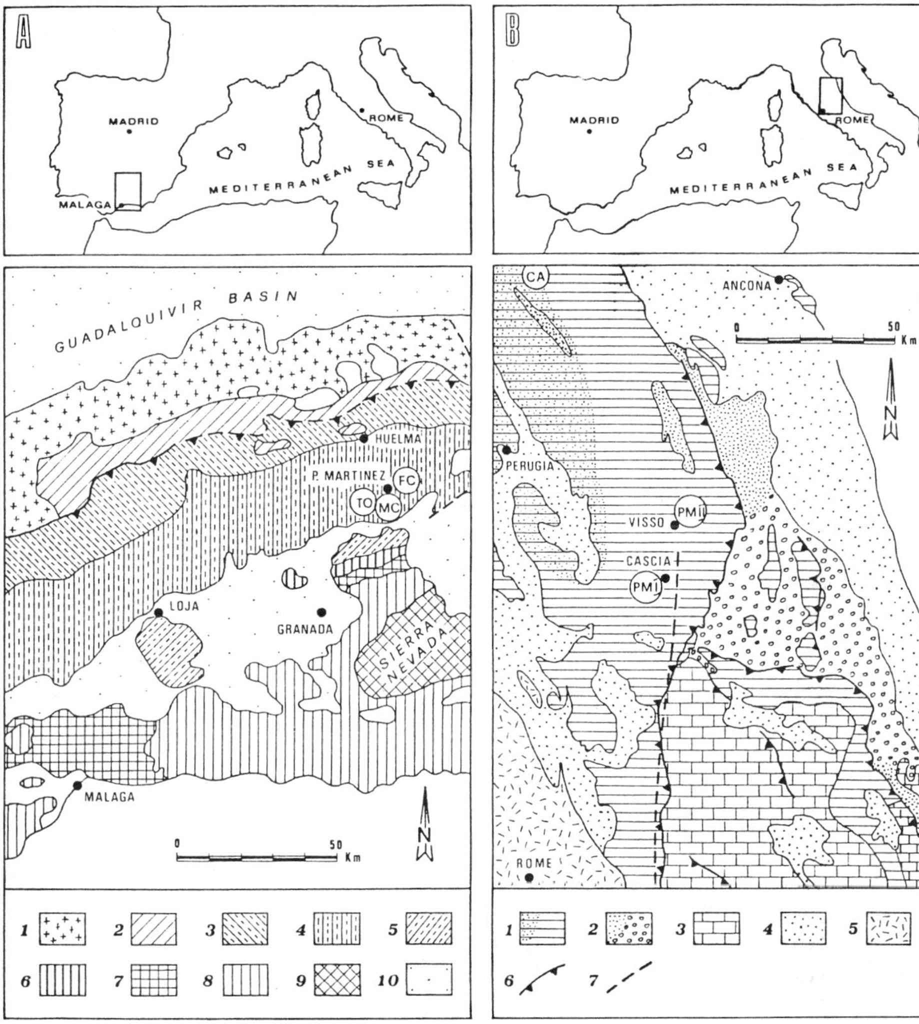

Fig.1. Geological sketch maps of Betic Cordillera and Umbro-Marchean area and location of the sections studied. A): TO: Torre Cardela section; MC: Molino de Cobo section; FC: Fuente Caldera section. B): CA: M. Cagnero section; PM II: North Visso section; PM I: Collecastellano section.
A)- 1) Guadalquivir allochtonous units; 2) Intermediate units; 3) External Subbetic; 4) Median Subbetic; 5) Internal Subbetic; 6) Rondaides or Betic dorsal; 7) Malaguide; 8) Alpujarride; 9) Nevado Filabride; 10) Upper Miocene-Pliocene-Quaternary.
B)- 1) Mesozoic and Cenozoic pelagic successions and flysches in the Umbro-Marchean basin; 2) Miocene-Pliocene late-orogenic successions; 3) Latium-Abruzzi carbonate platform; 4) Marine and continental Plio-Pleistocene post-orogenic successions; 5) Vulcanites of Tuscan-Latium system; 6) Overthrusts; 7) Fault.

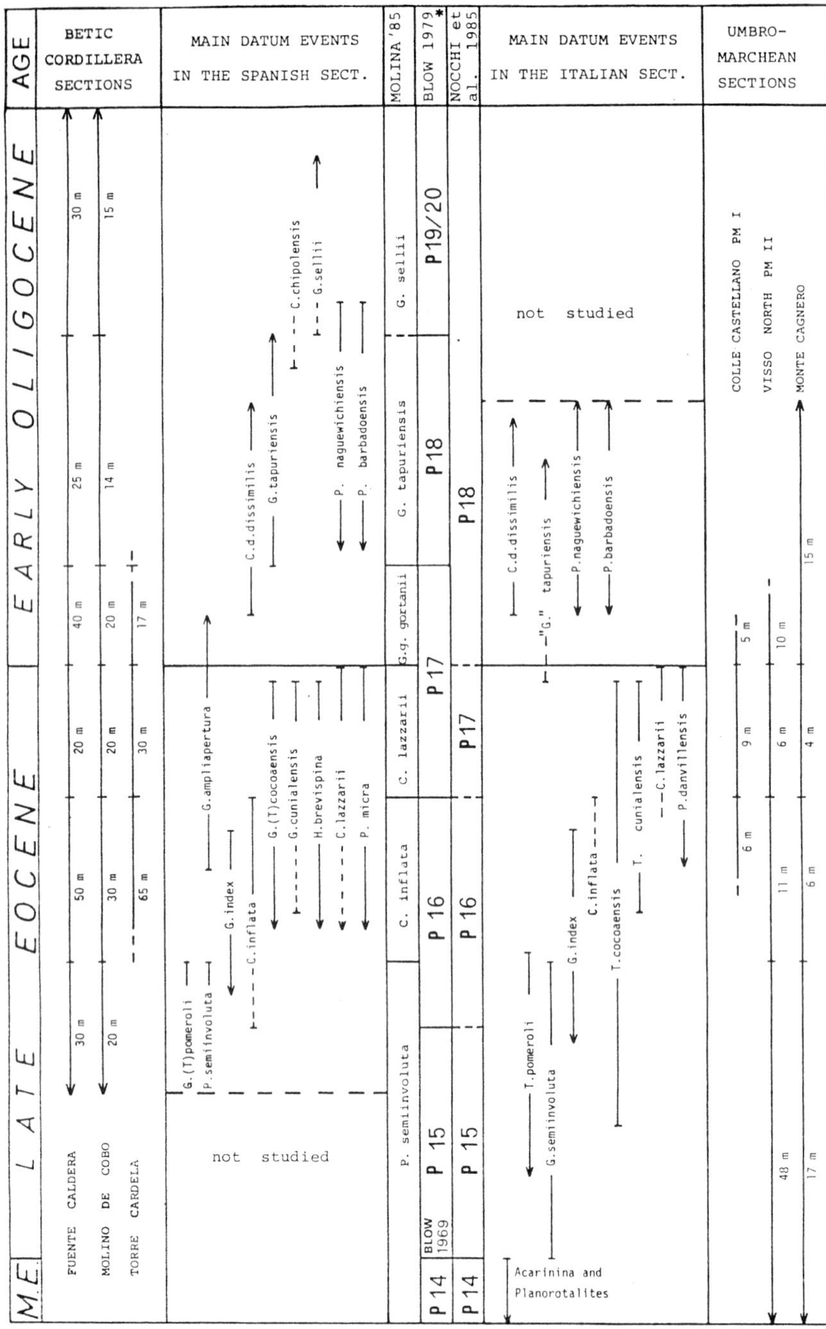

Fig.2. Main planktonic fòraminifera events in the Spanish and Italian sections. The dashed line indicates that the taxon is rare.
* As reported by Berggren et al.,(1983).

Eocene boundary to just below the Lower Oligocene/Middle Oligocene boundary, and thus encompasses Blow's (1969) biozones P14 to P20. The Middle Eocene/Upper Eocene boundary has not been studied in the Italian sections. A synthesis of our data and a summary of the main planktonic foraminiferal events are indicated in fig.2. This figure shows the biozones from the Italian and Spanish sections, as well as the stratigraphic thickness of each biozone. These biozones are correlated to those defined by Blow (1969), which were plotted as having the same thickness, thus distorting their chronostratigraphic extent.

A comparison of the same biostratigraphic interval between the Spanish and Italian sections reveals that the Subbetic sequences are thicker then the Umbro-Marchean ones. The thickness ratio, calculated between LO of Globigerinatheka semiinvoluta and the Eocene/Oligocene boundary, ranges from 5/1 to 8/1. Consequently the biostratigraphic events are more diluted in the Spanish sections and the microfauna are better preserved in these more argillaceous sequences.

The LO of G. semiinvoluta is the first event that was considered because it can be indirectly used for identifying the P15/P16 zonal boundary. The identification of this boundary is based on the FO of Cribrohantkenina inflata (Blow, 1969). In the Spanish sections the latter taxon overlaps for few meters with G. semiinvoluta, but it is rare in the lower portion of its range while in the upper part is more common. Cribrohantkeninas are rare or absent in the Umbro-Marchean sections except at the top of the Upper Eocene. Molina et al., (this volume) in the Late Eocene retained a lower G. semiinvoluta Zone corresponding to the total range of the nominal taxon, according to the Bolli's (1957a) original definition, overlain by the C. inflata Zone, the interval from the extinction of G. semiinvoluta to the extinction of G. inflata, which, consequently, results to be shorter than Blow's (1969) Zone P16. Based on the Spanish data, in the Umbrian sections the P15/P16 boundary was located just below the LO of G. semiinvoluta.

The next event in this study is the FO of Turborotalia cunialensis (Globorotalia cunialensis sensu Molina) that occurs just above the LO of G. semiinvoluta. This event is less evident in the Spanish sections than in the Italian sections. It should be noted that this event always is within P16.

Globigerinatheka index (Globigerapsis index sensu Molina) is common from the base of the sequences in both area. Its extinction is distinct and occurs after the appearance of T. cunialensis and before the extinction of C. inflata. This event always falls within P16 and G. index and T. cunialensis coexist in each section examined. The disappearance of G. index is accompanied by a very strong decrease in abundance of all Globigerinathekae. In the Italian sections, only rare forms of Globigerinatheka tropicalis continue up to the Eocene/Oligocene boundary, where they become extinct.

The P16/P17 boundary is defined by the LO of C. inflata that occurs above the LO of G. index.

At the top of the Upper Eocene there are further extinctions of the most typical Eocene forms, but they are not simultaneous. The extinction of the Turborotalia cerroazulensis group coincident with the LO of Hantkenina brevispina in Spain, occurs before LO of Hantkeninids and Pseudohastigerinids with subacute margins and greater then 150 microns in diameter. These last events are considered to characterize the Eocene/Oligocene boundary.

The P17/P18 boundary is not clearly defined. In the Umbro-Marchean sections, we have specimens attributed to "Globigerina" tapuriensis at the Eocene/Oligocene boundary while in Spain this taxon, in agreement with Blow (1979), appears above the boundary. After the Eocene/Oligocene boundary the FO of Catapsydrax d. dissimilis occurs in both areas.

Cassigerinella chipolensis has not been found in the lowermost Oligocene in either area. In Spain, this form appears close to the boundary P18/P19 which is later than has been reported by Bolli (1957b).

The small Pseudohastigerinids, P. naguewichiensis and P. barbadoensis, which are present since the top of the Upper Eocene, continue into the Lower Oligocene, while larger forms (greater then 150 microns), named P. micra by Molina et al. (this volume) and P. danvillensis by Nocchi et al. (this volume) disappear at the Eocene/Oligocene boundary.

The comparison of deep water benthic foraminifera from the two areas has revealed some similarities. Benthic microfauna are always subordinate to the planktonic one. Nuttallides truempyi is present but discontinous at the top of Eocene, and disappears before the Eocene/Oligocene boundary. In the Fuente Caldera section, the latter taxon disappears simultaneously with the T. cerroazulensis group; in the Italian sections N. truempyi disappears slightly before the T. cerroazulensis group.

After the disappearance of the T. cerroazulensis group, we find the FO of Anomalinoides pompilioides and the LO of Alabamina dissonata. After the LO of C. inflata there is in both areas, the occurrence of small forms of Bolivinids (Bolivina antegressa gr., B. floridiana and B. obscuranta) which characterize the fractions smaller than 150 microns. After the boundary P15/P16 Aragonia, represented mainly by A. janoscheki, disappear although they are already rather scarce since P15. After the FO of T. cunialensis we find the occurrence of large agglutinated foraminifera such as Cyclammina acutidorsata and Haplophragmoides sp.. Differences between the two areas concern the Uvigerinids, which are more abundant in the Italian sections after the Eocene/Oligocene boundary, while in Spain they remain subordinate to other forms. In the Upper Eocene the Uvigerinids are represented by U. rippensis, U. multistriata and U. havanensis; after the boundary U. eocaena is common. In the Upper Eocene of the Italian sections Cibicidoides and Heterolepa are more prevalent then Anomalinoides and Lenticulina, whereas the opposite is found in the Spanish sections.

The deep water assemblages in Spain are much rarer, however, they exhibit some similarity to those of the Italian basin considered in this work.

The shallow water microfaunas, which are found in the calciruditic and calcarenitic levels, are sometimes scattered in the marls, thus recording the presence of similar neritic source areas adjacent to the basins in both areas. The assemblages are generally characterized by Discocyclinidae and red algae. At the top of the Eocene, Asterigerina rotula and Gypsinidae become more common, while Nummulites are rare. In both areas, rare Cretaceous forms are sometimes present close to the Eocene/Oligocene boundary.

Upper Eocene shallow water microfaunas continue upward in to the portion of the Oligocene studied here. Rotalidae (mainly Pararotalia spp.) are the common forms at the base of the Oligocene in both areas. In the Fuente Caldera and Molino de Cobo sections of Spain, the first occurrence of Lepidocyclina is above the interval studied here, near the boundary between P20 and P21, before the FO of Globigerina angulisuturalis.

CONCLUSIONS

The comparison between the Subbetic and Umbrian areas has allowed us to recognize biostratigraphic events common to both areas occurring in the same vertical order as shown in fig.2.

Isochronous extinctions of G. index and Discoaster barbadiensis and D. saipanensis have been recognized by Nocchi et al. (this volume) and by Molina et al. (this volume). In absence of C. inflata these latter events were used to identify with same uncertainty the P16/P17 boundary in Italy. The comparison carried out in this work clearly shows that the events mentioned above occur below the LO of C. inflata. Therefore, the P16/P17 boundary must be drown above the LO of G. index and not equated to it.

In the Spanish sections Hantkeninidae are common or abundant and continously present, while in the Italian sections an increase in Hantkeninidae occurs after the disappearance of G. index. Moreover, it has been confirmed that the extinction of T. cerroazulensis group occurs slightly before the Hantkeninidae extinction, represented mainly by Cribrohantkenina lazzarii in both areas. The last occurrence of Pseudohastigerina with a diameter greater then 150 microns corresponds to the Hantkeninidae extinction which indicates the Eocene/Oligocene boundary.

The Subbetic sections and some Umbrian sections contain calcarenitic intercalations with reworked skeletal debris coming from a carbonate platform. The similar faunal content points out that in the Late Eocene there was an adjacent neritic environment with a high organic activity in both areas. In the lower planktonic biozones of the Oligocene Discocyclinidae are dominant. Chapmanina and Pellatispira, which are taxa characteristic of the Upper Eocene, have been found in both the Spanish and Italian sections while Lepidocyclinidae are absent and appear much later in the Late Oligocene.

REFERENCES

Baumann, P., 1970. Mikropaläontologische und stratigraphische untersuchungen der obereozänen-oligozänen Scaglia in Zentralen Apennin (Italien). Eclogae Geol. Helv.,63(3):1133-1211.

Baumann, P. and Roth, P.H., 1969. Zonierung des Obereozäns und Oligozäns des Monte Cagnero (Zentralapennin) mit planktonischen Foraminiferen und Nannoplankton. Eclogae Geol.Helv.,62(1):303-323.

Berggren, W.A., Kent, D.V. and Flyn, J.J., in press. Paleogene geochronology and chronostratigraphy. In:N.J.Snelling (Editor), Geochronology and Geological Record Geol.Soc.London,Spec.Pap.,1983.

Blow, W.H., 1969. Late middle Eocene to Recent planktonic foraminifera biostratigraphy. In: Proc.1st Plank.Conf.Geneva,1967,199-422.

Blow, W.H., 1979. The Cainozoic Globigerinida. In:Leiden E.J.Brill; I,II and III.

Bolli, H.M., 1957a. Planktonic Foraminifera from the Eocene Navet and S. Fernando formation of Trinidad, B.W.I. U.S.Nat.Mus.Bull.;215:155-198.

Bolli, H.M., 1957b. Planktonic Foraminifera from the Oligocene-Miocene Cipero and Lengua formations of Trinidad, B.W.I. U.S.Nat.Mus.Bull.:215:97-131.

Bolli, H.M., 1972. The genus Globigerinatheka Broennimann. Jour.Foram.Res.,2(3): 109-136.

Channel, J.E.T, D'Argenio, B. and Horvath, F., 1979. Adria, the African promontory in Mesozoic Mediterranean paleogeography. Earth Science Reviews,15:213-292.

Comas, M.C., 1978. Sobre la Geologia des Montes Orientales: Sedimentacion y evolucion paleogeografica desde el Jurasico al Mioceno inferior (zona Subbetica, Andalucia). Thesis, Univ.Bilbao:1-323.

Molina, E., Monechi, S. and . Description and biostratigraphy of the main reference section of the Eocene/Oligocene boundary in Spain: Fuente Caldera section. This volume.

Monaco, P., Nocchi, M. and Parisi, G., in press. Analisi stratigrafica e sedimentologica di alcune sequenze pelagiche dell'Umbria sudorientale dall'Eocene inferiore all'Oligocene inferiore. Bull.Soc.Geol.It., 1985.

Nocchi, M., Parisi, G., Monechi, S., Madile, M., Monaco, P., Napoleone, G., Orlando, M., Premoli Silva, I., Ripepe, M. and Bice, D.. The Eocene/Oligocene boundary in the Umbrian pelagic sequences, Italy. This volume.

PLATE 1

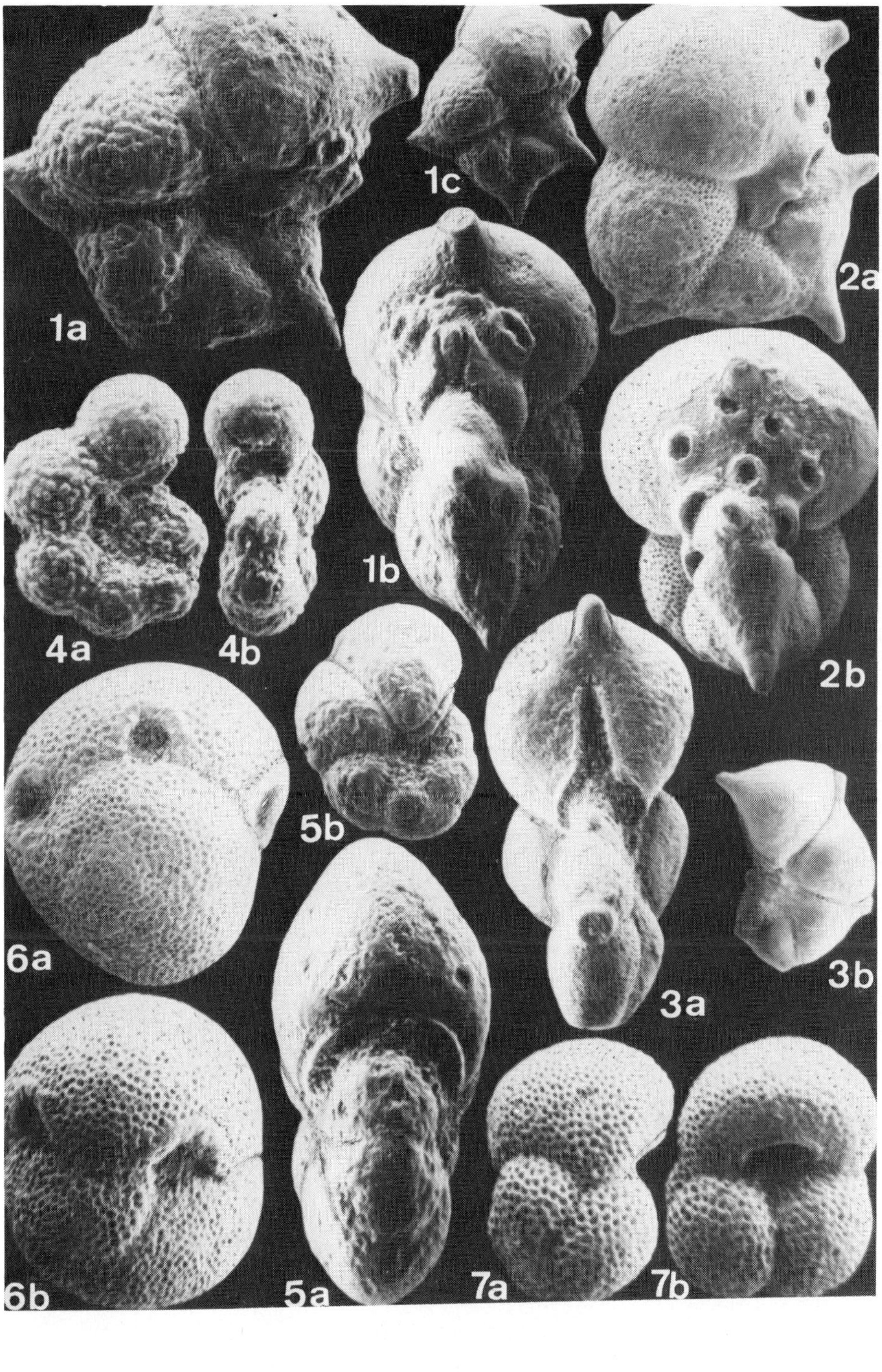

PLATE 2

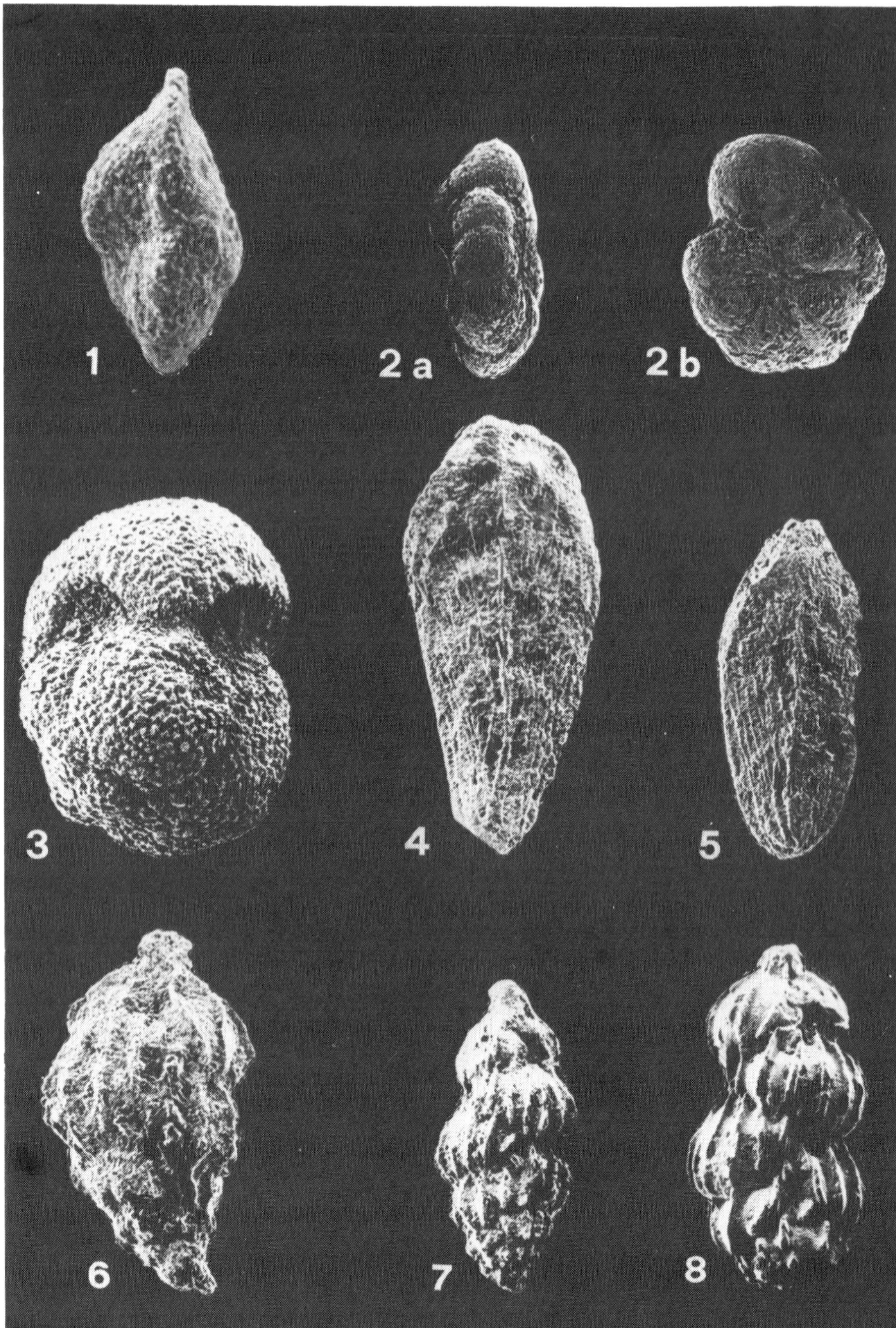

PLATE 1

1. Cribrohantkenina lazzarii (PERICOLI). Sample FC 13,8 Spain. Zone of C. lazzarii. x 100 (a,b), x 50 (c).
2. Cribrohantkenina inflata (HOWE). Sample FC 12 Spain. Zone of C.inflata. x 100.
3. Hantkenina brevispina CUSHMAN. Sample FC 7 Spain. Zone of P.semiinvoluta. x 100 (a), x 50 (b).
4. Pseudohastigerina naguewichiensis (MYATLIUK). Sample TC 9. Zone of G. g. gortanii. x 200.
5. Pseudohastigerina micra (COLE). Sample FC 12 Spain. Zone of C. inflata. x 200 (a), x 100 (b).
6. Porticulasphaera semiinvoluta (KEIJZER). Sample FC 7 Spain. Zone of P. semiinvoluta. x 100.
7. Globigerina tapuriensis BLOW & BANNER. Sample FC 17 Spain. Zone of G. tapuriensis. x 100.

PLATE 2

1. Turborotalia cunialensis (TOUMARKINE & BOLLI). Sample PMI 0 Italy. Zone P 16. x 100.
2. Pseudohastigerina danvillensis (HOWE & WALLACE). Sample PMI 26 Italy. Zone P 17. x 100 (a,b).
3. Globigerinatheka tropicalis (BLOW & BANNER). Sample PMI 7 Italy. Zone P 16. x 100.
4. Bolivina gr.antegressa SUBBOTINA. Sample PMI 26 Italy. Zone P 17. x 100.
5. Bolivina gr.antegressa SUBBOTINA. Sample PMI 26 Italy. Zone P 17. x 100.
6. Uvigerina sp. Sample PMI 37 Italy. Zone P 18. x 100.
7. Uvigerina multistriata HANTKEN. Sample PMI 37 Italy. Zone P 18. x 75.
8. Uvigerina cf multistriata HANTKEN. Sample FC 13,8 Spain. Zone of C. lazzarii. x 50.

ENGLISH REFERENCE SECTIONS WHICH SPAN THE EOCENE/OLIGOCENE BOUNDARY

D. CURRY[1] and E.A.HAILWOOD[2]
[1]Geology Dept., University College, Gower St., London WC1E 6BT (England)
[2]Oceanography Dept., Southampton University, Southampton S09 5NH (England)

INTRODUCTION

Sections spanning the Eocene/Oligocene boundary occur in England in the Hampshire Basin on the Isle of Wight and the nearby Hampshire coast. They are well-exposed in sea-cliff sections and have been studied for over 150 years. They are for the most part continental in origin and overlie a dominantly marine sequence without obvious major hiatuses which commences in the latest Palaeocene, the whole having been deposited in a gently subsiding basin. Criteria for the identification of the Eocene/Oligocene boundary are not yet internationally agreed but all workers would accept that, in the English sequences, this boundary must lie within the Solent Group of Daley and Insole (1984) (Headon, Osborne, Bembridge and Hamstead Beds of early authors). This unit has a total thickness of some 180 m and represents a time-span of about 5 m.y. Basic data in relation to the sequences are summarised in the notes which follow. The most important sections (see Figure 1) are at Whitecliff Bay (SZ6486), Bouldnor (SZ 3991) and Headon Hill (SZ306860) in the Isle of Wight; and Hordle Cliff (SZ264922), near Christchurch, Dorset. The two first of these sections have been selected (see Table 1) to typify the English succession.

STRATIGRAPHY

Lithology, sedimentology and structure

The sequences are of unconsolidated clays with subordinate fine sands and discontinuous beds of limestone. They are mostly horizontal, but locally are involved in the Isle of Wight monocline along the southern edge of the outcrop. For details see White (1915) and Daley & Insole (1984). Little has been published on the heavy minerals, clay minerals or geochemistry of the sequences.

Palaeomagnetism

Studies are in progress on the rock sequences illustrated in Table 1 under the direction of one of the present authors (EAH). The remanent magnetism of much of the sequence appears to be stable, and initial results suggest a dominant normal polarity throughout the Headon beds and a reverse polarity throughout much of the Bembridge Marls and the lower Hamstead Beds.

PALAEONTOLOGY

The mollusc and ostracod faunas indicate that the sequences were for the most part laid down in fresh water. The only fully marine sequence is that of the Brockenhurst Beds (Middle Headon Beds) and (possibly) the top of the Hamstead Beds. Restricted marine faunas are present in most of the remainder of the Middle Headon Beds, the Bembridge Oyster Beds and the Upper Hamstead Beds. However indications of mesohaline conditions are present at levels scattered throughout the sequences (see schematic indications in Table 1).

Mollusca

The marine fauna of the Brockenhurst Beds is matched most closely in the Grimmertingen sands of Belgium and the Latdorf beds (type Lattorfian) of Germany (Curry 1967). That of the Upper Hamstead Beds resembles that of the type Sannoisian of the Paris Basin. The succession of faunas of continental molluscs can be matched at several localities in France (Rey 1966).

Ostracoda

See Keen (1977, 1978). About 40 species are known from the Middle Headon Beds and 15 each from the Bembridge Oyster Beds and Upper Hamstead Beds. There are good possibilities of correlation with continental Europe.

Mammals

Over 90 species in all are known, from six main levels (Hooker & Insole 1980). Comparison with French localities shows that the fauna of the Lower Headon Beds is somewhat older than that of Euzet; that from the Upper Headon/ Osborne Beds may equate to that of La Debruge; the Bembridge Limestone fauna equates to that of Montmartre (Haute Masse), and the Lower Hamstead Beds yield elements of the Ronzon fauna.

Nannoplankton

See Aubry (1983). The Brockenhurst Beds have yielded *Isthmolithus recurvus*, *Discoaster saipanensis* and *D. tani*, and so have been referred to Zone NP19-20. Nannoplankton have not been recorded at levels above the Middle Headon Beds.

Palynology

See Machin (1971), Châteauneuf (1980). The sequences are in a facies favourable for preservation, and pollen is common. Dinoflagellates have been found at a few levels and characterise the *clathrata* and *gochti* Zones.

Seeds and leaves

Remains, especially of water-plants, occur throughout the succession.

Charophytes

See Feist-Castel (1977), Riveline (1984). About 30 species have been recognised and a succession of four floral zones has been defined based on the English successions (see Table 1).

CONCLUSION

English sections across the Eocene/Oligocene boundary are very accessible; the strata are little altered and are very fossiliferous, with few or no important hiatuses. The sequences are not suitable for the definition of an international boundary standard, however, because of the small representation of marine beds. Nevertheless the presence of rich sequences of mammals and charophytes and (potentially) of pollen in indisputable chronological sequence makes the English area a good reference section against which to correlate the, mostly scattered, occurrences of continental beds in Western Europe in general.

REFERENCES

Aubry, M.P., 1983. Biostratigraphie du Paléogène épicontinental de l'Europe du nord-ouest. Etude fondée sur les nannofossiles calcaires. Docum. Lab. Géol. Lyon, 89, 317p.

Châteauneuf, J.J., 1980. Palynostratigraphie et Paléoclimatologie de l'Eocène supérieur et de l'Oligocene du Bassin de Paris. Mém.B.R.G.M., 116, 424p.

Curry, D.,1967. Problems of correlation in the Anglo-Paris-Belgian Basin. Proc. Geol. Assoc. London, 77, 437-467.

Daley, B. and Insole, A., 1984. The Isle of Wight. Geol. Assoc. Guide, 25, 36p.

Feist-Castel, M., 1977. Evolution of the charophyte floras in the Upper Eocene and Lower Oligocene of the Isle of Wight. Palaeontology, 20, 143-157.

Hooker, J.J. and Insole, A.N., 1980. The distribution of Mammals in the English Palaeogene. Tertiary Res., 3, 31-45.

Keen, M.C., 1977. Ostracod assemblages and the depositional environments of the Headon, Osborne and Bembridge Beds (upper Eocene) of the Hampshire Basin. Palaeontology, 20, 405-445.

------, 1978. The Tertiary - Palaeogene. In: R.H. Bate and J.E.Robinson (Editors). A stratigraphical index of British Ostracoda. Seel House Press, Liverpool, pp. 385-450.

Machin, J., 1971. Plant microfossils from Tertiary deposits of the Isle of Wight. New Phytol., 70, 851-872.

Rey, R., 1966. Malacologie continentale oligocène dans l'ouest de l'Europe. Rev.sci. Bourbonnais (1966), 53-129.

Riveline, J., 1984. Les charophytes du Cénozoique (Danien à Burdigalien) d'Europe occidentale. Implications stratigraphiques. Thèse, Paris, 1011p.

White, H.J.O., 1915. The geology of the country near Lymington and Portsmouth. Mem. geol. Surv. G.B., 78p.

-------, 1921. A short account of the geology of the Isle of Wight. Mem.geol. Surv. G.B., 201p.

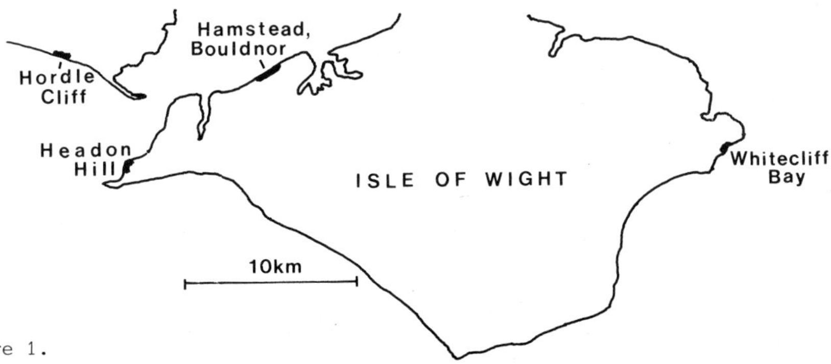

Figure 1.

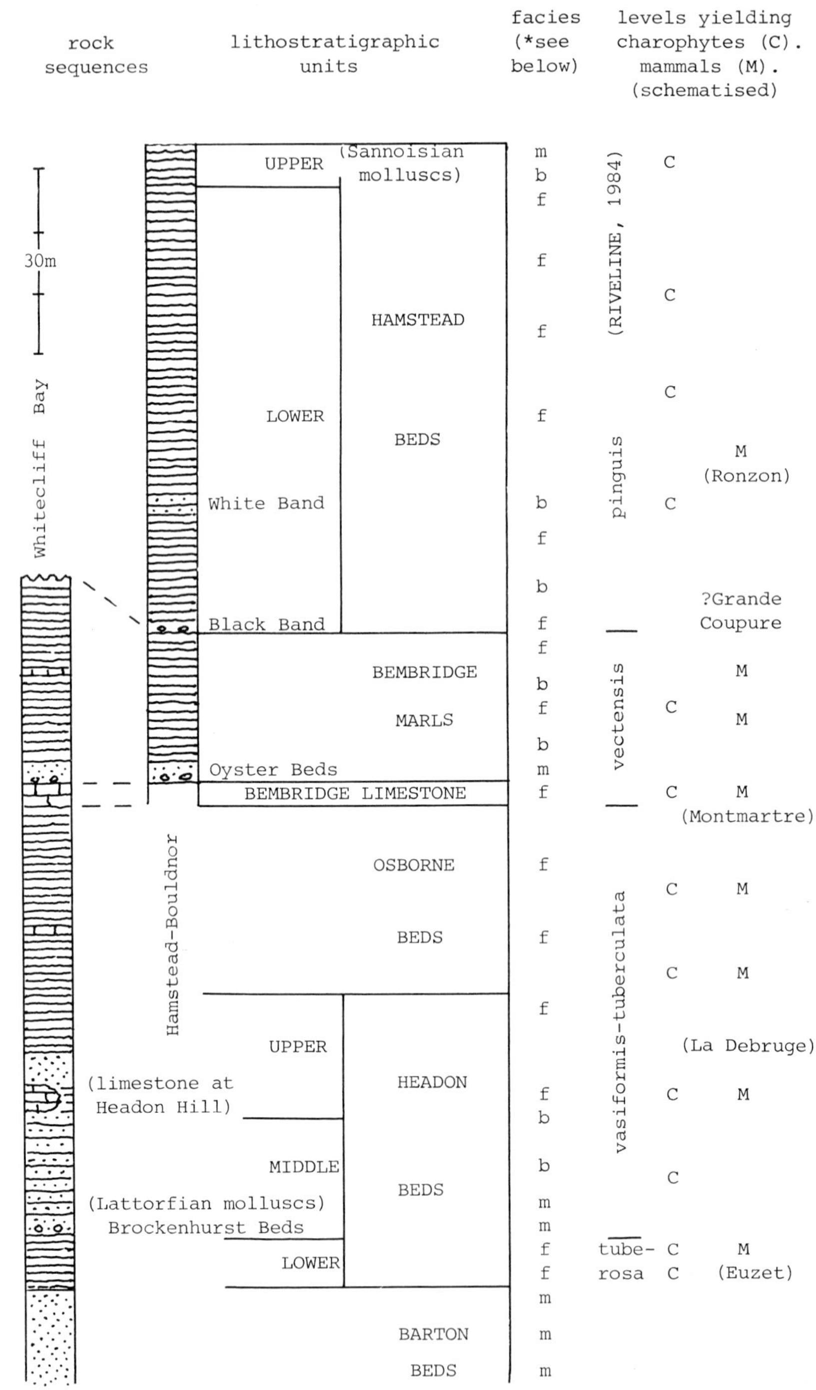

*m (marine), b (brackish), f (freshwater).

TABLE 1

CORRELATIONS AND BIOSTRATIGRAPHIC EVENTS AT THE EOCENE-OLIGOCENE BOUNDARY IN FRANCE AND IN EUROPE

By Claude CAVELIER +

INTRODUCTION

The Late Eocene and Early to Middle Oligocene sedimentary series, with which volcanic rocks are locally associated, are well represented in France, whether in the broad sedimentary basins of Paris and Aquitaine, in the external parts of the young orogenic belts of the Alps and the Pyrenees, or again in the numerous tensional troughs that affect in particular eastern France - those of Alsace, Limagne, Bresse, Bas Dauphiné, Provence and Languedoc, but also occur in the west, as at Rennes and Saint Maixent.

Because of the extremely varied character of the sedimentary environments and facies represented, ranging from deep open marine to frankly continental, in a geographic domain extending from the northern margin of Tethys to the southern confines of the "northern" seas, an important part is played by the French series, which are generally highly fossiliferous and well catalogued, in establishing correlations between southern and northern Europe (Cavelier, 1979).

Many sections that cross the Eocene-Oligocene boundary are known, not only in the Paris Basin, where the Ludian and Stampian stages have been defined, but also in Alsace, in the Rennes basin, in Limagne, in Aquitaine and in the external Alpine domain, from the Alpes Maritimes to Haute Savoie. Nevertheless, even in the Alps and in Aquitaine, where the deposits are marine, none can claim, in our present state of knowledge, to act as a standard reference for the Eocene-Oligocene boundary.

CORRELATIONS

The biostratigraphic content of the various Middle Eocene to Oligocene deposits in France, more and more closely inventoried, has enabled scales of reference to be established for the principal faunal and floral groups, whether marine or continental. Since the author's compilation was presented (Cavelier, 1979) these scales have been largely completed for the calcareous nannoplankton (Aubry-Bergreen, 1983), the pollens, spores and dinoflagellates (Chateauneuf, 1980), the planktonic foraminifera (Toumarkine, 1983), the charophytes (Feist-Castel, 1976; Riveline, 1984), the large mammals (Brunet, 1979), the rodents (Vianey-Liaud, 1979), the Artiodactyla (Sudre, 1977), the Creodonts

+ Direction de la Géologie - Service géologique national - B.R.G.M. - BP 6009
45060 ORLEANS - France

(Lange-Badré, 1979), the Marsupials (Crochet, 1980), the Insectivora and Chiroptera (Sigé, 1976), the Serpents (Rage, 1976) and the Chelonians (de Broin, 1977). Together, these works have defined certain age attributions and correlations between French and other European basins, but they have above all shown the importance and defined the characteristics of the proposed Eocene-Oligocene boundary at the base of the Stampian (Cavelier, 1979) and, incidentally, those of the break between the Bartonian and the Priabonian-Ludian. The groupe français d'Etude du Paléogène has, moreover, finally accepted this break as the boundary between the Middle and Late Eocene (Cavelier, 1984), following the example of the specialists in the intertropical Paleogene.

The abundance of continental deposits in France and western Europe, in many places intercalated with marine deposits, as in the Paris Basin, Aquitaine, Alsace and the western Alps, and the generally unreworked and highly evolved nature of their fossil content (e.g. mammals and charophytes), or its excellence as a climatic indicator (spores and pollens) have led to their being widely used for correlation between marine deposits of "northern" and Tethyan affinities. In this respect, attention must be drawn to the recent results of the faunal study of La Montagne du Charbon, in Savoie (Herb et al., 1984), which correlate the late Ludian, attested by mammals of Saint Capraise age and by Charophytes, with the late Priabonian, dated by Nummulites and by molluscs of the "*Cerithium diaboli*" association.

BIOSTRATIGRAPHIC EVENTS MARKING THE EOCENE-OLIGOCENE BOUNDARY

All the animal and vegetable groups underwent modification, in some instances very considerable ones, during the Late Eocene and Early Oligocene, to the extent that Middle Oligocene populations have in general very little in common with those of the Middle Eocene, or even Late Eocene.

Not only the qualitative aspect of these modifications, at the level of families, genera and species, but also their breadth is remarkable, so that the principal characteristic of the Oligocene populations is their limited diversity and great homogeneity compared with the exuberance and wide variability of those of the Middle and Late Eocene.

In other terms, the passage from the Eocene to the Oligocene is marked essentially by disappearances. These occurred by stages, certain of which are well marked not only in time - at the end of the Bartonian, during the Priabonian, at the beginning of the Stampian and during the early Stampian - but also in space. In fact, a paleolatitudinal gradient can be seen, particularly well defined at the base of the Stampian, so that in western Europe the Eocene-Oligocene boundary is much more sharply defined in the "northern" than in the Tethyan regions, and it is of course in the tropical domain that the phenomenon is least well marked.

It has been clearly demonstrated (Cavelier, 1979; Cavelier et al., 1981) that the disappearances during the period straddling the Eocene-Oligocene boundary were related to a global lowering of the average temperatures of the oceans, which occurred in several stages, culminating at the very beginning of the Stampian. It has also been shown that the cooling of the ocean waters at this period was coupled with a sharp cooling of the terrestrial climate, which was responsible for significant changes in the vertebrate fauna (the "Grande Coupure") and in the flora, among which the temperate elements known as "Arcto-Tertiary" became dominant in middle latitudes.

The negative climatic aspect of the events that affected both land and sea around the Eocene-Oligocene boundary imposed a severe breaking effect on evolution. The rate of appearance of new species was generally very low to nil in the majority of groups, and restricted to the more resistant elements (Mammals). Populations were smallest at the Eocene-Oligocene boundary fixed at the base of the Stampian (Cavelier et al., 1981). Persistence was greatest in the tropical domain, where species which formerly occurred in higher latitudes took refuge, as examplified by the migration, at the beginning of the Stampian, of residual elements of the Eocene Latdorfian fauna to the northern Tethys.

Nevertheless, whether in the regions influenced by the Tethys or in middle latitudes, it can be seen that the break corresponding to the boundary at the base of the Stampian and above the Priabonian is sufficiently well defined to have been noticeable to the geologists who constructed the regional stratigraphic columns: the base of the Stampian correlates in practice with the base of the "Latdorfiano" in Italy, the Khadumian in Russia, the Vicksburgian in the southeastern USA, the Zemorrian in California and the Whangaroan in New Zealand, to name only some of the more important examples.

Although clearly visible on the regional scale, the biostratigraphic events that mark the limit between the Eocene and the Oligocene are less readily discernible on the global scale, due essentially to their negative nature (disappearances) and to diachronisms caused by the late Eocene episodes of cooling. This is the case for the disappearance of the Discoaster barbadiensis and D. saipanensis which occurred much earlier in high and middle latitudes than in tropical regions, so also for all the Nummulites and Discoasters, no longer present in middle latitudes at the end of the Eocene, whereas they persisted and continued to evolve into the Oligocene in tropical zones (Cavelier, 1975, 1979).

Because of the climatic zoning that pertained throughout the world in the Paleogene and became accentuated at the Eocene-Oligocene boundary it seems vain to look for global biostratigraphic events marking this limit - the reverse, it appears, of the imprint left by natural phenomena in the fields of chemiostratigraphy and magnetostratigraphy.

Among the biostratigraphic events that can be most widely recognized in the tropical domain and its northern margin are:

Open sea/ocean

The disappearance of the Hantkenines and the *Turborotalia* gr. *cerroazulensis* in general later than that of the rosette shaped Discoasters (*D. saipanensis* and *D. barbadiensis*), which was more or less synchronous with the appearance of *Cassigerinella chipolensis* and even *Globigerina tapuriensis*.

Platform

The disappearance of the Discocyclines and numerous species of Nummulites, the *Lockartia*, *Fabiania*, *Orbitolites*, the genus *Spiroclypeus* gr. *granulosus* and *Pellatispira madaraszi* among others, followed, in the early Stampian, by the appearance of *Nummulites intermedius* and *N. vascus*, of *Peneroplis armorica* and relative forms. Among the Mollusca, forms of the genus *Ampullinopsis* appeared widely from Indonesia to the east coast of the USA (*A. crassatinus*, *A. vicksburgensis*) in the earliest Stampian, and among the Echinoderms representatives of the genus *Parmulechinus* (Scutellidae) appeared in Tethyan Europe and North Africa. Similarly, among the Dinoflagellates, *Wetzeliella gochti* appeared at the opening of the Stampian.

Continental

The Vertebrates, in particular the Mammals, show greatly renewed development at the Eocene-Oligocene boundary in western Europe, where it is well documented (the "Grande Coupure" of Stehlin, see Rage, 1984), in Asia, and again, though with distinctive characteristics due to their isolation, in North America. In Europe the composition of the terrestrial floras was regionally affected, at the end of the Priabonian and in particular at the opening of the Stampian, by the broad southward transgression of the "Arcto-Tertiary" species, or the Turgai association, that largely supplanted the "paleotropical" species. A similar phenomenon is known from North America.

In western Europe this modification of floras at the Eocene-Oligocene boundary takes the form of the appearance of several markers in the spore and pollen spectra, such as *Boehlensipollis hohli*.

In the fresh water and littoral margin environments the European fauna (Mollusca) and flora (Charophytes) show a similar evolution with, in particular, the extinction, at the Eocene-Oligocene boundary, of species persisting since the Middle Eocene and the early part of the Late Eocene, and on the contrary the persistence of part of the species which originated in the second half or at the end of the Late Eocene, such as *Nystia duchasteli*, *N. plicata*, and *Pseudocyrena convexa*, or again *Harrisichara tuberculata*, *Rhabdochara stockmansi*, etc..

CONCLUSIONS

Deposits close to the Eocene-Oligocene boundary at the base of the Stampian in France have been more and more closely calibrated, thanks mainly to intercalations of continental beds, containing mammals, molluscs and charophytes in particular, that are free of reworking and above all independent of the paleogeographic traits that affect the Tethyan and "northern" maritime domains. It has thus been possible to correlate the sequences with each other and with those of neighbouring regions, so enabling this boundary to be recognized and its characteristics in Europe to be determined (Cavelier, 1979).

As far as biostratigraphy is concerned, the Eocene-Oligocene boundary has the characteristics of a break that occurred not instantaneously, but during a brief period of time. Most of the species which disappeared at this time had appeared in the Middle Eocene and in the early part of the Late Eocene, whereas those which survived it were a part, varying in size according to group, of the "young" forms that originated in the second half of the Late Eocene. The appearance of new species was exceptional during the critical period but occurred slightly or palpably after it, in the "basal" or Early Stampian.

These negative phenomena (disappearances and lack of evolution) together constitute a great break that forms a boundary between the Eocene and the Oligocene comparable to, though lesser in degree than that which separates the Cretaceous from the Tertiary.

It can be seen, nevertheless, that the negative biostratigraphic events (disappearances) of the Eocene-Oligocene boundary had in fact been in preparation since the end of the Middle Eocene, and lasted through the beginning of the Oligocene, during which period evolution occurred in successive stages that were probably related to climatic "jumps".

This observation enables us to conclude that the events of the Eocene-Oligocene boundary were not, sensu stricto, catastrophic, and to look for their origin in the coincidence of various climatic and tectonic phenomena that affected both continents and oceans and whose driving force was probably located in the oceans.

BIBLIOGRAPHY

AUBRY, M.P., 1983. Biostratigraphie du Paléogène épicontinental de l'Europe du Nord-Ouest. Etude fondée sur les nannofossiles calcaires. Docum. Lab. Géol. Lyon N° 89, 317 p.

BROIN F. de, 1977. Contribution à l'étude des Chéloniens. Chéloniens continentaux du Crétacé du Tertiaire de France. Mém. Mus. nation. Hist. nat., n.s., C, N° 38, 366 p.

BRUNET M., 1979. Les grands Mammifères, chefs de file de l'immigration oligocène et le problème de la limite Eocène-Oligocène en Europe. Fondation Singer-Polignac édit., 281 p.

CAVELIER C., 1972. L'âge priabonien supérieur de la zone à Ericsonia subdisticha (Nannoplancton) en Italie et l'attribution des Latdorf Schichten allemands à l'Eocène supérieur. Bull. BRGM, 2ème série, section 4, N° 1, p. 15-24.

CAVELIER C., 1975. Le diachronisme de la zone à Ericsonia subdisticha (Nannoplancton) et la position de la limite éocène-oligocène en Europe et en Amérique du Nord. Bull. BRGM, 2ème série, section 4, N° 3, p. 201-225.

CAVELIER C., 1979. La limite éocène-oligocène en Europe occidentale. Mém. Sciences géologiques, N° 54, 280 p.

CAVELIER C., 1984. La limite Eocène moyen-Eocène supérieur en France. 27ème Congrès géol. intern. Moscou. Abstracts vol IX, part 2, p. 13.

CAVELIER C., CHATEAUNEUF J.J., POMEROL C., RABUSSIER D., RENARD M., VERGNAUD-GRASSINI C., 1981. The geological events at the Eocene/Oligocène boundary. Pal. Pal. Pal. 36 (3/4) p. 223-248.

CHATEAUNEUF J.J., 1980. Palynostratigraphie et paléoclimatologie de l'Eocène supérieur et de l'Oligocène du Bassin de Paris (France). Mém. BRGM N° 116, 360 p.

CROCHET J.Y., 1980. Les Marsupiaux du Tertiaire d'Europe. Fondation Singer-Polignac édit. 279 p.

FEIST-CASTEL M., 1976. Les Charophytes dans le Paléogène du Sud de la France (Provence, Languedoc, Aquitaine). Etude systématique et biostratigraphique. Thèse Montpellier, 82 p.

HERB R., HUGUENEY M., LANGE-BADRE B., WEIDMANN M., 1984. Données nouvelles sur les Mammifères et les Nummulites de l'Eocène supérieur subalpin (Synclinaux du Charbon et d'Entrevernes, Bauges, Savoie et Haute-Savoie). Géobios N° 17, fasc 2, p. 221-234.

LANGE-BADRE B., 1979. Les Créodontes (Mammalia) d'Europe occidentale de l'Eocène supérieur à l'Oligocène supérieur. Mém. Mus. nation. Hist. nat., n.s., C, 42, 249 p.

RAGE J.C., 1976. Paléontologie, phylogénie et paléobiogéographie des Serpents. Thèse, Paris, 144 p.

RAGE J.C., 1984. La "Grande Coupure" éocène/oligocène et les herpétofaunes (Amphibiens et Reptiles); problème du synchronisme des évènements paléobiogéographiques. Bull. Soc. Géol. France, (7), t XXVI N° 6, p. 1251-1257.

RIVELINE J., 1984. Les Charophytes du Cénozoïque/Danien à Burdigalien d'Europe occidentale. Implications stratigraphiques. Thèse, Paris, 2 vol, 417 + 517 p.

SIGE B., 1976. Les insectivores chiroptères du Paléogène moyen d'Europe dans l'histoire des faunes de Mammifères sur ce continent. Paléobiologie continent., 7, p. 1-25.

SUDRE J., 1977. Les artiodactyles de l'Eocène moyen et supérieur d'Europe occidentale. Systématique et évolution. Thèse. Montpellier, 257 p.

TOUMARKINE M., 1983. Les Foraminifères planctoniques de l'Eocène moyen et supérieur des régions tropicales à tempérées chaudes. Thèse, Paris, 219 p.

VIANEY-LIAUD M., 1979. Evolution des Rongeurs à l'Oligocène en Europe occidentale. Palaeontographica, A, 116, p. 136-236.

THE KALLO WELL AND ITS KEY-POSITION IN ESTABLISHING THE EO-OLIGOCENE BOUNDARY IN BELGIUM

E. STEURBAUT
Laboratorium voor Paleontologie, State University Ghent, Krijgslaan 281/S8, B9000 Gent (Belgium)

The Eo-Oligocene boundary-sequence in Belgium is represented by an alternation of clays and fine sands (a_1, S_1, a_2, S_2, a_3 and S_3), known as the "Kallo complex" (GULINCK, 1969a, b). This sequence, encountered in boreholes of northern Belgium, was established in the Kallo well, 10 km west of Antwerp, where it is best preserved and most complete (map-sheet 15/2, co-ordinates x = 144.86, y = 217.84). JACOBS (1978) redefined the "Kallo complex" in the Meetjesland region, some 50 km west of Kallo. It was subdivided in a lower Meetjesland Formation and an upper, thinner Zelzate Formation comprising the Bassevelde Sand and the Watervliet Clay Members. The presence of those units in the Kallo well has been established only recently (STEURBAUT in press; see also table). In this paper, STEURBAUT redefines the Zelzate Formation to include the Bassevelde Sand Member (= S_3), the Watervliet Clay Member (= a_4), and the Ruisbroek Sand Member (= S_4) as a new subdivision.

The Kallo well boundary-sequence is of shallow marine origin and characterized by very low deposition rates and several breaks in sedimentation. Until now, nothing has been published on the heavy minerals, clay minerals, geochemistry or palaeomagnetism of this sequence, but various palaeontological aspects have already been described. The units intermediate between the Wemmel Sand Member and the Boom Clay Member were investigated for foraminiferids (DROOGER, 1969:interval 87 to 178 m; HOOYBERGHS, 1967:interval 92 to 110 m), dinoflagellates (CHATEAUNEUF, 1980:4 samples in the interval 86 to 160 m; DE CONINCK, work in progress) and pollen and spores (ROCHE & SCHULER, 1980). Calcareous nannofossils were studied by MARTINI (1969) (4 levels in the interval 82.70 to 174.00 m). The nannofossil results were discussed by CAVELIER (1979). Regarding the Eo-Oligocene boundary in Belgium, three major theories have been proposed. The first was formulated by GULINCK (1969a) on the base of sedimentological features and was recently adopted

TABLE

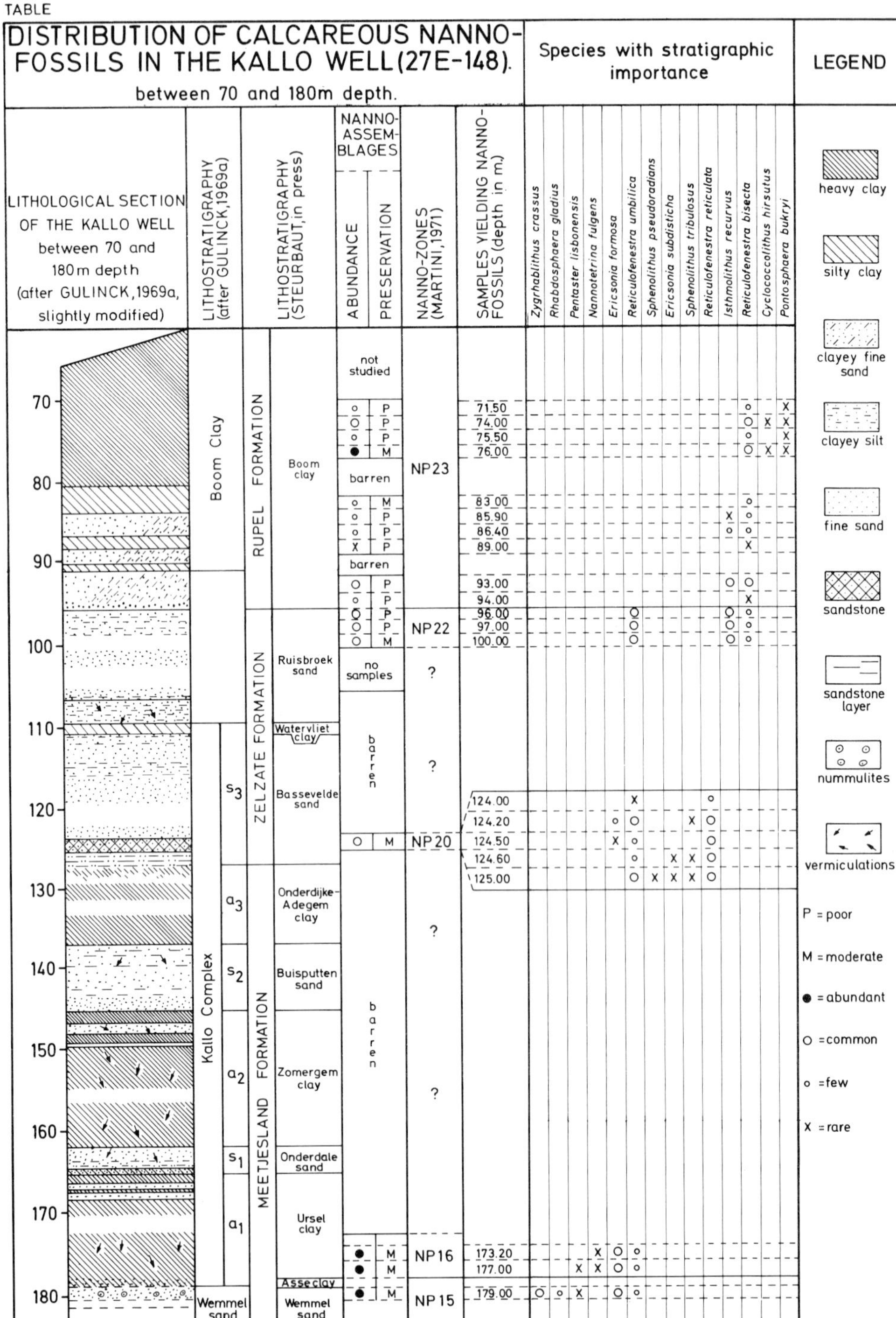
DISTRIBUTION OF CALCAREOUS NANNO-FOSSILS IN THE KALLO WELL (27E-148).
between 70 and 180m depth.
Species with stratigraphic importance
LEGEND
LITHOLOGICAL SECTION OF THE KALLO WELL between 70 and 180m depth (after GULINCK, 1969a, slightly modified)
LITHOSTRATIGRAPHY (after GULINCK, 1969a)
LITHOSTRATIGRAPHY (STEURBAUT, in press)
NANNO-ASSEMBLAGES
ABUNDANCE
PRESERVATION
NANNO-ZONES (MARTINI, 1971)
SAMPLES YIELDING NANNO-FOSSILS (depth in m.)
Zygrhablithus crassus
Rhabdosphaera gladius
Pentaster lisbonensis
Nannotetrina fulgens
Ericsonia formosa
Reticulofenestra umbilica
Sphenolithus pseudoradians
Ericsonia subdisticha
Sphenolithus tribulosus
Reticulofenestra reticulata
Isthmolithus recurvus
Reticulofenestra bisecta
Cyclococcolithus hirsutus
Pontosphaera bukryi
70
80
90
100
110
120
130
140
150
160
170
180
Boom Clay
Kallo Complex
s3
a3
s2
a2
s1
a1
Wemmel sand
RUPEL FORMATION
ZELZATE FORMATION
MEETJESLAND FORMATION
Boom clay
Ruisbroek sand
Watervliet clay
Bassevelde sand
Onderdijke-Adegem clay
Buisputten sand
Zomergem clay
Onderdale sand
Ursel clay
Asse clay
Wemmel sand
not studied
barren
no samples
NP23
NP22
NP20
NP16
NP15
?
71.50
74.00
75.50
76.00
83.00
85.90
86.40
89.00
93.00
94.00
96.00
97.00
100.00
124.00
124.20
124.50
124.60
125.00
173.20
177.00
179.00
heavy clay
silty clay
clayey fine sand
clayey silt
fine sand
sandstone
sandstone layer
nummulites
vermiculations
P = poor
M = moderate
● = abundant
O = common
o = few
X = rare

and slightly refined by GAEMERS (1984). According to these authors, the Zomergem Clay Member (= a_2) is the lateral equivalent of the Grimmertingen Sand Member and represents the basal part of the Lower Oligocene. Both other interpretations are supported by palaeontological data and suggest that the Grimmertingen Sand Member correlates with the Bassevelde Sand Member (= S_3). The age of both units, however, is still controversial. CAVELIER (1979) and CHATEAUNEUF (1980) suggested an Eocene age, while others (e.g. MARTINI, 1969 and ROCHE & SCHULER, 1980) proposed an Oligocene age.

The re-investigation of the Kallo well calcareous nannoflora (STEURBAUT in press, see also table) provides additional evidence for the identification of the Eo-Oligocene boundary in Belgium. It shows that during the interval Late Middle Eocene to Middle Oligocene a progressive impoverishment in nannospecies occurred, mainly due to a progressive decrease in temperature of the surface waters. A major change in nannoflora is recorded between the Late Eocene Bassevelde Sand Member (= NP 20) and the Early Oligocene Ruisbroek Sand Member (= NP 22). Between the middle part of the former, at -124 m, and the middle part of the latter, at -106 m, 18 m of decalcified clays and clayey sands occur, straddling the Eo-Oligocene boundary. This boundary lies within the Zelzate Formation, but, because of decalcification no precise limit can be drawn. Lithologically however, there is a break in sedimentation between the Watervliet Clay Member and the Ruisbroek Sand Member at -109 m (presence of a fine gravel layer), which might be connected indirectly with the Eo-Oligocene transition.

Correlation of the different members of the Zelzate Formation with the Lower Tongrian deposits of Eastern Belgium remains speculative. On the basis of its nannoflora, the Grimmertingen Sand Member has to be placed in the interval comprising the middle part of the Bassevelde Sand Member to the middle part of the Ruisbroek Sand Member. To what it corresponds exactly can be established only through the detailed analysis of the dinoflagellate associations (J. DE CONINCK, work in progress).

REFERENCES

Cavelier, C., 1979. La limite éo-oligocène en Europe occidentale. Sci. Géol. Mém. 54, 280 pp.

Chateauneuf, J.J., 1980. Palynostratigraphie et paléoclimatologie de l'Eocène supérieur et de l'Oligocène du Bassin de Paris. Mém. B.R.G.M., 116, 357 pp.
Drooger, C.W., 1969. Microfauna close to the Eocene-Oligocene boundary in the boring at Kallo. In : Le Sondage de Kallo (au nord-ouest d'Anvers). Toelichtende Verhand. Geol. Kaart Mijnkaart België, 11:9-27.
Gaemers, P.A.M., 1984. Fish otoliths from the Bassevelde Sand (Late Tongrian) of Ruisbroek, Belgium, and the stratigraphy of the Early Oligocene of Belgium. Meded. Werkgr. Tert. Kw. Geol., 21(1):13-57.
Gulinck, M., 1969a. Le passage Oligocène-Eocène dans le sondage de Kallo et le Nord de la Belgique. In : Colloque sur l'Eocène, Paris mai 1968. Mém. B.R.G.M., 69(III):193-196.
Gulinck, M., 1969b. Coupe résumée des terrains traversés au sondage de Kallo et profil géologique NS passant par Woensdrecht-Kallo-Halle. In : Le Sondage de Kallo (au nord-ouest d'Anvers). Toelichtende Verhand. Geol. Kaart Mijnkaart België, 11:3-7.
Hooyberghs, H.J.F., 1976. Remarks on the Eocene/Oligocene boundary with some preliminary results of the study of Eocene/Oligocene planktonic foraminifera in Belgium. Bull. Belg. Ver. Geol., 85(3):117-126.
Jacobs, P., 1978. Litostratigrafie van het Boven-Eoceen en van het Onder-Oligoceen in Noordwest België. Geol. Survey Belgium, Prof. Paper 1978/3, 151, 92 pp.
Martini, E., 1969. Calcareous nannoplankton from the Kallo well. In : Le Sondage de Kallo (au nord-ouest d'Anvers). Toelichtende Verhand. Geol. Kaart Mijnkaart België, 11:39-41.
Roche, E. & Schuler, M., 1980. Etude palynologique du "Complexe de Kallo". Geol. Survey Belgium, Professional Paper 1980/8, 178, 13 pp.
Steurbaut, E., in press. Late Middle Eocene to Middle Oligocene Calcareous nannoplankton from the Kallo well, some boreholes and exposures in Belgium and a description of the Ruisbroek Sand Member. Bull. Belg. Ver. Geol., 8 tabl., 2 pl.

EOCENE/OLIGOCENE BOUNDARY IN WESTERN GERMANY

F. GRAMANN, H. HAGN, S. RITZKOWSKI, V. SONNE

NORTHWEST GERMANY

Important stratigraphic gaps are widespread in NW-Germany close to the Eocene-Oligocene Boundary. But there are also regions with more or less undisturbed marine sedimentation.

Most of the complete sections have been observed in the eastern part of Lower Saxony (see Fig. 5). In the outcrop area, in the Helmstedt region, about 30 m of decalcified glauconitic sands of the Gehlberg Formation are intercalated between Annenberg Formation of NP15 and NP16 and Silberberg Formation with NP21 nannoflora. Further to the north, in the Wolfsburg-Gifhorn-Hankensbüttel area and in the well Wettenbosten Z 1, beds with Turborotalia cerroazulensis cerroazulensis or NP19/20 Nannoflora have been observed as latest marine Eocene. Slightly older beds with markers of Chiasmolithus oamaruensis Zone (NP18) have been recorded from the Eldingen 1 well. At least in part of this area, this Late Eocene is covered by finegrained marine sediments with Silberberg Formation fauna and NP21 nannoflora, as first observed from Hankensbüttel Süd 32 well (see Fig. 4). Silberberg Formation is regarded as lateral equivalent of the Latdorfian Stage due to faunal and nannoflora studies. Marine beds with NP21 nannoflora are known as transgressive near-shore sediments in the Osnabrück-Bünde region, where a complete sequence of marine Oligocene rocks includes Septaria clay formation and terminates with Chattian Stage calcarenites. Other regions of Lower Saxony, especially in the northwestern part, show transgressive Rupelian Clay (Septaria Clay Formation), usually with NP23 nannoflora, on top of much older Tertiary, very often dated as Late Middle Eocene NP15 or NP16.

The basal Neuengamme Sand Member of the Septaria Clay Formation, if developed, is usually followed by Rupelian Clay of NP23 nannoflora. Modern investigations suggest a diachronous age shift in the Wendland area in the northeast of Lower Saxony. So Neuengamme Sand Member and even the lowermost part of the Rupelian clay proper seem to commence there with NP22 nannoflora, instead with the NP23 assemblage as usual.

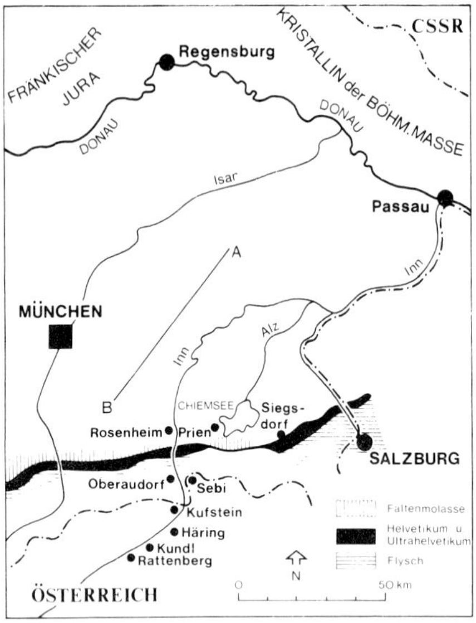

Fig. 1

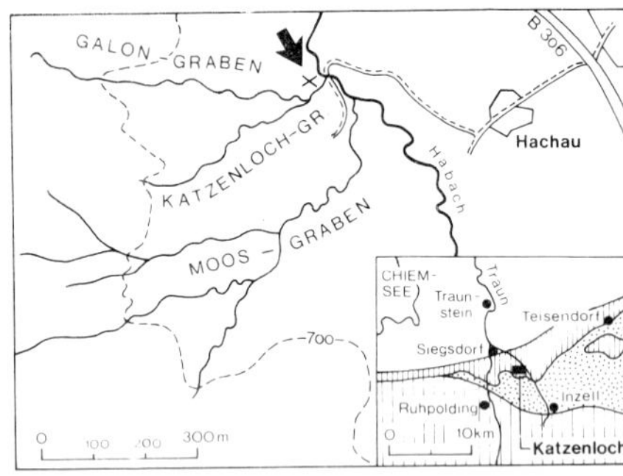

Fig. 2

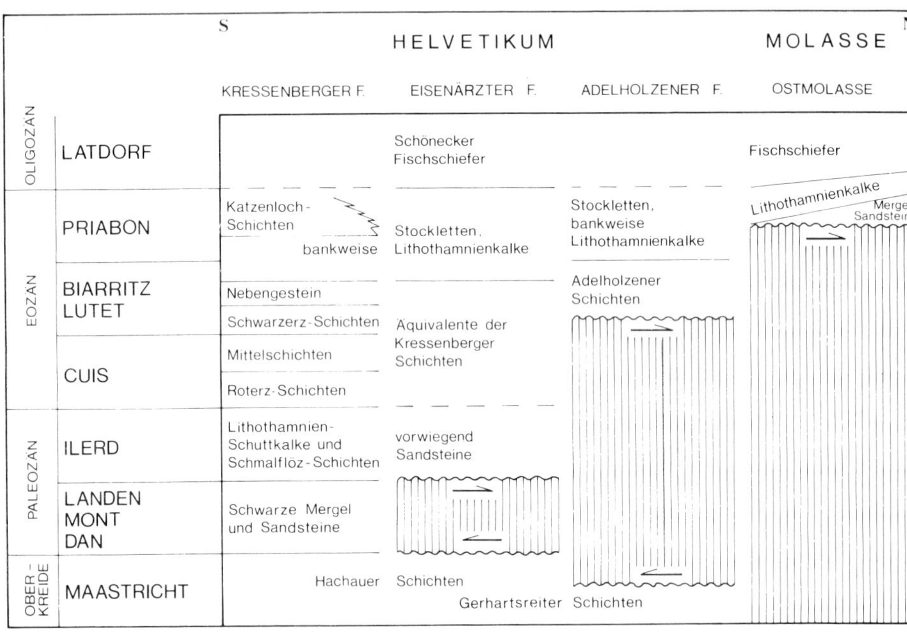

Fig. 3

Fig. 1: Structure-zones of the Bavarian Alps and their foreland.

Fig. 2: Sketch-map of locality "Katzenloch" in the Eastern bavarian Alps.

Fig. 3: Cretaceous to Oligocene stratigraphy of the Helvetic and Molasse zones in Bavaria.

HESSIAN DEPRESSION

In the Hessian Depression the Tertiary sedimentation starts in Eocene time with continental formations: the weathering horizon of the Mardorf ore is followed by the lignites of Borken and a kaolinitic clay sequence. They are overlain by limnic and brackish-water sediments of the Melania clay formation and by the marine clay of Rupelian age.

The following zones of calcareous nannoplankton have been proved (Müller, 1971; Martini, oral comm.):
NP22 in the Melania clay horizon D,
NP23 in the lower and middle part of the Rupelian clay,
NP24 in the upper part of the Rupelian clay and the Chattian.

The lignite of Borken has been dated as upper Lutetian due to a Lophiodon cf. cuvieri Watelet (Tobien, 1961). According to Russell et al. (1982) a late Eocene age (Auversian, Marinesian) could be taken into consideration. Micromammals of the vertebrate zone of Frohnstetten were described by Tobien (1971, 1972) from the level C of the Melania clay formation from Neustadt and Nordshausen.

All charophytes which have been found by Feist-Castel and Riveline in the Melania clay formation belong to the zone of Bembridge (Feist-Castel , 1977). Gyrogona caelata and G. wrighti appear already in the zone of Verzenay. Harrisichara tuberculata, Sphaerochara subglobosa and Psilochara div. sp. allow a comparison with the Bembridge marls. Therefore they indicate the zone of H. tuberculata and Steph. vectensis according to Riveline (1983).

The ostracodes of the brackish water beds of the Melania clay formation (i.e. level C) are similar to those of the Latdorf beds of NW-Germany: Hazelina indigena Moos, Hermanites triebeli Stchepinsky, Hermanites camelus Moos, Hermanites camelus turgidus Moos, Schuleridea (A.) perforata cognata Moos. Therefore it can be concluded that the Latdorfian sea has penetrated into the Hessian depression. Melania clay formation and Latdorf beds have the same age. Furthermore Hazelina indigena is a characteristic fossil of Keen's (1978) ostracode zone 12 from the upper Bartonian Brockenhurst bed.

Numerous small benthic foraminifera appear in the level B, C and D of the Melania clay formation. They are associated with dinocysts. They indicate the marine influence which is assumed to come from the Latdorfian sea in the North.

The "Borken Bild" has been defined by the microflores of the lignite of Borken, the "Heskem Bild" by a lignitic bed of the Me-

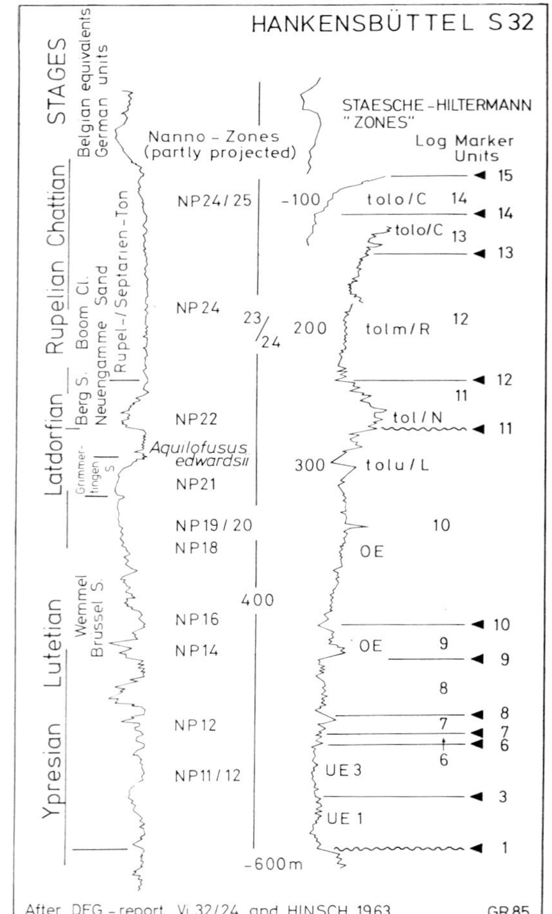

Fig. 4: Stratigraphy of Hankensbüttel S 32 - well.

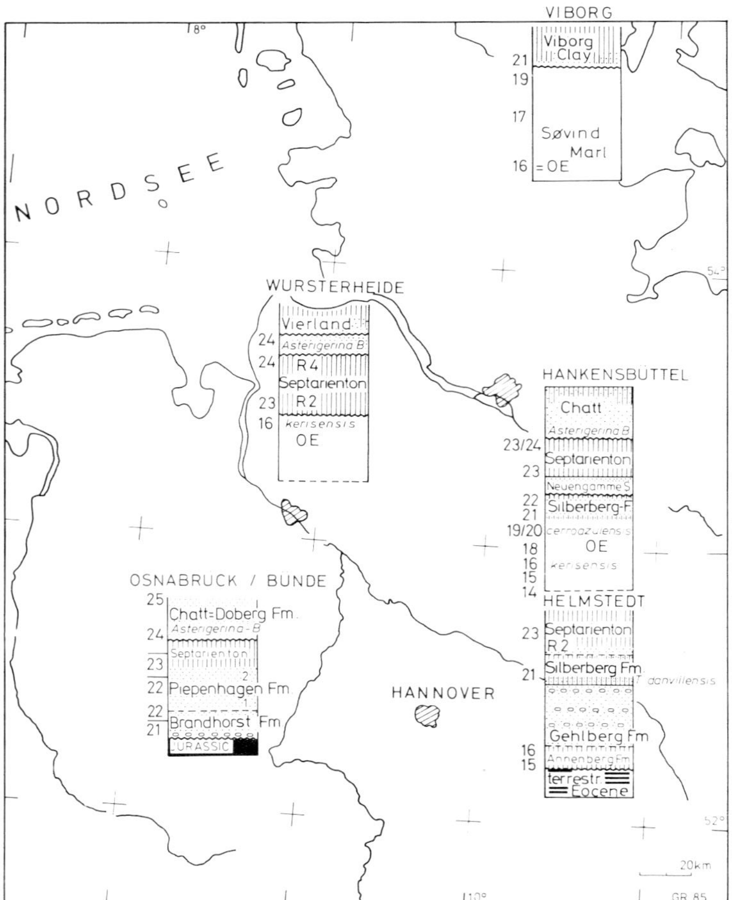

Fig. 5: Reference sections of Eocene-Oligocene sequences in Lower Saxony and southern Denmark.

lania clay formation. Chateauneuf (1980: 926) compares the kaolinitic clay of Borken with the Bartonian. On the other hand the microfloras of level B, C and D of the Melania clay formation are considered to be equivalent to those of the marnes supragypseuses from the Paris Basin (v.d. Brelie, 1969; Schuler, in prep.). In contrast level D shows an increase in Coniferae, Araliaceae and castanoids Cupuliferae (Chateauneuf, Schuler, in prep.) like the marnes vertes de Romainville (Sannoisian) of the Paris Basin. In the uppermost bed of the Melania clay formation ("level E") and in the Rupelian clay Boehlensipollis hohli has been proved by Schuler for the first time.

Up to now there is only one radiometric (K-Ar) age from glauconites of the base of the Rupelian clay: 29.8 $\pm$ 0,5 My (Kreuzer and Ritzkowski).

According to the original definition (Beyrich, 1854) the Eocene/ Oligocene boundary should be drawn at the base of the Latdorf stage. In the Hessian depression this boundary lies at the base of the Melania clay formation. The Ludien/Stampian boundary of the Paris Basin might be drawn between level C and D of the Melania clay formation.

UPPER RHINE GRABEN AND MAINZ BASIN

In the Upper Rhine Graben and Mainz Basin correlation by means of planktonic fossil zonation is only possible in the Foraminifera Marls (NP23) and Middle Pechelbronn Beds (NP22). The age of the remaining formations cannot be defined precisely.

The Eocene Basis Clay is -- at least to a large part -- older than Lutetian. However it might also have been formed in younger stages.

The Lymnaea Marls may have been deposited during Middle and Late Eocene and Early Oligocene (Latdorfian). Likewise sedimentation of the Lower Pechelbronn Beds could extend from the Late Eocene up into Early Oligocene. Since Limnaea Marls and Lower Pechelbronn Beds have not been found in the Mainz Basin the Eocene Basis Clay theoretically could also have been formed in the Early Oligocene here as well.

The brackish-marine Middle Pechelbronn Beds are dated by calcareous nannoplankton (NP22). They contain a rich ostracode fauna (Stchepinsky, 1960;, Triebel, 1963; Malz and Triebel, 1970; Malz, 1973), furthermore phytoplankton, Characean and Bryozoan communities. In some beds a rich planktonic foraminifera fauna is found.

It is uncertain wheather the Upper Pechelbronn Beds still belong to Zone NP22, but it appears probable. They are found exclusively in the Upper Rhine Graben.

The marine Foraminifera Marls occur discordantly in the Mainz Basin as in the Upper Rhine Graben.

THE BAVARIAN ALPINE FORELAND

The most valuable information on the Eocene/Oligocene boundary in the Bavarian alpine foreland is supplied by the southern segment of the Helvetic Zone (see Fig. 1). In the Limestone Alps and near the Untersberg (Salzburg) deposits of Late Priabonian and of Early Oligocene only occur as allochthonous pebbles.

In the lower part of the Katzenloch Graben (see Fig. 2) the Eocene/Oligocene boundary can be placed at the contact between the Stockletten and Schönecker Fischschiefer (Hagn et al., 1981: 122-125). The globigerina-bearing marls of the Stockletten are characterized by Globigerina eocaena Gümbel, G. ampliapertura Bolli, G. gortani praeturritilina (Todd). The ostracode genus Agrenocythere (det. Malz) indicates a great water depth. The rich calcareous nannoplankton (det. Martini) can be assigned to Zone NP19 (Isthmolithus recurvus-Zone).

As a result of tectonic reduction Zone NP20 cannot be proved.

The Schönecker Fischschiefer is a striking 30 m thick member, which is also very similar to Molasse deposits. The foraminiferal fauna is primarily composed of globigerinids. Among the Metazoan swimming organisms predominate. Pteropods (Limacina) cover many bedding surfaces. Fish remains of Elasmobranchian and Actinopterygian fauna rich in species and specimens (det. Pfeil) are extraordinarily frequent. The abundant calcareous nannoplankton (det. Martini) contains among others Reticulofenestra insignata Roth and Hay, R. umbilica (Levin) and Sphenolithus pseudoradians Bramlette and Wilcoxon. The fauna and flora can be placed in the Latdorf Stage (Zone NP21).

The Globigerina Marls of Priabonian age can be differentiated in terms of facies, fauna and by the great depth of deposition from the Latdorfian Fischschiefer. These differences are due to tectonic events in the alps at the Eocene/Oligocene boundary.

Equivalents to the Schönecker Fischschiefer are found in the foreland of the Grünten (Allgäu) and near Teisendorf. Thus, there is a connection with Glarner Fischschiefer of the Swiss Alps and the Menilit Slate of the Carpathians.

The close relationship between the Helveticum and the Foreland Molasse, which results from the stepwise shifting of the basin axis to the North, is visible in the Lithothamnian limestone. The so-called "Granitmarble" of the northern Helveticum and the similar red-algal limestones of the Foreland Molasse with Discocyclina and Gypsina linearis (Hanzawa), which have been encountered in numerous oil drillings, are placed in the Late Eocene, whereas the northern equivalents without these faunal elements extend into the Early Oligocene.

REFERENCES

Benedek, P.N.v. and Müller, C., 1976. Die Grenze Unter-/Mittel-Oligozän am Doberg bei Bünde/Westfalen. Phyto- und Nannoplankton. N. Jb. Geol. Paläont. Mh. 1976: 129-144.
Buchholz, P., 1984. Sedimentation and Diagenese des Lithothamnienkalkes der ostbayerischen Molasse. Diplomarbeit, Braunschweig (unpublished), : 146 pp.
Chateauneüf, J.-J., 1980. Palynostratigraphie et paléoclimatologie de l'Eocène supérieur et de l'Oligocène du Bassin de Paris (France). Mém. B.R.G.M., 116: 360 pp.
Feist-Castel, M., 1977. Evolution of the Charophyte floras in the Upper Eocene and Lower Oligocene of the Isle of Wight. Palaeont. Ass., 20: 143-157.
Hagn, H. et al., 1981. Die Bayerischen Alpen und ihr Vorland in mikropaläontologischer Sicht. Geologica Bavarica, 82: 408 pp.
Kreuzer, H. and Ritzkowski, S., 1982. NDS 217. In: G.S. Odin (Editor), Numerical dating in Stratigraphy. pp. 894-895.
Malz, H., 1971. Ostracoden aus dem Sannois und jüngeren Schichten des Mainzer Beckens, 3: Ehemalige "Cytheridea"-Arten und -Verwandte. Notizbl. hess. L.-Amt Bodenforsch., 101: 188-201.
Müller, C., 1971. Nannoplankton-Gemeinschaften aus dem W-deutschen Mittel-Oligozän. Notizbl. hess. L.-Amt Bodenforsch., 99: 43-53.
Riveline, J., 1983. Proposition d'une échelle zonale de Charophytes pour le Tertiaire (Danien à Burdigalien) de l'Europe occidentale. C.R. Acad. Sci. Paris, 296: (18.04.1983), sér. II, 1077-1080.
Russel, D.E., 1982. Tetrapods of the Northwest European Tertiary Basin. Geol. Jb., A 60: 5-74.
Spiegler, D., 1985. Gliederung des nordwestdeutschen Tertiärs (Paläogen und Neogen) aufgrund von planktonischen Foraminiferen. In: H. Tobien (Editor), The Northwest German Tertiary Basin (in press).
Tobien, H., 1961. Ein Lophiodon-Fund (Tapiroidea, Mamm.) aus den niederhessischen Braunkohlen. Notizbl. hess. L.-Amt Bodenforsch. 89: 7-16.
Tobien, H., 1971. Mikromammalier aus dem alttertiären Melanienton von Nordhessen, Teil 1: Marsupialia, Insectivora, Primates. Notizbl. hess. L.-Amt Bodenforsch., 99: 9-29.
Tobien, H., 1972. Teil 2: Rodentia, Biostratigraphie, Biostratonomie. Notizbl. hess. L.-Amt Bodenforsch., 100: 7-40.

UPPER EOCENE AND OLIGOCENE IN YUGOSLAVIA

R. PAVLOVEC
Odsek za geologijo, Univerza E. Kardelja, 61001 Ljubljana,
Askerceva 12, pp 311, YU.
K. DROBNE
Institut za paleontologijo ZRC SAZU, 61001 Ljubljana, Novi trg 5,pp 311, YU.
L. SIKIC
Geoloski zavod, 41000 Zagreb, Sachsova 2, YU.

A very brief description of beds is given containing certain fossils found close to the Eocene-Oligocene boundary. It must be mentioned that the transition of beds at the boundary E/O have not yet been described with certainty.

At the western margin of the Pannonian basin the Late Eocene beds at Makole and on Ravna Gora (mountain) were deposited disconformably on older rocks (fig. 1) (Šikić et al.,1976, Drobne et al., 1979). They consist of shallow reef facies. Also, the Oligocene beds on the western border of the Pannonian basin disconformably overlie older beds. Their basal layers consist of conglomerates or reef and near-reef deposits. Less developed are fresh-water beds. In the area of Slovenia most probably existed a connection between the sea in Transdanubia and in northern Italy (Cimerman, 1979, Drobne et al., 1985a, 1985b, Šikić, 1985).

In northern Bosnia (Majevica Mountain), on the margin of the Pannonian basin, the transition from the Late Eocene into the Oligocene most probably consists of flysch beds. South of Belgrade are known Late Oligocene beds with a Zizyphus zizyphoides flora. In Macedonia (Ovce polje, Tikveš and Kumanovo basins), sandstones and flysch were deposited in the Late Eocene (fig. 1). In these beds occur numerous corals (Lemoine, 1977). Oligocene beds attain considerable thicknesses. Alternate marine, fresh water and continental deposits occur.

In the Outer Dinarides along the Adriatic coast were deposited Late Eocene and Oligocene clastites. In the youngest Late Eocene or at the transition between the Eocene and Oligocene occurs Globigerina corpulenta. The presence of the Oligocene has been proved at Dugi otok island in off-shore boreholes (Jenko and Bistričić, 1978). The molasse Promina beds in Dalmatia are partly Late Eocene and partly Oligocene. In them are found Nummulites fabianii, Discocyclina pratti and Globigerina corpulenta (Krašeninikov et al., 1969, Muldini-Mamužić, 1971). In Hercegovina these beds have not been studied well enough yet. In them is an abundant macrofauna (fig. 1).

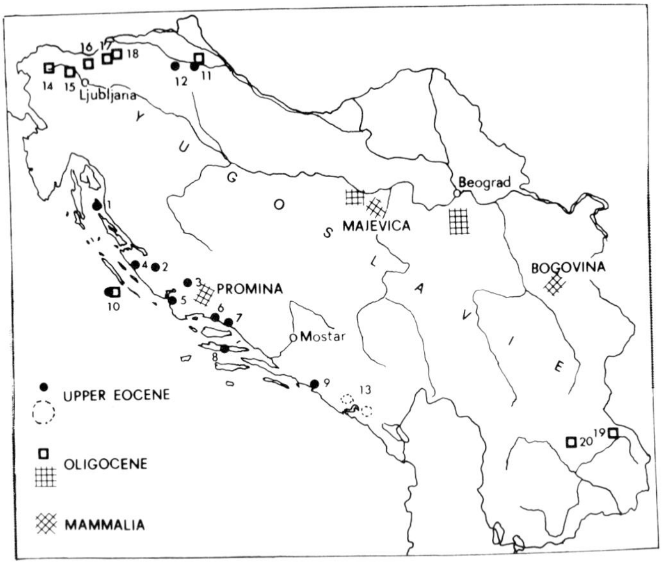

Fig. 1. Localities of foraminifera of Upper Eocene and Oligocene age. Explanation of numbers describing localities on Table 1 and 2.

Table 1

UPPER EOCENE	Rab 1	Ravni Kotari 2	Drniš 3	Zadar 4	Šibenik 5	Split 6	Omiš 7	Jelsa 8	Dubrovnik 9	Dugi otok 10	Ravna gora 11	Makole 12	S/SW Crna Gora 13
Nummulites fabianii	+	+	+	+	+	+	+				+	+	+
N. vascus incrassatus													+
Operculina alpina							+				+		
Discocyclina pratti		+	+	+			+						
Halkyardia minima	+			+			+				+		
Borelis vonderschmitti											+	+	
Chapmanina gassinensis											+	+	
Haddonia heissingi											+		
Asterigerina rotula											+	+	
Pellatispira madaraszi												+	
Gyroidinella magna											+	+	
Sphaerogypsina globula											+	+	
Globigerina corpulenta		+	+	+		+	+	+	+				
G. officinalis			+	+									
G. venezuelana								+					
G. gortanii										+			
Catapsydrax dissimilis			+	+									
Globorotalia cocoaensis							+						
Hantkenina suprasuturalis			+	+				+	+				

Table 2

OLIGOCENE	Dugi otok 10	Ravna gora 11	Bohinj 14	Poljšica 15	Kamniška Bistrica 16	Nova Štifta 17	Okonina 18	Crna Skala 19	Ovče polje 20
Cyclammina acutidorsata		+							
Tritaxia szaboi		+							
"Peneroplis" prisca			+	+		+			
Asterigerina bimammata			+	+		+	+		
Halkyardia maxima			+	+		+			
Sphaerogypsina globula			+	+	+	+			
Chapmanina gassinensis			+			+			
Nummulites fichteli			+	+	+			+	
N. vascus incrassatus			+					+	
N. vascus vascus						+			
N. germanicus			+			+	+		+
Planoperculina complanata				+		+			
Globigerina selli	+								
G. ampliapertura	+	+							
Catapsydrax dissimilis	+								
Globigerapsis tropicalis	+								
Globorotalia crassata	+								
G. cocoaensis	+								
G. opima opima	+	+							
Parvamussium duodecimlamellatum		+							

The most important smaller and larger foraminifera from Yugoslavia of Upper Eocene (1) and Oligocene (2) age.

In Montenegro have been determined Nummulites fabianii and Operculina alpina in Late Eocene clastites (Pavić, 1970). In the easternmost part of Yugoslavia in the Carpatho-Balkanids coal, oil shales and other economically important rocks are characteristic for the Late Eocene and Oligocene beds.

Mammal remains are extremely rare in the Paleogene of Yugoslavia. The remains described up to now mostly belong to the group of anthracotherids (Anthracotheriidae). Anthracotherium magnum and Microbunodon minus have been discovered in the coal basins of Bogovina (eastern Serbia), Ugljevik (northern Bosnia), Promina (Dalmatia). In the Promina beds Prominatherium dalmatinum was found. In Middle Oligocene beds of Slovenia Anthracotherium illyricum and Meninatherium telleri were found (fig. 1).

REFERENCES

Cimerman, F., 1979. Oligocene Beds in Slovenia. In: K. Drobne (Editor), 16th European Micropal. Coll., 65-70, Ljubljana.

Drobne, K., Pavlovec, R. and Drobne, F., 1979. /Characteristics of Microfossils in the Old Paleogene an the Western Border of the Pannonian Basin/. Zbornik radova, IV. Znan. skup za naftu JAZU, 155-172, Tab. 1-4, Zagreb.

Drobne, K., Šikić, L., Pavlovec, R., Cimerman, F. and Drobne, F., 1985a. /Upper Eocene and basal Oligocene beds on the western margin of the Pannonian basin/. Discussiones Paleontologicae, 31: 115-124, Budapest.

Drobne, K., Pavlovec, R., Drobne, F., Cimerman, F.and Šikić, L., 1985b. /Some larger foraminifera from the Upper Eocene and the basal Oligocene beds in north Slovenia/. Geološki glasnik, 28 (II): 77-88. Pl. 1-15, Sarajevo.

Jenko, K. and Bistričić, A., 1978. /Contribution to the knowledge of the biostratigraphic relationship of the Upper Paleogene, Neogene and Quarter in the Dugi otok basin/. Zbornik radova, IX. kongres geol. Jugoslavije, 25-39, Sarajevo.

Krašeninikov, V., Muldini-Mamužić, S. and Džodžo-Tomić, R., 1969. Signification des Foraminifères planctoniques pour la division du Paléogène de la Yougoslavie et la comparaison avec les autres régions examinées. Geol. vjesnik, 21: 117-147, Zagreb.

Lemoine, M., 1977. Etude d'une collection d'algues corallinacées de la région de Skopje (Yougoslavie). Rev. micropal., 20: 10-43, Paris.

Muldini-Mamužić, S., 1971. Biostratigraphy of Paleogene beds in the broader Zadar area. Nafta, 22/4-5, (simp. Zadar): 262-269, Zagreb.

Pavić, A., 1970. /Paléogène marin du Montenegro/. Inst. Geol. de Montenegro, Titograd, 192 pp, annex I:V, 20 pls.

Šikić, L., 1985. The Rupelian in the area of Ravna gora (NW Croatia) based on foraminiferous fauna. Geol. vjesnik, 38, in print.

Šikić, L., Šimunić, A., Pikija, M. and Šimunić, Al., 1976. /The Upper Eocene on Mt. Ravna gora (NW Croatia)/. Geol. vjesnik, 29: 191-197, Zagreb.

From the Abstracts of the IGCP Symposium, Proj. 174. Mostar, October 1984, several data on authors were taken: K. Drobne, D.S. Mihajlović, B. Milaković, N. Pantić, R. Pavlovec, L. Šikić, V. Temkova. Papers will be published in: Geološki glasnik, 28, spec. publ. Gradja 2, Sarajevo, 1985, 204 pp.

THE EOCENE/OLIGOCENE BOUNDARY IN HUNGARY

A. NAGYMAROSY, T. BÁLDI, M. HORVÁTH
Etvös University, Dept. of Geology, Museum Krt. 4, H.1088, BUDAPEST, Hungary

Though the position of the Eocene-Oligocene boundary in Hungary has been multilaterally examined, yet it is a hotly debated question. The main cause of the discussion lays in two important motives :

1) There is no unambigous agreement in the special literature of this problem, which biostratigraphical or chronostratigraphical event must be used for defining the E/O boundary and which environmental change resulted in changing the vertical distribution of living organisms. Some authors suspect a sudden "revolution", others a slow "evolution" of the environmental factors at this boundary.

2) Only a minor part of the biostratigraphical datums at the E/O boundary can be traced in the transitional E/O sequences in Hungary. The effect of the altered conditions does not appoint a well recognizable, sharp level.

The late Priabonian - Early Oligocene evolution of the Paleogene basin in Hungary was described in details by Kazmer, 1983, Baldi, 1984, and Baldi et al, 1984. The Priabonian transgression from the SW flooded over the Transdanubian Central Mountains and North Hungary until the Bükk Mts. The Buda Marl represents the uppermost part of the Late Eocene sequence. The offshore sediment consists of marl, calcareous marl and allodapic limestone intercalations with graded texture. The material of the allodapic limestone derives from the shallow margins of the basin, so containing the skeletons of shallow water dwelling organisms at bathyal depositional depth.

The Buda Marl is overlain by the Early Oligocene Tard Clay Formation, whose lower, marine member demonstrates an increased extension of the sea and additional deepening of the sedimentary basin between the Bükk Mts. and the so called Buda-line/paleotectonical line between Budapest and Romhány in a SSW-NNE direction, see Báldi, 1984. The fact of the transgression is in contradiction with the global eustatic regressional trend, showing the mobile position of the Hungarian Paleogene basin between orogenic belts.

In the higher part of the Lower Tard Clay the frequent laminated intercalations, the mixed faunas and the increased quantity of terrigenous material indicates a varying salinity and developing euxinic conditions. At the top of the lower member two mass-extinctions of Spiratellas (Pteropodas) were observed together with the endemic, brackish water Cardium lipoldi assemblage. The upper member of the Tard Clay (NP 23 nannoplankton zone) was deposited in a totally euxinic environment.

The sedimentation in the west side of the Buda line ended after the earliest Oligocene and the following denudation covered up the continuous sequences across the E/O boundary in Transdanubia. Thus, continuous sedimentation persisted only east of the Buda line, in North Hungary.

Detailed investigations were carried out on the material from several drill-notes and outcrops : petrographic parameters, macro- and' microfossils, ostracods, plant remains were examined and Marton, 1981, measured the magnetic anomalies of the Kiscell-1 (Budapest) borehole profile. Summarizing the results of these investigations the following important statements can be made :

- The fixing of the E/O boundary by using planktonic organisms is rather uncertain in Hungary. The index fossils are rare because of their climatic dependence, and the permanent reworking makes the extinction datums obscure.
- One possibility is to set the E/O boundary at the top of the NP 20 nannoplankton and the Globorotalia cerroazulensis planktonic foraminifera zone. This horizon occurs in the uppermost strata of the compact, calcareous Buda Marl in Hungary (see fig. 1). The figure also shows, that these zone boundaries fall into the C 15 N magnetostratigraphic chron, i.e. these extinction datums are earlier in Hungary than in the tropical areas (see Cavelier, 1979). The later occurence of some index nannofossils in this profile may be caused by reworking.
- The last occurences of Nummulites fabianii fall into the NP 21 nannoplankton zone and P 17/18 plankton foraminifera zone, regularly. It may be due to the persistence of N. fabianii throuth NP 20/21 boundary on the one hand, and to the reworking of the older non-consolidated shallow water sediments into the younger layers of the deep basins.
- A significant increase in the abundance of the cold water nannoplankton species Zygrhablithus bijugatus, Lanternithus minutus and Isthmolithus recurvus at the NP 20/21 boundary reflects the influence of cold water masses, though the general aspect of the nannofloras and foraminiferal assemblages did not change. The carbonate content - as a function of bioproduction - also does not show a sudden drop, but it decreases gradually upwards.
- The main event in Hungary, that changed the sedimentation and the distribution of living organisms essentially was the separation of the Paratethys from the world ocean and the developing of a series of euxinic basins (see Baldi, 1984). The Spiratella horizons at the NP 21/22 boundary and in

KISCELL (KL) -1. BOREHOLE

T. Báldi, M. Horváth, A. Nagymarosy, P. Varga 1980

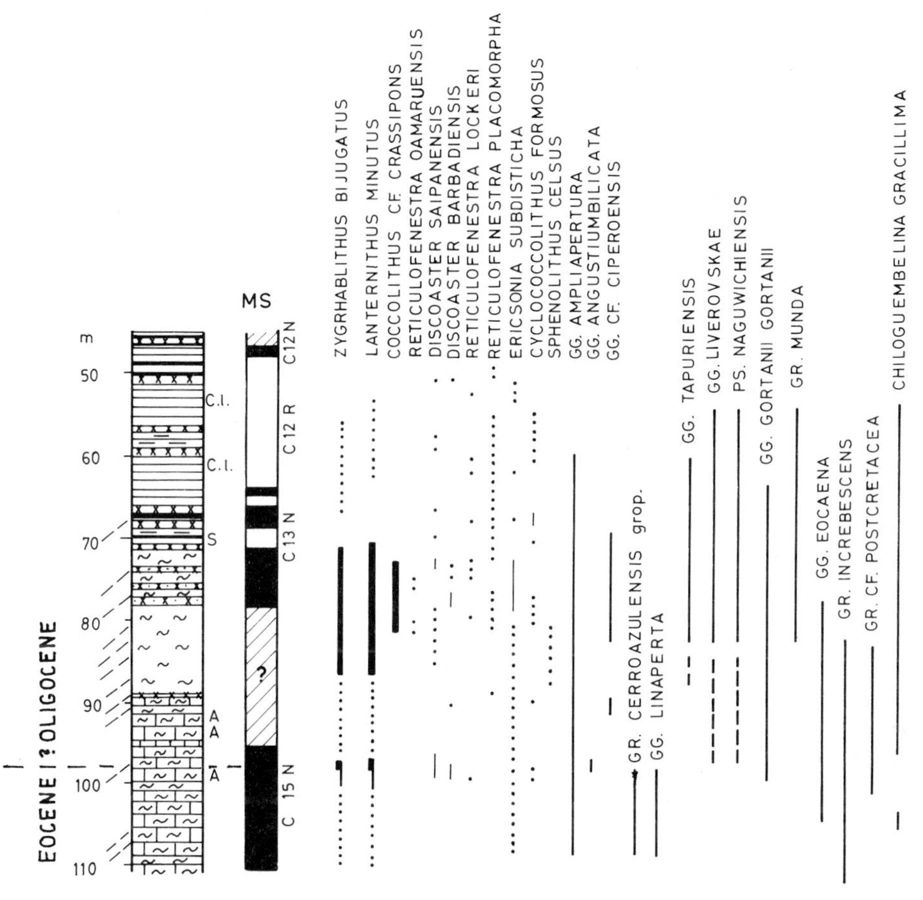

The log of the Kiscell-1 (Budapest) borehole, with the distribution of some important planktonic organisms. The calcareous Buda Marl is overlain by the marls, clayey marls and laminated clays of the lower member of the Tard Clay formation . Key to the symbols : A = Variamussium fallax, S = Spiratella horizon, C. l. = Cardium Lipoldi assemblage, MS = magnetostratigraphic chrons.

the P 18 zone can be regarded as the first sign of euxinic conditions. The appearance of the endemic Cardium lipoldi fauna and Transversopontis fibula Gheta nannoplankton species at the NP 22/23 boundary already reflects the total separation of the euxinic-brackish water basin.

The evidences from E/O sequences in Hungary do not prove sudden and sharp changes at the boundary. Our investigations outline a gradual evolution of the environmental factors, which affected only a minor part of the living organisms by extinctions. Conditions differing essentially from that of the Eocene have been established and became permanent only in the higher part of the Early Oligocene.

REFERENCES

Báldi, T., 1984. The terminal Eocene and Early Oligocene events in Hungary and the separation of an anoxic, cold Paratethys. Ecl. geol. Helv., 77 : 1-27.

Báldi, T., Horváth, M., Nangymarosy, A., Varga, P., 1984. The Eocene-Oligocene boundary in Hungary. The Kiscellian stage. Acta Geol., Hung., 27 : 41-65.

Cavelier, Cl., 1979. La limite Eocene-Oligocene en Europe occidentale. Sci. Geol. Mém., 54 : CNRS, 280 pp.

Kázmér, M., 1983. Microfacies of Upper Eocene carbonate rocks in Budapest, Hungary. TEE Meeting in Visegrad, Hungary. Abstracts, Manuscript : 25-32.

Márton, P., 1981. Report on the paleomagnetic studies of Oligocene core samples, borehole Kl-1. Manuscript, Budapest.

THE EOCENE-OLIGOCENE BOUNDARY FROM EASTERN RHODOPES, BULGARIA

A. GORANOV[1], G. ATANASOV[2], E. BELMUSTAKOV[2]

[1]Geological Institute, Bulgarian Academy of Sciences, Sofia (Bulgaria)

[2]Faculty of Geologt, University of Sofia "Clement Ohridsi" (Bulgaria)

The Paleogene from Eastern Rhodopes in Bulgaria is a part of the sediments of a Tethyan gulf which covers also large areas in Turkish and Greek Thracia (Doust and Arikan, 1974; Kopp, 1965).

The block disintegration of the Rhodope crystalline basement which began during the Paleocene (Atanasov and Goranov, 1984) governed the diverse development of the Paleogene basin in individual parts and the differences in the sedimentation regime in them. Synsedimentary medium-acid and acid volcanic activity was also an important factor in the Paleogene evolution. These two factors determine the varied character of the Eocene-Oligocene boundary (Atanasov et al, 1972). Several types are distinguished :

1. Transgressive transition

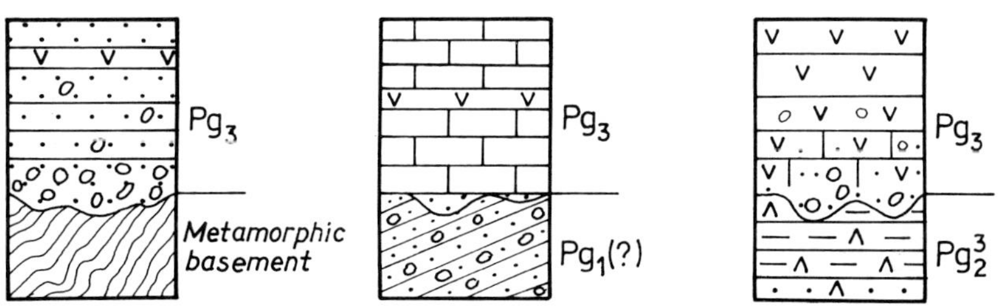

2. Sharp lithologic transition

3. Gradual lithological transition

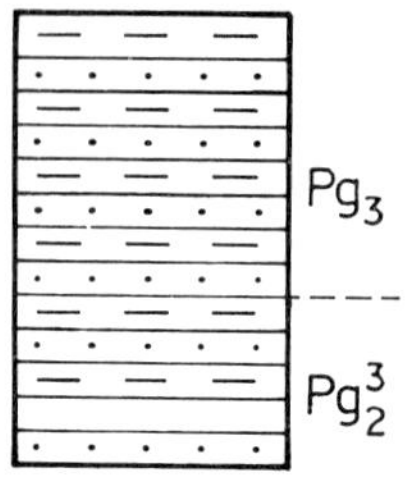

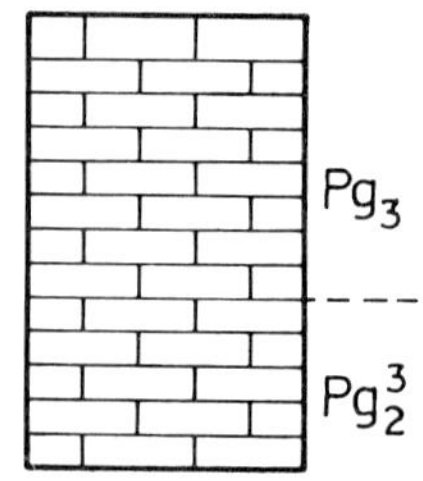

These types of boundaries are divided on the basis of three main features : structural, lithologic and faunistic. The structural features are expressed in the transgressive position of the Oligocene upon a basement of different age (including also Precambrian metamorphic rocks) as well as in local angular unconformities. The lithologic variations are fixed in abrupt changes of contrasting types of sedimentary rocks and in the first occurences of acid volcanism which usually marks the beginning of the Oligocene. The Eocene-Oligocene boundary is well defined faunistically (including also the places with lithologic transition). It is fixed above sediments with Nummulites fabianii and below sediments with Nummulites intermedius, above sediments containing g. Discocyclina and below sediments which do not contain the latter. The lowermost Oligocene sediments contain index from : Ampullinopsis crassatina (Lamarck), Globularia (Crommium) angustata (Grateloup), Miltha aliformis (Mayer), Levicardium comatulum (Brongniart).

All these data show that the boundary between the Eocene and the Oligocene is a period of rather intensive paleogeographic(readjustment) of the East Rhodope Paleogene depression.

REFERENCES

Atanasov, G. Goranov, Al. and Belmustakov, Em. 1972. Notes sur le problème de la limite entre le Priabonien et l'Oligocène dans les Rhodopes de l'Est. Annuaire de l'Université de Sofia, Faculté de Géologie et Géographie,v. 64, 1, Géologie; 41-48 (in Bulgarien).

Atanasov, G.D. and Goranov, A.G., 1984. On the paleogeography of the Eastern Rhodopes. Comptes rendus de l'Académie bulgare des Sciences, t. 37, N 6, Géologie; 783-784.

Doust, H., Arikan, Y., 1974. The Geology of the Thrace basin. Second Petroleum Congress of Turkey, Procedings, January, Ankara, 119-136.

Kopp, K.O., 1965. Das Tertiär Zwischen Rhodope und Evros. Geologie Thrakiens III. Labor. de Geologie de l'Université, 46, Athenes, 316-362 (in German).

ON THE BOUNDARY EOCENE - OLIGOCENE IN NE BULGARIA

V. SAPOUNDJIEVA
Géological Laboratories, Nassar Str. 16, 1113 Sofia

Oligocene sedimentsin NE Bulgaria follows as a narrow band the outlines of the present Black Sea coast north of Bjala. Surface outcrops are found in Kamchija Valley and Momino Highland.

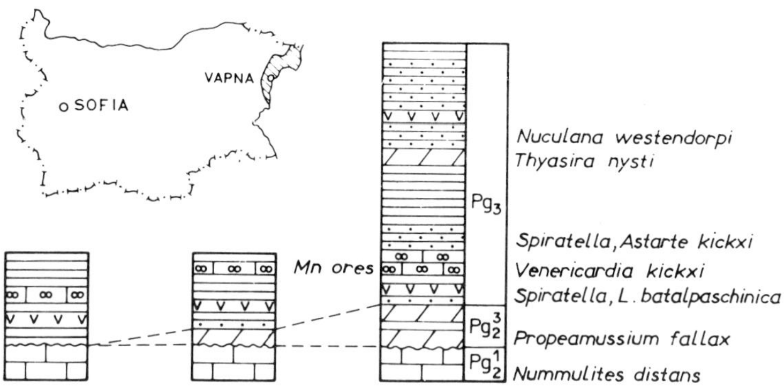

Fig. 1

The distribution of Oligocene sediments in NE Bulgaria marks the location of a Paratethyan gulf. In its peripheral parts the Oligocene locally overlies different stages of the Eocene. Thus in most cases the Eocene-Oligocene boundary is lithologically distinct and may be fixed by well-logging diagrams, petrographic studies and paleontologic data.

The Eocene marls include the index species Propeamussium fallax Korob., characteristic of the Upper Eocene, and Limopsis scalaris Sow., Venericardia hortensis (V. de Regny), Thyasira ignota Korob., T. dagestanica Korob.

The lithologic variations upwards in the section are insignificant and the marls with P. fallax gradually grade into silty clays with Spiratella (Oligocene age). The intermediate ten meters of sediments only contain spicules, fish scales and rare, small brachiopods.

The Oligocene deposits are a monotonous sequence made up mainly of clays with irregular content of carbonates to carbonate-free silty rocks. In the lower part there are tuffs and Mn mineralizations. In the middle part the clays are purer, carbonate-free with pyrite and glauconite. Fish scales, spicules and carbonized plant remains also occur. The lithologic and paleontologic data witness periodical intoxications with H_2S and hampered gass exchange.

The Spiratella beds contain the characteristic Lucina batalpaschinica Korob. , a single L. tenuistriata Heb., Pterolucina heberti (Desh.), Dentalium novaci Koen., and D. acutum Heb.. Above them follows a Mn-bearing horizon with Venericardia orbicularis tuberculata Münst., V. kickxi Nyst, Nuculata comta Goldf. Upwards a bed with Spiratella and Astarte kickxi Nyst, Lucina gracilis Nyst, Dentalium kickxi Nyst, etc ... occurs once again. Above the overlaying "marl horizon" only rare Nuculana westendorpi (Nyst) and a tooth of Odontaspis actussima Ag. have been found.

According to the faunistic features, the Oligocene sediments in NE Bulgaria show affinities with the Paratethys, being similar to the Borisphensk Formation, including the Rubanian and Nikopol Subformations (Veselov, 1979). They correlate well with the base of Tard Clay in Hungary (Baldi, 1984) and Mera Beds in Transilvania (Rusu, 1983). They contain also mollusks similar to those from the Northern bioprovince specifically to Rupelian. Beside the latter character, the beginning of a cold climate is witnessed also by the occurrence of radiolarite-diatomitic interbeds and by the palynological content.

In Upper Thracia the Oligocene beds with Nucula comta also belongs to the Paratethys. These threefold subdivision (Kojumdgieva, Sapoundgieva, 1981) are well correlated with the known phases at the development of the Paratethys.

REFERENCES

Báldi, T.,1984. The terminal Eocene and Early Oligocene events in Hungary and the separation of an anoxic, cold Paratethis. Eclogae geol. Helv. 7; 1-27.

Kojumdgieva, E., Sapondgieva, V.,1981. Biostratigraphie de l'Oligocène du bassin de la Haute Thrace d'après les Mollusques. Geologica Balc. 11,4; 93-114.

Rusu, A., 1983. Remarks on Oligocene chrono-and biostratigraphy in Transylvania (Romania). An. Inst. geol. geophiz. 59; 229-238.

Veselov, A., 1979. To the accurate definition of the stratigrafical correlation of the Oligocene-Lower Miocene border-marking horizons of the Eastern and Central Paratethys. Ann. Geol. Pays Hellen, hors série 3; 1243-1252.

EOCENE/OLIGOCENE BOUNDARY IN ROMANIA. PRESENT-DAY STATE OF INVESTIGATION

GEORGE BOMBITA
Institute of Geology and Geophysics, Bucharest (Romania)

INTRODUCTION

The present note is a summary of most recent investigations regarding two representative sections at the Eocene/Oligocene boundary in Romania. It offers complementary data, considering that the reader interested in the study of this boundary in the Romanian Carpathians has also become acquainted with our first note (Bombita and Rusu, 1981 - a review of data available by the end of 1979).

This second information is also exclusively of biostratigraphic nature. It concerns both Transylvania Basin and the Outer Zone of East Carpathian Flysch (Fig. 1).

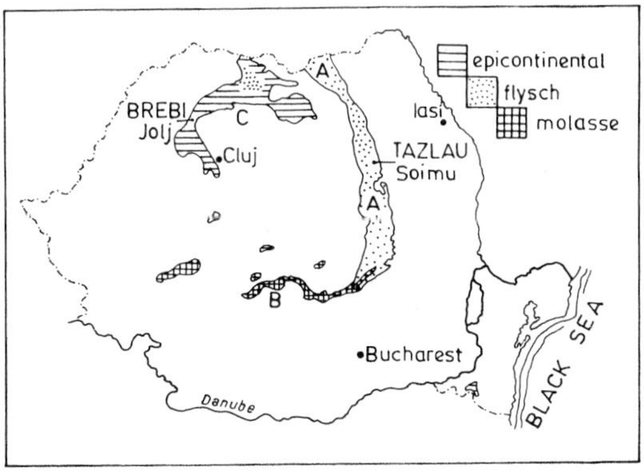

Fig. 1. Areas of Paleogene outcrop in Romania : A, East Carpathians; B, "Nummulitique Gétique"; C, Transylvanian Basin. Location of the two representative sections : Brebi-Jolj in Transylvania and Tazlau-Soimu in the East Carpathians.

In 1854 Beyrich introduced the notion of Oligocene related to the trilogy Septarienton sandwiched between the Magdeburg sands and the Cassel sands with their stratotypes in Germany and Belgium, that is in Northern Europe.

In Transylvania the Oligocene was first used by Austrian geologists : Hauer and Stache (1863) had timidly applied this new name and pointed to some lithological resemblance between the Septaria clay and the shale bearing fish remnants at Ileanda; in 1879 and 1887 Hofmann established the classical division of Paleogene formations in Transylvania using firmly the notion Oligocene.

Later on, Koch (1894) and his paleogenists successors during this century made detailed stratigraphic studies and enriched substantially the paleontologic inventory.

In the East Carpathians the Oligocene was used based on a petrographic analogy too. Cobalcescu (1883) assigned this Series the rock sequence in mainly euxinic facies which succeeds to the Eocene flysch, by considering the resemblance between the "Menilitic Sistema" and Ileanda Beds in Transylvania.

The scarcity of fossil remnants has long time been discouraging. In exchange, the lithostratigraphy has been accelerated either by the attraction to the study of complicated alpine structures or by the interest aroused by oil investigations.

H. Renz (1939) delimited the so-called "Uebergangsniveau Eozän/Oligozän" with double lithological affinities. I. Atanasiu (1943) recognized within this interval of continuous and graded transition some mixed, alternating Oligocene-Eocene members which were later thoroughly differentiated, described and named by Stoica (1944), Dumitrescu (1952) and Ionesi (1971).

The last decades are characterized by a real progress achieved by biostratigraphic studies.

It is to mention that the Paleogene evolution of the Carpathian territory includes, at the end of the Priabonian, a "moment" of connection with the large Mesogean domain, when the sedimentation processes and, implicitly, the pelagic microorganism contents were directly influenced by the ocean.

This short paleogeographic link is represented in Transylvania by the Brebi Marls and in the East Carpathians by the Globigerina Marls. As it will be shown below, both stratigraphic units yielded essential material to the recent biostratigraphic studies.

According to present-day conception, the radical change of lithology at the Eocene/Oligocene boundary corresponds to the parting of the Carpathian area from the Old Mediterranean Sea and its joining the north European basins.

In Romania the Oligocene is of generally regressive character.

TRANSYLVANIA

In the north-western sector of the intermountainous basin of Transylvania the micropaleontologic inventory at the Eocene/Oligocene boundary is well known, so the problems of zonation, boundaries and synchronization are not too controversial for the time being.

The most representative section is located in the Salaj district, east of the Brebi village situated on a asphalted secondary road south of the national road 28 almost at mid-distance between the towns of Zalau and Jibou. The outcrop from which Iva and Rusu (1982) collected and studied a suite of samples is some 700 m north of the road, on the right slope of Jolj valley.

Here, at their stratotype, the Brebi Marls crop out on almost all their thickness (60 m) between Nummulites fabianii level and the Hoia Beds (Fig. 2), and consist of grey-yellow-cream marls with siltic and calcareous layers, splintered, whitish and dusty when dry.

The formation is fossiliferous containing : molluscs, echinoids, bryozoans, ostracods, micro and macroforamifera, nannoplankton.

About 40 planktonic foraminifer taxons belong to two biozones (Iva and Rusu, l. c.) (Fig. 2) : Priabonian Globorotalia cerroazulensis zone (1, Glr. c. cocoaensiss subzone + 2, Glr. c. cunialensis subzone) and Lower Oligocene Globigerina ampliapertura/Pseudohastigerina micra zone. The Eocene/Oligocene boundary is marked by the disappearance of Glr. cerroazulensis group and its replacement by Glg. ampliapertura.

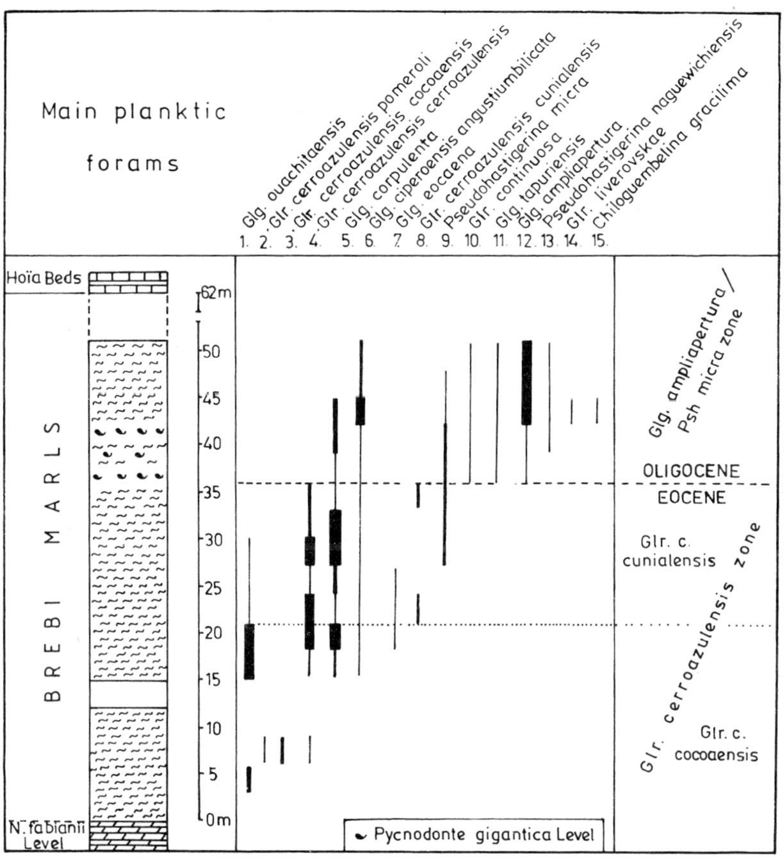

Fig. 2. Eocene/Oligocene boundary in Transylvania : Brebi village, Jolj valley (after Iva and Rusu, 1982, simplified).

Within the uniform sequence of Brebi Marls this boundary is also marked by the mollusc Pycnodonte gigantica located in the upper third of the Priabonian.

The latest study of nannoplankton (Gheta, 1984) confirmed at the Pycnodonte gigantica level the disappearance of Cyclococcolithus formosus. The last occurrence of this species is considered as the upper boundary of NP 21 zone (Ericsonia subdisticha). Above the Pycnodonte gigantica, species belonging the NP 22 zone (Helicopontosphaera reticulata) as for exemple Reticulofenestra retiformis, Transversopontis zigzag, Cyclicargolithus rupeliensis etc. occur.

The equivalence Lattorfian = Upper Priabonian and, implicitly, placing of the Eocene/Oligocene boundary between NP 21/NP 22, respectively P 17/P 18 zones are solutions successfully applied in Transylvania by Rusu (1977). In his opinion it is not advisible to place the Eocene/Oligocene boundary in Transylvania between the zones NP 20 and NP 21 because it would mean to place in Oligocene the Brebi Marls, that is a good part of the Nummulites fabianii zone.

EAST CARPATHIANS

The investigations regarding the Eocene/Oligocene boundary in the East Carpathians Outer Flysch followed a more sinuous way. This was obviously due to the peculiar sedimentary conditions in the Carpathian geosyncline, characterized by mainly detrital-terrigenous sequences of turbiditic genesis.

The lithostratigraphic succession at the Eocene/Oligocene boundary was concisely presented in another note (Bombita and Rusu, 1981).

The Upper Eocene formations are represented by the quasiheteropic trilogy Podu Secu Beds to the west, Plopu Beds, Bisericani Beds to the east the top of which consists of the well known member of Globigerina Marls, less than 50 m thick, with interbedding of Lucacesti Sandstone.

The Lucacesti Sandstone is in fact the first element which, from facies point of view, may be considered of "Oligocene" type. Being present or absent in places, it has led to a certain confusion as regards the lithostratigraphy.

The Globigerina Marls are overlain by the Linguresti Beds. These are represented by slate shale, marl or clay, grey-brown in colour and slightly bituminous, 5-25 m thick in the Outer Flysch, which may have or not interbedding of Fierastrau Sandstone confounded with the Lucacesti Sandstone.

The Lower Menilite (fetid opal cherts), which grades upwards into Bituminous Marls (5-50 m thick) and then into the Lower Dysodilic (papiraceous) Shale, marks the final settlement of typical Oligocene Carpathian facies.

Two sections situated in the Vrancea and Bistrita half-windows were studied by O. Dicea and Marieta Dicea (1980). The authors placed the Eocene/Oligocene boundary between the nannoplankton zones NP 20 and NP 21, that is below the Ericsonia subdisticha zone. On one of the two sections this index species occurs in reduced amounts in the Linguresti shales (with Fierastrau sandstones).

According to Ionesi (1982) and Florea (1982) the Eocene/Oligocene boundary would be placed between zones NP 21 and NP 22 which would occur within Lower Menilite. This conclusion was based on the presence of Globigerina assemblage characteristic of the Globigerapsis index zone within the pelitic joints among sandstone layers which are western lateral equivalents of the Lucacesti Sandstone; on a population of N. fabianii and Ass. exponens (!) in the Lucacesti Sandstone; and on a nannoplankton assemblage of the NP 21 zone (including Reticulofenestra oamaruensis, Isthmolithus recurvus and Ericsonia ? subdisticha) in the lower half of Lower Menilite.

A section outcropping next to Tazlau village, on the Soimu brook and yielding an interesting nannoplankton content has been recently studied by Micu and Gheta (1984, in press) (Fig. 3).

The locality of Tazlau, in the Neamt district, may be reached from the national road 15 between the towns of Piatra Neamt and Buhusi. A by-road deviates south of the Roznow locality and after 23 km reaches the Tazlau village.

The outcrop lies on the left slope of Soimu brook, at its confluence with the Tazlau valley and at some 6 km upstream the Tazlau village. Here all the formations implied in the Eocene/Oligocene boundary crop out continuously.

The study by Micu and Gheta points out two significant elements : 1, the Globigerina Marls has a trizonal nannoplankton content (NP 21 + NP 22 + NP 23); 2, the reworked nature of some nannoplankton species of the Eocene Carpathians flysch is mentioned for the first time and, as a consequence, the past tendency of considering them older than they are in reality.

The first 15-17 m of the Globigerina Marls (Fig. 3), overlaying the Bisericani Beds, contain the last specimens of Cyclococcolithus formosus. The disappearance of this species marks the boundary between the zones NP 21 and NP 22. On the following 10 m of Globigerina Marls Chiasmolithus alatus and Cruciplacolithus tarquinis appear and simultaneously Reticulofenestra umbilica disappears. Do to this fact the authors assign the top of the Globigerina Marls to NP 22 zone (Helicopontosphaera reticulata).

The last level (4 m) of the Globigerina Marls contain Sphenolithus distentus a representative of NP 23 zone (Sphenolithus predistentus), also present in overlying Lower Menilite and Bituminous Marls.

According to Micu and Gheta, from biostratigraphic point of view the Oligocene starts in the median part of the Globigerina Marls. The authors note that representatives of the Globigerina gortanii, Glg. tripartita and Glg. tapuriensis planktonic microforaminifera assemblage P 18 in Blow) are also present in the Globigerina Marls usually assigned to Upper Eocene.

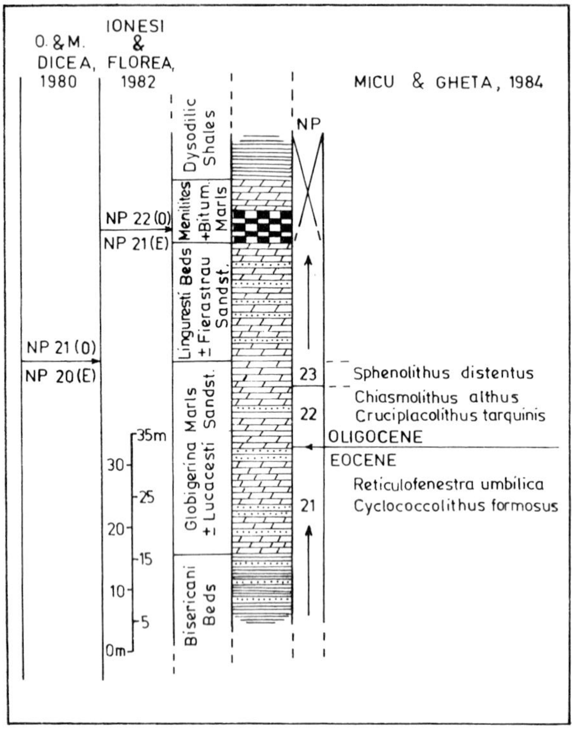

Fig. 3. Eocene/Oligocene boundary in the East Carpathians : Tazlau village, Soimu brook (after Micu and Gheta, 1984, adapted).

Here are three studies the results of which are not concordant yet. The micropaleontologic inventory in the East Carpathian and, above all, its critical analysis become tasks for the years to come.

REFERENCES

Atanasiu, I., 1943. Les faciès du flysch marginal dans la partie moyenne des Carpates Moldaves. Ann. Inst. Géol. Roum., XXII : 149-176. Bucarest.

Bombita, G., and Rusu, A., 1981. New data on the Eocene/Oligocene boundary in the Romanian Carpathians. Palaeogeogr., Plaeoclimat., Palaeoecol. 36 : 213-222. Amsterdam.

Cobalcescu, Gr. 1883. Studii geologice si paleontologice asupra unor taramuri tertiare ... Mem. geol., Sc. milit. Iasi, 161 pp., Bucuresti.

Dicea, O. and Dicea Marieta, 1980. Stratigraphic correlations on nannoplankton basis in the external flysch of the East Carpathians.D.S., Inst. Géol. - Géoph., LXV : 111-126. Bucarest.

Dumitrescu, I., 1952. Etude géologique de la région d'entre Oituz et Coza. Ann. Inst. Géol. Roum., XXIV : 195-271. Bucarest.

Florea, Fl., 1982. Contributions à la connaissance du nannoplankton du flysch paléogène ... Trav. Sess. Sci. "G. Cobalcescu" : 289-296. Iassy.

Gheta, N., 1984. The Eocene of Transylvania. A new biostratigraphic interpretation. D.S., Inst. Géol.-Géoph., LXIX/3 : 95-106. Bucarest.

Hauer, Fr. R. v., and Stache, G., 1863. Geologie Siebenbürgens, Braunmüller, 636 pp. Wien.

Hofmann, K., 1887. Geologische Notizen über die kristallinische Schieferinsel von Preluka and über das nördlich and südlich anschliessende Tertiärland. Jahresb. d. kgl. ung. geol. Anstalt f. 1885 : 31-61. Budapest.
Ionesi, L., 1971. Le flysch paléogène du bassin de la vallée Moldova. Acad. RSR, 250 pp. Bucarest.
Ionesi, L., 1982. Sur la limite Eocène-Oligocène dans le flysch externe carpatique. Trav. Sess. Sci. "G. Cobalcescu" : 349-365. Iassy.
Iva Mariana and Rusu, A. 1982. La limite Eocène-Oligocène en Transylvanie d'après les foraminifères planctoniques. D.S., Inst. Géol.-Géoph. LXVI : 157-160. Bucarest.
Koch, A., 1894. Die Tertiärbildungen des Beckens der siebenbürgischen Landestheile. I, Paläog. Abth., Mitt. a.d. Jahrb. d. kgl. ung. geol. Anst., X/6 : 177-399. Budapest.
Micu, M. and Gheta, N., in press. Eocene-Oligocene boundary in Romania on calcareous nannoplankton basis. D. S., Inst. Géol-Geoph., LXX/4. Bucarest.
Renz, H. H., 1939. Zur Geologie der Ostkarpathen in der Region von Dofteana, Doftanita und vorderem Slanictal. Geol. Rapport n° 995. Câmpina, Romania.
Rusu, A., 1977. La stratigraphie des dépôts oligocènes du nord-ouest de la Transylvanie ... Ann. Inst. Géol.-Géoph., LI : 69-224. Bucarest.
Stoica, C., 1944. Paleogenul din Valea Sibiciului (Jud. Buzau). Rev. Muz. Miner. Univ. Cluj (-Sibiu), VIII/1 : 64-85. Timisoara. Romania.

TERMINAL EOCENE GEOLOGICAL EVENTS IN TURKEY

SUNGU L. GÖKÇEN
Çukurova Üniversity, Department of Geology, Adana-Turkey

INTRODUCTION

Results of the geological studies in Late Eocene-Early Oligocene sequences of Turkey revealed that the terminal Eocene events in the country are closely related to the "Alpine Pyrenean Phase" and three main types of sedimentation could be classified as follows: (i) a continuous marine sedimentation occurs throughout the Late Eocene and Early Oligocene in the intermontaine flysch basins, (ii) in the tectonically active evaporitic basin marine Eocene sediments gradually pass into transitional and continental environments of Oligocene/Mio-Pliocene ages, (iii) the transgressive Oligocene rests unconformably on the eroded surface of Middle Eocene or older rocks in the orogenic belts.

EOCENE-OLIGOCENE BOUNDARIES

In northern Thrace the Eocene-Oligocene boundary can be easily followed along the foothills of the Istranca Massive. In this region the Nummulites gizehensis and Nummulites fabianii bearing limestones and sandstones of Middle-Upper Eocene are overlain by Nummulites intermedius and Nummulites vascus containing marly limestones of Lower Oligocene (GÖKÇEN 1975; KESKİN 1974). In the southern border of the region, contrary to the north, the turbiditic fining upward sequence of Keşan Formation of Upper Eocene-Lower Oligocene age represents the thickest marine sedimentation (ATAMAN and GÖKÇEN 1975). This formation is divided into two mapable units on the basis of lithology, stratigraphy and sedimentary features. The lower unit-Çınarlıdere Member-of Upper Eocene, composed of rhythmic turbidites intercalated with volcanics. The Lower Oligocene Upper unit-Şaplıdere Member-is composed of three types of turbidite alternations. The enclosed figure represent the geological details of the Keşan Formation being one of the best Eocene-Oligocene transitions in the country. In western Anatolia, in the Muğla and Denizli areas, the flysch type clastic sediments occur in the uppermost levels of Eocene as well as the Oligocene. These sequences are mainly composed of sandstone, mudstone and marl alternations. Deposits of the Burdur-Acıgöl area in the same region, containing Nummulites intermedius in the Oligocene part of the sequence, represent the characteristic continuous Eocene-Oligocene sedimentation. In the Antalya region, the Beydağları and Lyciennes Nappes of Late Eocene sandy flysch or marine carbonate deposits contain the late occurence of Globorotalia cerroazulensis and conformably overlain by Early Oligocene thick sandy conglomerates (BIZON et al. 1974; KOÇYİĞİT 1984; POISSON et al. 1984). The Bolkardağları and Ereğli-Ulukışla basins of SW and S Anatolia, are shallow marine Eocene deposits conformably overlain by evaporites and lacustrine limestones (DEMİRTAŞLI et al. 1984; OKTAY 1982). In the Erzincan-Refahiye area of Eastern Anatolia, Eocene shales containing Globigerinatheka semiinvoluta and Globigerina eocanea are concordantly overlain by sandstone, mudstone and limestones of Oligocene, bearing Hemicyprideis oertli and Hemicyprideis moyesi (GÖKÇEN 1974).

In the Çankırı-Çorum and the Kastamonu basins of Anatolia the continuous sedimentation occurred during the Early Eocene-Oligocene (NORMAN et al. 1980; YILMAZ 1980). Preceeding accumulation of shallow marine carbonates and clastics of Eocene; conglomerates, sandstones and mudstones of Oligocene were deposited in an alluvial environment. In the Ankara-Haymana Basin the Middle Eocene sequence is composed of slumps, debris flows and turbidites containing Assilina granulosa, Nummulites atacicus and Globigerinatheka index. Later the area was uplifted and eroded throughout the Upper Eocene and Oligocene (GÖKÇEN 1976). Around Kırıkkale, near Ankara, the marine conditions ended after the Middle Eocene, when a gradual uplift started with a regression. Reddish conglomerates and marls, lacustrine limestones, gypsiferous shales and lagoonal deposits of Oligocene covered the older formations (NORMAN et al. 1980). In the Sivas basin the 2700 m thick Eocene-Oligocene sequence is composed of thick bedded turbidites and olistostromes of slump origin being Eocene flysch, overlain disconformably by sandy and calcareous molassic Middle Oligocene deposits (GÖKÇEN and KELLING 1985).

Northeast of Elazığ-SE Turkey-transgressive Oligocene sequence composed of conglomerate, sandstone, marl and nummulitic limestone layers, overlie unconformably the Upper Cretaceous Ophiolitic Melange. Nummulites vascus, and Nummulites intermedius are the most characteristic species of this unit (SİREL et al. 1975). The Ağrı-Karaköse and Muş region-E Turkey-Oligocene sandstones, marly limestones containing Nummulites intermedius overlie unconformably the flysch sediments of the Middle Eocene.

CONCLUSION

It could be concluded that in the northern, central and southern as well as some parts of eastern Anatolia, the Late Eocene flysch sediments were uplifted, eroded and overlain disconformably by gypsum intercaleted biogenic limestones, red beds and conglomerates of shallow marine to continental origin. The continuous marine sedimentation from Late Eocene to Early Oligocene in Turkey appears typically in the northern and southern edges of Turkish Thrace and in the borders of the Taurus Orogenic Belt. Eventually, throughout the country, these events occurred and are affected by the Alpine Pyrenean orogenic phase.

REFERENCES

ATAMAN, G. and GÖKÇEN, S.L., 1975. Determination of source and paleoclimate from the comparison of grain and clay fractions in sandstones: a case study. Sediment. Geol., 13: 81-107.

BIZON, G., BIJU-DUVAL, B., MONOD, O., POISSON, A., ÖZER, B. and ÖZTÜMER, E., 1974. Nouvelles precisions stratigraphiques concernant les bassins Tertiaires du sud de la Turquie (Antalya, Mut, Adana). Revue.Inst.Français.Petr., 29/3: 305-323.

DEMİRTAŞLI, E., TURHAN, N., BİLGİN, A.Z. and SELİM, M., 1984. Geology of Bolkar Mountains.Proc.Int.Symposium on the Geology of the Taurus Belt.(Ankara-1983), M.T.A. Publ., pp: 125-142.

GÖKÇEN, N., 1975. Age and lateral variations in environmental conditions of the Pınarhisar Formation, north-northeastern Turkish thrace. Proc.Congress of Earth Sciences on the occasion of the 50th anniversary of Turkish Republic (Ankara-1973), pp: 131-145.

GÖKÇEN, S.L., 1974. Erzincan-Refahiye bölgesi sedimanter jeolojisi 1: olistolit, türbidit ve olistostrom fasiyesleri. Hacettepe Fen Müh.Bil.Derg., 4: 179-205.

GÖKÇEN, S.L., 1976. Sedimentology and provenance of resedimented deposits in part of the Haymana Basin-Central Anatolia 1.Yerbilimleri, 2: 161-200.

GÖKÇEN, S.L. and KELLING, G., 1985. Oligocene deposits of the Zara-Hafik region (Sivas, Central Turkey): evolution from storm-influenced shelf to evaporitic basin. Geologische Rundschau, 74/1: 139-153.

KESKİN, C., 1974. The stratigraphy of the northern Ergene basin. Proc.Second Petroleum Congress of Turkey. (Ankara-1974), pp: 137-163.

KOÇYİĞİT, A., 1984. Tectono-stratigraphic characteristics of Hoyran Lake region (Isparta Bend). Proc.Int.Symposium on the Geology of the Taurus Belt. (Ankara-1983), M.T.A. Publ., pp: 53-68.

NORMAN, T.N., GÖKÇEN, S.L. and ŞENALP, M., 1980. Sedimentation pattern in Central Anatolia at the Cretaceous-Tertiary Boundary. Cretaceous Research, 1/1: 61-84.

OKTAY, F.Y., 1982. Stratigraphy and Geological evolution of Ulukışla and its surroundings. Geol.Soc.Turkey Bull., 25/1: 15-24.

POISSON, A., AKAY, E., DUMONT, J.F. and UYSAL, Ş., 1984. The Isparta Angle: a Mesozoic paleorift in the Western Taurids.Proc.Int. Symposium on the Geology of the Taurus Belt. (Ankara-1983), M.T.A. Publ., pp: 11-26.

SİREL, E., METİN, S. and SÖZERİ, B., 1975. The stratigraphy and the micropaleontology of the marine Oligocene of Palu (NE of Elazığ). Geol.Soc.Turkey Bull., 18/2: 175-180

YILMAZ, O., 1980. Lithostratigraphic Units and Tectonics of Northern part of the Daday-Devrekâni Massif (Western Pontides, Turkey). Yerbilimleri, 5-6: 101-135.

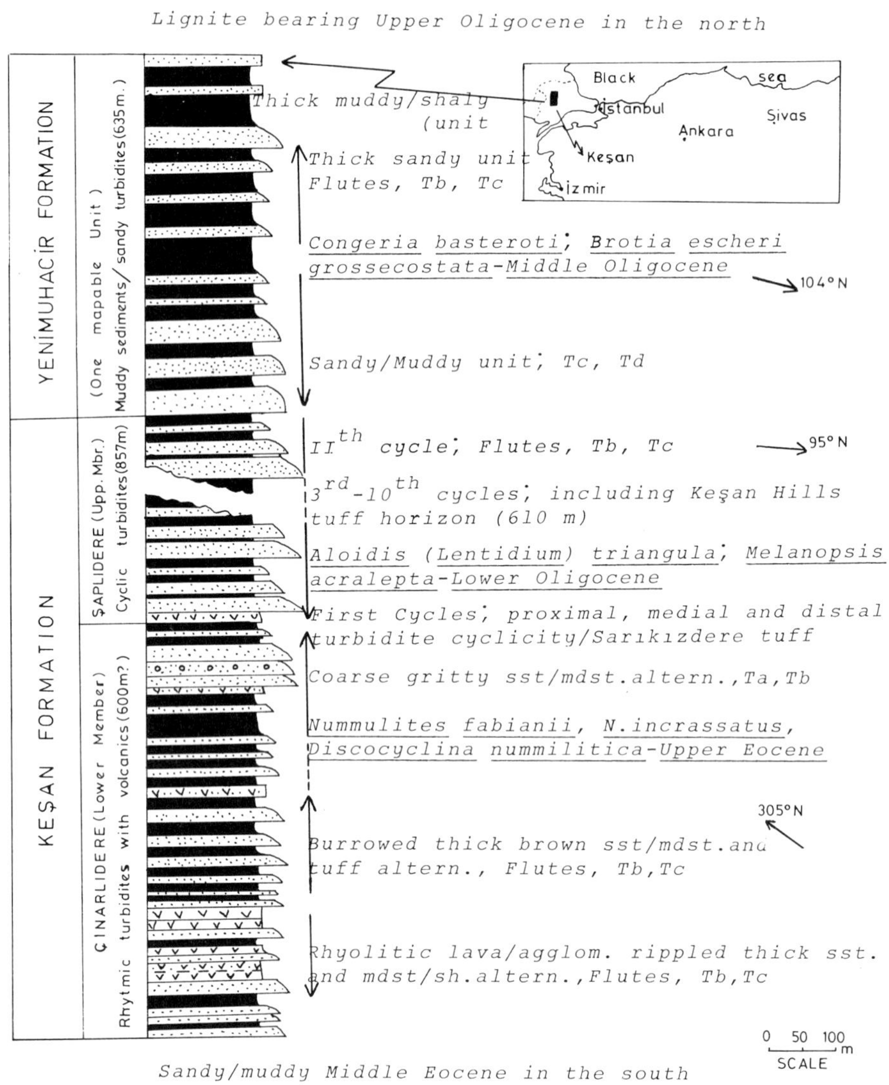

Fig. 1 . Simplified composite sedimentary geological columnar section of the Keşan-Yenimuhacir (Eocene-Oligocene) Formations of the Western Turkish Thrace (After ATAMAN and GÖKÇEN 1975, p.83, Fig.2).

SECTION LANDZHAR (USSR, SOUTH ARMENIA)

V.A.KRASHENINNIKOV, S.M.GRIGORIAN, Yu.A.MARTIROSIAN, A.E.PTUCHIAN, N.I.ZAPOROZHETS

On the southern slope of the Lesser Caucasus the Paleogene is widely distributed in the Yerevan-Ordubad structural-facies zone (the right bank of the Araks River), being composed of thick carbonate, clayey, tufogeneous-sedimentary, and partly sandy-argillaceous rocks. The taxonomic composition of planktonic foraminifers, nummulites and mollusks indicates that this territory was attributed to the Southern (Mediterranean) paleobiogeographic province.

The most complete section of Upper Eocene and Lower Oligocene sediments is near the village of Landzhar (Biralou), approximately at 100 km eastward of Yerevan. Here the transition from the Middle Eocene, the *Truncorotaloides rohri* zone to the Upper Eocene is very gradual.

Upper Eocene deposits and those of the Lower Oligocene *Globigerina tapuriensis* zone are represented by a pile of pale-grey and greenish-grey marls and calcareous clays with rare and thin interbeds of calcareous sandstones; thickness is about 100 m. In the lower part there occurs a member of more compact pale-grey limestones and marls with nummulites (layers with *Nummulites millecaput*), 10 m thick. In the Eocene light marls are predominant, whereas calcareous clays of a darker colour prevail in the Oligocene. Nevertheless, the Eocene - Oligocene boundary is not pronounced lithologically. This interval is characterised by a complete outcropping. Only at the contact of the *Globigerina tapuriensis* and *Globigerina sellii* zones is a member of rocks, 7 to 10 m thick, is not exposed.

The Lower Oligocene *Globigerina sellii* zone is represented by a different complex of deposits - alternation of yellow sandstones with subordinate interlayers of slightly carbonate yellowish arenaceous clays; thickness is 150 m. At the base there is a member of arenaceous limestones and sandstones with nummulites (thickness 13 m).

When judged by planktonic foraminifera, the Upper Eocene is subdivided into three zones:

The *Globigerapsis semiinvoluta* zone with rare specimens of the index-species and abundant *G.tropicalis*, *G. index*, *Globigerina corpulenta*, *G. tripartita*, *G. galavisi*, *G. subtriloculinoides* , *Globorotalia centralis*, *G. cerroazulensis*, *Hantkenina suprasuturalis*, *Pseudohastigerina micra*;

The *Globorotalia cocoaensis* zone with the numerous species given above (only *Globigerapsis semiinvoluta* is missing). Along with them, *Cribrohantkenina inflata*, *Globorotalia cocoaensis* and rare *G.cunialensis* were identified.

assemblage incorporating numerous *Globigerina corpulenta*, *G. galavisi*, *G. tripartita*, *G. ouachitaensis*, *G. praebulloides*, *Pseudohastigerina micra*, rarer *Globorotalia centralis*, *Globigerina gortanii* and sporadic *Hantkenina suprasuturalis* and *Globorotalia cunialensis*. They are accompanied by rare, recognized for the first time *Globigerina officinalis*, *G. prasaepsis*, *G. ampliapertura*, *Globorotalia permica*. Representatives of *Globigerapsis* and *Cribrohantkenina*, as well as *Globorotalia cerroazulensis* and *G. cocoaensis* are missing.

The following zones were identified in the Lower Oligocene by means of planktonic foraminifers:

The *Globigerina tapuriensis* zone with abundant and perfectly preserved *G.tapuriensis*, *G. ampliapertura*, *G. prasaepis*, *G. officinalis*, *G. praebulloides*, *G. galavisi*, *G. tripartita*, *G. ouachitaensis*, *G.angustiumbilicata*, *Cassigerinella chipolensis*, *Globorotalia nana*, *G. permicra*, *G. gemma*, *Pseudohastigerina micra*, *Ps. barbadoensis*, *Chiloguembelina cubensis*;

The *Globigerina sellii* zone,incorporating together with the above mentioned species, sporadic specimens of this index species. The foraminiferal assemblage is impoverished due to arenaceous sediments prevailing here.

Considering nannoplankton, the Upper Eocene deposits correspond to the *Discoaster barbadiensis* zone, where, along with the index species, there occur *D.salivanensis*, *D. tani*, *D. nodifer*, *Reticulofenestra bisecta*, *Corannulus germanicus*, *Lanternithes minutus*. The appearance of sporadic *Isthmolithus recurvus* enables to single out lower *Chiasmolithus oamaruensis* and the upper *Isthmolithus recurvus* subzones, the boundary between which coincides with the base of the *Globorotalia cocoaensis* zone. The boundary between the *Discoaster barbadiensis* and *Ericsonia subdisticha* zones where *Discoaster barbadiensis* and *D. saipanensis* disappear, is at 3-4 meters above the base of the *Globigerina tapuriensis* zone. Consequently determination of the Eocene-Oligocene boundary by means of plankotic foraminifers and nannoplankton is practically identical.

Large foraminifers are associated with two stratigraphic levels. In the lower part of the *Globigerapsis semiinvoluta* zone (Upper Eocene) there occur *Nummulites millecaput*, *N. fabianii*, *N. striatus*, *N. incrassatus* and *discocyclines* (the *Nummulites fabianii* zone). In the base of *Globigerina sellii* zone (Lower Oligocene) *Nummulites intermedius*, *N. fabianii retiatus*, *N. incrassatus*, *N. vascus initialis* were identified (the *Nummulites intermedius* zone).

The alteration of palynoflora along the section proceeds very gradually. The levels of these alterations do not coincide with zonal boundaries identified by planktonic foraminifers. The first palynocomplex characterizes the *Globigerapsis semiinvoluta* zone and the lower *Globorotalia cocoaensis* zone. Content of angiosperm pollen varies from 5 to 24%; invariably present are subtropical *Quercus*, *Castanopsis*, *Palmae*, *Nissa*. *Rhus*. Magnoliacea, and *Myrica*.

Content of gymnosperm pollen is relatively low: *Pinus* - below 50%, *Cedrus* - 20-50%, *Ephedra* - up to 20%, Taxodiaceae - up to 12%. Content of subtropical spores varies from 8,5 to 16%.

The second palynocomplex corresponds to the upper *Globorotalia cocoaensis* zone - the lower *Globigerina tapuriensis* zone. The amount of angiosperm pollen decreases to 2 %. The content of *Pinus* pollen reaches its maximum (up to 70 - 75%) with reduction of *Cedrus*, *Ephedra* and Taxodiaceae pollen. Almost complete reduction of angiosperm pollen and spores with appreciable increase of gymnosperms reflects the cooling of the Latest Eocene - Earliest Oligocene with the maximum in the *Globorotalia centralis* - *Globigerina gortanii* zone and the basal *Globigerina tapuriensis* zone, though the climate remained subtropical, transitional to temperate. The Eocene-Oligocene boundary is practically not expressed by a change of palynofloras, only the content of *Pinus* pollen at the base of the *Globigerina tapuriensis* zone reduces from 74 to 55%.

The third palynocomplex originates from the upper *Globigerina tapuriensis* zone, where the diversity and amount of angiosperm pollen (*Fagus*, oak, chestnut) increase (up to 24%), the total gymnosperm content being high (70-90%). The role of *Pinus* appreciably diminishes (down to 30-55%), with varying amounts of *Cedrus* pollen (8-27%), *Ephedra* (5-20%) and insignificant participation of Taxodiaceae (up to 5%).

The fourth complex characterizes the *Globigerina sellii* zone, where the content of angiosperm pollen increases up to 62%. Among spores Polypodiaceae and *Ligodium* are predominant. The total gymnosperm pollen content reduces to 10-27%, that of *Ephedra* - to 4-7%, and *Cedrus* -to 3%. The rise of angiosperm diversity in the *Globigerina sellii* zone can be regarded as an indication of some warming of the climate in the second half of the Early Oligocene.

The change of paleoenvironment, fauna and flora at the Eocene-Oligocene boundary (base of the *Globigerina tapuriensis* zone) over the South Armenia territory can be presented as follows: the tectonic regime remained tranquil; calcareous sediments were gradually substututed by argillaceous ones. Intensification of tectonic movements and appearance of thick piles of sandstones took place inside the Lower Oligocene (the *Globigerina sellii* zone). On the whole, planktonic foraminifers changed significantly, but this change took place at two stages: many specialized Eocene forms (*Globigerapsis*, *Cribrohantkenina*, the *Globorotalia cerroazuensis* group) disappear near the base of the *Globorotalia centralis* - *Globigerina gortanii* zone; above the top of this zone a typical Oligocene complex of foraminifers is widely distributed. Discoasters (*D.barbadiensis*, *D. saipanensis*) in the zone concerned are sporadic, their last specimens disappear in the basal layer of the *Globigerina tapuriensis* zone. The palynocomplexes testify to beginning of the cooling in the middle of the *Globorotalia cocoaensis* zone with its maximum in the *Globorotalia centralis* - *Globigerina gortanii* (Upper Eocene) zone, and the lower *Globigerina tapuriensis* zone (Oligocene). In the *Globigerina sellii* zone (Oligocene) the warming can be observed, but it does not reach the Upper Eocene values.

REFERENCES

Gabrielyan, A.A., 1964. Paleogene and Neogene of the Armenian SSR. Yerevan, Izdat. Akad. Nauk Arm. SSR, 229 pp. (in Russian).

Krasheninnikov, V.A., Ptuchyan, A.E., 1973. Relationship between the fauna of planktonic foraminifers and nummulites in Paleogene deposits of Armenia. Voprosy mikropaleontologii, v.16, Moscow, Nauka: 146-182. (in Russian).

Krasheninnikov, V.A., Muzylev, N.G., Ptukhyan, A.E., 1985. Stratigraphic subdivision of Paleogene deposits of Armenia by means of planktonic foraminifers, nannoplankton and nummulites (I. Key sections of Paleogene of Armenia). Voprosy mikropaleontologii, v.27, Moscow, Nauka: 130-169. (in Russian).

Saakyan-Gezalyan, N.A., 1957. Foraminifers of Tertiary deposits of the Yerevan basin. Yerevan, Izdat. Akad. Nauk Arm. SSR: 141pp. (in Russian).

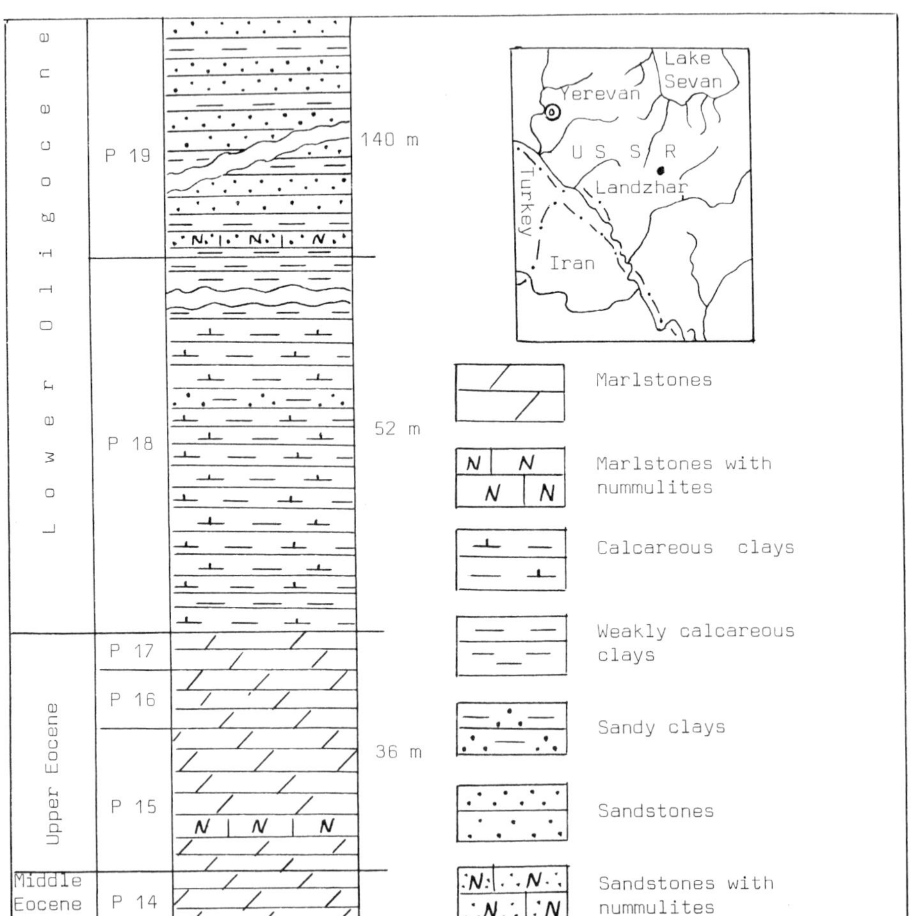

THE KUBAN RIVER SEQUENCE (USSR, NORTH CAUCASUS)

V.A.KRASHENINNIKOV

On the northern slope of the Caucasus the Paleogene deposits outcrop in a form of a continuous band dipping by an angle of 3-5° to the north. They crop out in the valleys of many rivers. A sequence on the left bank of the Kuban River near the town of Cherkessk is the best known.

The Upper Eocene deposits in this sequence conformably overlie brown and grey thin-layered, sometimes leaf-like marls with abundant fish scales of *Lyrolepis caucasica* (Middle Eocene). The sediments include numerous *Globigerina turcmenica, G. praebulloides, G. incretacea, G. subtriloculinoides, Pseudohastigerina micra, Acarinina rugosoaculeata, Chiloguembelina* sp. (the *Globigerina turcmenica* zone), and nannoplankton of the *Discoaster saipanensis* zone and of the lowermost part of the *Chiasmolithus oamaruensis* subzone. The sediments with such peculiar plankton were accumulated in the basin with abnormal hydrological environment (hydrosulphuric contamination); no benthic microfauna is present here. The brown leaf-like marls of the Kuma formation (or Buraya /Brown/ formation) extend from Bulgaria up to Caspian Sea and serve as a nice marking horizon.

Upper Eocene deposits are included in the Belaya Glina (White Clay) formation and considerably differ lithologically from the underlying sediments. They are represented by massive thick-layered clayey limestones and marls of white, light-grey and light-green colour about 100 m thick. The top of the formation incorporates a member (4-5 m) of grey-green marls with scattered pyrite.

By planktonic foraminifers the Upper Eocene is subdivided into two zones:

- the *Globigerapsis tropicalis* and large *Globigerina* zone, where *Globigerapsis tropicalis, G. index, Globorotalia centralis, Globigerina corpulenta, G. galavisi, Pseudohastigerina micra* were found. This zone corresponds to the *Globigerapsis semiinvoluta* and *Globorotalia cocoaensis* zones of the tropical realm. Identification of the latter ones is impossible here as a number of tropical species of planktonic foraminifers are absent due to paleoclimatic conditions.
- the *Globorotalia centralis - Globigerina gortanii* zone, where together with above-mentioned species were found *Globigerina corpulenta, G. galavisi* and relatively rare - *G. officinalis, G. ampliapertura, Globorotalia permicra* widespread in the Oligocene. The representatives of *Globigerapsis* are completely absent. The high-conic *Globigerina gortanii* Borsetti, 1959 is probably a syno-

nym of *Globigerina rubriformis* (Subbotina), 1953.

According to nannoplankton the Upper Eocene deposits of the Kuban river sequence correspond to the *Discoaster barbadiensis* zone subdivided into three subzones - *Chiasmolithus oamaruensis, Isthmolithus recurvus, Sphenolithus pseudoradians*.

Marls and limestones of the Belaya Glina (White Clay) formation contain numerous benthic foraminifers - *Bolivina antegressa, B. nobilis, B. budensis, Bifarina millepunctata, Neobulimina fraudulenta, Uvigerina costellata, U. jacksonensis, Bolivinoides reticulatus, Bulimina sculptilis, B. aksuatica, Vaginulina mexicana, Saracenaria arcuata, Frondicularia budensis, Siphonodosaria volgensis, S. exilis, S. spinescens, Nodosaria bacillum, Dentalina inornata, Marginulina behmi, M. fragaria, Lenticulina limbosa, L. laticostata, Valvulineria iphigenia, Cassidulina globosa, Planulina costata, Anomalina acuta, Cibicides pigmeus* and many others. The uppermost part of this formation (5-7 m) is particularly rich in benthic foraminifers; this member is called "layers with *Bolivina antegressa*".

The layers in question contain also different species of mollucs - *Variamussium fallax, Amussium semiradiatum, Palliolum meyeri, Spondylus buchi, Nuculana perovalis, Pholadomya subalpina,* etc. They are used to be considered as a fauna of Lattorfian type.

Oligocene deposits form the Maikop series and are represented by quite different sediments - dark noncarbonate clays. The boundary with the Belaya Glina (White Clay) formation is very sharp, but no traces of a hiatus or outwash have been found . The replacement of light marls and clayey limestones by dark clays takes place within a bed of about 1 meter thick.

The Lower Oligocene corresponds to the Khadum formation,35 m thick. It includes brown, dark grey and black leaf-like clays with fish remains. At the base of the formation the clays are weakly calcareous and upward the sequence becomes completely non-calcareous with jarosite spots.

The planktonic foraminifers are developed only in the basal calcareous clays - abundant *Globigerina officinalis, G. praebulloides, G. ampliapertura, G. prasaepis, G. ouachitaensis, Globorotalia permicra, G. gemma, Pseudohastigerina micra, Ps. barbadoensis*. They are indicative of the Lower Oligocene age of sediments, but taking into consideration the persistence of the sequence, the Khadum formation basal layers can be included in the *Globigerina tapuriensis* zone. Nannoplankton here is also poor - disappearance of *Discoaster saipanensis* and *D. barbadiensis* shows, that these deposits belong to the *Ericsonia subdisticha* zone. In the overlying non-carbonate clays nannoplankton is absent.

The lower part of the Khadum formation contains also molluscs - *Nucula compta, Corbula conglobata, Pseudamussium cossmani, Cassidaria buchi, Natica nysti,* etc.

The Khadum clays are overlain by the Ostracoda bed - white and grey marls with abundant thin-walled ostracods, up to 5 m thick. This bed is traced throughout the northern slope of the Caucasus (several hundred km) and serves as a good regional marking horizon.

Thus, the Kuban river sequence shows continuous succession of the Upper Eocene - Lower Oligocene deposits. At the Eocene - Oligocene boundary(top of the *Globorotalia centralis - Globigerina gortanii* zone by foraminifers, top of the *Discoaster barbadiensis* zone by nannoplankton) the process of sedimentation was sharply changed - calcareous sediments were replaced by non-carbonate ones, i.e. the lithological boundary coincides with the chronostratigraphic one. It should be noted that in the adjacent regions of the North Caucasus studies frequently fix a hiatus at the level under consideration. Fauna and flora at the Eocene - Oligocene boundary in this region were changed strongly, but this event was emphasized by a sharp change in facies. The latter does not permit the real evolution of fauna and flora to be traced.

REFERENCES

Krasheninnikov, V.A., Muzylev, N.G., 1975. Correlation between zonal scales by planktonik foraminifers and nannoplankton in Paleogene sequences of the North Caucasus. Voprosy mikropaleontologii, 18: 212-224 (in Russian).

Subbotina, N.N., 1953. Globigerinids, hantkeninids and globorotaliids. Trudy Vses. nauchno-issled. geol. rasv. inst., nov. ser., 76, 294 pp. (in Russian).

Subbotina, N.N., 1953. Upper Eocene lagenids and buliminids of the South of the USSR. Trudy Vses. nauchno-issled. geol. rasv. inst., nov. ser., 69, Mikrofauna SSSR, 6: 115-256 (in Russian).

Shutskaya, E.K., 1956. Stratigraphy of Paleogene lower horizons in the Central Forecaucasus by foraminifers. Trudy Inst. geol. nauk Akad. nauk SSSR, geol. ser., 164: 1-119 (in Russian).

Shutskaya, E.K., 1960. Lower Paleogene stratigraphy of the North Caucasus and Crimea. In: A.L.Yanshin and oth. (Editors), Paleogene deposits of the South of the USSR European part. Izd. Akad. nauk SSSR, Moscow, pp.207 220 (in Russian).

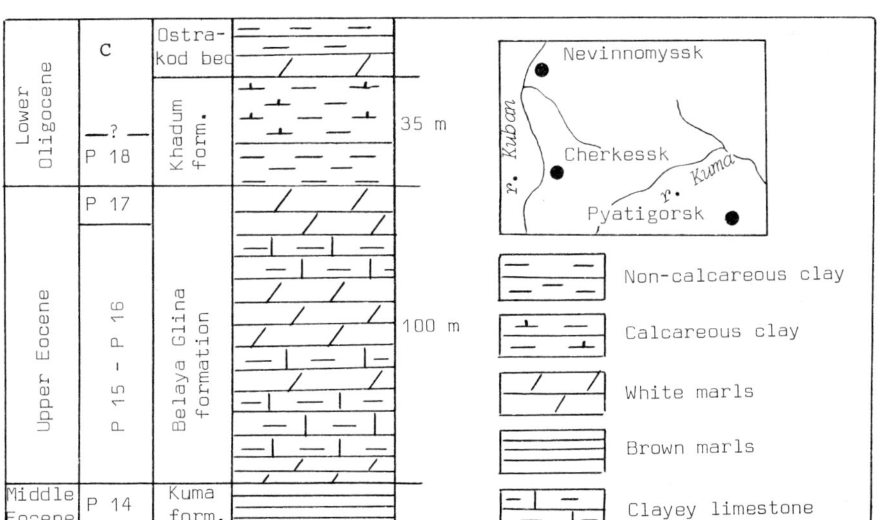

THE KIIN-KERISH SECTION (USSR, SOUTH-EASTERN KAZAKHSTAN, THE LAKE ZAISAN BASIN)

AKHMETIEV M.A., BORISOV B.A., EROFEEV V.S., TSEKHOVSKY Yu.G.

The Lake Zaisan basin is filled by Cenozoic lacustrine-alluvial deposits with a total thickness of up to 2 000 m. It separates the South Altai and Saur-Tarbagatai mountains (see fig.1).The Kiin-Kerish section with remains of representatives belonging to different groups[x] of continental biota was selected for description of Eocene-Oligocene boundary deposits. It occupies the north-east of the Lake Zaisan depression. The deposits of boundary interval compose two units (see Figs. 2,3). The lower Turangin formation (20 m) is represented by sands, siltstones and rarely clays. It is characterized by a limonitization, resulting from oxidation of sulphides scattered in the deposits. The formation's section is topped by the marking layer "A" - spotty-coloured pink siltstones (0.5-1.5 m). In the upper Tuzkabak formation (50 m) clays and silts predominate with lenses and interlayers of quartzic sands and calcareous sandstones. The specificity of the unit is expressed in an abundance of concretions and pseudomorphs of calcite with an admixture of siderite (10-15%) and goethite.

The Turangin formation with erosional surface overlies Paleocene - Middle Eocene (?) deposits and is subdivided into four members.

1. Sands with oblique waving or horizontal lamination 3-4 m. In the sands a jaw was found of the Paroxyclaenid *Kiinkerishella zaisanica* Gab. *et* Bir., and in the roof an incomplete skeleton *Zaisanamynodon borisovi* Bel.

2. Siltstones with clay interlayer (0.5 m) in the middle part. In the roof quartzic sands in places. In it was found a jaw of Amynodontidae ...... 4-5 m.

3. Alternation of siltstones and clays with fossil plants .... 7-8 m. The flora changes throughout the member. In the basal "Salvinia" layer remains of water plants predominate - *Salvinia mildeana* Goepp., *S. mirabilis* Iljinsk., *Nelumbo* sp., monocotyledones - *Zingiberites borealis* (Palib.) Iljinsk., dicotyledones - *Ulmus* sp. and *"Hibiscus" neuburgae* Iljinsk. Poor complex of pollen and spores with *Pinus*, Taxodiaceae, Cupressaceae, Palmae, *Corylus*, *Quercus*. Upward in the clays predominate *Alnus* ex gr. *subcordata* May, *Ulmus* spp., *Populus arnaudii* (Sap.) Iljinsk., *"Hibiscus" neuburgae* Iljinsk., cf. *Cupanites formosus*

---

[x] They were studied by Gabunia L.K., Biryukov M.D. and Beliaeva E.I. (large mammals), Shevyreva N.S. (rodents), Efimov M.V. and Chkhikvadze (reptiles and amphibians), Sychevskaya E.K. (fishes), Tolstikova N.V. (fresh-water mollusks), Iljinskaya I.A., Romanova E.V., Akhmetiev M.A., (plant megafossils), Rzhannikova L.N., Panova L.A., Sotnikova M.A., and Zaporozhets N.I. (pollen and spores).

Berry, Rare - *Lindera vassilenkoi* Iljinsk., *Zelkova zelkovifolia* (Ung.) Bužek *et* Kotlaba, *Myrica,Dryophyllum*,,leaves and fruits of Leguminosae *(Mimosites, Sophora)*, leaves of *Magnolia, Sterculia, Apocynophyllum*, Myrtaceae. In palyno-complex predominate *Alnus* and *Ulmus* (up to 60 %), Juglandaceae *(Juglans, Carya, Pterocarya, Platycarya)* (up to 10%), Hamamelidaceae *(Hamamelis, Fothergilla)* (up to 5%), *Zelkova, Planera, Celtis.* Single pollen of Palmae, *Salix, Myrica, Corylus, Ostrya, Quercus (Q. gracilis, Q. graciliformis, Q.* cf. *sibirica), Castanopsis, Magnolia, Liquidambar zaisanica* Pan., *Altingia, Laurus, Trochodendron,* Araliaceae. Pollen of gymnosperms - 8%: *Pinus* (up to 6%), *Ginkgo, Picea, Abies,* Taxodiaceae, Cupressaceae, spores of Polypodiaceae (1%). In the sample from the same level, but outside the flora-bearing lense - Taxodiaceae (up to 50%), *Tilia,* Liliaceae. Imprints of cycads and mollusks(*Contradens* cf. *compressopsis* Tolstik. and others), mammals: *Ardynomys glambus,* cf. *Anthracotherium, Eoentelodon* sp., *Gobiohyus* sp.).

4. Siltstones and clays with lenses of quartz and sands at the base. Abundant limonitization. Directly under the marking layer "A" - the lense of white flora-bearing clays .... 2-6 m

Predominant are the narrow- leaved *Dryophyllum curticellense* Sap.*et* Mar., *D.dewalquei* Sap. *et* Mar., *D. kryshtofovichii* Iljinsk., *Myrica acuminata* Ung., *Myrica* cf. *lignitum* (Ung.) Sap., *Populus arnaudii* Iljinsk., *P. kryshtofovichii* Iljinsk., rare *Taxodium dubium* (Sternb.) Heer, *Lindera vassilenkoi* Iljinsk., *Zelkova zelkovifolia* (Ung.) Bûžek *et* Kotlaba, *Ampelopsis schischkinii* Iljinsk., *Fraxinus* sp., *Astronium ninae* Iljinsk. *et* Akhmet. and others. Here occur also mollusks from ecozone *Contradens compressopsis (C. compressopsis* Tolstik., *Ptichorynchoidens longiusculus* Tolstik. and others.

The Tuzkabak formation includes 5 members:

5. Clays and siltstones with interlayers of sands and sandstones, In the roof - marking layer "B" - siltstones, black manganese ....4-6 m

In the lowermost part of member 5 were found rodents - Cylindrodontidae and Eomyidae gen. *et* sp. n..Upward in the section,in the member 5 and at the base are vertebrates: mammals-*Amynodon* sp.,*Brachyodus* sp., *Archaeomeryx* sp.,crocodiles - Alligatorinae, (?) *Tomistoma* sp., turtles *Chrysemys* sp. and *"Plastomenus" minusculus* Chkhik., vertebrae of salamanders and ooze fishes *(Amia barroisi* Leirich., *A. robusta* Priem.*)*.

6. Siltstones, sands and sandstones, carbonate, manganese .... 8-12 m

7.Clays spotty-coloured with lenses of carbonate sandstones and intercalating clays and siltstones ....15-16 m

Mammals: *Amynodon*sp ., *Amphechinus* sp., *Cadurcodon* cf. *zaisanensis* Bel., *Bothriodon* sp., Anthracotheriidae (cf. *Brachyodus*), *Entelodon* sp., *Artiodactyla* (cf. *Eumeryx,* cf. *Miomeryx)*, crocodiles - *Alligatorinae, Tomistoma* sp., turtles - *Chrysemys index* Chkhik., *"Plastomenus" minusculus* Chkhik., *Chelydropsis minax*

Chkhik.

8. Clays with ferruginous-carbonate concretions with interlayers of silts and sands. In the base - traces of erosional surface (in places) .... 10-12 m. In the middle part of the member - mollusks of ecozone *Rectidens asiaticus (R. asiaticus* Tolstik., *Ctenodesma subfoliaceae* Tolstik., *Margaritonopsis infans* Tolstik., *Nannonaia pucla* Tolstik., *Oxynaia subcultus* Mart. *et* Tolstik., *O. ferrata* Tolstik., *Rectidens kustovicus* Tolstik., *Physunio laeviusculus* Tolstik. and others.

In the upper part of the member the turtle armour *Trionyx zaisanensis* Chkhik. was found. At the boundary of the members 7 and 8, in the lens of white clays plants were found.

schischkinii Iljinsk., *Quercus* spp. (3 species), *Castanea antipovii* (Krysht) Iljinsk., *Alnus neuburgae* (Baik) Iljinsk ., *Populus arnaudii* Iljinsk., *Ailanthus confucii* Ung., more rare - *Liquidambar zaisanica* Akhmet., *Platanus vassiljevii* Iljinsk., *Eucommia palaeoulmoides Baik.*, *Pistacia zaisanica* Iljinsk., *Rhus kiinkerischica* Akhmet., *Astronium ninae* Iljinsk. *et* Akhmet., *Acer* cf. *tataricum* L. and others.

9. Clays, gypsum-bearing, varicoloured .... 6-10 m

They are overlain by sands and silts of the Upper Oligocene.

Due to the difficulties of correlation between the regional scale and the planktonic scales, the position of the Eocene-Oligocene boundary in the Kiin-Kerish section can be established only within an interval. The finds of such archaic forms as *Kiinkerishella*, *Zaisanamynodon*, *Gobiohyus*, *Eoentelodon* and a form close to *Anthracotherium* (the representative of Antracotheriidae), permit one to include the mb 1-3 in the Upper Eocene. The presence of "*Hibiscus*" *neuburgae* in combination with typical subtropical plants in the complex of the member 3 as well as finds in the palynocomplex of *Quercus gracilis* Boitz. are also indicative of an Eocene age. The appearance of temperate *Alnus*, *Ulmus* and others results from local conditions : closeness of mountains and the situation of the Zaisan basin bottom above sea level.

Vertebrates complex of member 6, with Amynodontidae, Anthracotheriidae *(Bothriodon, Brachyodus* and others , as well as flora complex belonging to the Turgai type of the member 7 are indicative of Oligocene age of enclosing rocks. This enables us to consider that the Eocene - Oligocene boundary lies in the interval of the members 4-5.

Conclusions.

1) In the section Kiin-Kerish, Eocene - Oligocene boundary interval includes members 4 and 5. As a preliminary variant, the boundary between the Eocene and Oligocene can be drawn at the contact between Turangin and Tuzkabak formations.

2) The nature of sedimentation in the boundary interval does not essentially change. There was only the change in the major paragenetic assemblages: the

clay-siderite-pyrite assemblage was replaced by a ferruginous-carbon one. No considerable hiatuses in sedimentation have been registered. Composition of the fossil complexes of organisms is also changing gradually.

3) The cooling of the territory of the Zaisan Basin at the end of Late Eocene and the beginning of Oligocene led to a gradual elimination of subtropic "Poltava" elements from the vegetation and their replacement by temperate "Turgai" elements accompanying a general reduction of moisture.

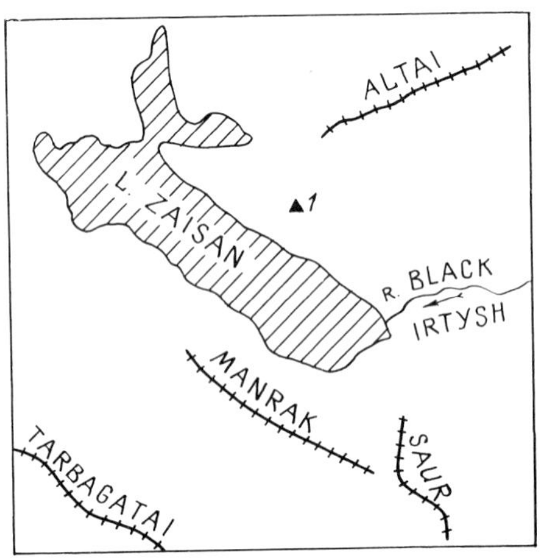

Fig.1 Index map, showing Kiin-Kerish
1 - Kiin-Kerish section

Fig.2 (below) Geological map (by Erofeev V.S.)
1 - clays of the Upper Paleocene-Eocene (P_1-P_2^2). Upper Eocene and Lower Oligocene deposits. 2,3 - Turangin formation (P_2^3-P_3^1): 2- member 1; 3- members 2-4; 4,5 - the Tuzkabak formation (P_2^3-P_3^1): 4- members 5-7; 5- members 8,9; 6- Upper Oligocene (P_3^2); 7- Quaternary deposits; 8-the marking horizon "A"; 9-faults; 10- stratigraphic contacts; 11- dip of strata; 12- localities of flora; 13- localities of vertebrates.

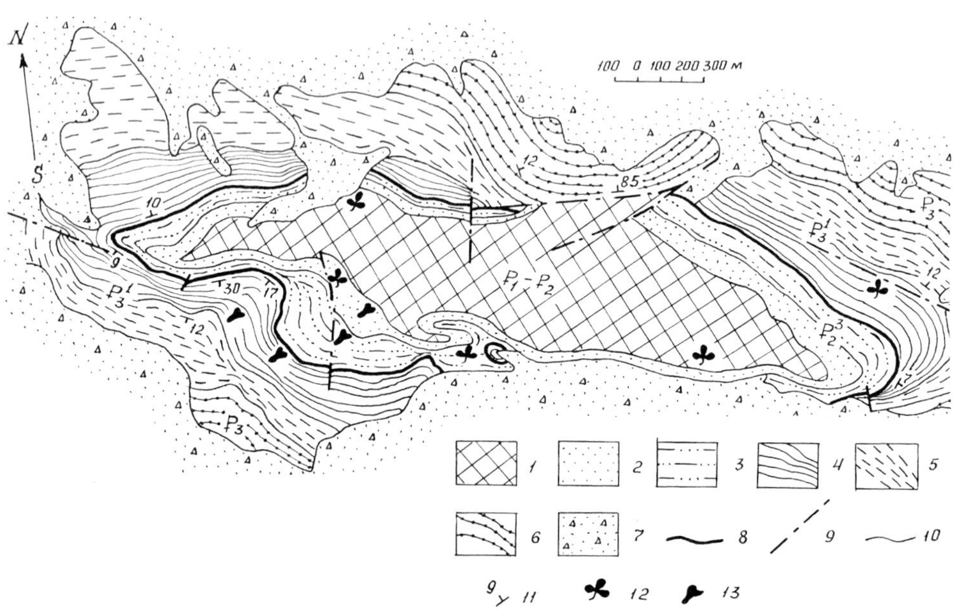

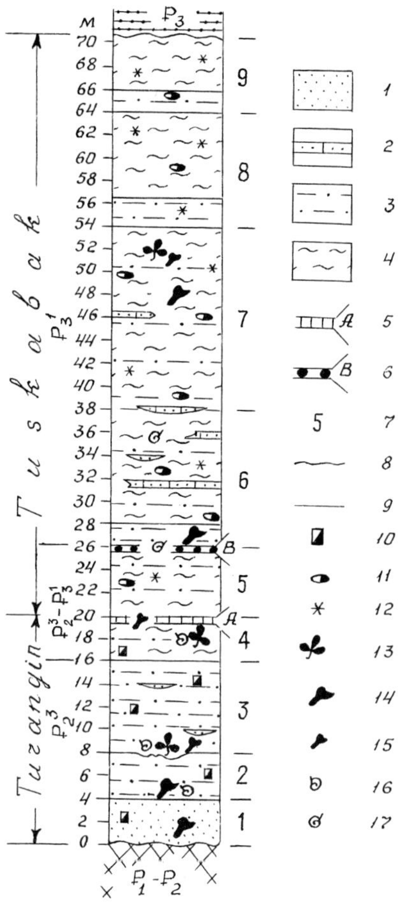

Fig. 3 Upper Eocene - Lower Oligocene Kiin-Kerish section.
1 - quartzic sands; 2 - quartzic sandstones with carbonate cement; 3 - silts 4 - clays; 5 - marking horizon "A"; 6 - marking horizon "B"; 7 - numbers of members in the section; 8 - contacts with errosional surface; 9 - conformable contacts; 10 - oxydizing sulphides; 11 - carbonate concretions; 12 - iron-manganese concretions; 13 - localities of fossil plants; 14,15 - major and isolated localities of vertebrates; 16,17 - mollusks of the ecozones *Contradens compressopsis* and *Rectides asiaticus*.

KARAGINSKY SECTION (USSR, KARAGINSKY ISLAND, EAST KAMCHATKA)

M.Ya.SEROVA

Marine and continental Upper Paleogene deposits are widespread within the north-western part of the Pacific Ocean (West Sakhalin , West and East Kamchatka and adjacent islands). They are present also in several depressions ofthe South-East Sakhalin, Koryak high-land, Anadyr-Penzha depression and on the Komandor Islands. Eocene-Oligocene boundary layers are represented here by predominantly marine terrigeno-volcanogenic sandy-clayey carbonate-free deposits up to several thousands km thick. In some regions they are coal bearing.

Late Paleogene deposits of the Far East region are rich in fossils-benthic foraminifers, mollusks, flora, spores, and pollen. However, they are practically free of calcareous plankton (planktonic foraminifers, nannoplankton). On the other hand, siliceous plankton is rather abundant but poorly preserved. Therefore, the benthic foraminifers and mollusks are the major groups for subdividing sequences and for interregional correlations. The complexes of benthic foraminifers are practically free of endemic forms and give the opportunity to correlate the biostratigraphic subdivisions of the Far East scale with the stage and zonal scale of California, elaborated on the basis of benthic foraminifers and mollusks which for the last few years was essentially corrected in definition of age by planktonic foraminifers (Lipps, 1967; Krasheninnikov, 1973, and others) and nannoplankton (Warren, Newell, 1976).

The Eocene-Oligocene boundary is drawn on the basis of completed wide correlations and a complex of biostratigraphic, climatological data and few measurements of absolute age: in West Kamchatka - between Kovachinsky and Amansko-Gankhinsky horizons, in the southern part of the Koryak high-land (the Ilpinsky Peninsula) - between the Kovachinsky and Aluginsky horizons, in East Kamchatka (the Karaginsky Island) - between the formation of the mountain Peresheek and Ilkhatunsky formation, on the coast of the Penzha bay - between the Irgirninsky and Rateginsky formations.

The Karagin Island sequence was interpreted as typical for the Eocene-Oligocene boundaries. Its brief description is given below.(Fig. 1).

The boundary deposits here are represented by three formations: the Cape Tons, the mountain Peresheek, and the Ilkhatunskaya one (Fig. 2).

The Cape Tons formation is represented by tuff-breccias and tuff conglomerates with interlayers of tuff-sandstones and tuff claystones including shells

Fig. 1

Index map

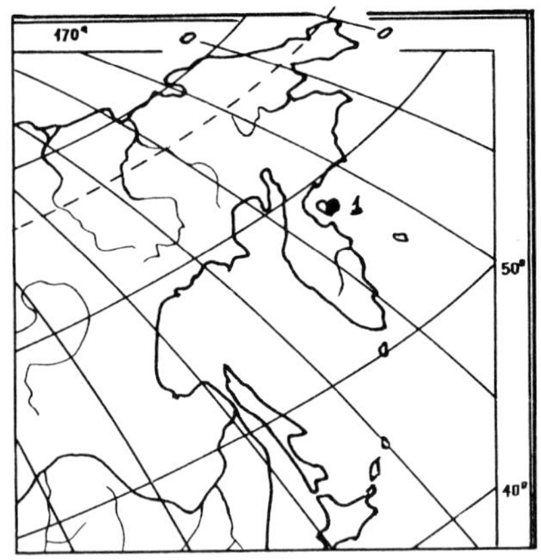

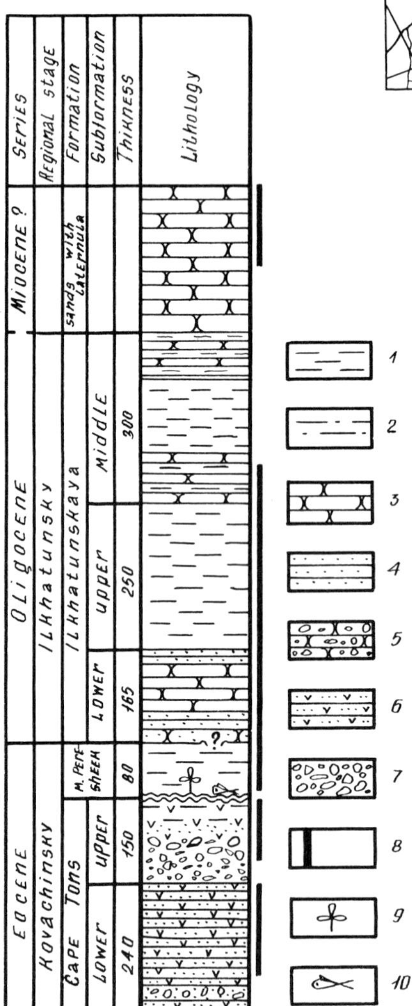

Fig. 2

Eocene - Oligocene stratigraphic column of Karaginsky Island

1 - mudstone
2 - siltstone
3 - sandstone
4 - clayey sandstone
5 - conglomerate
6 - tuffaceous sandstone
7 - tuff-breccia
8 - intervals of section sampled for Foraninifera
9 - flora fossils
10 - fish fossils

of mollusks and foraminifers.

The mountain Peresheek formation consists of dense sandy claystones with round loaf-like concretions, scattered fine pebbles and grit and mollusk shells. Upper part of the formation has interlayers of fine-platycalcareous claystones with abundant accumulations of fish remains. Thickness (visible) - 80 m.

The Ilkhatunskaya formation overlaying without visible unconformity the mountain Peresheek formation is represented by an alternation between sandstones, siltstones and claystones. The siltstone members are characterized by presence of calcareous concretions, and admixture of pebble and grit, as well as abundant mollusk shells with predominance of the *Yoldia* genus. Total thickness is 700-1000 m.

The boundary deposits of this sequence were described by three different foraminiferal complexes. The first (Upper Eocene) one is associated with Cape Tons formation, mountain Peresheek formation and lowermost parts (below 40 m) of the Ilkhatunskaya one. *Plectofrondicularia, Globulimina, Valvulineria, Caucasina* are predominant here. Typical species of the complex are: *Plectofrondicularia packardi packardi, P. packardi multilineata, P. smithi, Gyroidina condoni, H.obliquecamerates.* In the mountain Peresheek - *Uvigerina garzaensis nudorobusta, Bulimina sculptilis, Alabamina kernensis, Valvulineria involuta, Caucasina schwageri, Caucasina eocaenica kamchatica* are dominant.

The mollusks are rich in number of individuals and impoverished in systematic composition. Species composition does not practically change upwards except only changes in composition of the complex resulted from a change in sediment lithology. The finer (by composition of rocks) siltstones and claystones are characterized by abundant small finevalve scallops of the *Variamussium* type, fish remains and crustaceans, stems of sea algae. The gritstones are represented by fewer deep-water groups - *Yoldia, Lima, Mactra* and others. Typical are species: *Variamussium pillarense, V. omacusensis, V.* aff. *inollei, Lima (Acesta) amakusensis, Yoldia watasei* (Gladenkov, 1972). This last species is most abundant in the complex of the mountain Peresheek formation. The complex is close to the fauna of the Sakasegava formation in Japan (Serova *et al.*, 1975).

The foraminiferal and molluscan complex of the Ilkhatunskaya formation differs considerably in composition and structure from the complexes of the Cape Tons and mountain Peresheek formations. The most diversified foraminiferal assemblage characterizes the lower part of the formation (without lowermost 40 m). It is represented by more than 40 species of predominantly calcareous perforate foraminifers. The percentage of species passed from the underlying deposits is inconsiderable. New elements exceed 80%. The major background is composed of *Cibicides* and among them the species *Cibicides coalingensis* is most numerous. Besides this species there are present: *Cibicides americans, C. crassiseptus, C. hodgei, C. almaensis* together with *Eponides frizzelli, Melonis shimokinense,*

Cribroelphidium stelligerum, Lagena, costata, Nonionella tatumi, Siphonodosaria frizzelli, and also single *Bulimina sculptilis, Caucasina schwageri, Globulimina ilpinica, Cyclammina pacifica, C.incisa.* In this complex single *Pseudohastigerina micra* (Cole) is present also.

The upper, larger by thickness part of the Ilkhatunskaya formation is described by impoverished complex of agglutinated foraminifers: *Haplophragmoides latissimisuturalis, H. asagaensis, Asanospira carinata, Budashevaella deserta, C. incisa, C. concellata obesa, Martinotiella* aff. *communis* and others. The calcareous foraminifers in this part of the section are very rare.

The mollusks of the Ilkhatunskaya formation considerably differ from the complex of the mountain Peresheek and Cape Tons formations. Among *Yoldia* here together with *Yoldia watasei, Yoldia longissima* predominates, and in the upper member the species *Yoldia nitida* is abundant. They are accompanied by *Acila praedivaricata, A. gettysburgensis, Malletia* sp., *Laternula besshoensis, Macoma sumizuensis, Trominia japonica, Ancistrolepis* sp. (Gladenkov, 1972).

The greatest change in the mollusk and foraminiferal association occurs at the base of the Ilkhatunskaya Formation. It takes place slightly higher than the boundary of change in lithological composition of the rocks, i.e. above the boundary between the Peresheek Mountain and Ilkhatunskaya Formations.

The above described foraminiferal associations are alike both by systematic composition and by the structure fo the E/O boundary deposit complexes of the Pacific coast of North America with which they were correlated. The Aroio-el-Bulito sequence in California (Kleinpell, Weaver, 1963) is most interesting in this respect. This sequence near the type section of Refugian stage in the upper part of the Sakate formation contains nannofossils of the *Chiastmolithus camaruensis* subzone of Late Eocene *Discoaster barbadiensis* zone. In the formation Geviota, which by its benthic foraminifers belongs to Refugian stage, nannoplankton of the *Istmolithus recurvus* subzone of the *Discoaster barbadiensis* zone was found.

The Early Zemorrean is considered by nannoplankton as the Earliest Oligocene (Warren, Newell, 1976). In the Upper Eocene part of this section (the Geviota formation) as in the deposits of the Cape Tons and mountain Peresheek formations the major basis of the foraminiferal association is composed of the following species: *Bulimina sculptilis, Cibicides haydoni, Plectofrondicularia pachardi packardi, P. smithi. Cibicides americanus crassiseptus, C. hodgei, Eponides frizzelli* and others predominate in the lower part of Zemorrean stage (upper part of the Alegria formation), as well as in the Ilkhatunskaya formation.

The above given data show that within the Far East region at the Eocene - Oligocene boundary occur the most essential changes in benthic foraminiferal associations both in systematic composition and in the structure of the complexes, i.e. considerable decrease in amount of calcareous benthos and sharp increase in a number of agglutinated foraminifers with great diversity in systematic composition. A similar trend of the change in the biota is marked at the Eocene - Oligocene boundary in the adjacent areas of the Pacific province - North America, Japan and others. These changes are in close relationship with other geological events, such as climate cooling, which began at the end of the Eocene and markedly progressed in the Oligocene; volcanic activity, also started in the Middle Eocene. Sharp distrophic events embracing the whole region at Eocene - Oligocene boundary have not been registered. This is confirmed by absence of large hiatuses in sedimentation.

REFERENCES

Gladenkov, Yu.B., 1978. Marine Upper Cenozoic of the Northern regions. Trudy GIN, vol. 313, 194 pp.

Kleinpell, R.M., Weaver, D.W., 1963. Foraminiferal fauna from the Gaviota and Alegria formation. Univ. Calif. Publ. Geol. Sci., v. 43: 1-45.

Krasheninnikov, V.A., 1971. Stratigraphy and Foraminifers of Cenozoic pelagic sediments of the Pacific north-western part (based on the materials of deep-sea drilling). Voprosy mikropaleontologii, v. 14: 140-201.

Lipps, I., 1967. Planktonic Foraminifera intercontinental correlation and age of California Mid. Cenozoic microfaunal stages. I. Paleontol., v. 41: 994-999.

Serova, M.Ya., Borzunova, G.P., Shapiro, M.N., 1975. Paleogene of the North part of Koraginsky Island (East Kamchatka). Izv. Akad. Nauk, ser. geol., 11: 73-83.

Warren, A.D., Newell, J.H., 1976. (abs.) Nannoplankton biostratigraphy of the Upper Sacate and Gaviota formation, Aroyo el Bulito, Santa Barbara County, California. 51 Annual Meeting, Program Pacific Section, San Francisco: 19-21.

EOCENE - OLIGOCENE BOUNDARY IN MONGOLIA

D.DASHZEVEG, E.V.DEVYATKIN

The Central Asian regions including the South-East Gobi and adjacent areas of North China offer the best perspective for defining the position of the Eocene-Oligocene boundary in continental deposits. These regions were studied by expeditions of the American Museum of Natural History (Berkey, Morris, 1927), the Mongolian Paleontological Expedition of the USSR Academy of Sciences (Efremov, 1954), later by the Mongolo-Polish Paleontological Expedition (Kielan-Jaworowska, Dovchin, 1968), and starting from the middle 60-es by the Soviet-Mongolian Paleontological and Geological Expedition of the USSR Academy of Sciences and Mongolian Academy of Sciences (Dashzeveg, 1966; Janovskaja *et al.*, 1977; Devyatkin, 1981).

The South-East Gobi territory is a gently hilly plateau with moderate mountain massifs oriented to the south-east and separated by wide flat depressions. Eocene and Oligocene deposits filling these depressions are represented by lacustrine and lacustrine-alluvial facies - vari-coloured clays, quartzic sandstones, gritstones, pebble beds with interlayers of carbonate sediments and horizons of fossil soils. These deposits are fragments of lacustrine-alluvial plains spreading to the adjacent territory of China. Eocene and Oligocene deposits are about 120-140 m thick.

By now in the South-East Gobi 6 key sequences of the range between the Eocene and Middle Oligocene have been studied in detail (Fig.1), and the succession of their suites, beds and the enclosed fauna of mammals have been revealed (Table 1).

The lowermost horizons of the Eocene were recognized near the Dzamyn-Ude; further to the west towards Erghylin -Dzo the total Paleogene sequence extends up to the Middle Oligocene inclusively.

The Erghylin-Dzo suite sequences are the most complete and best studied. The Erghylin-Dzo escarpment is a single system of sequences with latitudinal orientation, traced along the plateau margin over more than 50 km. The history of research in stratigraphy and paleontology of these sections is given in the article by N.M.Janovskaya *et al.*, 1977. Recent interpretation of already known and new materials from the Erghylin-Dzo region is given below. Other sections are supplemental.

The Middle Eocene deposits of the Mergen sequence are represented by light-grey sands, red-brown sandstones and clays up to 10 m thick. They are overlain by washed-out light gravel sands and green-grey clays (up to 10 m thick) united into the Tsagan-Obo suite containing the mammal fauna of: *Lophialetes (?) minutus*, *L. expeditus*, *Eumys* sp., *Shamolagus* or *Gobilagus* sp., and *Triplophus (?) proficiens*. This fauna permits correlation of the Tsagan-Obo suite with the Irdyn-Manha Formation (China).

The lowermost part of the Upper Eocene is singled out as the Badrakh suite in Alag-Tsav sequence. The suite is represented by green and red clays,in the uppermost parts by gray sands with gravel up to 18 m thick. In the suite there were found: *Archaeomeryx optatus*, *Megalamynodon* sp., *Hyaenodon* sp.

The Khubsugul layers occupy the lowermost stratigraphical position in general sequence of Erghylin-Dzo suite. They are represented by red, horizontally layered clays (up to 16 m) with green interlayers not described paleontologically. They are overlain by the Dzangut beds which are well situated in the Khoer-Dzan sequence, where they are represented by greenish clays with white carbonate concretions at the base, replaced upwards by grey sandstones and fine gravels (up to 10 m thick). In Erghylin-Dzo region the Dzangut beds possibly drop out of the general sequence of the suite. The overlying Sevkhul beds are represented by quartzic light sands, loose conglomerates, replaced upwards by white and greenish carbonate, horizontally layered clays with sandstone interlayers. The total thickness of deposits is 15-25 m.

In a number of sections above the Sevkhul beds the red clays can be traced and are individualized as the Shavag beds (up to 4 m). They do not have a paleontologic basis but serve lithologically as a marking horizon.

In all sequences the Erghylin beds are overlain with outwash. They consist of alluvial coarse-grained yellow-brown cross-bedded gravel sandstones (up to 15-20 m). In these beds Early Oligocene key genera of mammals appear for the first time.

The Khetsu-Tsav beds are the youngest stratigraphic subdivision of the Erghylin-Dzo suite. They terminate the alluvial cycle and are represented by gravel and pebble conglomerates with inconsiderable thickness (from 1-5 to 10 m). In places they are separated from the underlying Sevkhul beds by horizons of fossil soils. In Khetsu-Tsav beds turtle remains predominate, *Ergilemys insolitus*, with few finds of *Cadurcodon ardynensis* teeth.

The Shand-Gol suite deposits terminate the Paleogene sequence of South-East Gobi. They are represented by red-brown clays and sandstones (up to 15 m) and contain remains of *Ardynia* cf. *praecox* and *Tsagonomys* sp.

Paleogene sequences of the adjacent North China territory are rather well stratigraphically and lithologically correlated with the studied sequences of South-East Gobi.

The Eocene-Oligocene boundary drawn in continental deposits is based on the principle of rapid development and widespread distribution of new mammal forms

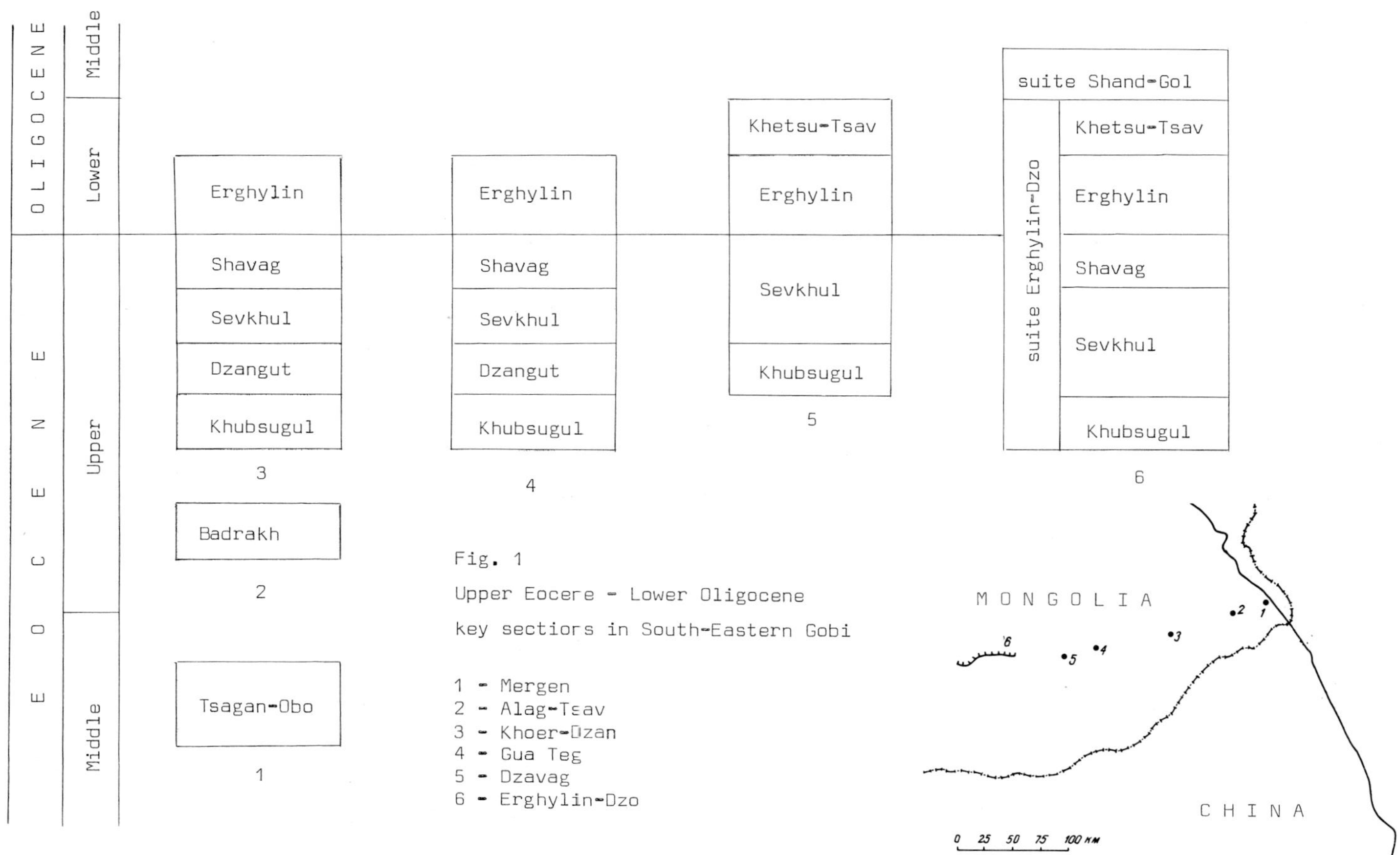

Fig. 1

Upper Eocere - Lower Oligocene

key sectiors in South-Eastern Gobi

1 - Mergen
2 - Alag-Tsav
3 - Khoer-Dzan
4 - Gua Teg
5 - Dzavag
6 - Erghylin-Dzo

TABLE 1

Distribution of mammalian fauna in the Upper Eocene and Oligocene key sections of South-Eastern Gobi, Mongolia

Stratigraphic subdivisions / Mammalia	Upper Eocene					Oligocene	
	suite Erghilin-Dzo						suite Shand-Gol
	Khubsugul	Dzangut	Sevkhul	Shavag	Erghylin	Khetsu Tsav	
Eucricetodon sp.					+		
Desmatolagus robustus			+				
Ardynomys olseni					+		
Tsaganomys sp.							+
Pterodon exploratus			+				
Pterodon (?) mongoliensis					+		
Pterodon sp.			+				
Hyaenodon eminus			+				
Hyaenodon incertus			+		+		
Hyaenodon gigas					+		
Hyaenodon sp.		+					
Ardynictis furunculus			+				
Cynodon sp.					+		
Nimravus mongoliensis					+		
Epimanteoceras robustus			+				
Parabrontops gobiensis			+				
Metatitan relictus			+				
Protembolotherium efremovi			+				
Embolotherium ergiliense					+		
Embolotherium andrewsi					+		
Embolotherium grangeri			+				
Embolotherium sp.		+					
Schizotherium avitum			+		+		
Eomoropus sp.					+		
Teleolophus magnus			+		+		
Deperetella sp.			+				
Colodon inceptus			+		+		
Ardynia praecox			+		+		
Ardynia cf. *praecox*							+
Prohyracodon meridionale			+				
Prohyracodon sp.		+					
Forstercooperia ergiliinensis					+		
Indricotherium (?)					+		
Amynodon lunanensis			+				
Cadurcodon ardynensis			+		+		
Gigantamynodon cessator			+		+		
Eoentelodon trofimovi			+				
Entelodon orientalis					+		
Bothriodon sp.					+		
Miomeryx altaicus			+		+		
Lophiomeryx angarae			+		+		
Lophiomeryx gobiae			+		+		

in a process of their adaptive radiation. On this basis we can suggest that the appearance of *Entelodon*, *Bothriodon*, *Brachyodus* and other genera in Paleogene sequences of Central Asia defines the lower boundary of the Oligocene.

In the Erghylin-Dzo sequence in Mongolia the essential change in ecological dominancy occurs between the Sevkhul and Erghylin layers. For the first time in Erghylin beds appear the genera of *Entelodon, Bothriodon, Cynodon, Nimravus*. They are accompanied by progressive forms of *Embolotherium ergiliense, Forstercooperia* sp. and others.

In North China the Lower Oligocene forms of mammals appear in Barun Sog formation and its stratigraphic analogues. In this horizon *Entelodon* sp., *(=Arhaetherium ordosius), Indricotherium, Cadurcodon ardynensis* and others were found. This formation is correlated with Erghylin beds from Mongolia.

Thus, the E/O boundary in the continental deposits of Central Asia is traced below the Erghylin beds (Mongolia) and the Barun Sog Formation (North China). It is overlain by outwash at the base of the Lower Oligocene deposits as well as by fine lacustrine clays and sands replaced in sequence by coarse alluvial pebble beds. Activation of tectonic movements and a tendency to aridification of the climate are probable in the Lower Oligocene.

REFERENCES

Berkey, Ch., Morris, F.K., 1927. Geology of Mongolia. V.2, New York; 475 pp.

Dashzeveg, D., 1966. New data on stratigraphy of Paleogene deposits of Eastern Mongolia. DAN SSSR, v. 168 (3): 647-649 (in Russian).

Devyatkin, E.V., 1981. The Cenozoic of Inner Asia (stratigraphy, geochronology and correlation). Trans. Joint Soviet - Mongolian Geol. Exped., v.27, Nauka, Moscow; 180 pp. (in Russian).

Efremov, I.A., 1954. Paleontological research in Mongolian People's Republic. Trans. Mong. Comm. Acad. Sci. USSR, Ser.59, Izdat. Akad. Nauk. SSSR, Moscow, pp. 3-26 (in Russian).

Kielan-Jaworowska, S., Dovchin, N., 1968. Narrative of the Polish-Mongolian Paleontological Expedition 1963-1965. Paleont. Polonica, v. 19: 7-30.

Janovskaja, N.M., Kurochkin, E.N., Devyatkin, E.V., 1977. Erghylin-Dzo locality - the stratotype of the Lower Oligocene in South-East Mongolia. Trans. Joint Soviet - Mongolian Paleont. Exped., v. 4, Nauka, Moscow, pp. 14-32 (in Russian).

THE BEST SECTION REPRESENTING THE EOCENE/OLIGOCENE BOUNDARY IN JAPAN

KUNIO KAIHO*, HISATAKE OKADA* and YOKICHI TAKAYANAGI**

The best section containing the Eocene/Oligocene boundary in Japan is a composite section exposed along the course of Kumanosawa and Yubari River in the Yubari area, central Hokkaido (Fig. 1). The Kumanosawa section represents the upper portion of the Poronai Formation whereas the Yubari River section comprises the top level of the Poronai Formation and the conformably overlying Momijiyama Formation. The upper portion of the Poronai Formation exposed along this combined section measures 480 m in thickness and consists principally of dark-gray massive siltstone intercalated with 5 cm to 10 m thick sandy tuff layers with scattered calcareous concretions. The Momijiyama Formation approximately 400 m thick is composed of volcanic sandstone and dark-gray tuffaceous sandy siltstone in the lower part (Jusanmairu Sandstone Member) and is tuffaceous sandy siltstone intercalated with fine tuff layers in the upper part (Fig. 2). The sequence is generally homoclinal with variable dips of between 12° and 60° and seems to represent a shallowing facies in a somewhat restricted inland basin of middle bathyal to outer neritic environments.

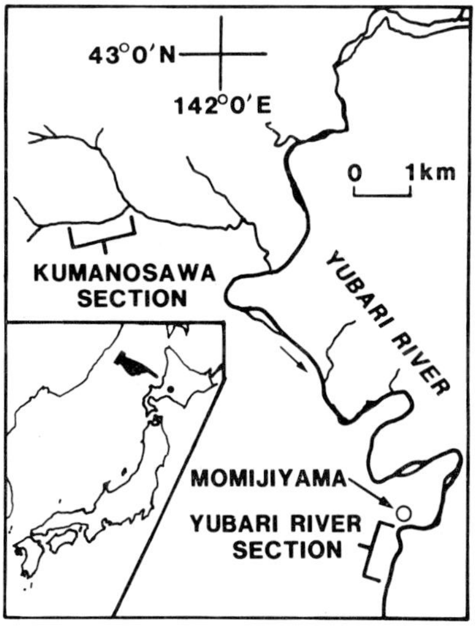

Fig. 1. Map showing the location of the Kumanosawa and Yubari River sections.

This sequence is rich in molluscs, crustaceans and benthic foraminifers whereas planktonic microfossils are rather scarce. It represents the upper two zones of the nine megafossil zones established in the Poronai Formation (Fig. 2), and the

*Department of Earth Sciences, Yamagata University
**Institute of Geology and Paleontology, Tohoku University

upper three zones of the six benthic foraminiferal assemblage zones recognized in the Poronai and Momijiyama Formations (Kaiho, 1984a, b). The palynological evidence indicate a cool-temperate climate with progressive cooling through this sequence.

A key nannofossil species Isthmolithus recurvus occurs exclusively in the upper part of the Poronai Formation indicating the I. recurvus Subzone (CP15b) of the latest Eocene and possibly a part of the Coccolithus subdistichus Subzone (CP16a) of the earliest Oligocene. The Momijiyama Formation is barren of nannofossils. Chiloguembelina cubensis characterizes the planktonic foraminiferal fauna in the upper part of the Poronai Formation. Although it commonly occurs in Oligocene sediments worldwide, its earlier range in the upper Eocene is an unsettled question. Globorotalia insolita which is an upper Eocene species of New Zealand is observed in the middle part (zone F) of the Poronai Formation and supports the age indicated by nannofossils.

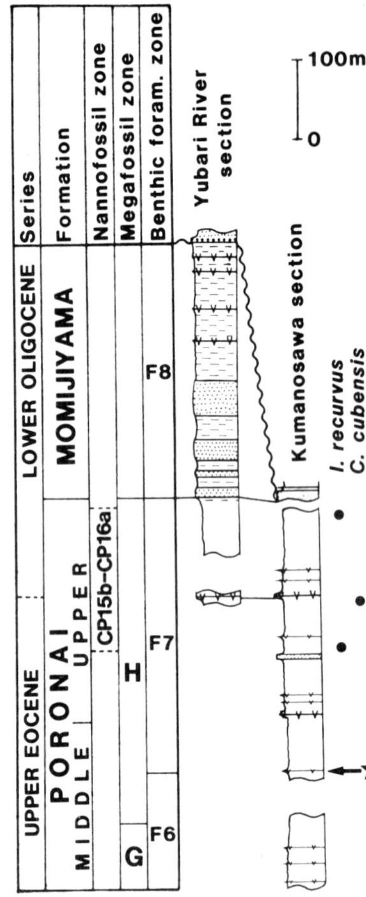

Fig. 2. Lithology and biozones of the Kumanosawa and Yubari River sections. : Explosive upward increase of calcareous foraminifers.

Although no radiometric data are available for this sequence, the middle part of the Nuibetsu Formation in eastern Hokkaido, which is correlatable with the Momijiyama Formation by means of benthic foraminifers, gives a K-Ar data of 32.6 ± 1.7 Ma.

REFERENCES

Kaiho, K., 1984a. Paleogene foraminifera from Hokkaido, Japan. Part 1. Lithostratigraphy and biostratigraphy including description of new species. Sci. Rep. Tohoku Univ., 2nd ser., 54: 95-139.

Kaiho, K., 1984b. Paleogene foraminifera from Hokkaido, Japan. Part 2. Correlation of the Paleogene system in Hokkaido. Ibid., 55: 1-74.

THE MAIN REFERENCE SECTION FOR THE EOCENE/OLIGOCENE BOUNDARY IN NEW ZEALAND

R.H.HOSKINS and H.E.G. MORGANS, New Zealand Geological Survey
PO Box 30368, Lower Hutt, New Zealand.

LOCATION

The main reference section in New Zealand is the Gibsons Beach section at Cape Foulwind, Westport, on the West Coast of the South Island. The section is 12 kilometres due west of the town of Westport and is accessible from a car park at the end of Cape Foulwind Road (see location map, fig. 1). A track leads down to the section which crops out in cliffs backing a beach. The sequence youngs to the east and is 100% exposed. Location can be aided by New Zealand Map Series 260 map "K29 Westport." The latest geological map of the area is by Nathan, 1976.

Two further Eocene/Oligocene boundary sections are being studied: (1) The Point Elizabeth section 75 km to the south-west of Cape Foulwind has 482 metres of mudstones containing a better planktic foraminiferal fauna deposited in more pelagic paleoenvironments, but has a problem of intermittent sand build-up on the beach platform. Seventy-nine samples for biostratigraphy and stable isotope stratigraphy are available for study.
(2) The Kakanui River mouth section, 230 km south-west from Christchurch, is a sequence of marls 24 metres thick. Twenty-five integrated magnetostratigraphic, biostratigraphic and stable isotope stratigraphic sites have been collected.

LITHOLOGY AND STRUCTURE OF CAPE FOULWIND SECTION

The 105 metre thick stratigraphic cliff section of massive mudstones and muddy siltstones begins with massive micaceous dark brown mudstones with intermittent discontinuous algal limestone bands. The middle of the section is poorly bedded with the upper section becoming more concretionary, more calcareous, sandier and eventually becoming well bedded. The strata have a 020^0 strike and a south-east dip of 12-19^0. Only one major fault with unknown throw has been observed.

MICROFOSSILS

Previous biostratigraphic and paleoecological studies carried out on this section include those by Srinivasan (1966, 1968), Srinivasan and Vella (1975) and MacGregor (1983).

One hundred and three biostratigraphic samples have been collected. Good benthic foraminiferal faunas with rare planktics suggest a shelf environment for the strata across the boundary.

Ranges of some foraminiferal key taxa near the Eocene/ Oligocene boundary in New Zealand

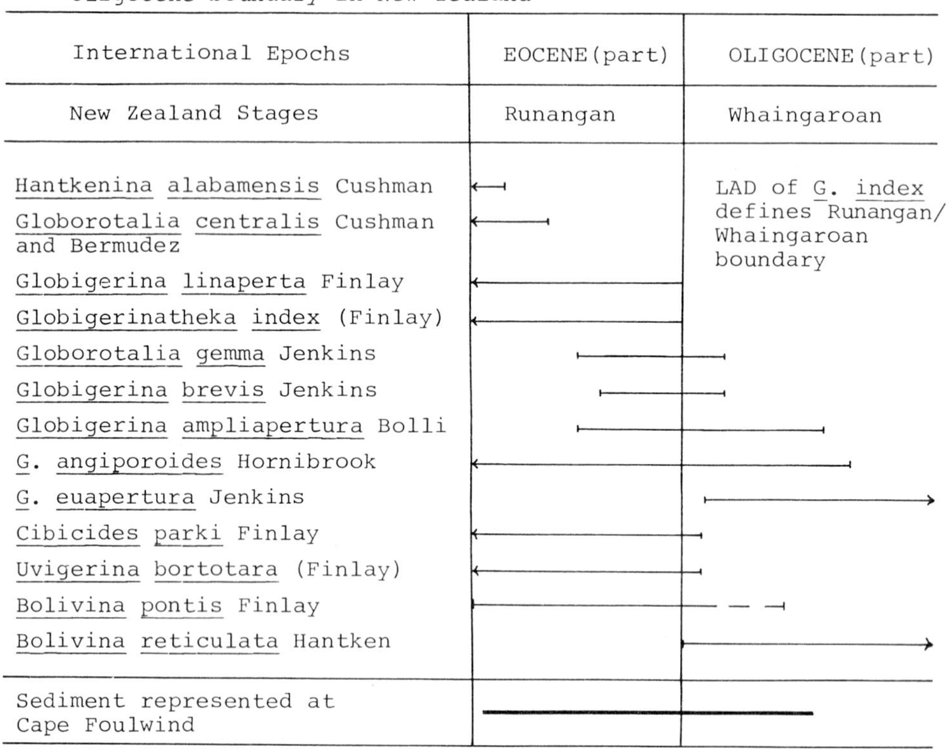

The last occurrence of the planktic foraminiferal species <u>Globigerinatheka</u> (<u>Globigerapsis</u>) <u>index</u> (Finlay) is used to define the Runangan/Whaingaroan boundary in New Zealand. The Eocene/ Oligocene boundary, if based on the last occurrence of <u>G</u>. <u>index</u> (Finlay), is about 55 metres above the base of the section and occurs within a massive non-bedded silty mudstone with continuous exposure and no lithological variation or angular discordance across the boundary.

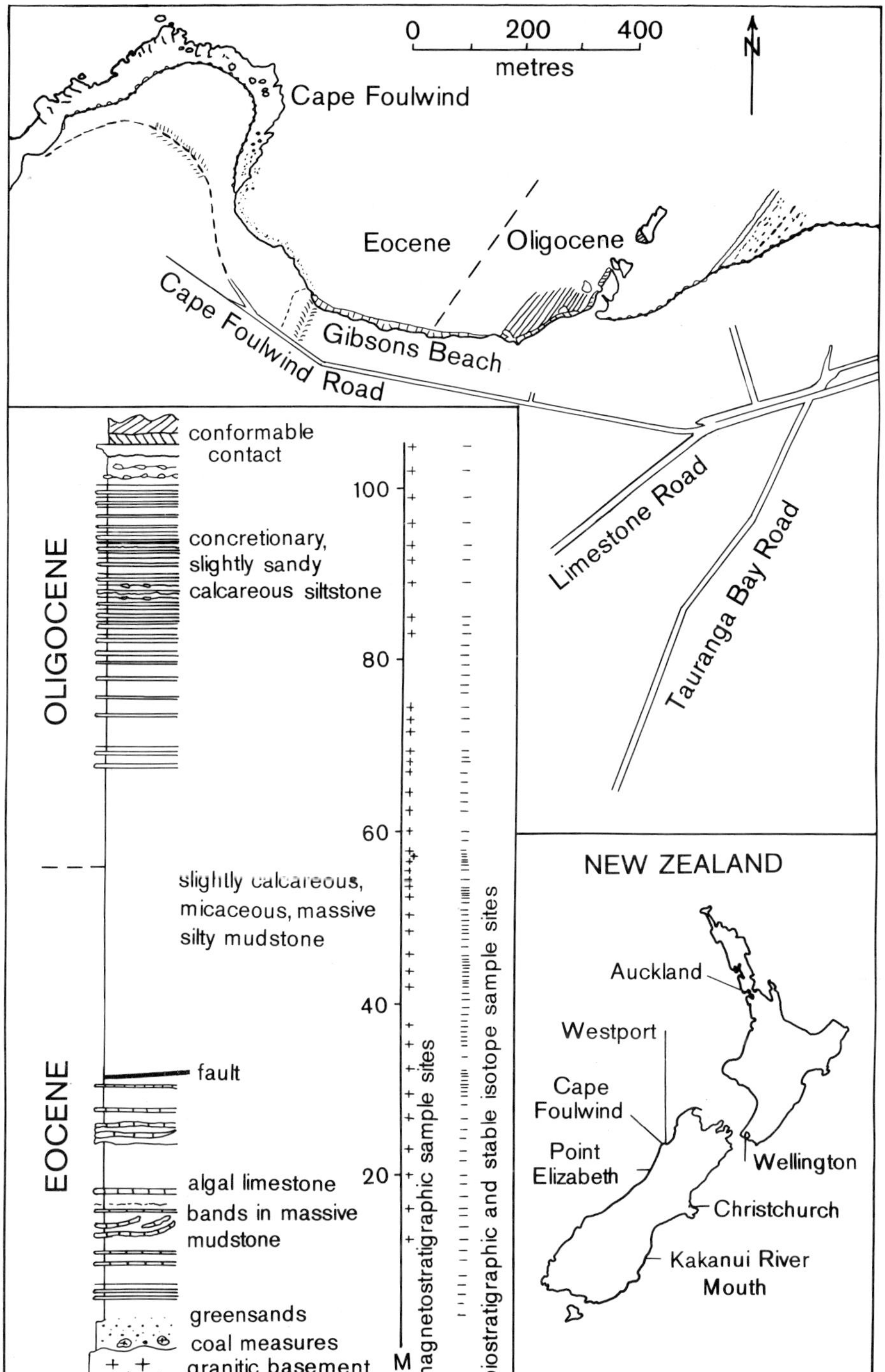

Fig. 1 Stratigraphic column for Cape Foulwind "Gibsons Beach" section and location map.

The last occurrence of the nannofossil Discoaster saipanensis Bramlette and Riedel, occurs 16 metres lower than the last occurrence of G. index (Dr T. Morgan pers. comm. Amoco Production Company, Tulsa, Oklahoma, USA).

GEOCHEMISTRY

One hundred and three samples have been collected for geochemistry/stable isotope stratigraphy. The only geochemistry presently available is an oxygen isotope study by Burns and Nelson (1981) on 11 samples originally collected by Srinivasan. The isotopic analyses were on benthic foraminifera with groupings at the family and generic level with some monospecific analyses. They interpreted the results to indicate a temperature deterioration through the late Eocene, with two distinct events in the late Eocene and the lowest temperature occurring in the earliest Oligocene. Temperatures then increased in the early Oligocene. The rates of temperature increase and decrease may have been similar.

MAGNETOSTRATIGRAPHY

No magnetostratigraphy is yet available. Thirty-nine sites were cored for magnetostratigraphy in November 1984 by a combined Japanese and New Zealand geological field party (see fig. 1).

RADIOMETRIC DATING

No radiometric dates have been taken from this reference section.

REFERENCES

Burns, D.A., Nelson, C.S., 1981. Oxygen isotopic paleotemperatures across the Runangan-Whaingaroan (Eocene/Oligocene) boundary in a New Zealand shelf sequence. New Zealand Journal of Geology and Geophysics 24(4): 529-538.

MacGregor, A.R., 1983. The Waitakere Limestone a temperate algal carbonate in the lower Tertiary of New Zealand. Journal of the Geological Society, London, 140: 387-399.

Nathan, S., 1976. Sheets S23/9 and S24/7 Foulwind and Westport (1st Ed.) Geological Map of New Zealand 1:25,000. Department of Scientific and Industrial Research, Wellington, New Zealand.

Srinivasan, M.S., 1966. Description of new species and notes on taxonomy of foraminifera from the upper Eocene and lower Oligocene of New Zealand. Royal Society of New Zealand, Transactions, Geology 3(17): 231-256.

Srinivasan, M.S. 1968. Late Eocene and early Oligocene planktonic foraminifera from Port Elizabeth and Cape Foulwind,New Zealand. Contributions from the Cushman Foundation for Foraminiferal Research 19(4): 142-159.

Srinivasan, M.S.,Vella, P. 1975. Upper Eocene-lower Oligocene benthonic foraminifera, Port Elizabeth and Cape Foulwind. New Zealand Journal of Geology and Geophysics 18(1): 21-37.

EOCENE/OLIGOCENE BOUNDARY, ADELAIDE REGION, SOUTH AUSTRALIA

J. MURRAY LINDSAY[1] and BRIAN McGOWRAN[2]

[1]South Australian Department of Mines and Energy, P.O. Box 151, Eastwood, South Australia, 5063. Published with the permission of the Director-General.
[2]Department of Geology and Geophysics, The University of Adelaide, G.P.O. Box 498, Adelaide, South Australia, 5001.

On the basis of foraminiferal biostratigraphy and as previously indicated (Lindsay, 1967, 1969, 1970, 1985), the Eocene/Oligocene (E/O) boundary is interpreted to occur within the lower part of the Port Willunga Formation, in coastal outcrop south of Adelaide at Port Willunga (Figs 1-3, 5-7) and at Port Noarlunga (Figs 2, 7); and correspondingly subsurface elsewhere in the eastern St Vincent Basin (Figs 2, 4).

Fig. 1. View of coastal exposures at Port Willunga, looking south. Port Willunga Formation (here latest Eocene and Oligocene) in lower part of cliff, dips gently southwest and is overlain with low-angle unconformity by flaggy Hallett Cove Sandstone (Late Pliocene), white earthy Burnham Limestone and slope-forming mottled clay and sand (Plio-Pleistocene), capped by middle Pleistocene calcrete. The basal (pale-colored) hard band of the Ruwarung Member of Port Willunga Formation forms the roof of the caves visible at centre: this is approximately the stratigraphic level of the interpreted E/O boundary.

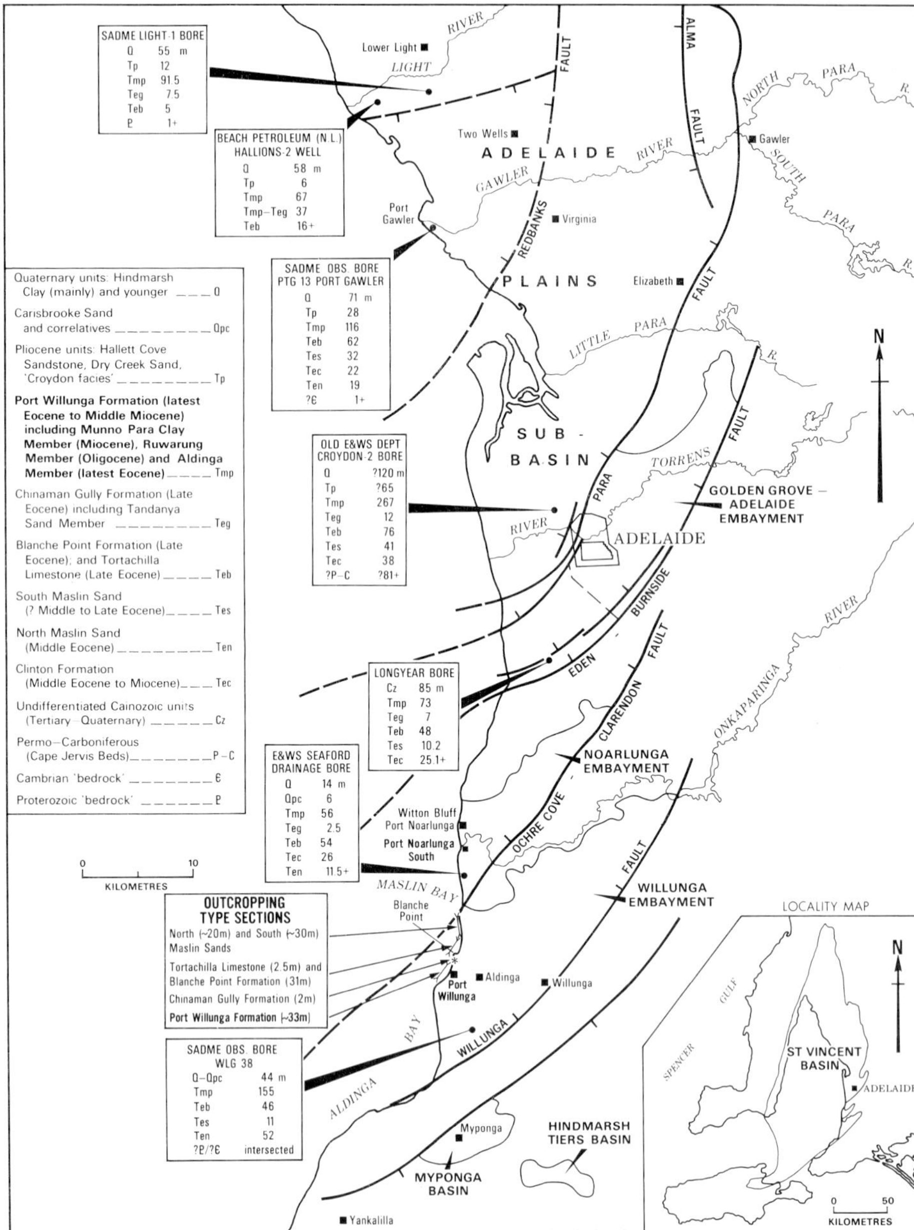

Fig. 2. Locality plan and regional stratigraphic setting of E/O boundary sections in outcrop of Port Willunga Formation at Port Willunga and Port Noarlunga, south of Adelaide, St Vincent Basin (after Lindsay, 1985).

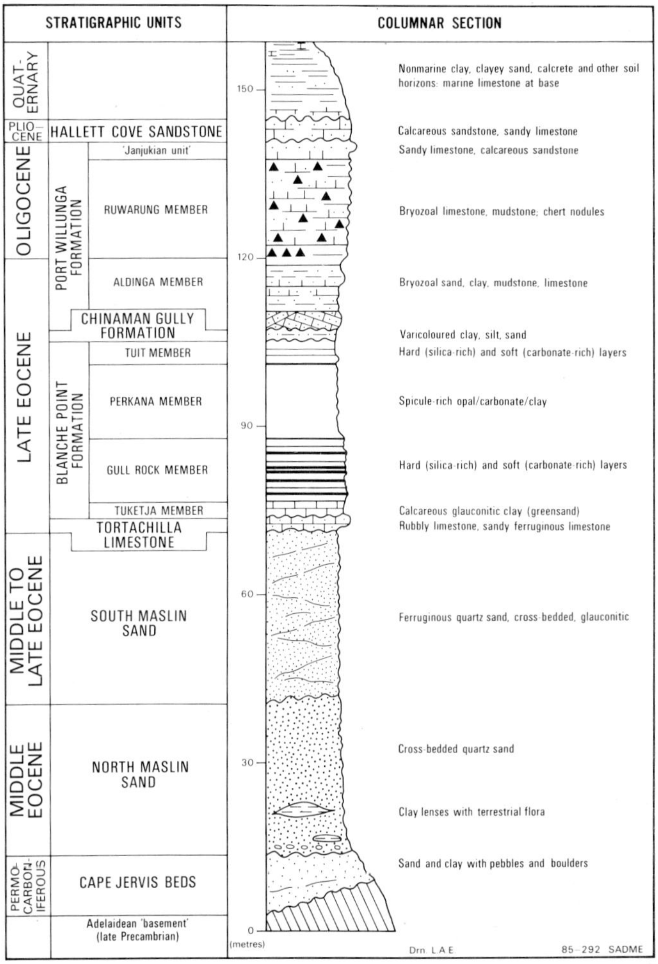

Fig. 3. Interpreted E/O boundary within stratotype Port Willunga Formation, in relation to composite columnar section of outcropping Tertiary units, Maslin Bay - Port Willunga, eastern St Vincent Basin, S.A. (after Lindsay _in_ Daily _et al._, 1976; McGowran and Beecroft, in press, a).

Coastal outcrops at the type section, Port Willunga, and at Port Noarlunga, have excellent exposure and accessibility. They display no discernible reworking, no major hiatus, and little deformation, in keeping with the tectonic setting of the St Vincent Basin near the passive margin of the Australian Plate. The basin has always been in temperate latitudes and somewhat restricted in access to the open ocean, so planktonic numbers are low (Fig. 6) and correlation with standard P. zones is generally indirect (Figs 5, 7).

Several foraminiferal datums suggest that the E/O boundary lies near the top of the Aldinga Member of Port Willunga Formation (Figs 4-7).

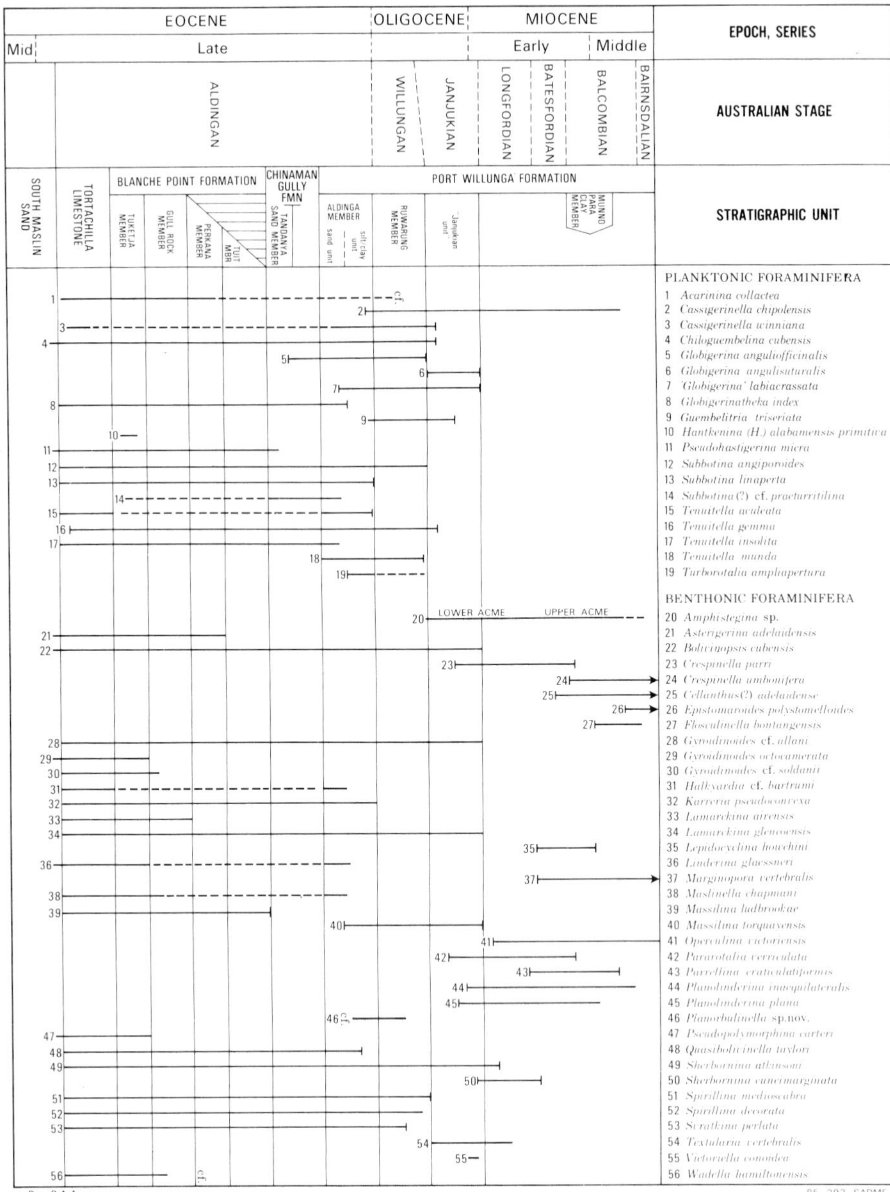

Fig. 4. Composite ranges of some foraminifera (mid-Tertiary outcrop and subsurface), eastern St Vincent Basin, S.A. (after Lindsay, 1985).

1. The St Vincent Basin has a good record of early Cassigerinella chipolensis (Cushman and Ponton), whose first appearance near the top of the Aldinga Member, in the presence of continuing C. winniana (Howe) (=C. eocaenica Cordey), seems to correlate most directly with the base of the Pseudohastigerina micra/C. chipolensis Zone (Bolli, 1966), a well-accepted E/O boundary marker.

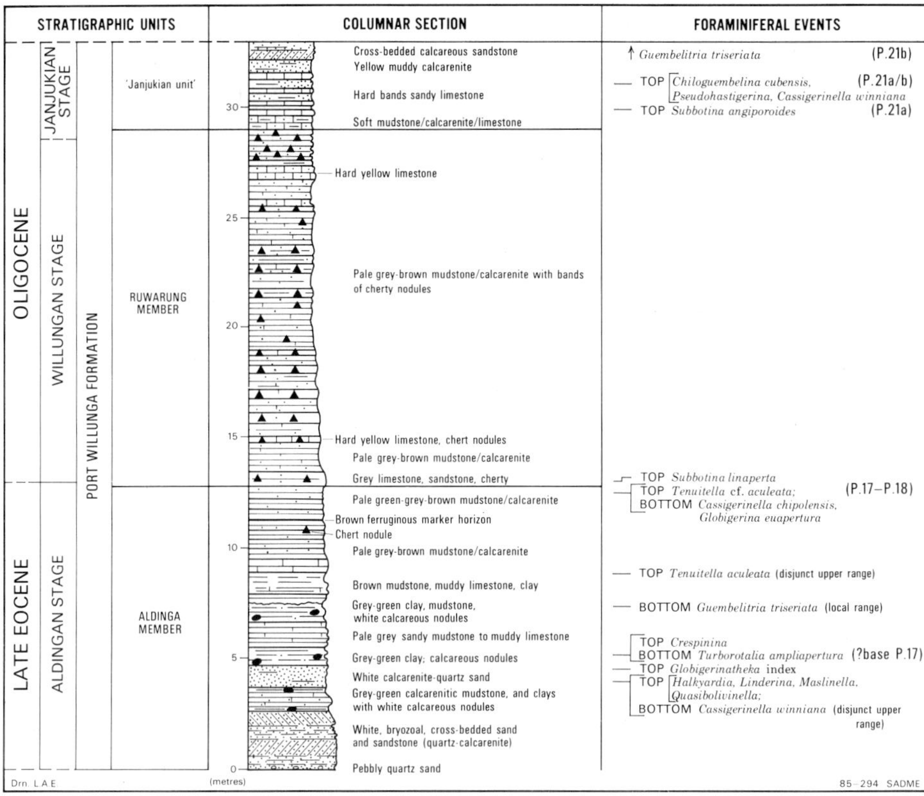

Fig. 5. Composite columnar section, type section of Port Willunga Formation, Port Willunga; stratigraphy, foraminiferal events, and correlations, in relation to interpreted E/O boundary (after Lindsay, 1967, 1985).

2. In such facies, the last appearance of carefully-discriminated Subbotina linaperta (Finlay) in populations of Subbotina, is considered (Lindsay, 1985) a better indicator of the E/O boundary in the Austral Region than the last appearance of rare and sporadic single-apertured Globigerinatheka index (Finlay), although the two are reported to disappear together at the top of the Runangan Stage in New Zealand, near the E/O boundary (Jenkins, 1966, 1971; Srinivasan, 1968; Hornibrook and Edwards, 1971).

3. The first appearance of Turborotalia ampliapertura (Bolli) in the presence of continuing T. increbescens (Bandy) occurs in the middle of the Aldinga Member at Port Willunga (Lindsay, 1967) (Fig. 5). The biostratigraphic context suggests that this appearance is time-correlative with that of Late Eocene age (Bolli, 1966) recorded at base Zone P.17 (Blow, 1969), and high in the N.Z. Runangan Stage (Jenkins, 1966, 1971; Srinivasan, 1968; Hornibrook and Edwards, 1971).

4. A benthonic fauna with Linderina, Crespinina, Halkyardia, Maslinella, and Quasibolivinella (Lindsay, 1967; cf. Quilty, 1981) makes its local final

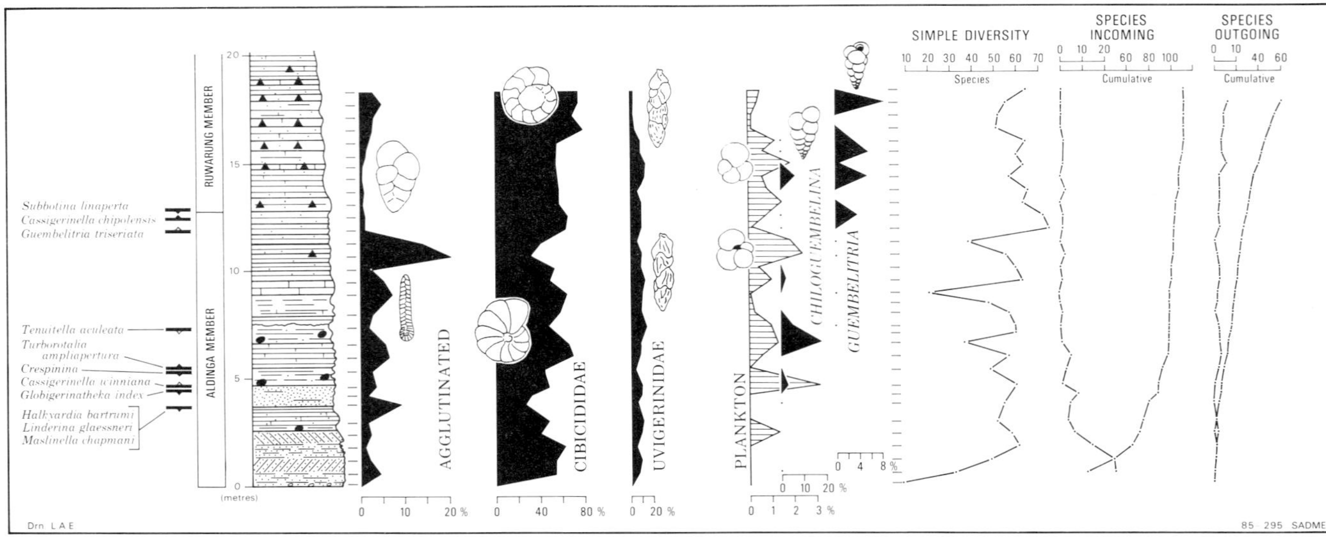

Fig. 6. Summary of quantified foraminiferal data in relation to E/O boundary section at Port Willunga. Note levels of samples, and variations in percentage scales. Datum levels at left: solid symbols are conventional. Hollow symbols: top T. aculeata is a well-marked local event, but not the final extinction (cf. Fig. 4); base C. winniana marks a migrational return, not a speciation (cf. Fig. 4); base Guembelitria is base of more or less continuous range (see below). Plots of agglutinated forms, Cibicididae, and Uvigerinidae, from McGowran and Beecroft (in press, a). Counts of plankton are from McGowran and Beecroft (1985). Although total plankton varies from low to vanishingly low, well-marked spikes of Guembelitria can be correlated with parallel spikes in more open neritic facies in the Otway Basin. Note that dots record the presence of rare Chiloguembelina and Guembelitria.

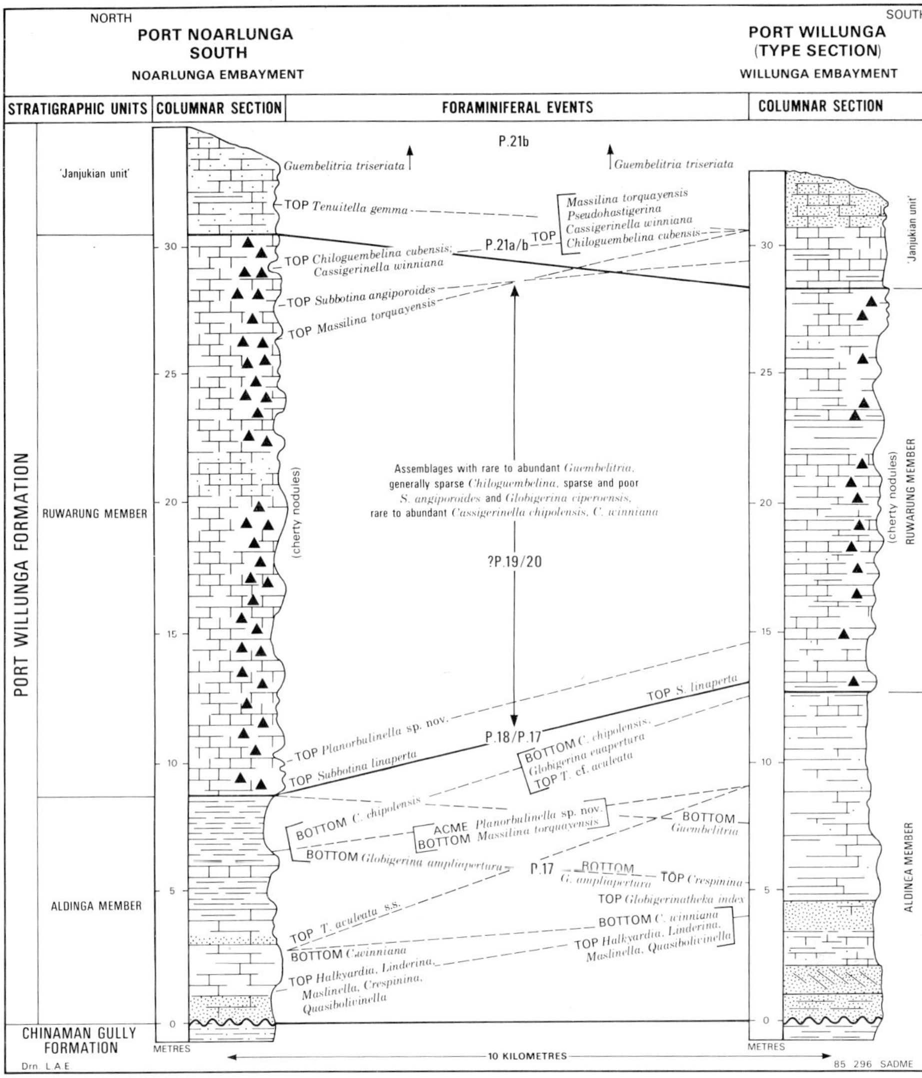

Fig. 7. Composite columnar sections showing correlations between Port Willunga Formation at Port Willunga and Port Noarlunga in relation to the interpreted E/O boundary.

appearance in the lower part of the Aldinga Member (Figs 5, 7), corresponding to the youngest of the Eocene warm episodes (extratropical excursions of larger benthonic foraminifera) noted by Lindsay (1976) and McGowran (1978, 1979). Linderina, at least, became extinct in the Indo-West Pacific Region prior to the end of the Eocene (Adams, 1970).

Species diversity, species incoming, and species outgoing, demonstrate

(Fig. 6) that any catastrophic 'Terminal Eocene Event' had little impact on the local inner-neritic community (McGowran and Beecroft, in press, b). The conformable change to nodular cherty lithology at about this time here (Ruwarung Member) may relate to the oceanic cooling, circulation changes, and mild eustatic sea-level fall, all reported at about the E/O boundary (Kennett and Shackleton, 1976; Vail and Hardenbol, 1979; Keigwin, 1980; Burns and Nelson, 1981; Loutit and Kennett, 1981), as may the local efflorescence of Guembelitria (Fig. 6; McGowran and Beecroft, 1985). Current local palynological biostratigraphy involving spores, pollen, and dinoflagellates (Harris, 1985) cannot resolve the boundary.

REFERENCES

Adams, C.G., 1970. A reconsideration of the East Indian Letter Classification of the Tertiary. Bull. Br. Mus. nat. Hist. (Geol.), 19(3).

Blow, W.H., 1969. Late Middle Eocene to Recent planktonic foraminiferal biostratigraphy. In: P. Bronnimann and H.H. Renz (Editors), Proceedings of the First International Conference on Planktonic Microfossils, Geneva, 1967. E.J. Brill, Leiden, 1:199-422.

Bolli, H.M., 1966. Zonation of Cretaceous to Pliocene marine sediments based on planktonic foraminifera. Boln inf. Asoc. Venez. Geol. Miner. Petrol., 9:3-32.

Burns, D.A. and Nelson, C.S., 1981. Oxygen isotopic paleotemperatures across the Runangan-Whaingaroan (Eocene-Oligocene) boundary in a New Zealand shelf sequence. N.Z. Jl Geol. Geophys., 24:529-538.

Daily, B., Firman, J.B., Forbes, B.G. and Lindsay, J.M., 1976. Geology. In: C.R. Twidale, M.J. Tyler and B.P. Webb (Editors), Natural History of the Adelaide Region. R. Soc. S. Aust., Adelaide, pp. 5-42.

Harris, W.K., 1985. Middle to Late Eocene depositional cycles and dinoflagellate zones in southern Australia. In: J.M. Lindsay (Editor), Stratigraphy, palaeontology, malacology: papers in honour of Dr Nell Ludbrook. Spec. Publ., S. Aust. Dept Mines and Energy, 5:133-144.

Hornibrook, N. de B. and Edwards, A.R., 1971. Integrated planktonic foraminiferal and calcareous nannoplankton datum levels in the New Zealand Cenozoic. In: A. Farinacci and R. Matteucci (Editors), Proceedings of the II Planktonic Conference, Roma, 1970. Edizioni Technoscienza, Roma, 1:649-657.

Jenkins, D.G., 1966. Planktonic foraminiferal zones and new taxa from the Danian to Lower Miocene of New Zealand. N.Z. Jl Geol. Geophys., 8:1088-1126.

Jenkins, D.G., 1971. New Zealand Cenozoic planktonic foraminifera. Paleont. Bull., Wellington, 42.

Keigwin, L.D. Jr., 1980. Palaeoceanographic change in the Pacific at the Eocene-Oligocene boundary. Nature, Lond., 287:722-725.

Kennett, J.P. and Shackleton, N.J., 1976. Oxygen isotopic evidence for the development of the psychrosphere 38 Myr ago. Nature, Lond., 260:513-515.

Lindsay, J.M., 1967. Foraminifera and stratigraphy of the type section of Port Willunga Beds, Aldinga Bay, South Australia. Trans. R. Soc. S.Aust., 91: 93-110.

Lindsay, J.M., 1969. Cainozoic foraminifera and stratigraphy of the Adelaide Plains Sub-Basin, South Australia. Bull. geol. Surv. S. Aust., 42.

Lindsay, J.M., 1970. Port Willunga Beds in the Port Noarlunga-Seaford area. Q. geol. Notes, geol. Surv. S. Aust., 36:4-10.

Lindsay, J.M., 1976. Tertiary history of South Australia - the foraminiferal record. 25th Int. Geol. Congr., Sydney, Abstr., 1:329-330.

Lindsay, J.M., 1985. Aspects of South Australian Tertiary foraminiferal biostratigraphy, with emphasis on studies of Massilina and Subbotina. In: J.M. Lindsay (Editor), Stratigraphy, palaeontology, malacology: papers in honour of Dr Nell Ludbrook. Spec. Publ., S. Aust. Dept Mines and Energy, 5:187-232.

Loutit, T.S. and Kennett, J.P., 1981. New Zealand and Australian Cenozoic sedimentary cycles and global sea-level changes. Bull. Am. Ass. Petrol. Geol., 65:1586-1601.

McGowran, B., 1978. Early Tertiary foraminiferal biostratigraphy in southern Australia: a progress report. In: D.J. Belford and V. Scheibnerova (Compilers), The Crespin Volume: essays in honour of Irene Crespin. Bull. Bur. Miner. Resour. Geol. Geophys., Aust., 192:83-95.

McGowran, B., 1979. The Tertiary of Australia: foraminiferal overview. Marine Micropaleontology, 4:235-264.

McGowran, B. and Beecroft, A., 1985. Guembelitria in the Early Tertiary of southern Australia and its palaeoceanographic significance. In: J.M. Lindsay (Editor), Stratigraphy, palaeontology, malacology: papers in honour of Dr Nell Ludbrook. Spec. Publ., S. Aust. Dept Mines and Energy, 5:247-262.

McGowran, B. and Beecroft, A., in press, a. Foraminiferal biofacies in a silica-rich neritic sediment, Late Eocene, South Australia. Palaeogeogr., Palaeoclimatol., Palaeoecol.

McGowran, B. and Beecroft, A., in press, b. Neritic, southern extratropical foraminifera and the Terminal Eocene Event. Palaeogeogr., Palaeoclimatol., Palaeoecol.

Quilty, P.G., 1981. Late Eocene benthic Foraminiferida, south coast, Western Australia. J.R. Soc. West. Aust., 64:79-100.

Srinivasan, M.S., 1968. Late Eocene and Early Oligocene planktonic foraminifera from Port Elizabeth and Cape Foulwind, New Zealand. Contr. Cushman Fdn foramin. Res., 19:142-159.

Vail, P.R. and Hardenbol, J., 1979. Sea-level changes during the Tertiary. Oceanus, 22:71-79.

EOCENE-OLIGOCENE BOUNDARY IN LIBYA

A.A. BUTT
Oasis Oil Cy of Libya, Geology Laboratory, P.O. Box 395, Tripoli.

There are two areas in northern Libya which comprise continuous succession across the Eocene-Oligocene boundary, namely, the Sirte Basin (the Augila formation of Late Eocene age and the Arida Formation of Early Oligocene age in subsurface sections) and the northeastern Libya (Darnah Formation of Late Eocene age and the Al Bayda Formation of Early Oligocene age). At both locations, the sequence is developed as a shallow-water facies, but the boundary can be better evaluated in the northeastern Libya on the basis of paleontological record of larger benthic formaminifera.

The geological work of Zert (1974) has brought to light the most recent stratigraphic details about the northeastern Libya, specifically the region of Jabal Al Akhdar (fig. location map). A continuous section through the Eocene-Oligocene boundary is developed as a shallow-water carbonate facies in Wadi Darnah section, 2 kilometers south of the town Darnah having a limited geographical extent (fig. 1). In Wadi Darnah the Darnah Formation (this formational name combines the Derna Limestone and the Slonta Limestone of Gregory, 1911, which are difficult to separate, essentially a nummulitic limestone, is about 140 meters thick and contains the following characteristic larger foraminifera in the middle and upper part, and some calcereous nannofossils identified by Hanzlikova (in Zert, 1974):

Nummulites gizehensis Forskal, N. gizehensis Lyelli d'Archiac and Haime, N. incrassatus de la Harpe, N. striatus Brugière, N. beaumonti d'Archiac, N. chavannesi de la Harpe, N. ex gr. fabianii Prever, N. pulchellus de la Harpe, Heterostegina (Grzybowskia) reticulata (Rutimeyer), Operculina Pyramidum Ehrenberg, Discocyclina varians (Kaufmann), D. pratti (Michelin). Calcereous nannofossils : Discoater distinctus Martini, D. kuepperi (Stradner), Lophodolithus reniformis Bramlette and Sullivan.

The Darnah Formation has been assigned a Priabonian age (Late Eocene) on the basis of foraminiferal evidence.

The Al Bayda Formation (former "Cyrene Formation" of Gregory, 1911, with two members, the Shabat Marl and Algal Limestone) is also developed as a calcareous facies overlying the Darnah Formation. It has been reported to contain the following important larger foraminifera indicative of Oligocene age : Nummulites cf. fichteli Michelotti, N. ex gr. hantkeni Engelberts, N. incrassatus de la Harpe, N. bouillei de la Harpe, Lepidocyclina (Eulepidina) dilatata (Michelotti), (Eulepidina) ephippiodes (Jones & Chapman).

In the present opinion, the faunal change across the boundary in a shallow neritic environment is characterized by changes within the Nummulites fauna, by the disappearance of the discocylinids and by a later appearance of Eulepidine and Lepidocyclinids. It is believed that the standard zonation across the boundary based upon the Nummulites fabianii/fichteli lineage can be used here (see Paleogene Stratigraphical Correlation Scale of Cavelier and Pomerol, 1985).

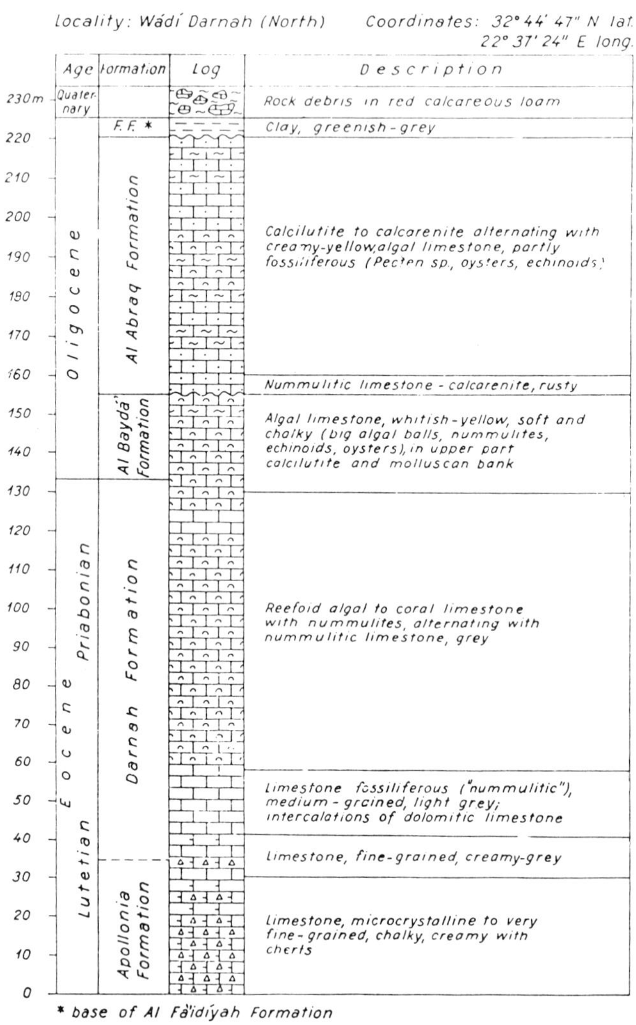

Fig. 1. Stratigraphic section along the Eocene-Oligocene boundary, 2 km South of the town Darnah after Zert (1974).

THE EOCENE-OLIGOCENE PASSAGE ZONE IN NORTHERN MOROCCO

Hugues FEINBERG - I.P.G. Paléomagnétisme. 4, Pl.Jussieu - 75230 PARIS

INTRODUCTION

The external zone of the Rif chain represents the ancient slope of the African plate. During the collision with Iberian and Alboran plates the Northern margin of Africa has been more or less destroyed and pushed Southwards. For that reason we can see now in the Prerifian zone witnesses of a continuous marine sedimentation during nearly all the Cenozoic.

Before the Upper Miocene tectonic climax these deposits have been lightly affected by tectonic movements during the Middle Eocene (Pyrenean phase) and also by paleogeographic changes towards the end of Oligocene but the Eocene/Oligocene boundary seems, up to now, a very quiet period. In this generally favourable environment for a detailed study of the Eocene-Oligocene boundary some difficulties however have to be taken into consideration:

The first concerns the effects of Miocene tectonic and tectono-sedimentary phenomena which gave a very complicated structure.

Secondary it appears very difficult to select reference sections, easy to reach and sample in a region devoid of road cuts and quarries.

GENERAL SETTING OF THE SECTIONS

The zone where Oligo-Miocene sections are the more frequent has a crescentic shape of nearly 200 km from east to west and about 30 km of width. About 40% of this region has been recently mapped in detail and regular geological maps (scale: 1/50 000) are available. For a rough selection of the sections the general map of the Rif chain at 1/500 000 (Suter,1982) is also very useful. The regional tectonic style corresponds to a "tiled structure" but the occurrence of sedimentary klippes is also frequent. In this context the Paleogene sediments

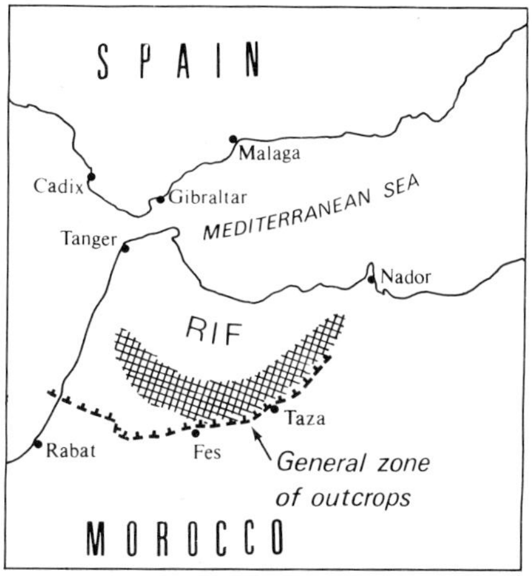

make units of various size lying on the Miocene with flat faulted contacts.
From a lithologic point of view the Eocene-Oligocene boundary is located in thick monotonous sequences of marls with interbedded sandstones which quantity could vary in a large way. In these sediments the location of the Eocene-Oligocene boundary is clearly indicated by planktonic foraminifera and calcareous nannofossils which are both abundant. North of the considered area, the Flysch type units (Numidian, Béni-Ider) are representing deep sea fan deposits and evidently contain very poor planktonic micro and nannofossils. On the other hand shelf deposits belonging to the Upper Eocene or Oligocene are completely unknown in the Rif chain.
Two sections respectively located North of FES and West from TAZA have already been described in the pointed out area (Feinberg,1984). We will now describe another section, easy to reach, and located N.-E. from TAZA.

SIDI MAATOUG SECTION

The section is located in the valley of the Msoun river close to the narrow road which joins the Secondary road n° 312 and the road from MSOUN to MEZGUITEM.
The Eocene-Oligocene transition outcrops in good conditions on the East flank of a large syncline cut by the Msoun valley. The narrowing of this valley related to basaltic outcrops (see Fig. 1) makes the place easy to recognize in the landscape. The beds are strongly dipping (45°) in the W-SW direction and the lithology is very uniform during the passage from Eocene to Oligocene.

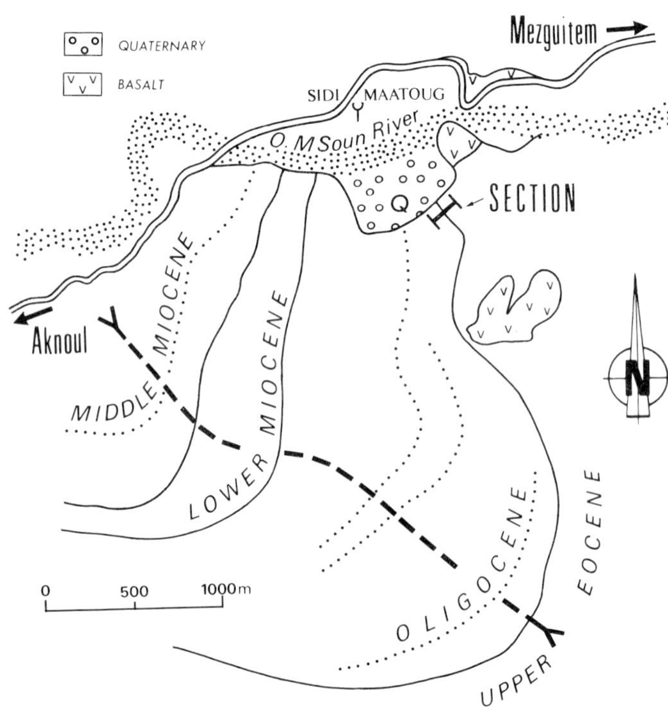

Fig. 1: Location of the S. Maatoug section. Solid lines represent geological contours; dotted and dashed lines correspond respectively to massive sandstone and synclinal axis.

The sequence is mainly marly, interbedded with thin layers (10-20 cm) of sandstone, each 3 or 4 meters. The color of the marls is blue or grey and the sandstones are brown. In the Oligocene unit the sandstones become more important and draw up lines of hill summits. This is indicated by dotted lines on the sketch map.

PRELIMINARY STUDIES

Preliminary sampling have been performed each five meters. The planktonic organisms are dominant in a moderatly deep environment of 600 to 1000 m.(W. Berggren personal communication). Near the boundary planktonic foraminifera show a gradual extinction:Hantkeninids first disappear followed by the Globorotalia of the cerroazulensis group. Examination of Nannofossils assemblages reveals that the extinction of Heliodiscoasterids occurs prior to the extinction of the last Eocene planktonic foraminifera. This fact has been pointed out in other sections of the Eocene-Oligocene passage sampled in North Africa (Magné,Perch-Nielsen et Feinberg,1982).

The Natural Remanent Magnetization of the marls varies from 0.5 to 2 10^{-6} UEM/cm^3.

CONCLUSION

A detailed study of the S. Maatoug section is needed. The moderate depth and the location of the Moroccan section in the subtropical zone at the junction between the Tethysian and Atlantic Oceans will provide information on both planktonic and benthic foraminifera and will give opportunity for accurate comparison with reference sections located in different environment.

In this way it will be probably possible to identify the effects of local control on the apparent ranges of the different fossil groups during the Eocene-Oligocene passage.

REFERENCES

Suter,G.,1980. Carte Géologique de la Chaîne Rifaine au 1/500 000. Notes Mém. Service Géol. Maroc, n° 245A.

Leblanc,D.,1982. Carte Géologique Aïn Bou Kellal - Msoun au 1/50 000. Notes Mém. Serv. Géol. Maroc, n° 256.

Feinberg,H.,1984. Les séries tertiaires des zones externes du Rif (Maroc). Notes Mém. Serv. Géol. Maroc, n° 315.

Magné,J.,Perch-Nielsen,K. et Feinberg,H.,1982. La limite Eocène/Oligocène dans les séries à faciès pélagique du Nord de l'Algérie et du Maroc. 9éme R.A.S.T. - Paris.

UPPER EOCENE AND LOWER OLIGOCENE STRATA IN SOUTHWESTERN ALABAMA, U.S.A.

Ernest A. Mancini, Geological Survey of Alabama and University of Alabama, Tuscaloosa, Alabama, U.S.A.

INTRODUCTION

The upper Eocene and lower Oligocene section of southwestern Alabama has been of particular interest to stratigraphers over the years because these strata represent one of the most complete exposed marine Eocene-Oligocene sections in North America. The Eocene-Oligocene contact is excellently preserved in a limestone quarry (St. Stephens Quarry), northeast of St. Stephens, Alabama, and along Little Stave Creek, north of Jackson, Alabama (Fig. 1).

St. Stephens Quarry (Lone Star Cement Company Quarry) is located on the west bank of the Tombigbee River, Washington County. The route to the quarry involves turning right (northwest) off County Highway 34 at St. Stephens United Methodist Church and Cemetery onto a paved road and then turning right (north) on a gravel road 0.1 mile from the junction of County Highway 34 and the paved road. The quarry gate is about 1.5 miles from the junction of the paved road and the gravel road. The St. Stephens Quarry measured section (Fig. 2) is about 0.25 mile from the west bank of the Tombigbee River in section 32, Township 7 North, Range 1 West, on the Saint Stephens, Alabama, 7.5-minute quadrangle.

The quarry is located near the Hatchetigbee anticline. The strata strike about N 70° W and dip 2 to 2.5° SSW (Glawe, 1967). The section exposed in the north quarry includes over 14 feet of upper Eocene (Priabonian) strata and over 25 feet of lower Oligocene (Rupelian) strata.

Little Stave Creek is a tributary of Stave Creek which flows southwestward to the Tombigbee River about 3 miles west of Jackson, Clarke County, Alabama. The route to Little Stave Creek involves turning left (west) off U.S. Highway 43, 0.35 mile from the junction of County Highway 10 and U.S. Highway 43, onto West Point Drive within the city limits of Jackson, and continuing to the end of West Point Drive about 0.3 mile. Park at the large oak tree at the dead end of West Point Drive and then walk along the gravel road and foot path to the southwest for about a mile until the path intersects Little Stave Creek at a large waterfall. The Little Stave Creek measured section (Fig. 3) continues upstream for over a mile in sections 19, 20 and 21, Township 7 North, Range 2 East, on the Jackson, Alabama, 7.5-minute quadrangle.

The strata exposed along Little Stave Creek dip to the east at 1.5 to 2°. The direction of dip is attributable to the Hatchetigbee anticline located to

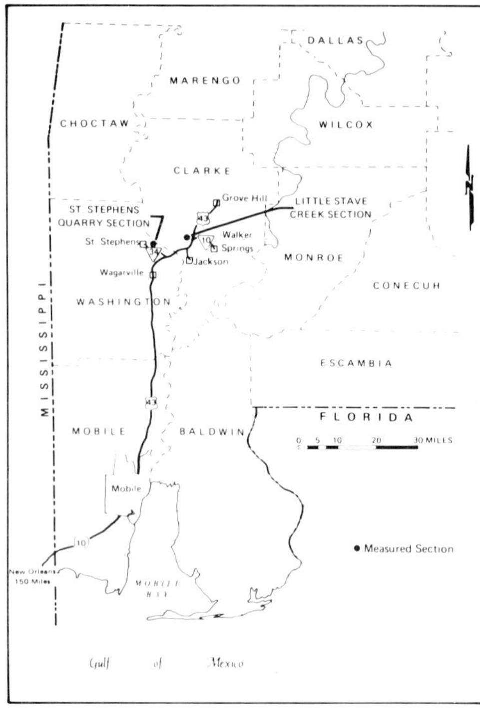

Fig. 1.--Location map for St. Stephens Quarry and Little Stave Creek measured sections, southwestern Alabama, U.S.A.

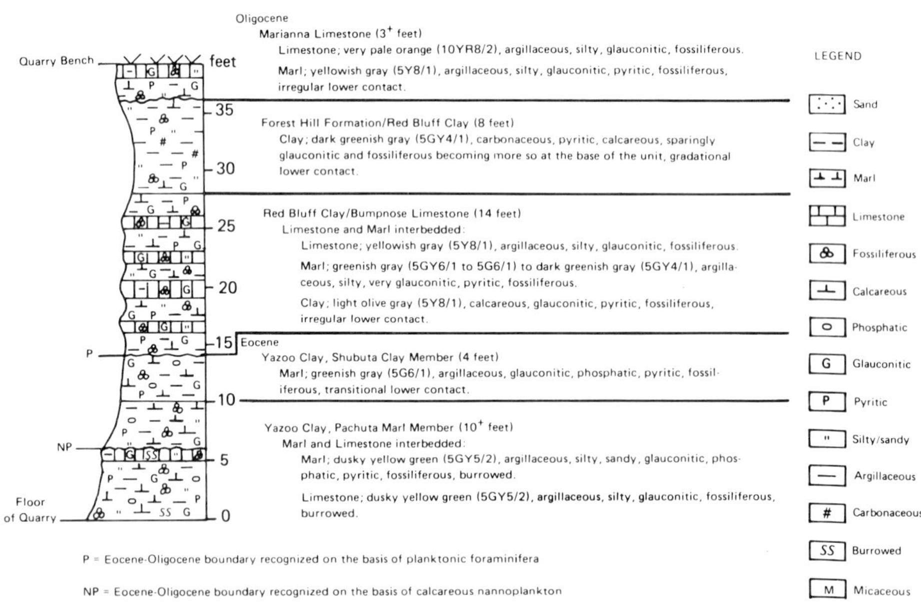

Fig. 2.--St. Stephens Quarry measured section, section 32, Township 7 North, Range 1 West, Washington County, Alabama.

the northwest (Bandy, 1949). Approximately 500 feet of surface displacement is evident on the Jackson Fault at this locality. Eocene-Oligocene strata are exposed on the upthrown side of the fault. The section includes over 61 feet of upper Eocene (Priabonian) strata and over 14 feet of lower Oligocene (Rupelian) strata.

The lithostratigraphic units exposed at St. Stephens Quarry and along Little Stave Creek in ascending order include the Moodys Branch Formation, the North Twistwood Creek Clay Member, Cocoa Sand Member, the Pachuta Marl Member, and the Shubuta Clay Member of the Yazoo Clay, the Red Bluff Clay/Bumpnose Limestone, the Forest Hill Formation/Red Bluff Clay, and the Marianna Limestone (Fig. 4).

PLANKTONIC FORAMINIFERAL BIOSTRATIGRAPHY

The Paleogene planktonic foraminiferal zonation utilized in this study was first established by Bolli (1957; 1966; 1972) and later modified by Stainforth et al. (1975) and Stainforth and Lamb (1981). This zonation has been used widely as an accepted biostratigraphic standard for warm water areas of the world, including the Gulf Coastal Plain region.

The Moodys Branch Formation at Little Stave Creek was assigned, in part, by Barker, in Blow (1979) to the upper Eocene (Bartonian) *Truncorotaloides rohri* Interval Zone (Fig. 4).

The North Twistwood Creek Clay Member, in part, and the Cocoa Sand Member of the Yazoo Clay at Little Stave Creek were placed by Barker, in Blow (1979) in the upper Eocene (Priabonian) *Globigerinatheka semiinvoluta* Interval Zone (Fig. 4). The presence of *Globorotalia cerroazulensis cerroazulensis* (Cole) and the absence of *Globorotalia cerroazulensis cocoaensis* Cushman and *Cribrohantkenina inflata* (Howe) in the North Twistwood Creek in southern Mississippi support this zone assignment.

The Pachuta Marl and Shubuta Clay Members of the Yazoo Clay at St. Stephens Quarry and Little Stave Creek have been assigned by Mancini (1979) and Waters (1983) to the upper Eocene (Priabonian) *Globorotalia cerroazulensis* (s.l.) Interval Zone (Fig. 4). *Hantkenina alabamensis* Cushman and subspecies of *Globorotalia cerroazulensis, including Globorotalia cerroazulensis cocoaensis,* occur in the Pachuta and Shubuta (Bandy, 1949; Deboo, 1965; Mancini, 1979; Waters, 1983). The last occurrences of *Hantkenina alabamensis* and subspecies of *Globorotalia cerroazulensis* that are considered autochthonous are in the Shubuta (Deboo, 1965; Mancini, 1979; Waters, 1983). Bybell and Poore (1983) suggested that some of the *Hantkenina alabamensis* found in the Shubuta at Little Stave Creek are reworked based on their study of the calcareous nannoplankton species that occur within the tests of these foraminifera. Also,

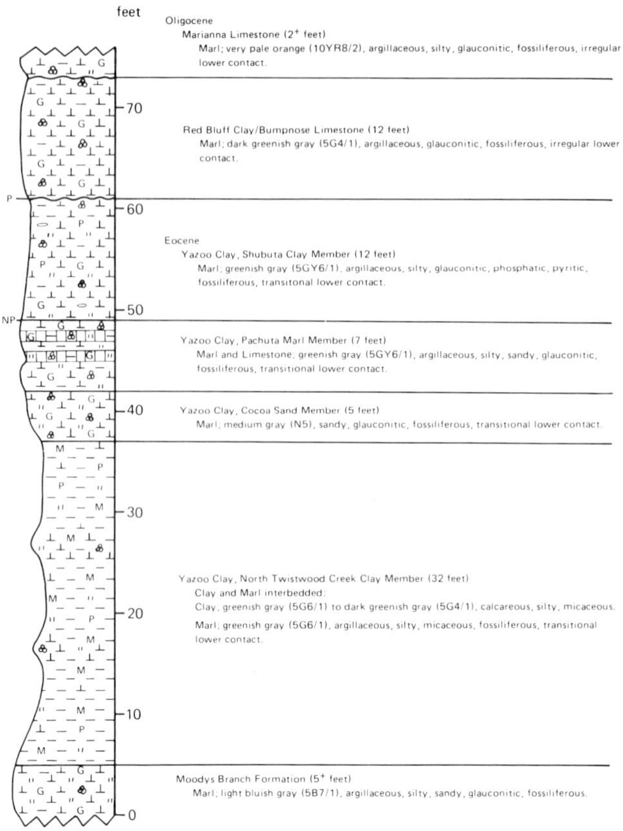

Fig. 3.--Little Stave Creek measured section, sections 19, 20 and 21, Township 7 North, Range 2 East, Clarke County, Alabama. See Fig. 2 for lithologic legend.

<table>
<tr><th colspan="2">FORMATION/MEMBER</th><th>PLANK. FORAM. ZONE</th><th>AGE</th></tr>
<tr><td colspan="2">Marianna Limestone</td><td rowspan="4">Pseudohastigerina micra Interval Zone</td><td rowspan="4">Rupelian</td></tr>
<tr><td colspan="2">Mint Spring Marl Mbr.</td></tr>
<tr><td colspan="2">Forest Hill Fm.</td></tr>
<tr><td colspan="2">Red Bluff Clay/ Bumpnose Limestone</td></tr>
<tr><td rowspan="4">Yazoo Clay</td><td>Shubuta Clay Mbr.</td><td rowspan="2">Globorotalia cerroazulensis (s.l.) Interval Zone</td><td rowspan="4">Priabonian</td></tr>
<tr><td>Pachuta Marl Mbr.</td></tr>
<tr><td>Cocoa Sand Mbr.</td><td rowspan="2">Globigerinatheka semiinvoluta Interval Zone</td></tr>
<tr><td>North Twistwood Creek Clay Mbr.</td></tr>
<tr><td colspan="2">Moodys Branch Fm.</td><td>Truncorotaloides rohri Interval Zone</td><td>Bartonian</td></tr>
</table>

Fig. 4.--Upper Eocene and lower Oligocene lithostratigraphy and planktonic foraminiferal biostratigraphy for southwestern Alabama.

Keller (1985) based on the abundance distribution of dominant planktonic foraminifera species occurring in the Yazoo Clay believed that the specimens of *Hantkenina* spp. and *Globorotalia cerroazulensis* subspecies found in the upper Shubuta are reworked. However, no sedimentologic evidence was observed in these sections to support the contention that the specimens found in the upper Shubuta are reworked. In addition, there is question as to the accuracy of unit thickness and/or sample location in the St. Stephens and Little Stave Creek sections published by Keller (1985).

The Red Bluff Clay/Bumpnose Limestone, Forest Hill Formation/Red Bluff Clay and Marianna Limestone have been assigned by Mancini (1979), Hazel et al. (1980) and Stainforth and Lamb (1981) to the lower Oligocene (Rupelian) *Pseudohastigerina micra* Interval Zone (Fig. 4). *Pseudohastigerina micra* (Cole) and *Globigerina ampliapertura* Bolli occur in the Red Bluff Clay/Bumpnose Limestone (Mancini, 1979; Waters, 1983) and in the Marianna Limestone (Stainforth and Lamb, 1981).

At St. Stephens Quarry (Fig. 2) and Little Stave Creek (Fig. 3), the Eocene-Oligocene boundary, as recognized on the basis of planktonic foraminifera, occurs at or near the top of the Shubuta Clay Member of the Yazoo Clay. The Eocene-Oligocene boundary is defined worldwide on the basis of the change in planktonic foraminiferal assemblages across this datum. Uppermost Eocene marine strata contain an assemblage consisting of *Hantkenina, Cribrohantkenina, Globorotalia cerroazulensis* subspecies, and *Globigerina* species having large tests (Stainforth et al., 1975). Lowermost Oligocene marine strata are usually characterized by an assemblage predominated by *Globigerina ampliapertura, Globigerina gortanii* (Borsetti), *Globorotalia increbescens* (Bandy), and *Pseudohastigerina* species. The Eocene-Oligocene boundary is drawn worldwide at the top of the *Globorotalia cerroazulensis* (s.l.) Interval Zone (Stainforth et al., 1975), which at St. Stephens Quarry and Little Stave Creek closely approximates the contact of the Shubuta Clay Member of the Yazoo Clay with the Red Bluff Clay/Bumpnose Limestone.

As recognized on the basis of calcareous nannoplankton, the Eocene-Oligocene boundary at St. Stephens Quarry could be placed at the disconformity in the Pachuta Marl Member (Fig. 2) and at the top of the Pachuta Marl Member of the Yazoo Clay at Little Stave Creek (Fig. 3). The last occurrences of *Discoaster barbadiensis* Tan Sin Hok, *Discoaster saipanensis* Bramlette and Riedel, and *Reticulofenestra reticulata* (Gartner and Smith) are reported by Bybell (1982) at these horizons. Cheetham (1963) using primarily cheilostome bryozoans also placed the Eocene-Oligocene boundary near the top of the Pachuta Marl Member in southwestern Alabama. Huddlestun and Toulmin (1965) based on macrofossils defined the boundary at the top of the Shubuta Clay Member of the Yazoo Clay.

Therefore, the Eocene-Oligocene boundary at St. Stephens Quarry (Fig. 2) based on planktonic foraminifera is 8 feet higher in the section than where the boundary could be recognized on the basis of calcareous nannoplankton. The epoch boundary at Little Stave Creek (Fig. 3) based on planktonic foraminifera is 12 feet higher in the section than where the boundary could be recognized on the basis of calcareous nannoplankton. The difference in elevation in placement of the Eocene-Oligocene boundary is not unusual. In fact, as reported by Gartner (1971), Stainforth and Lamb (1981), Poore et al. (1982), and Snyder et al. (1984), the boundary as defined by planktonic foraminifera occurs at a higher stratigraphic elevation in coreholes drilled in the Gulf of Mexico and the Atlantic Ocean than it would be if recognized on the basis of calcareous nannoplankton in these coreholes.

REFERENCES

Bandy, O. L., 1949. Eocene and Oligocene foraminifera from Little Stave Creek, Clarke County, Alabama. Bulletins American Paleontology, 32(131):210.

Bolli, H. M., 1957. The genera *Globigerina* and *Globorotalia* in the Paleocene-Lower Eocene Lizard Springs Formation of Trinidad, B.W.I. U.S. National Museum Bulletin 215, pp. 61-81.

Bolli, H. M., 1966. Zonation of Cretaceous to Pliocene marine sediments based on planktonic foraminifera. Asociacion de Venezolana Geologia, Mineria y Petroleo Boletin de Informacion, 9:3-32.

Bolli, H. M., 1972. The genus *Globigerinatheka* Bronnimann. Journal of Foraminiferal Research, 2:109-136.

Blow, W. H., 1979. The Cainozoic Globigerinida. E. J. Brill, Leiden, 1413 pp.

Bybell, L. M., 1982. Late Eocene to early Oligocene calcareous nannofossils in Alabama and Mississippi. Gulf Coast Association of Geological Societies Transactions, 32:295-302.

Bybell, L. M., and Poore, R. Z., 1983. Reworked *Hantkenina* specimens at Little Stave Creek, Alabama. Gulf Coast Association of Geological Societies Transactions, 33:253-256.

Cheetham, A. H., 1963. Late Eocene zoogeography of the eastern Gulf Coastal region. Geological Society of America Memoir 91, 113 pp.

Deboo, P. B., 1965. Biostratigraphic correlation of the type Shubuta Member of the Yazoo Clay and Red Bluff Clay with their equivalents in southwestern Alabama. Alabama Geological Survey Bulletin 80, 84 pp.

Gartner, S., Jr., 1971. Calcareous nannofossils from the JOIDES Blake Plateau cores and revision of Paleogene nannofossil zonation. Tulane Studies Geology Paleontology, 8(3):101-121, pls. 1-5.

Glawe, L. N., 1967. Stop 16--Lone Star Cement Company Quarry at St. Stephens Bluff on Tombigbee River, 2.2 miles northeast of St. Stephens, Washington County, Alabama. In: D. E. Jones (Editor), Geology of the Coastal Plain of Alabama. Geological Society of America 80th Annual Meeting, New Orleans, Louisiana, Alabama Geological Society Guidebook, University, Alabama, pp. 107-113.

Hazel, J. E., Mumma, M. D., and Huff, W. J., 1980. Ostracode biostratigraphy of the lower Oligocene (Vicksburgian) of Mississippi and Alabama. Gulf Coast Association of Geological Societies Transactions, 30:361-401.

Huddlestun, P. F., and Toulmin, L. D., 1965. Upper Eocene-lower Oligocene stratigraphy and paleontology in Alabama. Gulf Coast Association of Geological Societies Transactions, 15:155-159.

Keller, G., 1985. Eocene and Oligocene stratigraphy and erosional unconformities in the Gulf of Mexico and Gulf Coast. Journal of Paleontology, 59(4):882-903.

Mancini, E. A., 1979. Eocene-Oligocene boundary in southwest Alabama. Gulf Coast Association of Geological Societies Transactions, 29:282-286.

Poore, R. Z., Tauxe, L., Percival, S. F., Jr., and LaBrecque, J. L., 1982. Late Eocene-Oligocene magnetostratigraphy and biostratigraphy at South Atlantic DSDP site 522. Geology, 10(10):508-511.

Snyder, S. W., Muller, C., and Miller, K. G., 1984. Eocene-Oligocene boundary: biostratigraphic recognition and gradual paleoceanographic change at DSDP site 549. Geology, 12(2):112-115.

Stainforth, R. M., and Lamb, J. L., 1981. An evaluation of planktonic foraminiferal zonation of the Oligocene. Kansas University, Paleontological Contributions, Paper 104, 42 pp.

Stainforth, R. M., Lamb, J. L., Luterbacher, H., Beard, J. H., and Jeffords, R. M., 1975. Cenozoic planktonic foraminiferal zonation and characteristics of index forms. Kansas University, Paleontological Contributions, Article 62, 425 pp.

Waters, L. A., 1983. Correlation of upper Eocene and lower Oligocene strata in Mississippi and Alabama. Unpublished Master's Thesis, University of Alabama, 169 pp.

THE EOCENE-OLIGOCENE BOUNDARY IN MEXICO

P. SALMERON U.(*)

ABSTRACT

The Eocene-Oligocene boundary in Mexico is characterized by the extinction of the *Globorotalia cerroazulensis* Group and of the species of the genera *Hantkenina* and *Cribrohantkenina*. Two biozones were established in this boundary: the *Globorotalia cerroazulensis* s.l. Biozone of the Late Eocene, that underlies the *Cassigerinella chipolensis-Pseudohastigerina micra* Biozone of the Early Oligocene. Up to now, the only area where the Eocene-Oligocene boundary has been found conformably is the Tertiary Basins of Southeastern Mexico, especially the Nicapa and Mono Pelado Sections, located in the area between parallels 17°15'00" and 17°30'00" North and 93°25'00" and 93°35'00" West. In the rest of Mexico, this limit is unconformable or is not represented.

INTRODUCTION

The geological and biostratigraphic studies that are carried out mainly by Petroleos Mexicanos and the Instituto Mexicano del Petroleo, in order to support the oil exploration in the sedimentary basins with economic potential, have provided important information concerning the Eocene-Oligocene boundary.

The rocks of this stratigraphic level are widely distributed in the coastal plain of the Gulf of Mexico; on the other hand, their distribution is very limited in the Pacific Ocean slope.

Based on the study of planktonic fossils the author only found conformably the Eocene-Oligocene boundary in the subprovince belonging to the Tertiary Basins of Southeastern Mexico, mainly in the Nicapa and Mono Pelado Sections, which are located in the State of Chiapas; in the rest of Mexico this limit is unconformable or is not present.

BIOSTRATIGRAPHY

The following results concerning the Eocene-Oligocene boundary are based on the study of the planktonic fossils found in the Tertiary sedimentary basins of Mexico.

Salmerón (1971) studied the materials of the Nicapa Section, and established the conformity of the Eocene-Oligocene boundary in southeastern Mexico, based on the *Globorotalia cerroazulensis* s.l. Biozone of the Late Eocene and the *Cassigerinella chipolensis-Pseudohastigerina micra* Biozone of the Early Oligocene.

(*) Instituto Mexicano del Petróleo, Exploración.

The *Globorotalia cerroazulensis* s.l. Biozone was determined in the Late Eocene. In the upper part of this biozone, the species of the Group *Globorotalia cerroazulensis* and the species of *Hantkenina* and *Cribrohantkenina* were extinguished. The most important species of the planktonic foraminifera of this biozone are the following: *Hantkenina alabamensis*, *Cribrohantkenina bermudezi*, *Globigerinita dissimilis*, *Globigerinita unicava*, *Pseudohastigerina micra*, *Globigerina pseudoampliapertura*, *Globigerina venezuelana*, *Globigerina yeguaensis*. The bentonics species are: *Bulimina alazanensis*, *Osangularia culter*, *Globocassidulina subglobosa*, *Bathisiphon eoceanicus*, *Planulina wuellerstorfi*.

In the lower part of the Early Oligocene, the *Cassigerinella chipolensis*-*Pseudohastigerina micra* Biozone was determined, and it overlies the above mentioned biozone. The following species are commonly found in this biozone: *Globigerina pseudoampliapertura*, *Globigerina ampliapertura*, *Pseudohastigerina micra*, *Cassigerinella chipolensis*, *Cibicides pseudoungerianus*, *Cibicides mexicanus*, *Cibicides perlucidus*, *Ammospipirata mexicana*, *Uvigerina mexicana*, *Ammodiscus incertus* and *Nodosaria longiscata*.

When Salmerón (1972) studied the materials of the Mono Pelado Section, he established that the boundary between the Eocene and the Oligocene was represented by the *Globorotalia cerroazulensis* Biozone s.l. of the Late Eocene, which is overlaid by the *Cassigerinella chipolensis*-*Pseudohastigerina micra* Biozone of Early Oligocene age. The species of *Globorotalia cerroazulensis* s.l. and the species of *Hantkenina* and *Cribrohantkenina* were extinguished in the Late Eocene.

Salmerón (1973) also studied the materials of the Amates Section, in the same area, and found that the Eocene-Oligocene boundary is unconformable, that is, rocks of the Middle Eocene (*Orbulinoides beckmanni* Biozone) are overlaid by rocks of the Middle Oligocene (*Globigerina ampliapertura* Biozone).

Butterlin (1961) determined the Eocene-Oligocene boundary, based on the macroforaminifera of the Rio Palizada 2 well, which is located in the boundary between the Macuspana-Campeche Basin and the Plataform of Yucatan. He mentions the following species of Late Eocene age: *Asterocyclina minima*, *Lepidocyclina (Nephrolepidina) chaperi*, *Lepidocyclina (Neolepidina) macdonaldi*, *Lepidocyclina (Neolepidina) pustulosa*, *Helicolepidina nortoni*, *Nummulites floridensis*, *Nummulites trinitatensis*, and *Lepidocyclina (Eulepidina) favosa* (=*L.* (*E*). *undosa*) of the Oligocene.

Therefore, it is possible that the Eocene-Oligocene boundary mentioned by Butterlin is not conformable, because he does not report characteristic species in the lower part of the Lower Oligocene. On the other hand, *Lepidocyclina (Eulepidina) undosa* has a range up to the Early Miocene-Aquitanian? according Butterlin (1981).

Salmerón (1969) found the Eocene-Oligocene boundary unconformable in the Santa Isabel 1 well, which is in the vicinity of the Rio Palizada 2 well, that is, the Upper Eocene rocks (*Globorotalia cerroazulensis* s.l.) are overlaid by the Middle Oligocene rocks (*Globigerina ampliapertura* Biozone). The following foraminifera of Late Eocene age were found: *Globorotalia cerroazulensis* s.l., *Globigerina ampliapertura*, *Globigerina pseudoampliapertura*, *Pseudohastigerina micra*, *Globigerinita dissimilis*, *Hantkenina dumblei*, *Hantkenina alabamensis*, *Baggina cojimarensis*, *Rotaliatina mexicana*, *Osangularia mexicana*, *Bulimina semicostata*.

Associated with the above mentioned species, the following allochthonous of macroforaminifera were found, which although they are not characteristic of the depositional environment above mentioned, they are of the same age; the most significant forms are the following: *Helicostegina polygiralis*, *Helicolepidina nortoni*, *Sphaerogypsina pilaris*, *Lepidocyclina* (*Neolepidina*) *pustulosa pustulosa*, *Asterocyclina minima*, *Nummulites floridensis*, *Heterostegina ocalana* and *Fabiania cassis*.

In the Jonuta 1 well, which is located at 7900 m S 83°E of the Church of Jonuta in the State of Tabasco, Salmerón (1969) established that the Upper Eocene is overlaid unconformably by detritic rocks of the Lower Miocene. The rocks of the Upper Eocene are very recrystallized and sometimes poorly compacted limestones, in which the following species of macroforaminifera were identified: *Lepidocyclina* (*Neolepidina*) *pustulosa pustulosa*, *Fabiania cassis*, *Amphistegina parvula*, *Helicolepidina nortoni*, *Heterostegina ocalana*, *Asterocyclina minima* and *Operculina mariennensis*.

In the Mompuyil 2 well, which is located at approximately 2920 m S 25° E of the Mompuyil 1 well, this author found that the Upper Eocene outcrops on the surface, or sometimes it may be covered by Recent alluvial deposits. The Upper Eocene is determined by *Operculina mariennensis*, *Nummulites willcoxi*, *Nummulites floridensis*.

In the Tampico-Misantla Basin, located in the eastern-central part of Mexico, and in the Burgos Basin, in northeastern area, as well as in the Peninsula of the Baja California in the northwest, the Eocene-Oligocene boundary has been referred to as unconformable. The details concerning this boundary are a separate paper.

REFERENCES

Butterlin J, 1961. Grandes foraminíferos del Pozo Palizada No. 2. Municipio de Palizada Estado de Campeche. U.N.A.M. Inst. Geol. Paleontología Mexicana No. 10, 59 p., 11 láms.

Butterlin J., 1981. Claves para la determinación de Macroforaminíferos de México y del Caribe, del Cretácico Superior al Mioceno Medio. Instituto Mexicano del Petróleo, 97 p., 51 láms.

Salmerón P. 1969. Biostratigraphie du Tertiaire basée sur les foraminiferes dans la region située entre la fosse sédimentaire de Macuspana-Campeche et la presqu'ile de Yucatán, Mexique. Fac. Sc. Univ. Paris. These Doct. 3e Cycle, 184 p. 14 planches, 12 figs., vol. I et II. (unpublised).

Salmerón P. 1971. Estudio Bioestratigráfico de la Sección Nicapa y la Seccion II del subsuelo. Cenozoico area de Huimanguillo. I.M.P., 321 p. (unpublished).
Salmerón P. 1972. Estudio Bioestratigráfico de la Sección Mono Pelado. Cenozoico area Huimanguillo Zona Sur. I.M.P., 291 p. (unpublished).
Salmerón P. 1973. Estudio Bioestratigráfico del área comprendida entre los Estados de Chiapas, Veracruz y Tabasco. Resumen III. Conv. Geol. Nacional. Soc. Geol. Mexicana.

THE EOCENE-OLIGOCENE BOUNDARY IN THE BATH CLIFF SECTION, BARBADOS, WEST INDIES

E. MüLLER-MERZ and J.B. SAUNDERS

The study of a section of deep sea sediments that crosses the Eocene-Oligocene boundary and is exposed in Bath Cliff, Barbados, resulted in a paper published as Saunders et al. (1984). It was a co-operative project by the co-authors whose names are given in the bibliographic reference. The following report lists findings from this joint work that are particularly relevant to the epoch boundary. The data on which the conclusions are based will be found in the cited paper. The present summary enables us to make alterations to the original text figures some of which have become possible due to later findings reported in Sanfilippo (1985). Comparisons between radiolarian events near the boundary as seen in Bath Cliff are compared with their equivalents in cores from widely spaced Deep Sea Drilling sites by Riedel and Sanfilippo (this Volume).

The Bath sediments are indurated biogenic oozes with additional disseminated volcanogenic fragments and discrete ash beds. They span a time interval from the Middle Eocene *Podocyrtis goetheana* radiolarian Zone to the Early Oligocene *Theocyrtis tuberosa* radiolarian Zone (Fig. 1). The biogenic fraction is predominantly siliceous with only a minor calcareous component. The total carbonate fraction for the Late Eocene is only about 20%; the input of volcanic ash is a little above 8%, the remaining 72% consisting of the siliceous biogenic component + quartz and clay minerals.

The linear sedimentation rate for the Late Eocene interval gives a value of 27.1 m/m.a., uncorrected for compaction. This corresponds to a bulk mass accumulation rate of 3.58 $g/cm^2/10^3 yr$. This is approximately 18 times faster than the rate at DSDP Site 543 east of the toe of the Barbados accretionary prism, and 11 times faster than the rate at DSDP Site 149 in the Venezuela Basin. The expanded nature of the section suggests accumulation in an area of greater than usual siliceous biogenic productivity.

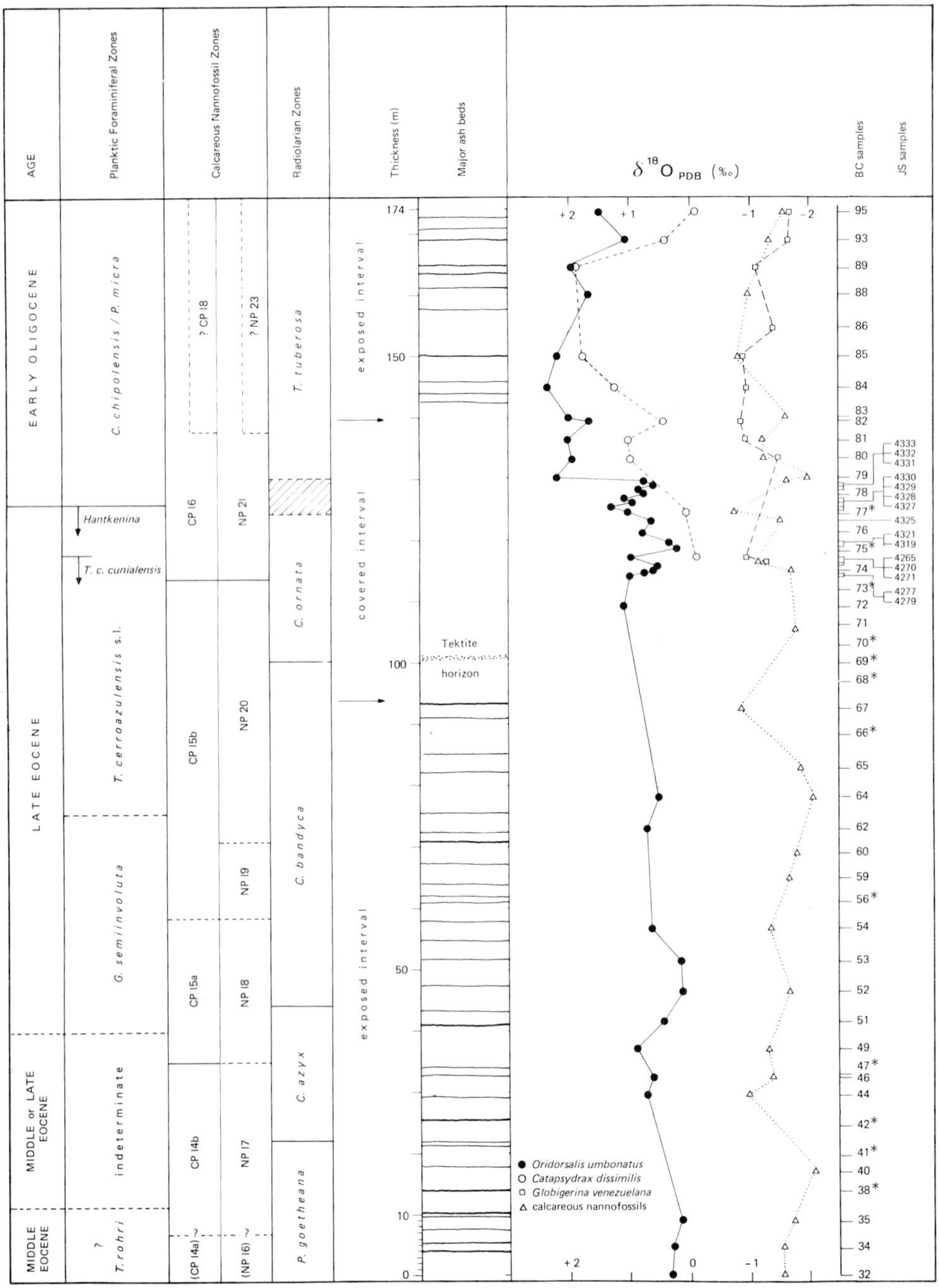

Fig. 1. Stratigraphic summary of the Bath Cliff section with biostratigraphic and oxygen isotope information plotted against a column showing the position of the microtektite horizon together with the more important ash beds.

The benthic foraminifera suggest a paleowater depth greater than 2000 meters, and probably more than 2800 meters. The calcareous fauna would indicate that the site of deposition was never far from the CCD. In the Eocene, the planktic/benthic ratio points to a loss of calcareous fauna varying from 90 to 100%. Shortly after the Eocene/Oligocene boundary, the site of deposition diverged from the CCD and this is shown by a higher planktic/benthic ration suggesting a loss of the calcareous fauna closer to 50%.

Figure 2 summarises the sequence of the more important events as they occur in the Bath Cliff section. Those that occur close to the epoch boundary are, in ascending order: last occurrence (LO) disc-shaped discoasters, LO *Nuttallides truempyi*; LO *Turborotalia cerroazulensis* s.l.; LO *Hantkenina*; evolutionary transition of the *Lithocyclia aristotelis* group to *L. angusta*. Following convention, we draw the boundary at the last occurrence of the genus *Hantkenina*.

The radiolarian studies show that the *Theocyrtis bromia* radiolarian Zone of the Late Eocene can be replaced by 3 new zones. From top to bottom these are: *Cryptoprora ornata* Zone, *Calocyclas bandyca* Zone and *"Carpocanistrum" azyx* Zone.

The nannofossil studies confirm that it is difficult to subdivide the Late Eocene in a low latitude oceanic section as the markers used for the standard zonations prefer high latitudes and/or hemipelagic environments. Regarding topmost Eocene, it was found that, even in the absence of the usual markers *Isthmolithus recurvus* and *Sphenolithus pseudoradians*, it is possible to assign samples to this level if they have disc-shaped discoasters but lack *Chiasmolithus grandis* and *Cribrocentrum reticulatum*.

The impoverishment of the calcareous fauna due to deposition close to the CCD has given a somewhat spotty distribution to the planktic foraminifera, particularly the marker species which are susceptible to solution. However, a small group of samples all of which contain *Hantkenina* and *Cribrohantkenina*, followed upwards by continuing richly calcareous samples without these marker forms has led us to draw the Eocene-Oligocene boundary here (sample 4328 on Fig. 1). The highest occurrence of the *Turborotalia cerroazulensis* group including *T. cerroazulensis cunialensis* (Toumarkine & Bolli) occurs approximately 8 m below (sample 4265) and, although the interval between suffered badly from solution, the absence of the

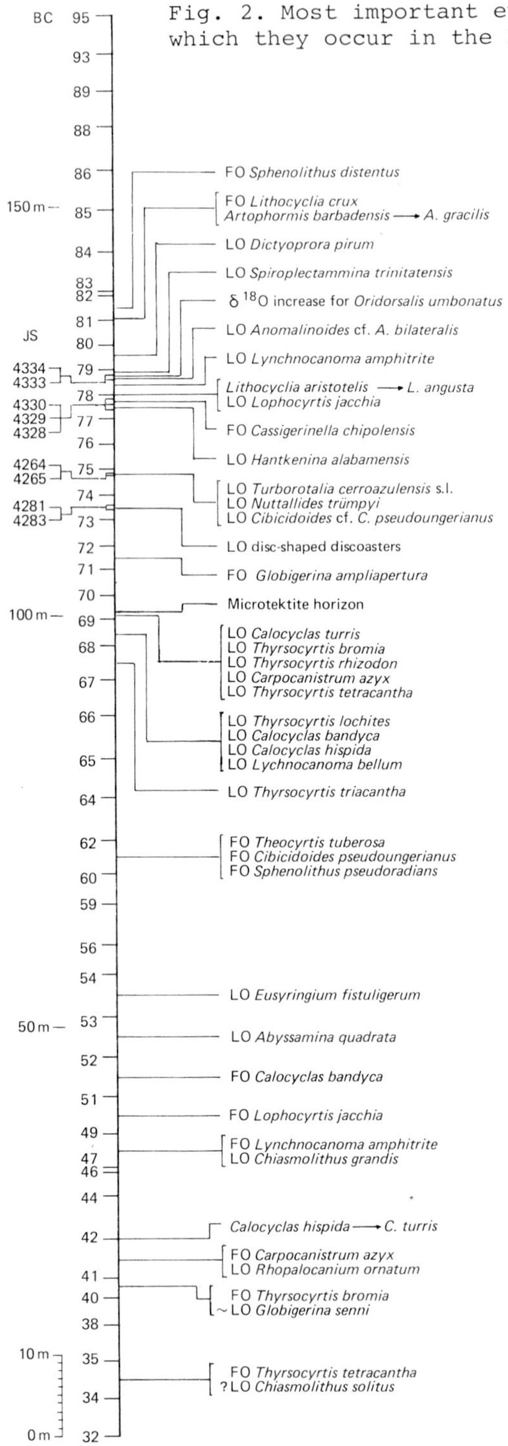

Fig. 2. Most important events in order in which they occur in the Bath Cliff section.

Turborotalia group from the samples containing the youngest occurrences of *Hantkenina* leads us to believe that *T. cerroazulensis* does become extinct before *Hantkenina* and *Cribrohantkenina* in the Bath Cliff section.

In the benthic foraminifera, there is a decrease in numbers of *Nuttallides truempyi* during the Late Eocene with an increase in numbers of *Cibicidoides pseudoungerianus* and *C. perlucidus*. The more diverse agglutinated fauna seen in the Late Eocene becomes a great deal poorer in the Oligocene. *Spiroplectammina trinitatensis* is the main constituent of the fauna in the Eocene but disappears just above the epoch boundary and therefore can be used as a local marker.

The oxygen-isotope record across the Eocene-Oligocene boundary suggests a drop in bottom water temperature of 5-6°C and of deeper surface water of 3-4°C with no measurable change in surface water temperature.

A layer with scattered microtektites occurs approximately 100 m above the base of the Bath Cliff section. Details of the occurrence, which is thought to be part of the 'North American' strewn field, are given in Sanfilippo et al. (1985). The microtektite-bearing interval is approximately 20 cm above the last occurrences of *Thyrsocyrtis tetracantha*, *Thyrsocyrtis rhizodon*, *Thyrsocyrtis bromia*, *Calocyclas turris* and *"Carpocanistrum" azyx* whose extinctions are virtually simultaneous. The intervening 20 cm of indurated sediment is thought to represent 13,000-14,000 years which contradicts any causal relationship between the extinctions and the microtektite event. However, the extinctions are coincident with an iridium-rich level that is described in the same paper. The microtektite layer falls slightly less than 26 m below the last occurrence of *Hantkenina* which would represent approximately one million years.

Throughout the time span of the section there is a gradual change in both floral and faunal composition. Although several events are clustered around the boundary, their separation suggests that they are not due to a catastrophic situation.

REFERENCES

Sanfilippo, A., Riedel, W.R., Glass, B.P., and Kyte, F.T., 1985. Late Eocene microtektites and radiolarian extinctions on Barbados. Nature, 314:613-615.

Saunders, J.B., Bernoulli, D., Müller-Merz, E., Oberhänsli, H., Perch-Nielsen, K., Riedel, W.R., Sanfilippo, A., Torrini, R., jr., 1984. Stratigraphy of the late Middle Eocene to Early Oligocene in the Bath Cliff section, Barbados, West Indies. Micropaleontology, 30: 390-425.

TERMINAL EOCENE EVENTS IN VENEZUELA

MARIA L. DIAZ DE GAMERO

The Cenozoic geologic history of Venezuela is usually subdivided into two complex sedimentary cycles, separated by an important tectonic event starting in the Late Eocene (González de Juana et. al., 1980).

This event is particularly marked in Western Venezuela where, after a brief but very widespread marine transgression during the late Middle Eocene, corresponding to the Orbulinoides beckmanni and Truncorotaloides rohri zones (Walton, 1966; Furrer, 1971; Hunter, 1974), there is no sedimentation until the Middle Miocene, with a short marine invasion in the Maracaibo Basin. The terminal Eocene geological events in this region are dominated by the initial emergence of the Merida Andes, the northeastern tip of the Andes range. Some sedimentary units of fluvio-lacustrine origin might have been deposited during this time in the area south of the present Lake of Maracaibo, which seems to have been a depression all through the Cenozoic, and indeed up until modern times (González de Juana et al., 1980). Outside this region, an important regional unconformity separates the Middle Eocene sequences from the Miocene and younger rocks in all Western Venezuela (see Fig. 1).

In a geologically different area, the Falcón Basin, situated in Northwestern Venezuela, the youngest Eocene sediments are of Late Eocene age, belonging to the Globigerinatheka semiinvoluta Zone (Hunter, 1974), and are found in an anticlinal structure in the southeastern part of the basin. Except for a few sandy intervals of Miocene age, the bulk of the sedimentation in the eastern Falcón Basin is composed of marls and clays deposited mostly at bathyal depths, with no obvious break in sedimentation. There have been no reports of early Oligocene faunas anywhere in this area, the oldest dated Oligocene being from the Globorotalia opima opima Zone (Diaz de Gamero, 1982). The interval from the latest Eocene to this level is apparently barren of calcareous foraminifera, the only microfaunas studied to date. It is still undecided whether there is a hiatus at the Eocene-Oligocene boundary, as in all the rest of Western Ve-

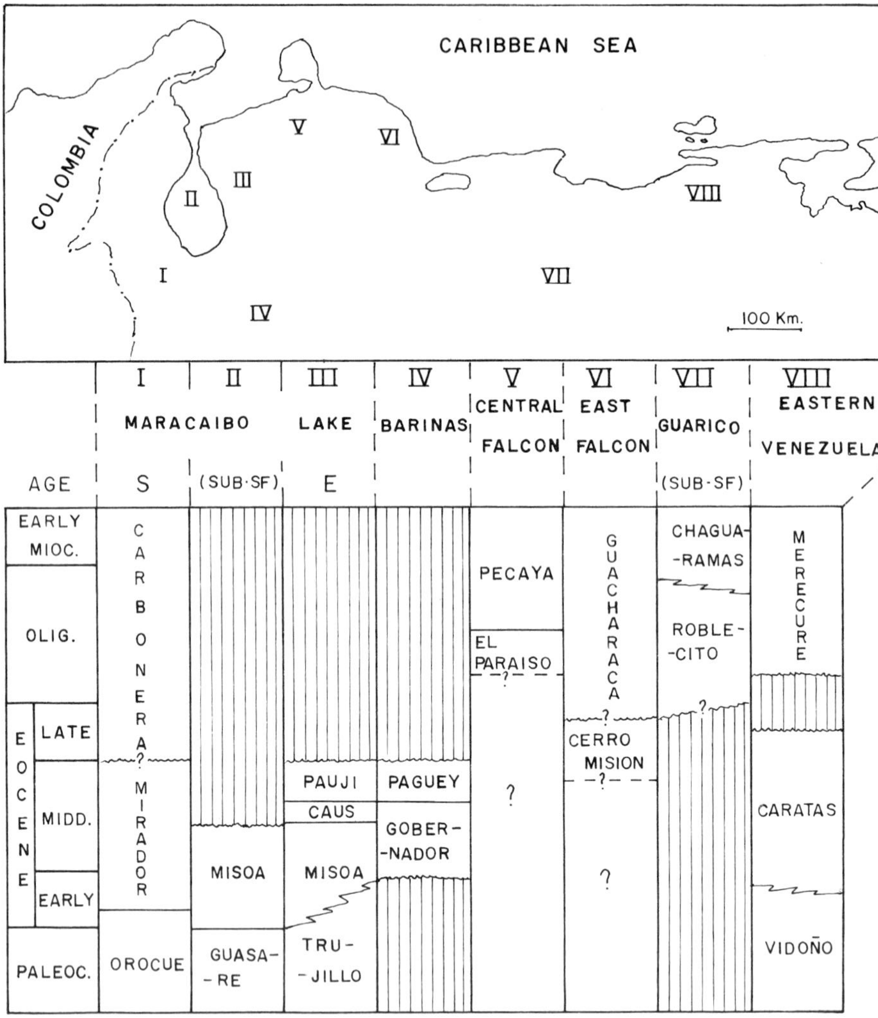

Fig. 1. Correlation chart of Eocene-Oligocene formations in Venezuela.

nezuela or, perhaps, the marine sedimentation was continuous, with a change in the physicochemical characteristics of the water mass.

Within the Eastern Venezuela Basin, the Eocene-Oligocene boundary is, again, essentially marked by a hiatus. The youngest Eocene sedimentation in the basin, dated by planktonic foraminifera, is of latest Middle Eocene age (Hunter, 1974), but the Globigerinatheka semiinvoluta Zone of the Late Eocene has been recognized locally (Furrer, personal communication, 1984). Unconformably above these sediments a thin clastic sequence, devoid of fossils, is followed by a marine shale with microfaunas belonging to the Globoortalia opima opima and Globigerina ciperoensis zones of the Oligocene (Lamb, 1964).

In short, with the possible exception of the deep marine sequence of the easternmost part of the Falcón Basin and some non marine sedimentation preserved locally in downfaulted or subsiding blocks, the Eocene-Oligocene boundary event was one of general uplifting and erosion in Venezuela, north of the stable craton of the Guayana Shield. The uplift was relatively short in Eastern Venezuela, but of a much longer duration, and accompanied by structural deformation in Western Venezuela.

REFERENCES

Diaz de Gamero, M.L., 1982. Estratigrafía y micropaleontología de la parte norte de la subcuenca de Agua Salada, Falcón oriental. Universidad Central de Venezuela, Unpublished, 158 pp.

Furrer, M.A., 1971. La edad de la Formación Paguey. IV Congr. Geol. Venezolano, Mem., I: 405-409.

González de Juana, C., Iturralde de Arozena, J.M. y Picard Cadillat, X., 1980. Geología de Venezuela y de sus Cuencas Petrolíferas. Ediciones FONINVES, Caracas, 2 vols., 1031 pp.

Hunter, V.F., 1974. The mid-Tertiary stratigraphic unit of the southern Caribbean area., In: Jung, P. (Editor). Contributions Dedicated to the Geology and Paleobiology of the Caribbean and Adyacent Areas. The Kugler Volume. Verhandl. Naturf. Ges., Basel, 84 (1): 172-190.

Lamb, J.L., 1964. The geology and paleontology of the rio Aragua surface section, Serranía del Interior, State of Monagas, Venezuela. Asoc. Venez. Geol. Min. y Petr., Bol. Inf., 7 (4): 111-123.

Walton, W.M., 1966. The Pauji and Mene Grande formations. Asoc. Venez. Geol. Min. y Petr., Bol. Inf., 9 (12): 325-337.

THE EOCENE/OLIGOCENE BOUNDARY IN DEEP SEA DEPOSITS

D. GRAHAM JENKINS

INTRODUCTION

Ever since the original definition of the Oligocene by Beyrich (1854), the placement of the Eocene/Oligocene boundary has been tied to ill-defined Stage boundaries in various parts of Europe (Berggren *et al.*, 1985 in press), and pressure to progress beyond this state of confusion led some workers to tentatively define the boundary without reference to the exact Stage boundaries in Europe. Thus, Bolli (1957) working in Trinidad placed the boundary at the extinctions of *Hantkenina*, and *Globorotalia cerroazulensis* (= *G. centralis* and *G. cocoaensis*), between the *G. cocoaensis* and *G. ampliapertura* Zones. In contrast, the boundary was positioned at the extinction of *Globigerinatheka index* in New Zealand between the two local stages Kaiatan and Whaingaroan by Finlay and Marwick (1940); the warmer water species used in Trinidad were either absent or had become extinct much earlier (Jenkins, 1971).

DEEP SEA DEPOSITS

In the 15 years of the *Glomar Challenger* phase of the Deep Sea Drilling Project (1968-1983) only a few continuous sections were found where the boundary could be identified by means of microfossils. Some sites were found to have been affected by dissolution (e.g. Site 77) while at other sites the boundary was within an unconformity (e.g. Site 608); the locations of these two, plus six other sites in the Pacific and Atlantic are given in Table 1.

TABLE 1 : Locations of DSDP Sites

Site	Latitude	Longitude
77	00° 28.90'N	133° 13.70'W
277	52° 13.43'S	166° 11.48'E
516	30° 16.59'S	35° 17.11'W
522	26° 06.80'S	05° 07.80'W
549	49° 05.29'N	13° 05.81'W
574	03° 59.24'N	134° 08.53'W
592	36° 28.40'S	165° 26.53'E
608	42° 50.21'N	23° 05.25'W

Pacific

At Site 277 in the Southwest Pacific, Shackleton and Kennett (1975) and Kennett and Shackleton (1976) showed that the beginning of a temperature drop of 4-5°C in bottom water coincided with the extinction of G. index. Further work on this site has shown that there is an important time-sequence of events: namely the extinction of G. index at a peak of species diversity, followed by rapid reduction in diversity and then the 4-5°C drop in bottom water temperature (Figure 1). Further north, at site 592 in the Tasman Sea there is an improvement in the calcareous nannofossil flora and the following sequence has been observed: the extinctions of Discoaster saipanensis, D. barbadiensis and C. reticulatum followed by the extinctions of Globanomalina micra, Globigerina linaperta, Globorotalia insolita and G. index, and then the drop in bottom and surface water temperatures (Kennett et al. in press). Although there is a similar sequence of fossil extinctions at site 593 a little further south in the Tasman Sea the drop in temperature is seen to begin at or is slightly before the extinction of G. index (Kennett et al. in press).

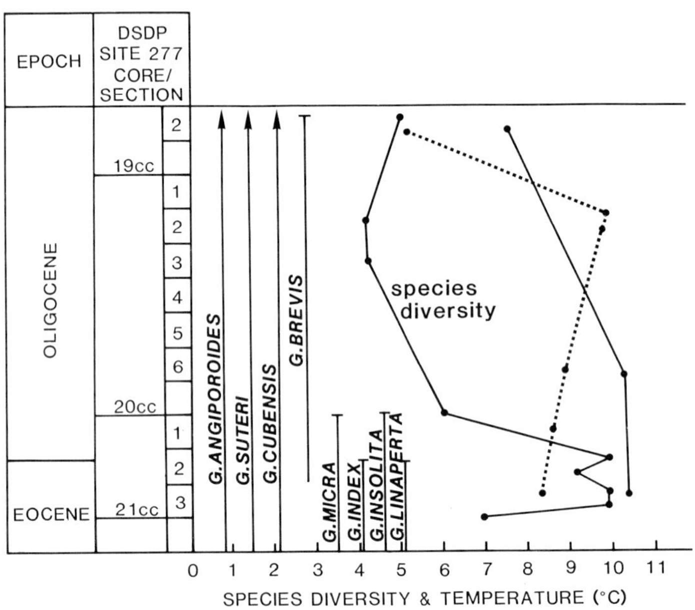

Figure 1. Species diversity, and paleotemperature based on oxygen isotope analysis, benthonic foraminifera (dotted) and planktonic foraminifera (solid line) at Site 277 (after Kennett and Shackleton 1976).

In the Eastern Equatorial Pacific Site 77 Keigwin and Keller (1984) placed the boundary slightly after the beginning of the drop in temperature, and at the nearby Site 574, Miller and Thomas (1985 in press) placed the boundary in a similar position.

It should be noted that not one of the sites in the Pacific has any paleomagnetic control across the boundary.

Atlantic

The best Eocene/Oligocene boundary in deep sea deposits was found at the South Atlantic Site 522, where there is both a good paleomagnetic and oxygen isotope record (Poore and Matthews 1984; Poore et al., 1982). Here the sequence of events is, anomaly 15, followed by the extinctions of Globigerinatheka, D. barbadiensis, G. cerroazulensis and Hantkenina, then the drop in temperature during anomaly 13 (Figure 2).

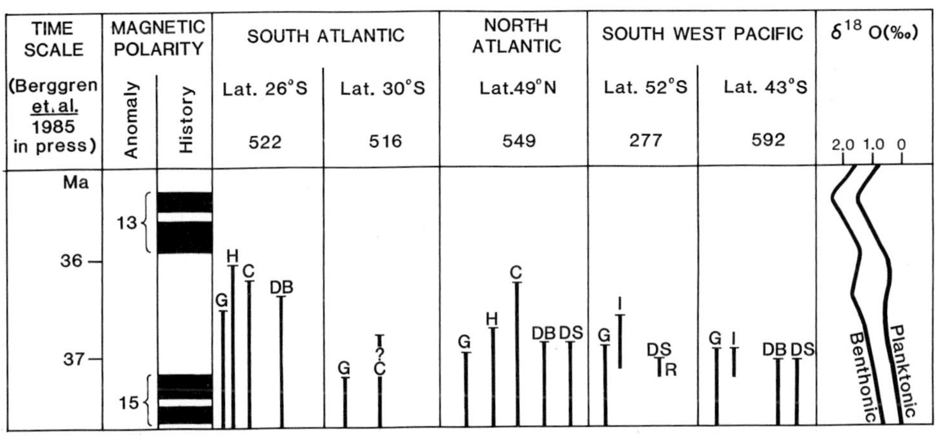

Figure 2. DSDP Sites in the Atlantic and Pacific; paleomagnetic and isotopic records from Site 522; there is some paleomagnetic control at Site 516 (Berggren et al. 1983) and none at Site 277, 549 and 592, but from the oxygen isotopic records the extinctions occurred before the drop in temperature during anomaly 13; C = G. cerroazulensis, G = Globigerinatheka, H = Hantkenina, I = G. insolita, DB = D. barbadiensis, DS = D. saipanensis, R = reworked?

At the North Atlantic Site 549 there is no paleomagnetic record but the sequence of events is: the extinctions of Globigerinatheka, D. saipanensis and D. barbadiensis, G. cerroazuelensis and Hantkenina followed by the drop in water temperature (Snyder et al. 1984).

EXTINCTIONS

According to Van Valen's rule, extinctions of related species within an adaptive zone occur as random events and over a long period of time the extinction rate is constant (Van Valen, 1973; Raup, 1975). Van Valen used Blow's

(1969) data to show that the extinction rate for the warm water planktonic foraminifera in the Middle Eocene-Pleistocene was 0.14 species per million years, and the extinction rate for all of the Cenozoic species in New Zealand is the same (Jenkins, 1985 in press). Any change in species diversity within a given time interval is dependent on 3 components: numbers of long ranging species, extinctions, and the appearance of new species. Analysis of New Zealand data across the Eocene/Oligocene boundary shows that the drop in species diversity (Figure 1) was not brought about by an increase in the rate of extinction, but was due to a drop in the number of initial appearances (Jenkins, 1973).

CONCLUSIONS

There is now a well established sequence of events across the Eocene/Oligocene boundary in deep sea deposits: starting at the bottom with anomaly 15, there follow the extinctions of species of calcareous nannofossils and planktonic foraminifera followed by the drop in temperature during anomaly 13. The sequence of extinctions appears to be latitudinally controlled by original paleotemperature and water-mass (Figure 2).

REFERENCES

Berggren, W.A., Aubry, M.P. and Hamilton, N. 1983. Neogene magnetostratigraphy of Deep Sea Drilling Project Site 516 (Rio Grande Rise, South Atlantic). Init. Rep. Deep Sea Drilling Project 72: 675-713.

Berggren, W.A., Kent, D.V., Flynn, J.J. 1985 in press. Paleogene Geochronology and the Geological Record. Geol. Soc. Lond. Spec. Publ.

Beyrich, E., 1854. Uben die Stellung der Hessischen Tertiarbildungen. Ber. Berh. Kgl. Preuss. Akad. Wiss. Berlin 1854: 640-666.

Blow, W.H., 1969. Late Middle Eocene to Recent planktonic foraminiferal biostratigraphy. In: P. Bronnimann and H.H. Renz (Editors), Proc. First Int. Conf. Plank. Microfossils, Geneva, 1967, 1: 199-421, E.R. Brill, Leiden.

Bolli, H.M. 1957. Planktonic foraminifera from the Eocene Navet and San Fernando Formations of Trinidad, B.W.I. U.S. Nat. Mus. Bull. 215: 155-172.

Finlay, H.J. and Marwick, J., 1940. The divisions of the Upper Cretaceous and Tertiary in New Zealand. Trans. Roy. Soc. New Zealand 70(1): 77-135.

Jenkins, D.G. 1971. New Zealand Cenozoic planktonic foraminifera. New Zealand Geol. Surv. Pal. Bull. 42, 278 pp.

Jenkins, D.G. 1973. Diversity changes in the New Zealand Cenozoic foraminifera. Journ. Foram. Res 3(2): 78-88.

Jenkins, D.G. 1985 in press. Biostratigraphic framework: planktonic foraminifera. In: Ciesielski, P. (Ed), Cenozoic paleoenvironment of the Southern Ocean. Amer. Geophys. Union.

Keigwin, L.D. and Keller, G. 1984. Middle Oligocene climatic change from equatorial Pacific DSDP Site 77. Geology 12: 16-19.

Kennett, J.P. and Shackleton, N.J. 1976. Oxygen isotopic evidence for the development of the psychosphere 38 Myr ago. Nature 260: 513-515.

Kennett, J.P. and von der Borch, C. et al. in press. Paleotectonic implications of increased late Eocene-early Oligocene volcanism: South Pacific DSDP sites. Nature.

Miller, K.G. and Thomas, E. 1985, in press. Late Eocene to Oligocene benthonic foraminiferal isotopic record, Site 574, equatorial Pacific. Init. Rep. Deep Sea Drilling Project,85.

Poore, R.Z. and Matthews, R.K. 1984. Late Eocene-Oligocene oxygen and carbon isotope record from South Atlantic Ocean, DSDP Site 522. Init. Rep. Deep Sea Drilling Project, 73: 725-735.

Poore, R.Z., Tauxe, L., Percival, S.F. and La Breque, J.L. 1982. Late Eocene-Oligocene magnetostratigraphy and biostratigraphy at South Atlantic DSDP Site 522. Geology 10: 508-511.

Raup, D.M. 1975. Taxonomic survivorship curves and Van Valen's law. Paleobiol. 1: 83-96.

Shackleton, N.J. and Kennett, J.P. 1975. Paleotemperature history of the Cenozoic and the initiation of Antarctic glaciation: oxygen and carbon isotopic analyses in DSDP Site 277, 279 and 281. Init. Rep. Deep Sea Drilling Project 29: 743-755.

Snyder, S.W., Muller, C. and Miller, K.G. 1984. Eocene-Oligocene boundary: biostratigraphic recognition and gradual paleoceanographic change at DSDP Site 549. Geology, 12: 112-115.

Van Valen, L. 1973. A new evolutionary law. Evol. Theory 1: 1-30.

EOCENE-OLIGOCENE BOUNDARY REFERENCE SECTIONS IN THE PACIFIC

GERTA KELLER

Dept. of Geological and Geophysical Sciences, Princeton University, Princeton, NJ 08544

In over 15 years of deep-sea drilling the Eocene-Oligocene boundary was recovered in relatively few DSDP sites in the Pacific Ocean located primarily in the equatorial region and the southwest Pacific (Figure 1). Of 16 deep-sea sites which recovered this boundary, four have primarily siliceous ooze sediments (Sites 69A, 70A, 166, 291) and indicate deposition at or below the calcium carbonate compensation depth (CCD). Poor carbonate preservation near the Eocene-Oligocene boundary affects six sites (Sites 77B, 305, 310, 315A, 462, 574) presumably due to fluctuating CCD and lysocline. Of the remaining six sites with good carbonate preservation, three are in the middle and high latitude southwest Pacific (Sites 277, 592, 593) and one of these (Site 593) has a 15 m sequence of lithified volcanoclastic breccia preceding the Eocene-Oligocene boundary. Of the remaining two sites core recovery is better in the middle latitude Site 592 than in the high latitude Site 277. Site 592 represents thus the only well-preserved reference section for the late Eocene to early Oligocene calcareous biostratigraphy in middle latitudes of the southwestern Pacific (Kennett et al, 1985).

In the equatorial region, of three deep-sea sites with good carbonate preservation, Site 167 has poor core recovery; Site 317B has good core recovery, but Blow (1969) and Bolli (1966) zones were found to be overlapping by Kaneps (1976). The third, Site 292, has been extensively studied and represents the best reference section for the low latitude Pacific Ocean. Deep-sea drilling sites which recovered late Eocene to early Oligocene sediments in the Pacific are indicated in Figure 1 along with Eocene-Oligocene sections studied in the Atlantic and Indian Oceans.

DSDP Site 292

Site 292 is located on Benham Rise in the West Phillippine Basin at 2943 m depth (15°49.11'N, 124°39.05'E). Upper Eocene to lower Oligocene sediments consist of clay-rich nannofossil chalk with about 2% radiolaria and thin beds of volcanic ash. Sediments are light to dark gray in color, moderately mottled and bioturbated. Percent carbonate is generally high averaging between 80-85% with the exception of short intervals at two microtektite horizons in Zones P16 and P15 where carbonate drops to 40%. Anomously high iridium was discovered at the lower microtektite horizon in Zone P15 (Keller et al, in preparation) at 38.2 Ma based on sediment accumulation rates.

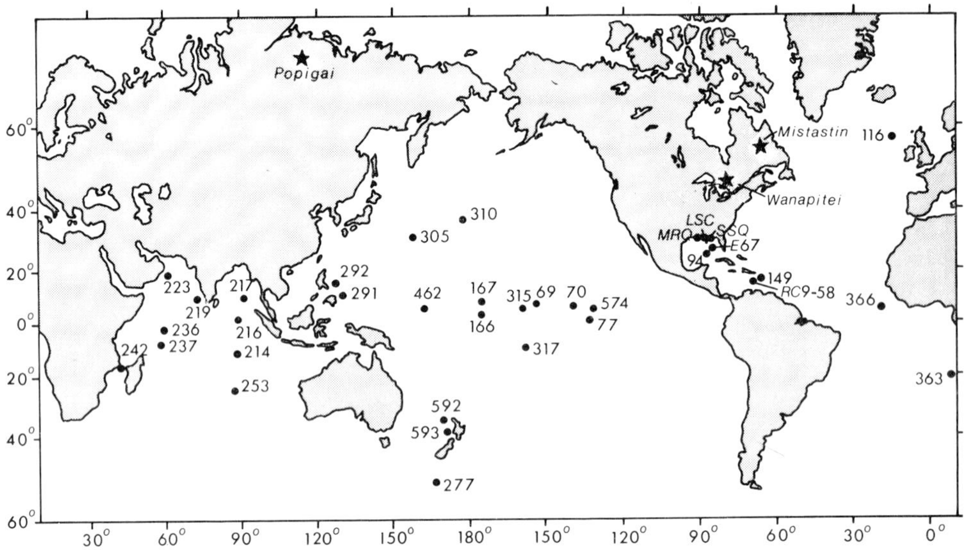

Fig. 1. Location map of Late Eocene to Early Oligocene DSDP cores and sections studied. Stars mark locations of impact craters of Late Eocene age, Popigai 39±9 Ma, Mistastin 38±4 Ma and Wanapitei 37±2 Ma (Grieve, 1982).

Glass and co-workers (1982, this volume) consider these two microtektite layers as the same layer as Zone P16 microtektites which they consider to represent upward reworking of the lower layer. However, it can be shown stratigraphically in terms of foraminifera and radiolaria that the lower layer at Site 292 is indeed older and of Zone P15 age and the upper Zone P16 layer corresponds in age to the Caribbean microtektite layer. Moreover, geochemically the Zone P15 microtektites are significantly different, primarily in their higher CaO and MgO contents, from the Caribbean microtektites (Keller et al, in preparation). For further discussion of this problem see Keller, this volume.

Site 292 has been extensively studied in terms of planktonic and benthonic foraminifera (Keller, 1983, Fig. 1, this volume; Corliss et al., 1984), calcareous nannofossils (Ellis, 1973), and stable isotopes (Keigwin, 1980). Radiolaria are also present, but no detailed analysis has been published since the initial report by Ling (1973).

All foraminiferal species extinctions and originations characteristic of the Late Eocene to Early Oligocene are present and appear to be isochronous when compared with other sites. Late Eocene occurrences, however, may be spotty due to frequent carbonate dissolution intervals. The nannofossil

Discoaster barbadiensis is last observed in Core 36cc and D. saipanensis in Core 36-2 (Ellis, 1973) and hence the CP15/CP16 boundary occurs before the E/O boundary as observed by the extinction of planktonic foraminifera (Globorotalia cerroazulensis, Gl. cocoaensis, G. cunialensis, and Hantkenina alabamensis) which is generally observed in the deep sea. Unfortunately, poor core recovery provides for an imcomplete Early Oligocene record in Site 292.

DSDP SITE 592

Site 592 is located on the southern Lord Howe Rise at 1098 m depth (36°28.40'S, 165°26.53'E). Late Eocene sediments consist of chalk and calcareous ooze with numerous volcanic ash layers. Planktonic foraminifera and nannofossils are well preserved. In the southwest Pacific middle to high latitudes the Eocene-Oligocene boundary is also marked by the extinction of the Globorotalia cerroazulensis - Gl. cocoaensis and Hantkenina groups when present, as in low latitudes. But, in addition, Globigerinathcka index, Globigerina linaperta and G. insolita also become extinct at the Eocene/Oligocene boundary and these species are the primary marker species in higher latitudes (Figure 2). In low latitudes sites Gr. index disappears earlier in the Globigerapsis semiinvoluta Zone whereas Globigerina linaperta ranges into the Oligocene, basal Globorotalia opima zone, and G. insolita is not recognized.

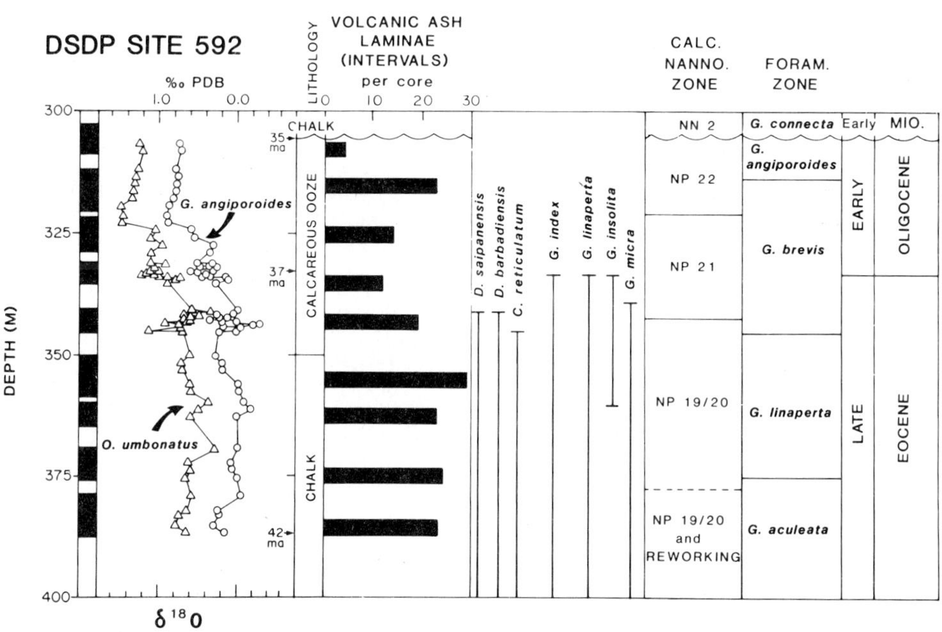

Fig. 2. Selected foraminifers and nannoplankton species ranges, volcanic ash intervals and oxygen isotope data of planktonic and benthonic foraminifera of Site 592 (from Kennett et al, 1985).

Similar to low latitudes, however, the nannofossils Discoaster saipanensis and D. barbadiensis disappear before the extinction of the foraminiferal Eocene-Oligocene boundary marker species (Kennett et al, 1985).

The differences in species ranges clearly indicate the necessity for separate high and low latitude reference sections and a means of calibrating them independent of microfossils which is possible for the Eocene-Oligocene boundary by using the oxygen isotope record. Thus, based on core recovery fossil preservation, biostratigraphic and stable isotope studies Sites 292 and 592 represent the best candidates for Eocene-Oligocene reference sections to date for the equatorial and southwest Pacific.

REFERENCES

Blow, W.H., 1969. Late middle Eocene to Recent planktonic foraminiferal biostratigraphy. In: Bronnimann, P. and Renz, H.H. (Eds.), Internatl. Conf. Plankt. Microfossils, 1st Proc. Leiden (E.J. Brill) : 199-421.

Bolli, H.M., 1966. Zonation of Cretaceous to Pliocene marine sediments based on planktonic foraminifera: Assoc. Venezolano Geol., Min. Petrol., Bol. Inform., 9 : 1-26. .

Corliss, B.H., Aubry M.-P., Berggren, W.A., Fenner, J., Keigwin, L.D. Jr. and Keller, G., 1984. The Eocene/Oligocene Boundary Event in the Deep Sea. Science, 226: 806-810.

Ellis, H.C., 1973. Calcareous nannofossil biostratigraphy - Leg 31, DSDP. In: Initial Reports of the Deep Sea Drilling Project, 31: 655-676.

Glass, B.P., DuBois, D.L. and Ganapathy, R., 1982. Relationship between an Iridium anomaly and the North American Microtektite Layer in core RC9-58 from the Caribbean Sea. Proceedings of the 13th Lunar and Planetary Sci. Conf., part 1, J. Geophys. Res., 87, supplement : A425-A428.

Kaneps, A.G., 1976. Cenozoic planktonic foraminifers, Equatorial Pacific Ocean, Leg 33, DSDP. In: Initial Reports of the Deep Sea Drilling Project, 33: 361-368.

Keigwin, L.D. Jr., 1980. Paleooceanographic change in the Pacific at the Eocene-Oligocene boundary. Nature, 287 (5784) :722-725.

Keller, G., 1983. Biochronology and paleoclimatic implications of middle Eocene to Oligocene planktonic foraminiferal faunas. Marine Micropaleontology 7: 464-486.

Keller, G., this volume. Late Eocene Impact Events and Stepwise Mass Extinctions.

Keller, G., D'Hondt, S., Keigwin, L.D., Onstott, T., Orth, C. and Shoemaker, E., in prep. Age and Geochemistry of Late Eocene Microtektite Horizons.

Kennett, J.P. and others, 1985 . Paleotectonic Implications of Increased Late Eocene-Oligocene Volcanism: South Pacific DSDP Sites. Nature, 316:507/511

Ling, H.Y., 1973. Radiolaria: Leg 31 of the Deep Sea Drilling Project. In: Initial Reports of the Deep Sea Drilling Project, 31: 703-762.

PART III

BIOSTRATIGRAPHICAL, GEOCHEMICAL, GEOPHYSICAL, SEDIMENTOLOGIC, TECTONIC AND COSMIC EVENTS

TERMINAL EOCENE EVENTS: PLANKTONIC FORAMINIFERA AND ISOTOPIC EVIDENCE.

BOERSMA, ANNE
Microclimates, Route 1, Box 405, Stony Point, New York 10980

PREMOLI SILVA, ISABELLA
Dipartimento di Scienze della Terra, Università di Milano

INTRODUCTION

The fact that planktonic foraminiferal assemblages from the late Middle Eocene are markedly different from those of the Early Oligocene has been known for over twenty years (Cifelli, 1967). Such differences may be described as the development of an Early Oligocene assemblage rich in simple, globigerina-shaped forms replacing an assemblage with a variety of different morphologies including the subquadrate acarininids, globorotalid-forms such as the turborotaliids, and the hantkeninids with their tubulospines. The aim of the IGCP Project No 174 was to demonstrate in detail how such changes took place.

To explain these events we have attempted first to document the types and sequence of terminal Eocene events among planktonic Foraminifera. Our approach has then been to compare synoptic measurements of estimated near-surface paleotemperatures and meridional and vertical thermal and carbon isotope gradients through the water column, to the spatial and temporal distribution patterns of those planktonic Foraminifera.

Despite the efforts of several scientists to document evolutionary patterns among the planktonic Foraminifera across the Eocene/Oligocene boundary, very few sections contained complete enough records for this task. Other sections, although relatively complete, were strongly affected by reworking and/or substantial dissolution. The few, relatively complete and utilizable sections, also summarized by Jenkins (this volume), include land sections from Barbados (Saunders et al., 1985), Australia (Lindsay and McGowran, this volume), New Zealand (Hoskins and Morgans, this volume), the southern Spanish Fuente Caldera (Molina, this volume), the Hungarian Buda marls (Nagymarosy et al., this volume), the Italian Umbrian basin (Nocchi et al., this volume), and various Deep Sea Drilling sections including Sites 516, 522, 592, 277, and 548 from Atlantic and Pacific Oceans. Such localities are well distributed through latitude and so should allow reconstruction of planktonic foraminiferal evolutionary events both spacially and temporally (Figure 1).

Detailed studies of the biogeographic patterns of Late Eocene planktonic Foraminifera (Haq et al., 1977; Boersma and Premoli Silva, 1985, in press) have demonstrated the latitudinal dependence of planktonic foraminiferal events and the diachronous ranges of several so-called biostratigraphic index species. Such species included: *Cribrohantkenina inflata*, *Globigerinatheka index*, *Subbotina linaperta*, and the *Turborotalia cerroazulensis* group, among others. The diachronous ranges of other taxa may result from their rarity in many samples and/or their very small size.

PLANKTONIC FORAMINIFERAL EVOLUTION

Planktonic foraminiferal originations and extinctions punctuate the entire period from the latest Middle Eocene into the Early Oligocene. Based on our composite range chart summarizing planktonic foraminiferal ranges through the Eocene at Deep Sea Drilling sites from the Atlantic (Haq et al., 1977) and from numerous land sections (Boersma and Premoli Silva, in press) including several discussed in this volume we estimated the degree of biotic overturn among planktonic Foraminifera. Because evolution tends to occur in pulses (Gould, 1977) we do not feel constrained to correct our events for the duration of each zone. Originations exceeded extinctions in Zone P16 of the Late Eocene. In Zone P15 extinctions and originations were nearly equal in number, 4 extinctions and 5 originations. Zone P17, associated with the Eocene/Oligocene boundary, was the only time when extinctions exceeded originations by 7 to 1 (Table 1).

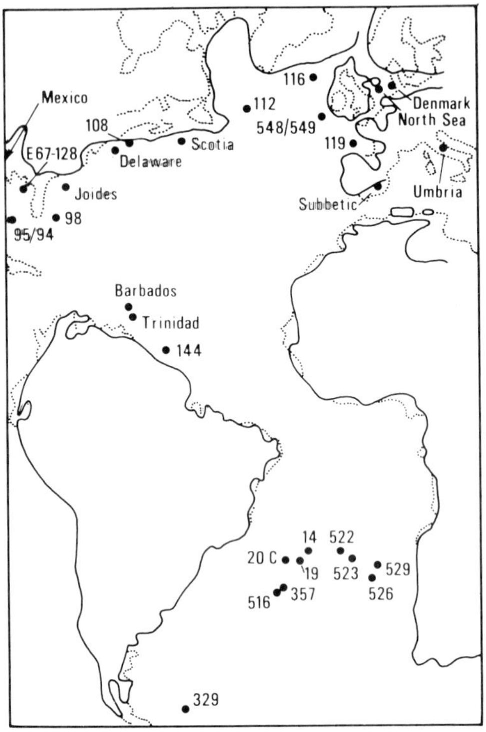

Fig. 1. Locations of sites included in this study (modified from Boersma and Premoli Silva, 1983).

Comparison of the numbers of extinctions, originations and the amount of evolutionary overturn between the zones of the Late and Middle Eocene demonstrates that except in Zone P16 evolutionary events were much fewer in number during the Late Eocene.

PLANKTONIC FORAM BIOZONES	DIVERSITY	ORIGINATIONS	EXTINCTIONS	OVERTURNS	OLIGOCENE REPRESENTATIVES
P14	56	13	18	31	22
P15	43	5	4	9	25
P16	51	12	9	21	35
P17	43	1	7	8	36
early P18	42	6	1	7	-

TABLE 1. Planktonic foraminiferal diversity, originations, extinctions and the numbers of events combined to be termed biotic overturn.

To a certain degree evolution slowed down during this time especially among the older Eocene groups. As indicated by Corliss et al., (1984) the Eocene/ Oligocene boundary was not a mass extinction event. In addition, the Late Eocene was not even a time of mass extinctions when compared with other intervals of the Middle Eocene or the Paleocene (Boersma and Premoli Silva, 1983).

We wish to stress that terminal Eocene events involve approximately equal numbers of originations and extinctions among the planktonic Foraminifera.

A large number of typically Oligocene planktonic Foraminifera first evolve in the Eocene; this evolution is accelerated in the Middle Eocene Zones P13-P14 with the appearance of the globoquadrinids, and species of *Catapsydrax*, *Globigerina*, *Subbotina*, and *Turborotalia (T. pseudoampliapertura* and *T. ampliapertura)*.

These originations took place largely in the tropical and subtropical belts; for example, these species have previously been recorded from Trinidad, Cuba, Mexico,and Barbados.Because in the latest Eocene and Oligocene the turborotaliids and globoquadrinids occupy similar latitudes to the mid Eocene morozovellids and acarininids, and also register some of the warmest paleotemperatures in the samples where they have been analyzed (Douglas and Savin, 1978; Boersma and Shackleton, 1977; Matthews and Poore, 1984) we assume that they replaced the morozovellids and acarininids as the low latitude, warm water-preferring index groups.

BIOGEOGRAPHIC PATTERNS OF LATE EOCENE PLANKTONIC FORAMINIFERA

By the beginning of the Late Eocene (Zone P14/P15 zonal boundary) the Early to Middle Eocene warm water planktonic Foraminifera had disappeared. They first contracted their ranges from middle to lower latitudes, decreasing in overall abundance, and then disappeared altogether. In many cases these forms disappeared nearly 0.5 m.y. earlier from the middle than from the lower latitudes (Boersma and Premoli Silva, 1983; in press). In the early Late Eocene, lower latitude faunas are dominated by forms such as *Globigerinatheka*, the complete group of *Subbotina*, *Pseudohastigerina*, and representatives of the *Turborotalia cerroazulensis* lineage, all forms previously typical of middle latitudes and of the gyre margins. Present also, although in very small numbers, are the globoquadrinids and turborotaliids which will become the low latitude index forms of the latest Eocene and earliest Oligocene. Higher latitude assemblages consist largely of high-spired subbotinids, *Catapsydrax*, *Globorotaloides*, and small globigerinids. The low-spired subbotinids(*S. eocaena* gr.) typical of high latitude regions in the earlier Eocene, had disappeared and by the Late Eocene *S. linaperta* s. str. was more typical of middle to lower latitude areas. Higher latitude index forms, such as *Catapsydrax* and *Globorotaloides*, nevertheless occur at all latitudes.

Several important events occur during the course of Zone P16. Evolutionary change was concentrated in the *Turborotalia cerroazulensis* lineage and among the cribrohantkeninids. Highest-spired forms, not present previously, first appeared in Zone P16. Biogeographically, the majority of genera, as well as many species, become cosmopolitan. Globigerinathekids, *Turborotalia cerroazulensis* s. str., globoquadrinids,hantkeninids, and *Turborotalia ampliapertura* penetrated as far north as Rockall Bank (DSDP Site 116), and *T. ampliapertura* even entered the North Sea. *Pseudohastigerina* appears to be confined to the western side of the Atlantic (DSDP Site 112),but its absence in the northeast at DSDP Site 116 may be an artifact of poor sediment preservation. The high-spired subbotinids, tenuitellids, *Catapsydrax*, *Globorotaloides*, biserial heterohelicids, and

Subbotina linaperta are present at all latitudes. This resulted in a sort of homogenization of faunas through latitude. Furthermore, the numbers of higher, middle and lower latitude index species occur in nearly equal proportions in many faunas at lower and middle latitudes,a further aspect of this faunas homogenization. Higher latitude faunas contained a large proportion of biserial heterohelicids after this time.

TABLE 2 BIOGEOGRAPHIC INDEX GROUP

PLANKTONIC FORAM-BIOZONE	LATITUDE LOW	MIDDLE	HIGH
Middle Eocene P11/P14	ornamented acarininids *Morozovellae* *Hantkeninae*	N. Atlantic biconvex morozovellids smooth acarininids *Planorotalites* *Acarinina spinuloinflata* S. Atlantic low-spired subbotinids smooth acarininids *Acarinina spinuloin-flata*	W.North Atlantic smooth acarininids *Globigerinatheka* subbotinids *Pseudohastigerina* E.North Atlantic high-spired subbotinids,biserial heterohelicids spinose acarininids
Late Eocene	*Globigerinatheka* turborotaliids *Cassigerinella* globoquadrinids	*Globigerinatheka* turborotaliids high-spired subbotinids	biserial heterohelicids *Catapsydrax* *Globorotaloides* tenuitellids

At lower latitudes the biserial heterohelicids and pseudohastigerinids occur alternately. When pseudohastigerinids are rare, biserial heterohelicids may be as abundant as they had become at higher latitudes.

By the end of Zone P16 *Globigerinatheka* decreased dramatically in abundance and they became a very minor component of the faunas. Conversely, hantkeninids, large-sized pseudohastigerinids and high-spired subbotinids increased in abundance at lower, and especially, at middle latitudes.

By almost the end of Zone P17 the *Turborotalia cerroazulensis* group became extinct. Shortly after, the very last *Globigerinatheka tropicalis* disappeared together with all hantkeninids, *Pseudohastigerina danvillenis*, and large-sized *P. micra*. This later group of events is equated to the Eocene/Oligocene boundary. The only definite origination during this time involved the appearance of primitive *"Globigerina" tapuriensis*. This form evolves in the very highest layer that can be attributed to the Eocene.

From these distribution patterns, the bioprovincial preferences of many taxa can be reconstructed, as shown in Table 2.

EVENTS IN COMPLETE EOCENE/OLIGOCENE BOUNDARY SEQUENCES

The actual sequence of these final Eocene events is usually difficult to observe due to the large numbers of hiatuses and dissolution episodes which chara cterize the Eocene/Oligocene transition. Where this crucial interval is preserved, it is characterized by strong fluctuations in the abundance of those groups which do cross the boundary. Two of the best examples of sequences considered to be complete are found in the Mediterranean region, an area which belonged to the subtropical climatic province. The sequence of events and abundances of planktonic Foraminifera from the Italian Umbrian region and from the Fuente Caldera section in Southern Spain are shown in Figure 2. Above the boundary faunas consist of about 30% biserial heterohelicids, about 20% high-spired subbotinids (*S. gortanii* and related forms) and 25% *Catapsydrax*. Except for *S. gortanii* which is very large, all other planktonic Foraminifera are small sized. This unique fauna, which has been found also at Site 522 in the South Atlantic and in the Gulf of Mexico (Boersma, 1985 in press), continues for a short interval in the earliest Oligocene. By Anomaly 13 time, coincident with a positive excursion in oxygen isotopes (Oberhänsli and Toumarkine, 1985, in press) the *S. gortaniii* group becomes less abundant and other species have returned to their more usual sizes. In these faunas *Catapsydrax*, biserial heterohelicids,

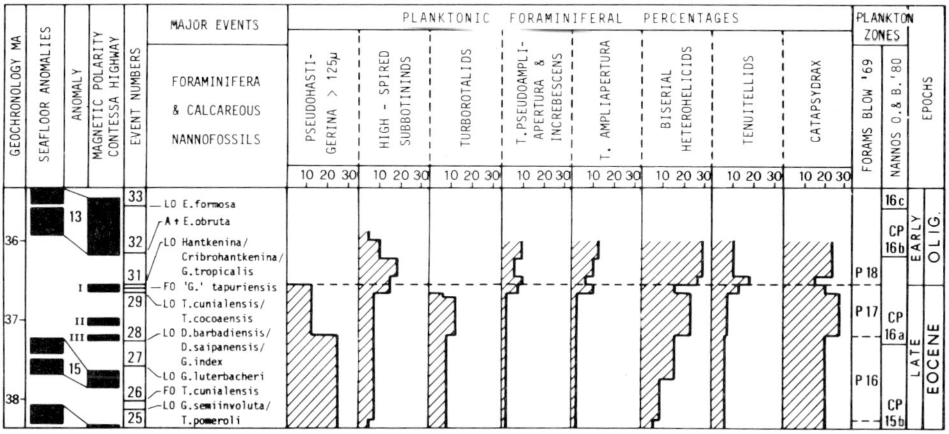

Fig. 2. Abundance of some planktonic foraminiferal groups across the Eocene/ Oligocene boundary from the Umbrian and Subbetic areas. Magnetic reversal sequence and calcareous plankton events after Nocchi et al., this volume.

and *Globorotaloides* are still common, while tenuitellids, turborotaliids (*T. ampliapertura*, *T. increbescens*, and *T. pseudoampliapertura*) increase in abundance. Pseudohastigerinids are represented by *P. barbadoensis* and *P. naguewichiensis*, rather than *P. micra* which had become very rare. Globoquadrinids and large globigerinids are only a minor component of these faunas.

TEMPERATURE STRATIFICATION PATTERNS OF PLANKTONIC FORAMINIFERA

The relative temperature stratification patterns of extinct planktonic Foraminifera are determined by their oxygen isotopic values which increase (are heavier) for forms dwelling in cooler waters. Measurement of a large number of Late Eocene to Early Oligocene age samples indicates the following relative temperature rankings from warmest to coolest (Table 3): biserial heterohelicids and pseudohastigerinids overlie turborotaliids and globigerinathekids, high-spired subbotinids, some small globigerinids, *Globorotaloides*, and *Catapsydrax*. Following the Eocene/Oligocene boundary extinctions which eliminated several of the thermophilic dwelling species, the relative temperature ordering was: *Cassigerinella*, biserial heterohelicids, the Oligocene turborotaliids, globoquadrinids, high-spired subbotinids, large globigerinids, small globigerinids, *Globorotaloides*, and *Catapsydrax*. These stratification patterns are difficult to confirm in the Early Oligocene because of the very small thermal contrast between the warmest and coolest species outside of the tropical zone, and because of confusing values which probably reflect a well-mixed near surface zone.

TABLE 3

PLANKTONIC FORAMINIFERAL STRATIFICATION ACCORDING TO $\delta^{18}O$ ORDERING CORROBORATED BY $\delta^{13}C$ ORDERING

	Middle Eocene	Early Late Eocene	Late Eocene	Base Oligocene
	Morozovella	*Globigerinatheka*	biser.heterohelicids	biser.heterohelicids
	Acarinina	turborotaliids	*Pseudohastigerina*	*Cassigerinella*
COOLER ↓	*Globigerinatheka*	*Hantkenina*	*Globigerinatheka*	turborotaliids
	Hantkenina	biser.heterohelicids	turborotaliids	*Globoquadrina*
	biser.heterohelicids	*Globigerina*	*G. index*	large globigerinids
	Subbotina	*Globorotaloides*	*Hantkenina*	h-spired subbotinids
	Globigerina	*Catapsydrax*	*Globigerina*	small globigerinids
	Globorotaloides		subbotinids	*Globorotaloides*
	Catapsydrax		*Globorotaloides*	*Catapsydrax*
			Catapsydrax	

CARBON ISOTOPIC STRATIFICATION OF PLANKTONIC FORAMINIFERA

During the Late Eocene planktonic foraminiferal species register relatively consistent carbon isotopic values. Among the planktonics the overall trend is towards less positive value and a decreasing gradient of values between the warmest and coolest preferring species. In Zone P15 the inter-specific gradient average about 2.77-1.92‰, by Zone P16 values are similar, about 2.77-1.92‰, but in Zone P17 values drop to 1.35-1.19‰ and the overall gradient decreases by about one-half. In Early Oligocene Zone P18 coeval with the oxygen isotope excursion, in areas not overlain by upwellings the carbon isotope values at the bottom become significantly more positive while the surface changes little, nevertheless resulting in a decrease in the surface to bottom carbon isotope gradient. Similar decreases in the carbon isotopic gradients among planktonic Foraminifera occur at the Cretaceous/Tertiary (Boersma et al., 1979) and at the Paleocene/Eocene boundaries (Shackleton and Hall, 1984; Boersma, 1984) and have been interpreted to represent a breakdown in oceanic primary productivity and carbon cycling. The very low inter-specific gradient at the termination of the Eocene and continuing through the Early Oligocene may represent a similar decrease in oceanic primary productivity.

As shown in Table 4 individual species consistently demonstrate higher or lower carbon isotope values. Recent and Pleistocene planktonic Foraminifera also demonstrate consistently high or lower carbon isotope values which correlate with a near surface or deeper habitat. One species, *Globorotalia scitula*, registers the lowest carbon isotope values and has been confirmed to inhabit an oxygen minimum layer (Shackleton and Vincent, 1978). Miocene biserial heterohelicids also register very low carbon isotope values and are interpreted to inhabit an oxygen minimum habitat (Resig and Kroopnick, 1982). Paleocene biserial heterohelicids also record lowest carbon isotope values and have been interpreted to inhabit the oxygen minimum zone in the Paleocene (Boersma et al., 1979; Boersma and Premoli Silva, 1983).

The carbon isotope ordering of species can also be confirmed by the oxygen isotope ordering. Species which demonstrate more positive carbon isotope values should also register lighter oxygen isotope indicating that they existed at warmer temperatures, shallower in the water column. Despite a high degree of variability in the Late Eocene data, these orderings are confirmed and further suggest that the relative carbon isotope values are not simply a function of disequilibrium precipitation of carbonate and/or the presence of symbionts (Hemleben and Spindler, 1983).

Assuming that high or low carbon isotope values of species represent some feature of their habitat, then Late Eocene species can be categorized as either high carbon isotope or low carbon isotope species. But what did high or low values of individual species signify? According to Broecker (1974) and later workers, the carbon isotopic composition of oceanic surface waters is modified by the cycling of light carbon through biologic systems. Higher in the water column where light and nutrients are more plentiful and primary production greater, light carbon is preferentially removed thus enriching the water in heavy carbon. Planktonic Foraminifera which secrete their tests in equilibrium with waters at these levels should give heavy carbon isotope values. Slightly deeper in the near surface zone where nutrient levels are lower, primary production is not occurring, and therefore light carbon is not being incorporated into biologic systems, but being returned to the water via their oxidation. Planktonic Foraminifera which secrete their tests at these levels record lighter carbon isotope values.

From the relative carbon isotope values of the Late Eocene planktonics we hypothesize the following habitat preferences; higher nutrient habitats are preferred by *Globigerinatheka* and to a lesser degree, by *Turborotalia*. The taxa of the *T. ampliapertura* group appear to like warm habitats, but appear to be pre-adapted to the lower nutrient levels at these warm temperatures thus fitting them for the lower productivity Oligocene ocean. Forms living in an oxygen minimum situation include the biserial heterohelicids, pseudohastigerinids, and possibly hantkeninids. The biserial heterohelicids and pseudohastigerinids appear to live in a warmer (shallower) oxygen minimum situation than do the hantkeninids.

The habitat of the biserial heterohelicids deserves special mention. These forms register the lowest carbon isotope values of all planktonics during most of the Paleocene, Early and Middle Eocene. However, as the carbon isotope values at the surface become more negative and the inter-specific gradients decrease, the biserial heterohelicids change their habitat. They move from intermediate to warmest temperatures and no longer register the lowest carbon isotope values. They demonstrate this same trend at the Cretaceous/Tertiary boundary when the carbon isotopic gradient among the planktonics also is very low (Boersma and Shackleton, 1981).

BIOGEOGRAPHY	OXYGEN RATIOS	CARBON RATIOS	SPECIES/GENUS
middle latitude gyre margin	low	high	*Globigerinatheka*
middle latitude to subtropics	intermediate	low	*Hantkenina*
middle latitude	variable	variable	biserial heterohelicids
low-middle latitude	medium low	medium high	*T. cerroazulensis* gr.
middle latitude slope species	intermediate	low-intermed-iate	high-spired subbotinids
middle latitude	low	low	*Pseudohastigerina*

TABLE 4. Biogeographic indications of 6 genera and/or species of Late Eocene planktonic Foraminifera plotted beside their characteristic oxygen and carbon isotopic ratios. Low oxygen isotope ratio species live at warm temperatures. Low carbon isotope ratio species may live in an oxygen minimum situation (Boersma et al., 1979). Biserial heterohelicids register lower oxygen isotope values in times of low carbon isotope gradients through the water column and/or among the planktonic Foraminifera. Conversely, during times of higher carbon isotopic gradients, biserial heterohelicids register intermediate oxygen isotope values. Carbon isotopic values of this group are lower in times of a high surface to bottom gradient in carbon isotopes, and higher during times of a low surface to bottom gradient.

CONCLUSIONS AND SPECULATIONS

We conclude that the key terminal Eocene to Early Oligocene events among the planktonic Foraminifera are:

1) Increasing abundance of biserial heterohelicids often alternating with that of *Pseudohastigerina* in fine fractions;
2) The evolution and spread of the highest-spired subbotininds in Zone P16;
3) The diminished adundance of *Globigerinatheka* at the top of Zone P16;
4) The homogenization of assemblages through latitude during Zone P16;
5) The extinction of *Turborotalia cerroazulensis* group in Zone P17 just prior to the Eocene/Oligocene boundary;
6) The first occurrence of *"Globigerina" tapuriensis* overlapping with the last Eocene taxa;
7) The extinction of *Globigerinatheka* and hantkeninids defining the Eocene/Oligocene boundary at the top of Zone P17;

8) The bloom of biserial heterohelicids and small-sized planktonics associated with large-sized *Subbotina gortanii* in earliest Zone P18.

We suggest that these events were the result of the modification and/or elimination of specialized habitats which Middle Eocene planktonic Foraminifera previously and preferentially occupied. In general the faunal homogeneity of the Late Eocene reflects the gradually decreasing thermal contrasts and lowering fertility of oceanic surface waters as the ocean cooled from the bottom. Near the beginning of the Late Eocene forms which occupied a low oxygen habitat but had large temperature (density) tolerances moved to the ocean surface indicating the development of at least moderately low oxygen conditions at the shallowest (warmest) levels occupied by the planktonic Foraminifera. The gradual cooling and spread throughout the ocean of the conditions of a shallow nearshore habitat suggested by the increasing import of the high-spired subbotinids, particularly at middle latitudes. The cool, low oxygen habitat previously occupied by the hantkeninids was eliminated altogether at the Eocene/Oligocene boundary. The warm, high nutrient near surface habitat of *Globigerinatheka* and of the *T. cerroazulensis* group was first restricted, then eliminated at the termination of the Eocene.

We have speculated elsewhere (Boersma and Premoli Silva, 1985, in press) that this faunal reorganization was the result of the transition from a well-stratified ocean cooling from the bottom and decreasing in fertility to one with homogeneous, well-mixed surface zone of low fertility. The decreasing fertility may be related to the lock-up of nutrients on the shelves during the Middle through Late Eocene high stand of sea level or, at the very end of the Eocene, by decreasing in upwelling (Fenner, this volume; Lipps, 1970, 1986). The evolution to a well-mixed ocean by the end of the Eocene eliminated specific oxygen-nutrient related water layers including the cool, deeper oxygen minimum and the subsurface oxygen maximum. In middle latitudes the near surface at the Eocene/Oligocene transition contained a low productivity and only moderately oxygenated zone overlying an essentially thermally uniform, well mixed, low nutrient and low oxygen subsurface. These oxygen lows were the result not of oxidation of organic matter, but a lack of ventilation. The decreased stratification of the surface zone may have developed, in part because of a decrease in a saline Tethyan Mediterranean outflow which had contributed to the stratification of Atlantic waters previously in the Eocene.

ACKNOWLEDGMENTS

The Authors are indebted to the Deep Sea Drilling Project for having provided the many samples utilized in this research. Warm thanks go to N.J. Shackleton, M. Orlando, M. Nocchi, G. Parisi, and P. Monaco. Financial support went to I.P.S. in a NSF grant to W.A. Berggren (Woods Hole Oceanographic Institution), and M.P.I. 40% 82/83 grant. The paper was read by M.B. Cita and J. Lipps to who our thanks are extended.

REFERENCE

Berggren, W.A., Kent, D.V. and Flynn, J.J., 1985. Paleogene geochronology and chronostratigraphy. In: Snelling, N.J., Ed., Geochronology and the geological record. Geol. Soc. London, Spec. Pubbl., in press.

Boersma, A., 1984. Oligocene and other Tertiary benthic foraminifers from a depth traverse down Walvis Ridge, Deep Sea Drilling Project Leg 74, Southeast Atlantic. Init. Repts. DSDP, 75/2: 1273-1300.

Boersma, A., 1985. Atlantic Paleoceanography using benthic Foraminifera across the Eocene-Oligocene transition. (This volume).

Boersma, A. and Premoli Silva, I., 1983. Paleocene planktonic foraminiferal biogeography and the paleooceanography of the Atlantic Ocean. Micropaleontology, 29/4: 355-381.

Boersma, A. and Premoli Silva, I., 1985. Eocene to Oligocene planktonic foraminiferal biogeography reflecting major climatic changes in the Atlantic. TERRA Cognita, 5/2-3: 117.

Boersma, A. and Premoli Silva, I., in press. Eocene planktonic foraminiferal biogeography and the paleooceanography of the Atlantic Ocean. Submitted to: Paleo., Paleo., and Paleo., Elsevier, Amsterdam.

Boersma, A. and Shackleton, N.J., 1977. Oxygen and carbon isotope record through the Oligocene, DSDP Site 366, Equatorial Atlantic. In: Lancelot, Y., Seibold, E., et al., Init. Repts. DSDP, 41: 951-962.

Boersma, A. and Shackleton, N.J., 1981. Oxygen and carbon isotope variations and planktonic foraminiferal depth habitats, Late Cretaceous to Paleocene. In: Thiede, J., Vallier, T.L., et al., Init. Repts. DSDP, 62: 513-526.

Boersma, A., Shackleton, N.J., Hall, M. and Given, O., 1979. Carbon and oxygen isotope variations at DSDP Site 384 (north Atlantic) and some paleotemperatures and carbon isotope variations in the Atlantic Ocean. In: Tucholke,B., Vogt, P., et al., Init. Repts. DSDP, 43: 695-717.

Broecker, W., 1974. *Chemical Oceanography*. Harcourt Brace Janovich, New York, 175 pp.

Cifelli, R., 1969. Radiation of Cenozoic planktonic Foraminifera. Syst. Zool., 18: 154-168.

Corliss, B.H., Aubry, M.P., Berggren, W.A., Fenner, J.M., Keigwin, L.D. and Keller, G., 1984. The Eocene/Oligocene Boundary Event in the Deep Sea. Science, 226: 806-810.

Douglas, R.G. and Savin, S.M., 1978. Oxygen isotopic evidence for the depth stratification of Tertiary and Cretaceous planktonic foraminifera. Marine Micropal., 39: 175-196.

Gould, S.J., 1977. Eternal metaforce of Paleontology. In: A. Hallam (Editor), Patterns of Evolution. Elsevier, Amsterdam, 1-26.

Haq, B.U., Premoli Silva, I. and Lohmann, G.P., 1977. Calcareous plankton paleobiogeography evidence for major climatic fluctuations in the Early Cenozoic Atlantic Ocean. Jour. Geophys. Res., 92: 386-398.

Hemleben, C. and Spindler, M., 1983. Recent advances in research on living planktonic foraminifera. Utrecht Micropaleont. Bull., 30: 141-170.

Hoskins, R.H. and Morgans, H.E.G., 1985. The main reference section for the Eocene/Oligocene boundary in New Zealand. (This volume).

Jenkins, D.G., 1985. The Eocene/Oligocene boundary in deep sea deposits. (This volume).

Lindsay, J.M. and McGowran, B., 1985. Eocene/Oligocene boundary, Adelaide Region, South Australia. (This volume).

Lipps, J.H., 1970. Plankton Evolution, Evolution, 24: 1-22.

Lipps, J.H., 1986. Extinction Dynamics in pelagic Ecosystems. In: D.K. Elliot (Editor), Dynamics of Extinctions. John Wiley and Sons, Inc. New York, in press.

Molina, E., 1985. Description and biostratigraphy of the main reference section of the Eocene/Oligocene boundary in Spain: Fuente Caldera section. (This volume).

Nocchi, M., Parisi, G., Monechi, S., Madile, M., Monaco, P., Napoleone, G., Orlando, M., Premoli Silva, I., Bice, D., 1985. The Eocene/Oligocene boundary in the Umbrian pelagic sequences (Italy). (This volume).

Nagymarosy, A., Baldi, T. and Horvath, M., 1985. The Eocene/Oligocene boundary in Hungary. (This volume).

Oberhänsly, H. and Toumarkine, M., 1985. The Paleogene oxygen and carbon isotope history of Sites 522,523, and 524 from the Central Southern Atlantic. In: Symposium on the South Atlantic Paleooceanography, Zurich, 1983. K.J. Hsu and H. Weissert Eds., Oxford Press, in press.

Poore, R.Z. and Matthews, R.K., 1984. Oxygen isotope ranking of late Eocene and Oligocene planktonic foraminifers: Implications for Oligocene sea surface temperatures and global ice volume. Marine Micropal., 9: 111-134.

Resig, J.M. and Kroopnick, P.M., 1983. Isotopic and distributional evidence of a planktonic habit for the foraminiferal genus *Streptochilus* Bronnimann and Resig, 1971. Marine Micropal., 8: 235-248.

Saunders, J.B., Bernoulli, D., Muller-Merz, E., Oberhansli, H., Perch-Nielsen, K., Riedel, W.R., Sanfilippo, A. and Torrini, R. jr., 1984. Stratigraphy of the late Middle Eocene to Early Oligocene in the Bath Cliff section, Barbados, West Indies. Micropaleontology, 30/4: 390-425.

Shackleton,N.J. and Hall, M., 1984. Carbon isotope data from Leg 74 sediments. In: Moore, T.C., jr., Rabinowitz, P.D., et al., Init. Repts. DSDP, 74: 613-620.

Shackleton, N.J. and Kennett, J.P., 1975. Paleotemperature history of the Cenozoic and the initiation of Antarctic glaciation. Oxygen and carbon isotope analyses in DSDP Sites 277, 279, and 281. In: Kennett, J.P., Houtz, R.E., et al., Init. Repts. DSDP, 29: 743-755.

Shackleton, N.J. and Vincent, E., 1978. Oxygen and Carbon isotope studies in Recent Foraminifera from the Southwest Indian Ocean. Marine Micropal., 3: 1-13.

Eocene/Oligocene Atlantic Paleo-Oceanography Using Benthic Foraminifera

Anne Boersma, Microclimates. Box 404 RR1, Stony Point, N.Y. 10980

Abstract

Benthic foraminiferal paleoenvironmental indices demonstrate that from late Eocene into earliest Oligocene time the Atlantic Ocean gradually became more corrosive and less fertile, from abyssal depths up to the slope. Development of glacial conditions in the early Oligocene (early in Anomaly 13 time) reversed the process; the CCD dropped, fertility increased, and higher levels of residual organic carbon were distributed from the slope to the lower bathyal zone.

Introduction

Through the past 10 years study of open ocean, Eocene/Oligocene age benthic foraminifera has been directed towards several key areas: 1) establishment of indices for interpreting past environmental conditions (Gernant and Kessling, 1966; Boersma, 1977, 1984b, 1985a,b,c; Grunig and Herb, 1980; Douglas and Woodruff, 1980 Tjalsma and Lohmann, 1983; Corliss and Keigwin, 1985;
2) determination of diversity and evolutionary overturn to assess the so-called catastrophic nature of the Eocene/Oligocene boundary event (Corliss, 1981; Tjalsma, 1983; Parker et al., 1984; Corliss et al., 1984); 3) establishing accurate biostratigraphic ranges of species whose endpoints have been associated with the Eocene/ Oligocene boundary (Proto Decima and Bolli, 1978; Schnitker, 1979; Corliss, 1979, 1981; Douglas and Woodruff, 1980; Tjalsma and Lohmann, 1983; Tjalsma, 1983; Clark and Wright, 1984; Boersma, 1984a, 1985b,c; Corliss et al., 1984; Saunders et al., 1985);
4) factor analytic study of species distribution patterns through space and time (Schnitker, 1979; Tjalsma and Lohmann, 1983);
5) recognition of environmental index faunas based on analogy to where (under what unique conditions) the species in the fauna were first described (Berggren and Aubert, 1976; Gradstein and Berggren,

1981; Tjalsma and Lohmann, 1983; Clark and Wright, 1984; Boersma, 1985a); 6) detailed 'event stratigraphy' closely tied to high resolution biostratigraphy and/or chronostratigraphy, involving benthics through the Eocene/Oligocene transitional interval (Monaco et al., 1984; Saunders et al., 1985; Boersma, 1985a,c); and 7) use of benthics in conjunction with stable isotopic analyses to interpret paleoclimate and paleo-oceanography (Shackleton and Kennett, 1975; Boersma and Shackleton, 1977; Vernaud Grazzini et al., 1980; Keigwin, 1980, 1983; Oberhansli, 1982; Miller and Curry, 1983; Oberhansli et al., 1984; Shackleton et al., 1984; Saunders et al., 1985; Boersma, 1985a,c; Corliss and Keigwin, 1985; Keigwin and Corliss, 1985. Discussion of all this data is beyond the scope of this paper; much of the recent general oceanographic interpretive work is summarized in Corliss and Keigwin, 1985. Therefore, only the major features of benthic biostratigraphy and the several environmental index species, some deriving from factor analytic studies, will be summarized here.

Biostratigraphy

Despite the large number of studies documenting the stratigraphic ranges of benthic foraminifera across the Eocene/Oligocene boundary, relatively few species' endpoints have been found to characterize this interval. Open marine bathyal to abyssal species which become extinct through this interval include, *inter alia*, *Plectofrondicularia lirata*, *Uvigerina rippensis*, *Alabamina dissonata*, *Abyssamina* spp., *Cibicidoides truncanus* = *Cibicidoides parki* and *Nuttallides truempyi*. Species which originate during this transitional interval are *Uvigerina pygmaea*, *Hopkinsina mioindex*, *Sphaeroidina bulloides*, *Rectuvigerina prisca*, *Laticarinina alto-camerata*, and *Pullenia bulloides*. Most of these evolutionary events are concentrated at the oxygen isotope excursion, now calibrated with magnetic Anomaly 13 in the early Oligocene (Tauxe et al., 1983; Oberhansli and Toumarkine, 1985;) rather than to the actual Eocene/Oligocene boundary located slightly below. Corliss (1981), Parker et al. (1984), and Corliss and Keigwin (1985) have therefore concluded that this is not a period of catastrophic extinctions among bathyal or abyssal benthic foraminifera and that a 26 my cyclicity of extinctions is not upheld at the Eocene/Oligocene boundary.

Paleodepth Index Species

Because of extensive migration through depth (Douglas, 1973) many benthic foraminiferal species cannot be used to interpret paleodepths except in the broadest terms. For example, *Nuttallides umbonifera*, usually an index for abyssal and lower bathyal depths (Tjalsma and Lohmann, 1983)in some areas migrates upslope to depths as shallow as 1,000 meters during glacial episodes and/or where an overturn displaces bottom waters and associated conditions in the sediment, to shallower depths (Boersma, 1985a, b, c). Nevertheless,there are some Eocene/Oligocene species with restricted depth ranges in the open ocean and these are shown in Table 1.

Table 1

Paleodepth Range	**Species**	Site/Reference
no>than 1,000 m	**Bulimina sculptilis, B. jacksonensis(<600 m)**	J2
	large costate uvigerinids, rectuvigerinids	(1)(2)
	Cibicidoides truncanus, C. parki, Cibicides spp.	(2)(3)
	large, ornamented lenticulinids and nodosarids	J2, 526593
	Rotaliatina, Palmula, Amphycornya, Dyocibicides spp.	(1)
abyssal	**Abyssammina spp., Glomospira spp.**	522,528,550
no > than lower	**Cibicidoides mexicanus, ornamented cibicidids**	116,549,593
bathyal	**Bulimina alazanensis, B. macilenta,**	
	B. glomarchallengeri	(1)(4)
	strongly evolute rotaloid species	(2)(4)

Table 1: depth indications of selected late Eocene/Oligocene benthic foraminifera in the Atlantic Ocean. References: (1)Boersma, 1984a; (2)Boersma, 1985b; (3)Tjalsma, 1983; (4) Tjalsma and Lohman, 1983.

Paleoenvironmental Index Species

Determination of most Eocene/Oligocene paleoenvironmental index species has been based on analogy with the environmental preferences of that taxon or a related group in the Recent or Pleistocene. Indices for various environmental variables currently being appraised and/or applied are shown in Table 2.

Table 2

Paleoenvironmental Feature	Species	Reference
high salinity	**Trifarina spp.**	(1)
cool, corrosive, high nutrient, oxygenated bottom water	**Nuttallides umbonifera**	(2)
low oxygen environment	**Bolivina spp.**	(3)
oxygen not less than 0.2 ml/l	**invertebrate activity**	(3)
high residual organic carbon in sediments	**Uvigerina spp.**	(3)
Tethyan ecophenotypes found in Atlantic sites	**Uvigerina bortotara, Cibicidoides parki**	(4)
	Heronallenia, Siphotextularia aurora	(5)(6)
	Brizalina reticulata, Rectuvigerina prisca	(4)
very low oxygen, corrosive sediment	**Gyroidina spp., Glomospira**	(7)
poor ventilation, non-corrosive	**Gyroidina spp. with Hoeglundina**	(7)
'old',nutrient rich watermass	**Uvigerina,Planulina,miliolids,Hoeglundina**	(2)
South Atlantic/ equatorial intermediate water	**Uvigerina spinulosa**	(8)
North Atlantic/ Gulf of Mexico intermediate water	**Uvigerina mexicana**	(8)
high nutrient, low oxygen around outfalls	**Bulimina, Bolivina spp.**	(9)
most tolerant of low oxygen	**Bolivina spp.**	(10)
low ph mineral, kaolinite	**Articulina, other miliolids**	(7)
terrigenous, not marine, organic matter	**cibicidids replace eponidids**	(11)
low sedimentation rate, corrosive	**Glomospira charoides**	(11)
oxygen minimum, not organic carbon maximum	**Brizalina aenariensis**	(12)
high rates of sediment accumulation	**Uvigerina spp**	(4)

Table 2: Eocene/Oligocene paleoenvironmental indications of benthic foraminifera primarily based on analogy to Recent distribution patterns. References: (1) Oberhansli, 1962; (2)Bremer and Lohman, 1982; (3)Douglas, 1982; (4)Boersma, 1985b; (5)Tjalsma, 1983; (6)Proto Decima and Bolli, 1977; (7)Cita and Zocchi, 1978; (8)Boersma, 1984b; (9)Seiglie, 1959; (10) Boltovskoy and Wright, 1976; (11) Poag, 1982; (12) Saunders et al., 1985).

Eocene/Oligocene Benthic Paleo-Oceanography

Determination of these index groups from a large number of open Atlantic ocean sections in many of which the stable isotopes also have been analyzed (Proto Decima and Bolli, 1977; Boersma, 1977; Schnitker, 1979; Corliss, 1981; Tjalsma and Lohmann, 1983; Tjalsma, 1983; Clark and Wright, 1984; Boersma, 1984a, 1985a,b,c in press; Corliss et al., 1984; Corliss and Keigwin, 1985 in press) suggests the following scenario in the open ocean from the latest Eocene (Zone P16/17) into the early Oligocene (Zone P17) and during the earliest Oligocene benthic oxygen isotope excursion (lower Anomaly 13 time, Zone P18) currently interpreted as indicating a glacial (Mathews and Poore, 1980).

Latest Eocene

Benthic faunas and sediment indicate extensive dissolution of Atlantic Ocean carbonates at depths from over 3,000 (Sites 550, 528, 522) to close to 1,800 (Site 357) meters paleodepth. In the abyssal eastern and western South Atlantic (Sites 522, 528, V20-220) faunas are low in diversity and numbers of benthics, very similar in species content, have index species which suggest low oxygen conditions and intense carbonate dissolution, and include very large sized, highly ornamented, redeposited shelf/slope benthics in pulses. Such faunas and sediment have been interrpeted to indicate the presence of a cool, corrosive (CO_2-rich) Antarctic -derived bottom water type. The redeposited shelf-slope species suggest the presence of vigorous mid ocean intermediate depth current activity and sediment undermining at slope depths along the Walvis Ridge and Rio Grande Rise. Abyssal faunas of the northeastern Atlantic in the Bay of Biscay(Site 550, V27-134) are strongly dissolved and are similar to those of the southeast Atlantic, but lack the redeposited slope benthics.

Lower bathyal faunas are much higher in diversity, benthics are larger sized and more numerous. At these depths in the southeast Atlantic(Site 529) dissolution is moderate, but mixing of sediment is intense. Hiatuses and extreme dissolution, accompanied by index species for corrosive bottom conditions, span most of the latest Eocene. At intermediate depths in the eastern South Atlantic(Sites 359, 526) faunas are rich in large sized benthics, include large numbers of high organic carbon index species, demonstrate some dissolution, but are terminated by a hiatus spanning the remainder of the Eocene. In the western South Atlantic, upper middle bathyal depth sites (Site 516,< 1,000 meters) are somewhat dissolved, but are considered continuous across the boundary. Faunas are rich in intermediate depth species typical of the Gulf of Mexico and Caribbean, but lack the high organic carbon index species. Site 516 faunas suggest that surface production and/or preservation of organic carbon at the bottom near from 500-1,000 meters paleodepth was greater, and subsequent current activity or dissolution was more vigorous in the southeastern than in the southwestern Atlantic, where dissolution and/or erosion was more effective at greater depths, closer to 2,000 meters. That is, in both basins dissolution and/or erosion were greater at the depths where southern-derived currents were flowing towards the equator. The eastern side of the ocean was also cooler through most of the Eocene (Shackleton and Boersma, 1981).

The carbonate record of the latest Eocene is missing in most of the slope to abyssal depth sites of the northwest and northernmost Atlantic (Tucholke and Vogt, 1979). At intermediate depths in De Soto Canyon in the northeastern Gulf of Mexico(E67-128) faunas include high abundances of radiolarians, sponge spicules, and high organic carbon benthic foraminiferal index species all of which suggest high surface water productivity and residual organic carbon. Pulses of low oxygen species which arrive just before the extinction of Hantkenina accompany a decrease in the high organic carbon species and in both the radiolarians and the sponge spicules. These faunal changes suggest either the development of an oxygen minimum situation in the canyon and/or oxidation of the organic carbon accompanying a slight decrease in surface water productivity at the very end of the Eocene. Moderate amounts of canyon channelling of shallow water species and mixing of sediment are also evident.

In the Bay of Biscay intermediate depth sites(Site 548) do not contain high abundances of high organic carbon species, but foraminifera are badly corroded and fragmented and very large numbers of invertebrate fragments are present. This was apparently an area of active predation, suggesting at least moderate levels of dissolved oxygen at the bottom. The presence of benthic morphotypes typical of the Tethys and of the Mediterranean slope basins suggests a lingering influence of the Mediterranean (Tethyan) outflow in Biscay near the end of the Eocene. Prior to the extinction of Hantkenina, however, the influx of a new intermediate depth index more typical of the equatorial Atlantic and eastern Caribbean, suggests that waters impinging at intermediate depths in Biscay may no longer be formed from the Tethyan outflow, but to the south in the equatorial convergence zone.

Earliest Oligocene

Following the extinctions of *Hantkenina* and *Turborotalia cerroazulensis cunialensis* dissolution becomes even more extreme at most sites, removing a large proportion of the coarse fraction fossils. Few sites contain a continuous record through this interval. Residues contain intermediate sized benthic and planktonic foraminifera, and large amounts of very small planktonic species. Benthics are smaller sized and less abundant at all but intermediate depths. High organic carbon index species decrease in abundance or disappear altogether from most open ocean sites except in areas of high sedimentation rates(Sites 277, 593) and/or high surface production(E67-128) as indicated by the planktonic siliceous fossils. High organic carbon index species occur in abundance only in slope and outer shelf desposits (J2, Vicksburg faunas) of this age.

At abyssal sites (522, V20-220) most carbonate is dissolved, the redeposited slope benthics become rare, benthics are smaller sized, rarer, and species tolerant of very low levels of oxygen or indicative of corrosion increase in abundance. Because these species are also dissolution resistant, interpretation of their increase is not clear.

At intermediate depths dissolution also increases. In the Gulf of Mexico(E67-128) redeposition decreases, the high organic carbon indices and the amounts of siliceous fossils decrease in abundance. The coarse fractions are noticeably dissolved, leaving mostly intermediate and small sized fossils. The index for corrosive conditions at the bottom, usually found at abyssal and lower bathyal depths, occurs for the first time at intermediate depths near 1,000 meters in the Gulf of Mexico (and in the high middle latitudes of the Tasman Sea). These faunas suggest a decrease in productivity and a shallowing of the carbonate lysocline and/or of watermasses rich in CO_2 to shallow depths, from 600-1,000 meters, just above the Eocene/Oligocene boundary.

Early Oligocene Glacial

Although the oxygen isotopic excursion signalling glacial conditions has been measured at a large number of sites, the earliest Oligocene is most often missing and the excursion simply tops the hiatus in most areas. At all sites containing this interval, preservation improves. Abyssal sediments of the southeastern and southwestern Atlantic(Sites 522, V20-220) regain coarse fraction fossils, size and diversity of benthics increase, and the corrosion and low oxygen index species are joined by a new faunal element which appears to have migrated downdepth from lower bathyal areas of the mid ocean ridges. A similar downdepth faunal migration occurred in the Bay of Biscay(Sites 400,401) at this time. Thus faunas and sediment both appear to reflect the drop in the CCD (Van Andel, 1975; Melguen, 1977) in both the North and South Atlantics with the development of glacial conditions.

At several lower and middle bathyal sites(Sites 529, 363) high organic carbon indices regain their import and suggest increased delivery and/or preservation of organic carbon at greater oceanic depths. Siliceous fossils increase again in the Gulf of Mexico De Soto Canyon where, however, with the development of the glacial, the sponge spicules are more common and suggest heightened productivity at the bottom rather than at the surface. Because the slight sea level drop at this time could be related to the glacial episode (Hardenbol, personal communication, 1985) , delivery of nutrients off the shelf may have stimulated production at the canyon bottom.

Summary

From the latest Eocene to the early Oligocene Atlantic Ocean benthic faunas indicate the gradual updepth expansion of cool, corrosive conditions from the abyssal areas to the lower bathyal zone, and finally into intermediate depths. Accompanying this expansion was a decrease in sediment erosion and redeposition to abyssal areas and an apparent decrease in the Mediterranean outflow. Delivery and/or preservation of residual organic carbon in the sediments also decreased through the latest Eocene, first at lower bathyal then at mid bathyal depths. By early Oligocene time high organic carbon indices were localized on the slopes and shelves. With the development of the early Oligocene glacial episode in Anomaly 13 time, the CCD dropped, sediment preservation improved throughout depth and most faunas moved downslope. Oceanic fertility increased with renewed upwelling in boundary currents and delivery of nutrients offshore, and higher levels of residual organic carbon again characterized the shelf/slope to the lower bathyal zone.

These fertility, dissolution, and circulation patterns represent the transition from a vigorously circulating, cooling ocean absorbing more CO_2 and thermally homogenizing from the bottom updepth with the decrease of the Mediterranean source of intermediate and/or deep water formation. Following this homogenization process, decreasing fertility resulted from decreased upwelling, the lock-up of nutrients onshore, and possibly from the alkalinity of surface waters. Accompanying the subsequent glacial, the drop in the CCD, and enhanced oceanic fertility, South and/or equatorial Atlantic source waters were compartmentalized in the South Atlantic and eastern North Atlantic; North Atlantic source water was compartmentalized in the western basin and produced the ubiquitous erosion surfaces of the far northern and western North Atlantic.

References

Beckman, J., H. Bolli, K. Perch Nielsen, F. Proto Decima, J. Saunders, and M. Toumarkine. 1981. Major calcareous nannofossil and foraminiferal events betwen the middle Eocene and early Miocene. Paleogeogr., Paleoclim., Paleoecol. 36:155-190.

Berggren, W. and J. Aubert. Late Paleogene (Late Eocene and Oligocene) benthonic foraminiferal biostratigraphy and paleobathymetry of Rockall Bank and Hatton-Rockall Basin. Micropaleo. 22: 307-326.

Boersma, A. 1977. Eocene to early Miocene benthic foraminifera, DSDP Leg 39, South Atlantic. Init. Repts. DSDP 39: 643-656.

Boersma, A. 1984a. Oligocene and other Tertiary benthic foraminifera from a depth traverse down Walvis Ridge, Deep Sea Drilling Project Leg 74. Init. Repts. DSDP 75:1273-1300.

Boersma, A. 1984b. <u>Handbook of Common Tertiary Uvigerina</u>. Microclimates Press, N. Y., 255pp.

Boersma, A. 1985a. Oligocene benthic foraminifera from North Atlantic Ocean sites 558 and 563; benthic foraminifera as water mass indices in the North and South Atlantic. Init. Repts. DSDP 82: 575-600.

Boersma, A. 1985b in press. Atlantic Ocean paleo-oceanography through the Eocene/Oligocene boundary interval based on benthic foraminifera. Paleogeogr. Paleobiol. Paleoecol.

Boersma, A. and N. Shackleton. 1977a. Tetiary oxygen and carbon isotope stratigraphy, DSDP Site 357 (southwestern Atlantic). Init. Repts. DSDP 39: 911-924.

Boersma, A. and N. Shackleton. 1977b. Oxygen and carbon isotope record through the Oligocene, DSDP Site 366, equatorial Atlantic. Init. Repts. DSDP 41: 957-962.

Boltovskoy, E. 1980. On the benthonic bathyal-abyssal foraminifera as stratigraphic guide fossils. Jour. For. Res.10:163-172.

Boltovskoy, E. 1981. Foraminiferos benthonicos del sitio 360 del Deep Sea Drilling Project (Eoceno medio-Plioceno inferior). Assoc. Geologica Argentina XXXVI: 389-423.

Bremer, M. and P. Lohmann. 1982. Evidence for primary control of the distribution of certain Atlantic Ocean benthonic foraminifera by degree of carbonate saturation. Deep Sea Res. 29: 987-998.

Cita, M. and M. Zocchi. 1978. Distribution patterns of benthic foraminifera on the floor of the Mediterranean Sea. Oceanolog. Acta 1: 445-462.

Clark, M. and R. Wright. 1983 Paleogene abyssal foraminifera from the Cape and Angola Basins, South Atlantic. Init. Repts. DSDP 73: 459-480.

Corliss, B. 1979. Response of deep-sea benthonic foraminifera to development of the pyschrosphere near the Eocene/Oligocene boundary. Nature 282: 63-65.

Corliss, B. 1981. Deep-sea benthonic foraminiferal faunal turnover near the Eocene/Oligocene boundary. Mar. Micropal. 6: 367-384.

Corliss, B. 1982. Linkage of North Atlantic and Southern Ocean deep-water circulation during glacial intervals. Nature 298: 458-460.

Corliss, B., M. Aubry, W. Berggren, J. Fenner, L. Keigwin, and G. Keller. 1984. Eocene/Oligocene boundary events in the deep sea. Science 226: 806-810.

Corliss, B. and L. Keigwin. 1985 in press. Eocene/Oligocene paleooceanography. AGU Review volume.

Douglas, R. 1973. Benthonic foraminiferal biostratigraphy in the central North Pacific, Leg 17; Deep Sea Drilling Project. Init. Repts. DSDP 17: 591-605.

Douglas, R. and H. Heitman. 1979. Slope and basin benthic foraminifera of the California borderland. SEPM Spec. Publ. 27:231-246.

Douglas, R. and F. Woodruff. 1981. Deep Sea Benthic Foraminifera. in Emiliani, ed., The Sea, v.7, Wiley Interscience Pubs., pp.1233-1327.

Gernant, R. and R. Kessling. 1966. Foraminiferal paleoecology and paleoenvironmental reconstruction of the Oligocene Middle Frio in Chambers County, Texas. Trans. Gulf Coast Assoc. of Geol. Soc. XVI: 131-158.

Gradstein, F. and W. Berggren. 1981. Flysch-type agglutinated Foraminifera and the Maestrichtian-Paleogene history of the Labrador and North seas. Mar. Micropaleo. 6: 211-268.

Keigwin, L. 1980. Paleooceanographic change in the Pacific at the Eocene-Oligocene boundary. Nature 287: 722-725.

Keigwin, L. and B. Corliss. 1985 in press. Stable isotopes in Eocene/Oligocene foraminifera. Geol Soc. Am. Bulletin

Melguen, M. 1978. Facies evolution, carbonate dissolution cycles in sediments from the eastern South Atlantic (DSDP Leg 40) since the early Cretaceous. Init. Repts. DSDP 40: 981-1025.

Miller, K. and W. Curry. 1982. Eocene to Oligocene benthic foraminiferal isotopic records in the Bay of Biscay. Nature 296: 347-352.

Monaco, P., M. Nocchi, and G. Parisi. 1984. Preliminary bio-and lithostratigraphy of the middle Eocene-lower Oligocene southern Umbrian sedimentary series. Int. Geol. Correl. Proj. Excursion (Gubbio-Visso-Cascia) Program:1-6.

Oberhansli, H., J. McKenzie, M. Toumarkine. 1983a. A paleoclimatic and paleooceanographic record of the Paleogene in the central South Atlantic (Leg 73, Sites 522, 523, 524). Init. Repts. DSDP 73: 737-747.

Oberhansli, H., A. Grunig, and R. Herb. 1983b. Oxygen and carbon isotope study in the late Eocene sediments of Possagno (northern Italy). Rev. It. Paleo. Strat. 89:377-394.

Poore, R. and R. Mathews. 1984a. Late Eocene/Oligocene oxygen and carbon isotope record from South Atlantic Ocean, Deep Sea Drilling Project Site 522. Init. Repts. DSDP 73: 725-735.

Poore, R. and R. Mathews. 1984b. Oxygen isotope ranking of late Eocene and Oligocene plantkonic foraminifera; implications for Oligocene sea surface temperatures and global ice volume. Mar Micropaleo. 9: 111-134.

Proto Decima, F. and H. Bolli. 1978. Southeast Atlantic DSDP Leg 40 Paleogene benthonic foraminifera. Init. Repts. DSDP 40: 571-634.

Schnitker, D. 1979. Cenozoic deep water benthic foraminifera, Bay of Biscay. Init. Repts. DSDP 48: 377-413.

Shackleton, N., M. Hall, and A. Boersma. 1984. Oxygen and carbon isotope data from leg 74 foraminifera. Init. Repts. DSDP 74: 599-612.

Strakov, K. 1979. La stratigraphic, paleoecologie, paleogeographie et les foraminiferes de l'Oligocene du nord-est de la Hongrie. Cahiers de Micropaleo. 3: 7-97.

Tjalsma, L. 1983. Eocene to Miocene benthic foraminifera from Deep Sea Drilling Project Site 516, Rio Grande Rise, South Atlantic. Init. Repts. DSDP 72: 731-755.

Tjalsma, L. and P. Lohmann. 1983. Paleocene-Eocene bathyal and abyssal benthic foraminfiera from the Atlantic Ocean. Micropaleo. Sp. Pub. 4.

Tucholke, B. and P. Vogt. 1979. Western North Atlantic; sedimentary evolution and aspects of the tectonic history. Init. Repts. DSDP 43:795-863.

Van Andel, T. 1975. Mexozoic/Cenozoic calcite compensation depth and the global distribution of calcareous sediments. Earth Plan Sci Lett. 26: 187-195.

Vernaud-Grazzini, C., C. Muller, C. Pierre, R. Letolle, and J. Peypouquet. 1979. Stable isotopes and Tertiary paleontological paleooceanography in the northeast Atlantic. Init. Repts. DSDP 48:475-491.

LARGER FORAMINIFERA AND EVENTS AT THE EOCENE/OLIGOCENE BOUNDARY IN THE INDO-WEST PACIFIC REGION

C. GEOFFREY ADAMS, Department of Palaeontology, British Museum (Natural History), Cromwell Road, London SW7 5BD, U.K.

JACQUES BUTTERLIN, Institut de Physique du Globe de Paris, Université Pierre et Marie Curie, 4 Place Jussieu, 75230 Paris, Cedex 05, France.

BIMAL K. SAMANTA, Department of Geology, University of Calcutta, 35 Ballygunge Circular Road, Calcutta 0900017, India.

INTRODUCTION

The changes undergone by the larger foraminifera at the Eocene/Oligocene boundary have long been known to be of considerable stratigraphical significance. Yet whether they represent purely palaeontological phenomena or signify a major geological event which had a marked - even catastrophic - effect on the faunas, has never been determined.

Since Tertiary larger foraminifera are used to date shallow-water carbonates, critical sequences of Late Eocene limestones in the Indo-West Pacific region are here reviewed with the object of establishing the nature of the faunal change (sudden or gradual) at the Eocene/Oligocene boundary and of linking it with any geological features of regional importance which may be apparent from the successions themselves. The evidence for mass extinction is considered in relation to possible physical causes, i.e., eustatic movements such as those postulated by Vail et al. (1977) and temperature variations deduced by Shackleton (1984). Shallow-water carbonates lend themselves to investigations of this kind since they (and their foraminiferal faunas) are particularly sensitive to physical changes in the environment.

SUCCESSIONS AND FAUNAS

Eocene and Oligocene limestones are numerous in the Indo-West Pacific region, and a great deal is known about their faunas. There is, however, very little published information on the boundary itself, and only at Melinau, Sarawak, have Eocene and Oligocene limestones been found at outcrop in an apparently

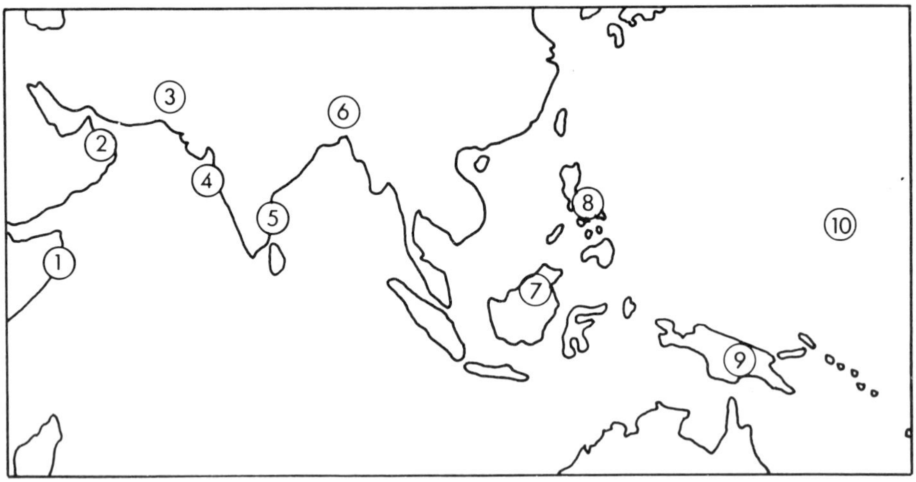

Fig. 1. Geographical location of stratigraphical sequences shown in Fig. 2.

unbroken sequence. Seemingly continuous successions across the boundary occur elsewhere, e.g. Meghalaya and in the deeper parts of the Bengal and Cambay basins, but these are not wholly developed in carbonate facies. An exception may be the complete, but very poorly fossiliferous, carbonate succession thought to straddle the boundary in the eastern part of the Bombay Offshore Basin.

The distribution of the principal carbonate successions known to span the Eocene/Oligocene boundary is shown in Fig. 1. There are, of course, others, but these either duplicate the information by the primary succession or are too poorly described to be considered here. The successions are correlated in Fig. 2, notes and references being provided in the Appendix. Age determinations are not necessarily those of the original authors, but are based on the scheme outlined below (Adams, 1970), which was itself derived from that proposed by van der Vlerk (1955).

Mid Oligocene. Association of *Lepidocyclina (Eulepidina* or *Nephrolepidina)* spp with *Nummulites fichteli* and/or *N. vascus*.

Early Oligocene. Occurrence of *N. fichteli* and/or *N. vascus* (or other small striate nummulites) with or without *Borelis pygmaeus* and *B. inflata*. *Lepidocyclina* absent.

Late Eocene. Occurrence of *Biplanispira* spp, *Pellatispira* spp, reticulate nummulites of the N. fabianii/retiatus group, striate nummulites such as N. chavannesi and N. pengaronensis, Spiroclypeus abdopustula and S. vermicularis, with or without Asterocyclina, Discocyclina, Fabiania and other genera indicating a general pre-Oligocene age.

Other taxa may also be used, but the presence of one or more of the above is usually considered to be essential for the recognition of each subepoch. This scheme, with few modifications, can be applied in the Mediterranean region, but in the Americas different criteria have to be employed. See Fig. 4.

The principal additions to knowledge of Tertiary carbonate successions and faunas since the last regional review (Adams, 1970) have come from oil and gas exploration around the coasts of India and from surface exposures in Meghalaya and Sarawak. This new information may be summarized as follows:

Mohan and Kumar (1985) have shown that only one supposedly complete succession of shallow-water carbonates is known to span the Eocene/Oligocene boundary in offshore areas of India, although continuous sequences certainly exist in deeper-water marls and clays. Throughout the Indian region (see "Notes on successions"), shallow-water carbonates are largely confined to zones P15 and P16. Larger foraminifera characteristic of Zone P17 are known with certainty only from N.E. India and, by inference, from Sarawak and the Marshall Isles. Elsewhere this zone is represented by a disconformity which in most areas extends down at least into Zone P16.

A review of the numerous samples colleced across the Eocene/Oligocene boundary in the Melinau Limestone (north faces of the Melinau and Medalam gorges), Sarawak, based on material described or referred to by Adams (1965, 1970) and on additional samples recently collected by the Malaysian Geological Survey, has confirmed that a thick sequence of algal-rich limestones separates the last age-diagnostic Eocene and first datable Oligocene faunas. The Early Oligocene faunas are unusual only in respect of the presence of Fallotella melinauensis (Adams), a species known also from the Bukit Sarang Limestone, some 240 km to the S.W. (Adams, 1964). In other respects the assemblages are similar to those found in the Oligocene part of the Kirthar Limestone, Gaj River, Pakistan (Khan, 1967).

The uppermost part of the Eocene at Melinau is developed in virtually a single carbonate facies and carries a restricted foraminiferal fauna. Only five age-diagnostic taxa have been found

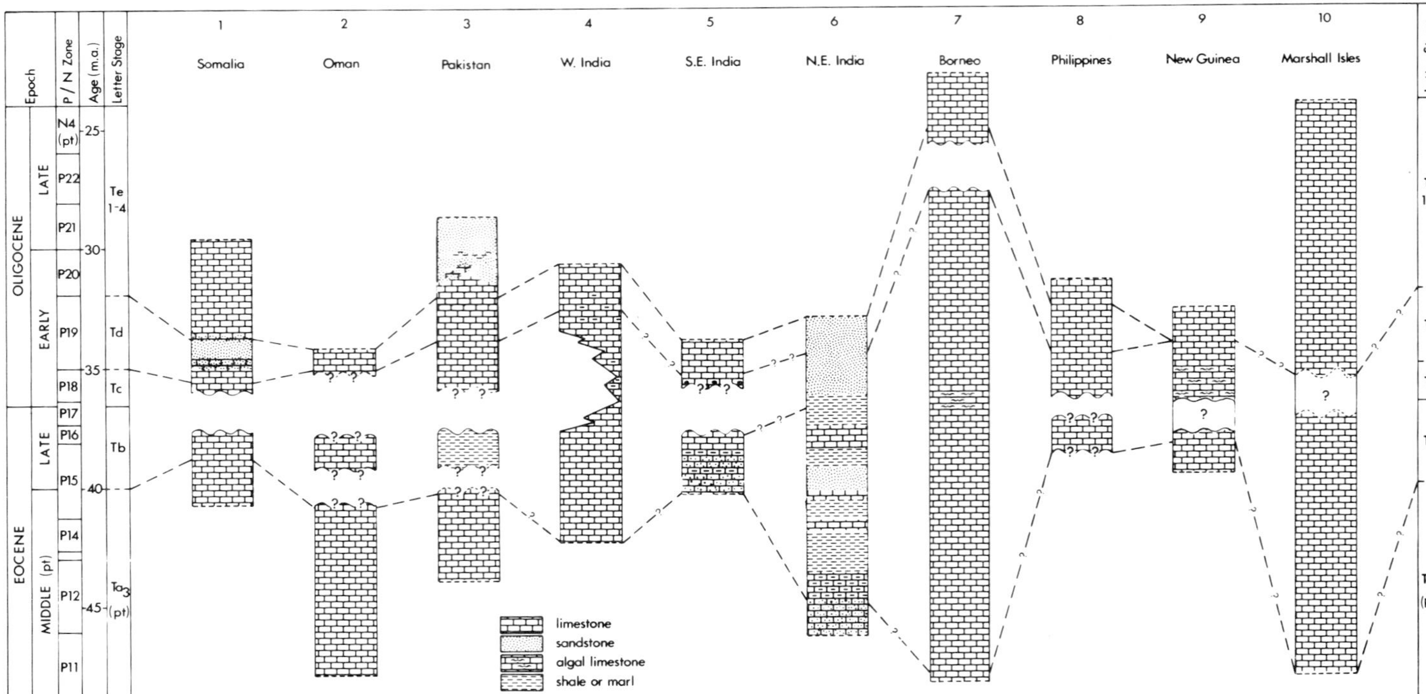

Fig. 2. Representative shallow-water carbonate successions spanning the Eocene/Oligocene boundary in the Indo-West Pacific region. Sections other than 7 and 9 are composite, schematic, and not exactly to scale. Section 6, although not wholly carbonate, is included because it has yielded important Eocene and Oligocene assemblages. Note widespread hiatus at boundary. Other disconformities not shown except in 7. Sources: 1. Azzaroli (1952, 1958); 2. Montenat et al. (1977) and unpublished data; 3. Khan (1967), Williams (1959) and unpublished data; 4. Western India, including Bombay Offshore, Mohan et al. (1982), Mohan and Kumar (1985), Pandey (1982), and Mohan (pers. comm.); 5. Mohan and Kumar (1985); 6. Meghalaya, Nagappa (1959), Samanta and Raychaudhari (1983), Mohan and Kumar (1985), Mohan (pers. comm.); 7. Melinau Limestone (Adams, 1965, 1970); 8. Hashimoto et al. (1977) and unpublished data; 9. Chimbu Limestone, Bain and Binnekamp (1973); 10. Eniwetok, Cole (1958a).

in the highest beds, and these cannot be regarded as fully representative of the region as a whole. It is nevertheless clear that Discocyclina, Fabiania saipanensis, Pellatispira, Spiroclypeus vermicularis and at least one species of Nummulites, disappeared within a short space of time, and that the succeeding Oligocene fauna appeared gradually - apparently as new elements were introduced into the area.

A survey of shelf carbonates in other parts of the region has revealed that Eocene limestones are always terminated by disconformities, unconformities, marked lithological changes, or faunal evidence for shallowing (Fig. 2). Only at Melinau and perhaps Chimbu, New Guinea, is there any strong indication that deposition might have been continuous across the boundary, and in both areas the limestones above the youngest datable Eocene are algal. There may, of course, be places in Kalimantan and Irian Jaya where continuous carbonate sequences include the boundary, but if so they have not yet been adequately described. There are certainly a few areas, e.g., southern Java (Baumann et al., 1972) and off Tarapur, western India (Mohan et al., 1982; Mohan and Kumar, 1985) where shallow-water Eocene limestones are succeeded by deeper water marls, but they have never been adequately described or dated, and their significance remains uncertain. Mohan and Kumar (1985) mention a miliolid dominant zone, possibly in the basal Oligocene part of Well G, Dahanu Depression, off Tarapur.

Collation of all the published evidence suggests that the ranges of age-diagnostic Late Eocene and Early Oligocene taxa are as shown on Fig. 3. A comparison of this figure with the range chart for the Americas (Fig. 4), shows that although the faunal change was apparently equally dramatic in both regions, different taxa were involved. It should be noted that the faunal changes at about the end of P16 times were also considerable.

A regional faunal discontinuity can be demonstrated for the Americas and an actual physical break shown to occur. Olsson et al. (1980) have demonstrated the existence of an hiatus in sediments extending along the eastern seaboard from Long Island to

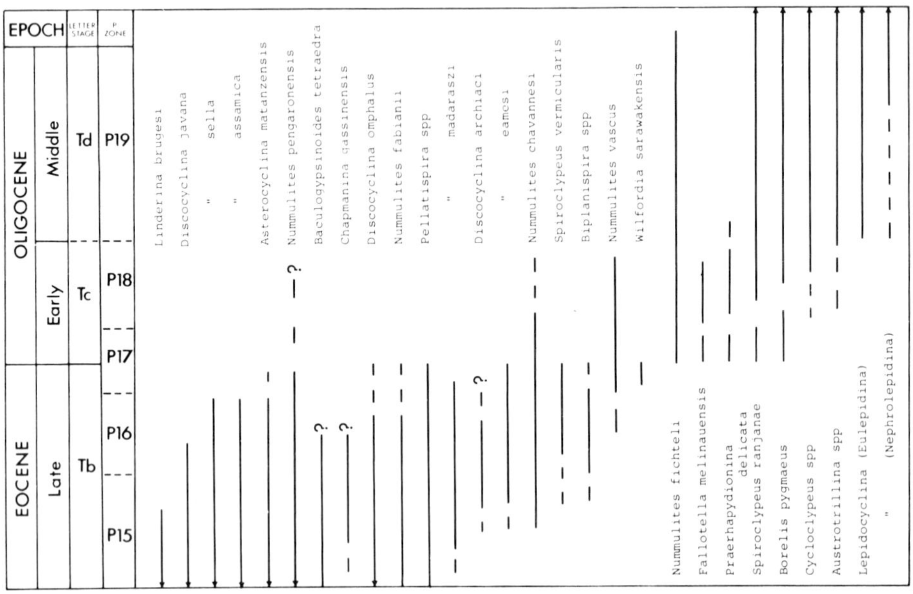

Fig. 3. Chart showing the ranges of some age-diagnostic taxa in the Indo-West Pacific region. The scale of the faunal discontinuity is evident. Solid lines mark known ranges; broken lines indicate uncertainty. (CGA & BKS)

Carolina, and Hunter (1978) has shown that a similar break exists throughout the southern Caribbean. The faunal discontinuity indicates that this hiatus must also occur throughout the southern part of the United States.

No continuously deposited succession of shallow-water carbonates across the Eocene/Oligocene boundary has been described from the Middle East, although Bizon et al. (1972: Figs 1, 2) indicated that such a sequence might exist to the N.N.W. of Estehbanat, Iran. Similar sequences have not, however, been reported from other parts of this area.

The only place in the Mediterranean region where deposition seems to have been continuous across the Eocene/Oligocene boundary is Priabona in northern Italy. Here, Eocene limestones with larger foraminifera and bryozoa pass upwards into algal limestones within which the typical Oligocene fauna gradually appears. See Barbin and Bignot (this volume).

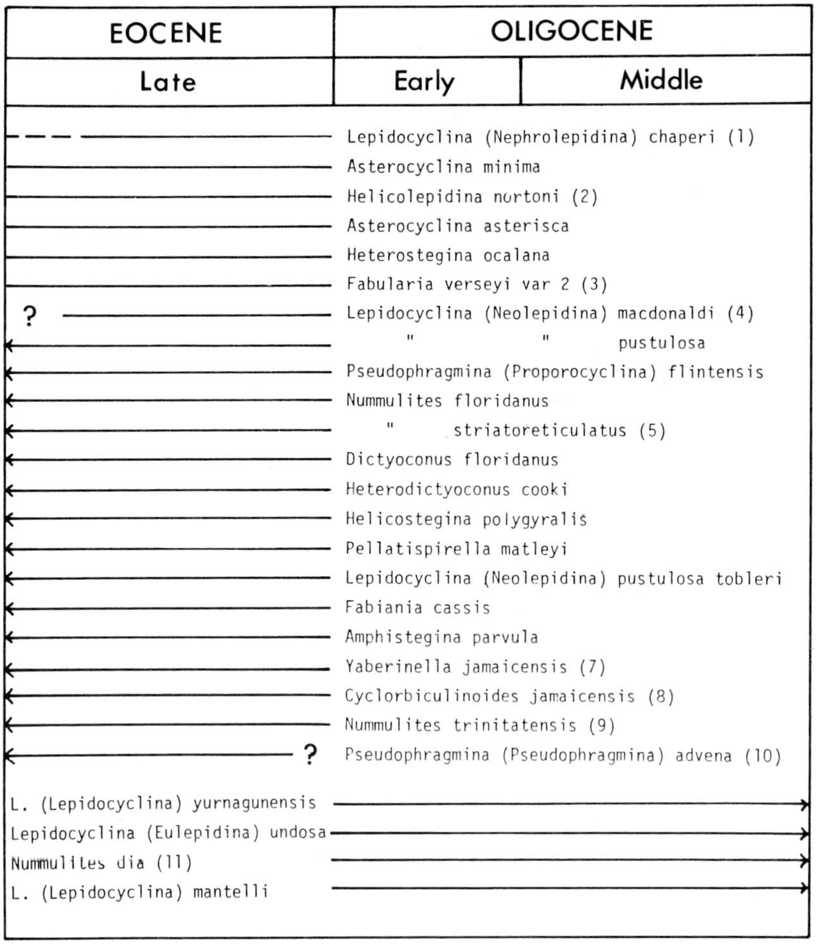

Fig. 4. Ranges of some age-diagnostic larger foraminifera in the Americas.

This arrangement of species reflects (a) the probability that they became extinct at the end of Eocene times, and (b) their known occurrences in the Upper Eocene. It should be noted that L. (L.) mantelli is relatively rare in Lower Oligocene strata.

(1) There is some doubt about the first appearance level of L. (N.) chaperi. According to Robinson (1974, 1977) Heterodictyoconus americanus disappeared a little above the first appearances of some variants of Fabularia verseyi which he places at the top of the Middle Eocene (beginning of Zone 4). For Robinson the first appearance of L. (N.) chaperi is much higher in Jamaica although McFarlane (1977) indicates an "overlap of the stratigraphical ranges" of H. americanus and L. chaperi, and Butterlin (unpublished) has confirmed this association in Haiti. Moreover, Woodring (1957: 17-20) reported L. (N.) chaperi with Globigerinoides mexicanus and G. ouachitensis in Panama, an association which characterizes the zone of Globigerinatheka semiinvoluta (P15-P16 pars).

(2) Cole (1962) placed *Helicolepidina nortoni* in synonymy with *H. spiralis*, a view not shared by Butterlin (1981).
(3) *Fabularia verseyi* has been reported only from Jamaica.
(4) Certain authors (Robinson, 1968 - Jamaica; Pecheux, 1984 - Mexico; Andreieff, pers. comm. - Haiti) consider that *L. (Np.) macdonaldi* first appeared in the upper Middle Eocene (Zone P13 or P14).
(5) Pecheux (1984) regarded *N. striatoreticulatus* as a junior synonym of *N. willcoxi*, but they are frequently found in association and are very difficult to distinguish.
(6) *Helicocyclina paucispira* was considered a junior synonym of *Helicostegina polyralis* by Cole (1960b), but this has been contested by some later authors, e.g. Caudri (1975).
(7) *Yaberinella jamaicensis* is known only from Jamaica, Costa Rica (Butterlin, unpublished) and Panama. Robinson (1968: 192) considered this species (and its synonym, *Y. trelawnyensis* Yaughan) to occur only in the Middle Eocene of Jamaica, but it has also been found in the Upper Eocene of Panama (Cole, 1952) and, together with *L. (Np.) macdonaldi*, in Costa Rica (Butterlin, unpublished).
(8) *C. jamaicensis* has been recorded only from Jamaica.
(9) If Cole (1958b) was correct in regarding *N. kugleri* as a synonym of *N. trinitatensis*, it ranges to the top of the Eocene in Trinidad and Haiti (Butterlin, unpublished), but not in Florida (Cole, 1945).
(10) *P. (P.) advena* poses a problem of definition owing to synonymies proposed by Cole (1969) [cf. Butterlin, 1967]. Its stratigraphical distribution is to this extent uncertain. Cole (1969) thought that it had the same range as *P. (P.) flintensis* (P12-P17).
(11) Some authors (e.g., Pecheux, 1984 and Andreieff, 1985) consider *N. dia* to be a synonym of *N. panamensis*. (JB)

POSSIBLE EVENTS

The major change which everywhere affected larger foraminifera at or towards the end of Eocene times seems to have been marked by a distinct depositional hiatus in most areas of shallow-water carbonate sedimentation. The cause could have been a global fall in sea level which exposed many inner shelf carbonates to erosion. Only in those places where local subsidence kept pace with, or exceeded, the eustatic fall, could limestones have been deposited continuously across the boundary. Vail et al. (1977) postulated a terminal Eocene fall on seismic evidence, and tentatively estimated its order as 35 m. This would scarcely have been enough to have had a catastrophic effect on the foraminiferal faunas. On the other hand, a fall of 50-70 m would have exposed a sufficient area of shelf carbonates to produce a global disconformity and to alter their faunal composition.

Severe climatic deterioration would also produce a marked faunal change, and some authorities believe that this occurred during the Early Oligocene. Shackleton (1984) has graphed the calculated temperature values for Tertiary surface waters at two points in low and mid latitudes, but his overall figures, although in keeping with most earlier estimates, are rather low. His postulated average temperature of around 18°C for low latitude surface waters would not maintain the present-day distribution of tropical larger foraminifera, much less would it account for the expansion observed at times during the Tertiary. Most living larger foraminifera need mean annual surface temperatures in excess of 18°C to support reproduction (Hornibrook, 1968; Murray, 1973), and it is noteworthy that Tertiary forms achieved their maximum latitudinal distribution (Alaska to south of New Zealand) in Mid to Late Eocene times when temperatures were presumably higher than they are today. It is difficult to see how they could have been some 10°C lower as suggested by Shackleton.

Cooling of the oceans would be unlikely to produce sudden extinctions on a large scale. Rather would they tend to be progressive, extending gradually from high to low latitudes and affecting first one taxon then another. In this connexion, it is worth noting that the general cooling trend postulated by Shackleton (1984) for the greater part of the Oligocene appears to be inconsistent with the increasing diversity seen amongst the larger foraminifera of Early to Mid Oligocene times.

Two tentative conclusions may be drawn from this review. First, that extinctions affecting the larger foraminifera at the Eocene/Oligocene boundary constituted a sudden event the world over, although the number of taxa involved is difficult to determine owing to the paucity of continuously deposited carbonate successions across the boundary. Second, that the extinctions could have been caused by a global fall in sea level which produced the depositional hiatus in shallow-water carbonate sediments now traceable throughout the circumtropical region. The eustatic fall might have been associated with climatic deterioration and cooling of the surface waters of the oceans.

The possibility that the faunal change observed at the Eocene/Oligocene boundary may be an artifact resulting from our limited knowledge of Zone P17 faunas has to be recognized. But the existence of an hiatus in all but a few of the known carbonate sequences in the Indo-West Pacific region, together with the similar changes seen in the American (Fig. 4) and Mediterranean (Barbin and Bignot, this vol.) successions and faunas, strongly suggests the effect of a geological event with world-wide repercussions.

APPENDIX Notes on important carbonate successions (Fig. 2).

The successions listed below are believed to be representative of the Indo-West Pacific as a whole. For locations see Fig. 1. It should be noted that only the offshore Indian sequences and those from Meghalaya have been dated in part by planktonic foraminifera.

Somali Republic

The relevant successions were described by Azzaroli (1952, 1958), and although he (1958: 125) stated that there was no apparent break between the Eocene and Oligocene, his faunal lists indicate the presence of a major disconformity which cuts out most, if not all, of the Upper Eocene. Identifications of the key taxa (*Nummulites fabianii* and *Pellatispira tudensis*) are questionable, and associated species (e.g., *Orbitolites complanatus*, *Dictyoconoides kohaticus* and *Dictyoconus africanus*) strongly suggest that this part of the sequence should be assigned to the Middle Eocene.

Arabia (Oman)

The coastal limestones are known to include Middle Eocene strata overlain by reefal limestones of Mid Oligocene to Early Miocene age (Montenat et al., 1977). Exposures of Late Eocene limestone also occur but are not in continuous sequence with the older and younger carbonates.

Pakistan

Khan (1967) showed that the type section of the Kirthar Limestone (Gaj River, Sind) includes an Early Oligocene foraminiferal fauna which unconformably overlies Lutetian-basal Auversian limestone yielding both larger and planktonic foraminifera. Only in a few places in Pakistan, e.g. Zinda Pir, western Punjab, are Late Eocene sediments with *Pellatispira madaraszi* and *Baculogypsinoides tetraedra* known to occur (Eames, 1952; Samanta, 1978). In no sequence has continuous carbonate sedimentation across the boundary been demonstrated, although in some localities (e.g., Mardan River, Jahan area) Late Eocene shales are interposed between older Eocene and younger Oligocene limestones (unpublished).

Western India

Onshore successions in Kutch, Surat and Broach are incomplete and always show a break between the youngest exposed Eocene and the oldest Oligocene or Miocene. In Kutch the Middle Eocene Fulra Limestone is overlain disconformably by the Maniyara Fort Formation of Early Oligocene age (Biswas and Raju, 1971). In Surat and Broach the uppermost Eocene limestones contain Baculogypsinoides tetraedra, Nummulites fabianii, and Pellatispira crassicolumnata (Samanta, 1978). In the offshore area, Zones P15 and P16 are well represented by shallow-water carbonates, but P17 is known mainly from deeper-water sediments which occur in the southern part of the Cambay Basin. Overlying N. fichteli-bearing limestones are not dated in terms of plankton and may thus be P18 or P19 in age (Mohan and Kumar 1985). Some specialists (e.g., Mohan and Kumar, 1985) believe that uppermost Eocene carbonates in the southern part of the Cambay Basin yield N. retiatus and N. fabianii, but this has not yet been confirmed.

Southeastern India

No complete succession is known on land, and although the offshore sequence in the Krishna-Godavari Basin is probably complete in areas of deep-water sedimentation, it is not known to be continuous across the boundary in wells which penetrated shallow-water limestones.

The only shallow-water carbonate succession so far known to include the boundary is that encountered in a Well in the southern part of the Cauvery Basin (Ramnad-Palk Bay Depression) where, a limestone, at least 40 m thick, has yielded a well-diversified larger foraminiferal fauna. Although the boundary was not cored, it is clear from the faunal evidence that an hiatus occurs (Mohan and Kumar, 1985). Indeed, the contact between Eocene and Oligocene sediments is now believed to be unconformable, despite a suggestion to the contrary by Raju (1971). An offshore well in this area penetrated only 5 m of strata tentatively assignable to the Late Eocene (Early P17). See Mohan and Kumar (1985).

N.E. India and Bengal

The highly fossiliferous limestones of the Prang Formation are followed by shales and thinly bedded limestones of the Kopili Formation (Nagappa, 1959). Planktonic and larger foraminifera typical of zones P14 to P17 have been recorded (Samanta, 1971). Zone P17, known only from wells, contains Discocyclina (Samanta, 1971). The Kopili Formation passes upwards into the deltaic "Barail Group". According to Mohan (CGA: pers. comm.) the Eocene/

Oligocene boundary is marked by a well-diversified planktonic foraminiferal fauna, above which N. fichteli occurs. One well in the Bengal Basin (Mohan and Kumar, 1985) shows that Discocyclina locally ranges slightly higher than Pellatispira.

Java and Sumatra

Few marine Eocene sequences are known from these islands (van Bemmelen, 1949), and no described shallow-water carbonate successions include the boundary. Baumann et al. (1972) reported marls overlying Discocyclina limestones in southern Java, and these presumably indicate a local deepening of the sea towards the end of Eocene times.

Borneo

Although van Bemmelen (1949: 139) reported Tc limestones overlying Tb carbonates in three areas of Kalimantan, the sequences concerned seem never to have been described. The succession at Melinau, Sarawak, was described by Adams (1965) who reported a thick development of shallow-water, algal-rich limestone (lacking age-diagnostic foraminifera) between the youngest datable Eocene and oldest definite Oligocene beds. Subsequent work on the succession in the Medalam gorge, some 10 km to the N.E. of the Melinau section, has suggested that these algal limestones may be thinner (20-50 m only) than was estimated originally. Unfortunately, there is no planktonic control over this part of the sequence. In both the Melinau and Medalam successions the Oligocene fauna appears gradually, beginning with the occurrence of Nummulitis fichteli Michelotti and Borelis inflata (Adams) and/or B. pygmaeus Hanzawa.

Philippines

Recent reviews by Hashimoto et al. (1977) and Hashimoto (1981) indicate that no sequence of carbonates is known to span the boundary in this area although both Late Eocene and Early Oligocene limestones occur. The Lutak Hill Limestone could be important for regional correlation (see Adams, 1984).

New Guinea

Visser and Hermes (1962) indicated that the Eocene/Oligocene boundary occurs in the New Guinea Limestone, but that its nature is not known. It may, however, be that the succession seen in the Chimbu Limestone of the Central Highlands (Rickwood, 1955; Bain and Binnekamp, 1973) is typical. Only about 12 m of Late Eocene limestone has been recognised in the Chimbu succession, but it is separated from the earliest datable Oligocene by some 60 m of algal limestone devoid of age-diagnostic fossils.

Marshall Islands

The only known sequence thought possibly to be continuous across the E/O boundary in the western Pacific is that described by Ladd et al. (1953) and Cole (1958) from the Marshall Isles. Boreholes on Eniwetok Atoll penetrated 93 ft (28.4 m) of carbonates of indeterminate age between the lowest recognisable Late Oligocene (Te_{1-4}) in drill-hole F-1 and the highest definite Eocene in hole E-1.

Other areas

The boundary must also be present in Saipan (Hanzawa, 1957; Cole, 1957); the Moluccas (Bursch, 1947); Misool (van Bemmelen, 1949), and Fiji (Cole, 1960a), to name but a few of the places where Eocene and Oligocene limestones have been recorded in close proximity, but in none of these areas have samples been collected from continuous sequences.

ACKNOWLEDGEMENTS

We wish to acknowledge the generous assistance of Dr S.K. Biswas, Mr Madan Mohan, and Mr P. Kumar of the Petroleum Institute, Oil & Natural Gas Commission, Dehra Dun, India, and of the Director and staff of the Malaysian Geological Survey. Without their active help (particularly in obtaining new data, reviewing and publishing borehole evidence, and in collecting additional material) this paper could not have appeared in its present form. Mr Mohan also commented helpfully on the typescript. Thanks are also due to Mrs L.M. McCormick for drafting the figures.

REFERENCES

Adams, C.G. 1964. The age and foraminiferal fauna of the Bukit Sarang Limestone, Sarawak, Malaysia. Ann. Rept. Geol. Surv. for 1963: 156-162.

_____ 1965. The Foraminifera and Stratigraphy of the Melinau Limestone, Sarawak, and its importance in Tertiary Correlation. Q. Jl. geol. Soc. Lond., 121: 283-338.

_____ 1970. A reconsideration of the East Indian Letter Classification of the Tertiary. Bull. Brt. Mus. nat. Hist. (Geol). 19(3): 87-137.

_____ 1984. Neogene Larger Foraminifera, Evolution and Geological Events in the Context of Datum Planes. In: Ikebe, N. and Tsuchi, R. (Editors), Pacific Neogene Datum Planes. University of Tokyo Press: 44-67.

Andreieff, P. 1985. Stratigraphic range of Caribbean Larger Foraminifera from Oligocene to Pliocene: state of knowledge in 1985. In: Mascle, A. (Editor), Géodynamique des Caraibes. Edit. Technip, Paris, pp 99-100.

Azzaroli, A. 1952. I Microforaminiferi della serie Carcàr (Eocene Medio e Superiore) in Somalia e la Loro Distribuzione Stratigrafica. Palaeontogr. Ital. 47 (N.S.)17 for 1950: 99-141.

_____ 1958. L'Oligocene e il Miocene della Somalia Stratigrafica, Tettonica Palaeontologia (Macroforaminiferi, Coralli, Molluschi) Paleontogr. Ital. 52 (N.S.)22: 1-143.

Bain, J.H.C. and Binnekamp, J.G. 1973. The Foraminifera and stratigraphy of the Chimbu Limestone, New Guinea. Bull. Bur. Min. Res. Geol. and Geophys. 139: 1-12.

Barbin, V. and Bignot, G. (this vol.). La limite Eocene-Oligocene dans la coupe-type de Priabona d'apres les microfacies.

Baumann, P., Oesterle, H., Suminta and Wibisono 1972. The Cenozoic of Java and Sumatra. Proc. Indon. Petrol. Assoc. 1st Ann. Conv., pp 31-42.

Bemmelen, R.W. van 1949. The Geology of Indonesia. Vol. 1A. Government Printing Office. The Hague. 732 pp.

Bizon, G., Bizon, J.J. and Ricou, L.E. 1972. Étude stratigraphique et paléogéographique des formations tertiaires de la Région de Neyriz (Fars interne, Zagros iranien). Rev. Inst. fr. Pétrol. 27(3): 369-405.

Biswas, S.K. and Raju, D.S.N. 1971. Note on the rock-stratigraphic classification of the Tertiary sediments of Kutch. Q. Jl. Geol. Min. Met. Soc. India, 43(3): 177-180.

Bursch, J.G. 1947. Mikropaläontologische Untersuchungen des Tertiärs von Gross Kei (Molukken). Schweiz Paleont. Abh. 65: 1-69.

Butterlin, J. 1960. Geologie generale et regionale de la Republique d'Haiti. Inst. Htes Etudes Amer. Latine, 194 pp.

_____ 1967. Revision d'espèces du genre Pseudophragmina H. Douvillé, 1923. Eclog. geol. Helv., 60(2): 543-51.

_____ 1981. Claves para la determinación de macroforaminiferos de Mexico y del Caribe, del Cretacico supérior al Miocene medio. Inst. Mexicano del Petroleo, 219 pp.

Caudri, C.M.B. 1975. Geology and Palaeontology of Soldado Rock, Trinidad. Pt 2, The Larger Foraminifera. Eclog. geol. Helv., 68(3): 533-89.

Cole, W.S. 1945. Stratigraphic and paleontologic studies of wells in Florida, N°4. Bull. Flor. geol. Surv., No. 28: 160 pp.

_____ 1952. Eocene and Oligocene larger foraminifera of the Panama Canal Zone and vicinity. Prof. Pap. U.S. geol. Surv., No. 244, 41 pp.

_____ 1957. Larger Foraminifera of Saipan Island. Prof. Pap. U.S. geol. Surv. 280-I: 321-360.

_____ 1958a. Larger Foraminifera from Eniwetok Atoll Drill Holes. Prof. U.S. geol. Surv., 260-V: 743-784.

_____ 1958b. Names and variation in certain American larger foraminifera, particularly the Camerinids, No. 2. Bull. Am. Paleont., 38: 261-84.

_____ 1960a. Upper Eocene and Oligocene Larger Foraminifera from Viti Levu, Fiji. Prof. Pap. U.S. geol. Surv. 374-A: 1-7.
_____ 1960b. Revision of Helicostegina, Helicolepidina and Lepidocyclina (Polylepidina). Contr. Cushm. Found. Foram. Res., XI (2): 57-63.
_____ 1962. Periembryonic chambers in Helicolepidina. Ibid. XIII (4): 145-52.
_____ 1963. Tertiary Larger Foraminifera from Guam. Prof. Pap. U.S. geol. Surv. 403-E: 1-28.
_____ 1969. Names and variation in certain American larger foraminifera, particularly the Eocene pseudophragminids, No. 4. Bull. Amer. Paleont., 56, No. 248, 55 pp.
Eames, F.E. 1952. Contributions to the study of the Eocene of W. Pakistan and W. India. A: The geology of standard sections in Western Punjab in the Kohat District (part 2). Q. Jl geol. Soc. Lond., 107: 159-200.
Hanzawa, S. 1957. Cenozoic Foraminifera from Micronesia. Mem. geol. Soc. Am. 66. 63 pp.
Hashimoto, W. 1981. Supplementary notes on the geological development of the Philippines. Contr. Geol. and Palaeont. S.E. Asia 218: 171-190, 12 tables.
_____, Matsumaru, K., Kurihara, K., David, P.P. and Rice, G.R., 1977. Larger foraminiferal assemblages useful for the correlation of the Cenozoic marine sediments in the mobile belt of the Philippines. Contr. Geol. and Palaeont. S.E. Asia. 183: 103-123.
Hornibrook, N. de B. 1968. Distribution of some Warm Water Benthic Foraminifera in the N.Z. Tertiary. Tuatara 16(1): 11-15.
Hunter, V.F. 1978. The mid-Tertiary stratigraphic unit of the southern Caribbean area. Verh. Naturf. Ges. Basel, 84(1): 172-190.
Khan, M.H. 1967. Time-hiatus at the Oligocene/Eocene boundary in Sind. Pakistan Jl Scient. ind. Res. 10: 287-295.
Ladd, H.S., Ingerson, E., Townsend, R.C., Russell, M. and Stephenson, H.K. 1953. Drilling on Eniwetok Atoll, Marshall Islands. Bull. Am. Assoc. Petroleum Geologists, 37(10): 2257-2280.
McFarlane, N.A. 1977. Some Eocene and Oligocene faunas from Central Jamaica. Mem. Segundo Congr. Latino-Americ. geol., Caracas, 1973, t.III: 1393-1411.
Mohan, M. 1983. Geohistory Analysis of Bombay High Region. Bull. O.N.G.C. 20(2): 23-46.
_____, Kumar, P., Narayanan, V., Govindan, A., Soodan, K.S. and Singh, P. 1982. Biofacies, Paleoecology and Geological History of Bombay offshore Region. Bull. O.N.G.C. 19(1): 1-15.
_____, and Kumar, P. 1985. Eocene Oligocene Boundary Events in India O.N.G.C., Dehra Dun (in press).
Montenat, C. and Blondeau, A. 1977. Premier aperçu Tertiaire d'Oman (Péninsule arabique orientale). Bull. Soc. géol. France (7) XIX (No. 6): 1285-95.
Murray, J. 1973. Distribution and Ecology of Living Benthic Foraminiferids. Heinemann, London. 274 pp.
Nagappa, Y. 1959. Foraminiferal biostratigraphy of the Cretaceous-Eocene succession in the India-Pakistan-Burma region. Micropaleontology, 5(2): 145-192.
Olsson, R.K., Muller, K.G. and Ungrady, T.E. 1980. Late Oligocene transgression of middle Atlantic Coastal Plain. Geology, 8: 549-554.
Pandey, J. 1982. Chronostratigraphic correlation of the Neogene sedimentaries of Western India Shelf, Himalayas and Upper Assam. Spec. Publ. Pal. Soc. India, 1: 95-129.

Pecheux, J.F. 1984. Le Sénonien supérieur-Tertiare du Chiapas (S.E. Mexique) et ses Macroforaminifères. These 3éme cycle, Univ. Nice, Lab. Micropal. & Géol. marines, 154 pp.
Raju, D.S.N., 1971. Upper Eocene to Early Miocene planktonic foraminifera from the subsurface sediments in Cauvery Basin, South India. Jb. Geol. B.A., 17: 7-68.
Rickwood, F.K. 1955. The geology of the Western Highlands of New Guinea. J. geol. Soc. Aust. 2 (1954): 63-82.
Robinson, E. 1968. Stratigraphic ranges of some larger Foraminifera in Jamaica. Trans. 4th Caribb. geol. Conf., Port of Spain, Trinidad, 1965: 189-94.
_____ 1974. Some larger Foraminifera from the Eocene limestone at Red Gal Ring, Jamaica. In: Contributions to the geology and palaeobiology of the Caribbean and adjacent areas. Verhandl. Naturf. Gesellsch., Basel, 84(1): 281-92.
_____ 1977. Larger imperforate foraminiferal zones of the Eocene of Central Jamaica. Mem. Segundo Congr. Latino-Americ. Geol., Caracas, 1973. t.III: 1413-21.
Samanta, B.K. 1969. Taxonomy and stratigraphy of the Indian species of Discocyclina (Foraminifera). Geol. Mag. 106: 115-129.
_____ 1971. Early Tertiary stratigraphy of the area around Garampani, Mikir-North Cachar Hills, Assam. Jl. Geol. Soc. India 12(4): 318-327.
_____ 1978. The Middle-Upper Eocene boundary in the Indian region. Q. Jl. Geol. Min. Soc. India, 50: 137-149.
_____ 1984. The genus Biplanispira Umbgrove (Foraminiferida) and its occurrence in India. Geol. Mag. 121, (4): 311-318.
_____ and Raychaudhuri, A.K., 1983. A revised lithostratigraphic classification of the Cretaceous - Lower Tertiary shelf sediments of eastern Khasi and Jainta Hills, Meghalaya. Q. Jour. Geol. Min. Met. Soc. India, 55(3): 101-129.
Shackleton, N.J., 1984. Oxygen isotope evidence for Cenozoic climatic change. In: P.J. Brenchley (Editor), Fossils and Climate, pp 27-34.
Vail, P.R., Mitchum. R.M. Jr. and Thompson, S., 1977. Seismic stratigraphy and Global changes of Sea level, Part 4: Global Cycles of Relative Changes of Sea Level, pp 83-97.
Visser, W.A. and Hermes, J.J., 1962. Geological results of exploration for oil in Netherlands New Guinea. Verhand. Kon. Ned. geol. mijn, Genoot. 20. 265 pp.
Vlerk, I.M. van der 1955. Correlation of the Tertiary of the Far East and Europe. Micropaleontology 1: 72-75.
Woodring, W.P., 1957. Geology and Paleontology of Canal Zone and adjoining parts of Panama: Geology and description of Tertiary Mollusks (Gastropods: Trochidae to Turritellidae). Prof pap. U.S. geol.Surv., 306A, 145 pp.

RADIOLARIAN EVENTS AND THE EOCENE-OLIGOCENE BOUNDARY

W.R. RIEDEL and ANNIKA SANFILIPPO

Because radiolarians are not present in the stratotypes of stages used to define the Eocene-Oligocene boundary, this microfossil group cannot be used independently to distinguish late Eocene from early Oligocene. This epoch boundary can be recognised only via the calcareous microfossil groups. Recently it has become possible to place these calcareous microfossil boundaries within a context of radiolarian events. We here describe the radiolarian stratigraphy around the Eocene-Oligocene boundary at three widely separated Deep Sea Drilling Project sites (no.94 in the Gulf of Mexico, no.462 in the western tropical Pacific and no.366 in the eastern tropical Atlantic), and compare them with the succession at Bath Cliff, Barbados.

Although these DSDP sites are the best available for this study, in that they sample apparently continuous sequences of calcareous-siliceous sediments spanning the desired stratigraphic interval, two of them have substantial deficiencies. At Site 94 the Eocene-Oligocene boundary occurs in a coring gap (Worzel, Bryant et al., 1973). And at Site 462 many older microfossils are reworked into younger sediments (Larson, Schlanger et al., 1981) - but analysis of the relative abundances of stratigraphically significant species allows us to "see through" this obscuring factor. At Site 366 (Lancelot, Seibold et al., 1978), the Paleogene radiolarian assemblages have a colder-water aspect than at the other tropical sites. This presents some difficulties for correlations, but on the other hand provides information helpful for establishing links with higher latitudes. The section on Barbados has none of these drawbacks, and moreover accumulated more rapidly than most deep-sea sequences, thus permitting higher stratigraphic resolution (Saunders et al., 1985; Sanfilippo et al., 1985).

In the correlation diagram, events are shown as Tm (latest occurrence of morphotypes), Bm (earliest occurrence of morphotypes), and evolutionary transitions indicated by arrows. Approximate thicknesses of sequences are indicated by the meter-levels near the top and bottom of each box. Vertical dimensions of the lettered rectangles show the interval of uncertainty in the placement of each event; their transverse dimensions are a matter of drafting convenience. Solid lines connect events that are confidently located, and dashed lines indicate events less confidently positioned. For example, event C (latest occurrence of morphotypes of Lithocyclia crux) is recognized more confidently at Sites 94 and 366 than in the other two sequences, but at Site 94 it occurs in a long gap between cores. This same coring gap may be responsible for the apparently anomalously early transition from Artophormis barbadensis to A. gracilis (event B) at Site 94. Although the younger morphotypes predominate in the two samples examined below this gap, this predominance may oscillate back and forth between the ancestral and descendant form in the part of the sequence not cored. The Eocene-Oligocene boundaries based on foraminifera and nannofossils (events H and I, respectively) are plotted according to the interpretations in the DSDP Initial Reports cited.

The data on which radiolarian correlations of the DSDP sites are based are presented in the tables, and those for Bath Cliff can be found in our cited publications. Abundances are given as percentages of the total radiolarian assemblages coarser than 44 or 63 microns. Numbers in italics represent estimates derived from our or H.P. Foreman's published records of abundances in terms such as "common", "few", "rare", etc.

The correlation diagram shows that there are four radiolarian events (D-G, the last not confidently located in Site 462) approximately coincident with the Eocene-Oligocene boundary defined on the basis of foraminifera. We know of no radiolarian events coincident with the boundary defined on the basis of calcareous nannofossils, but a tight group of five upper limits of taxa occurs about half a million years before it (using the sediment accumulation rate estimated by Saunders et al., 1985) - events J-N on the correlation diagram.

Lancelot, Y., Seibold, E., et al., 1978. Init. Repts. Deep Sea Drilling Project v.41. U.S.Govt. Printing Office, Washington.
Larson, R.L., Schlanger, S.O., et al., 1981. ibidem v.61.
Sanfilippo, A., Riedel, W.R., Glass, B.P., and Kyte, F.T., 1985. Late Eocene microtektites and radiolarian extinctions on Barbados. Nature, 314: 613-615.
Saunders, J.B., Bernoulli, D., Müller-Mertz, E., Oberhänsli, H., Perch-Nielsen, K., Riedel, W.R., Sanfilippo, A., Torrini, R., jr. 1984. Stratigraphy of the late Middle Eocene to Early Oligocene in the Bath Cliff section, Barbados, West Indies. Micropaleontology, 30: 390-425.
Worzel, J.L., Bryant, W., et al., 1973. Init. Repts. Deep Sea Drilling Project v.10. U.S. Govt. Printing Office, Washington.

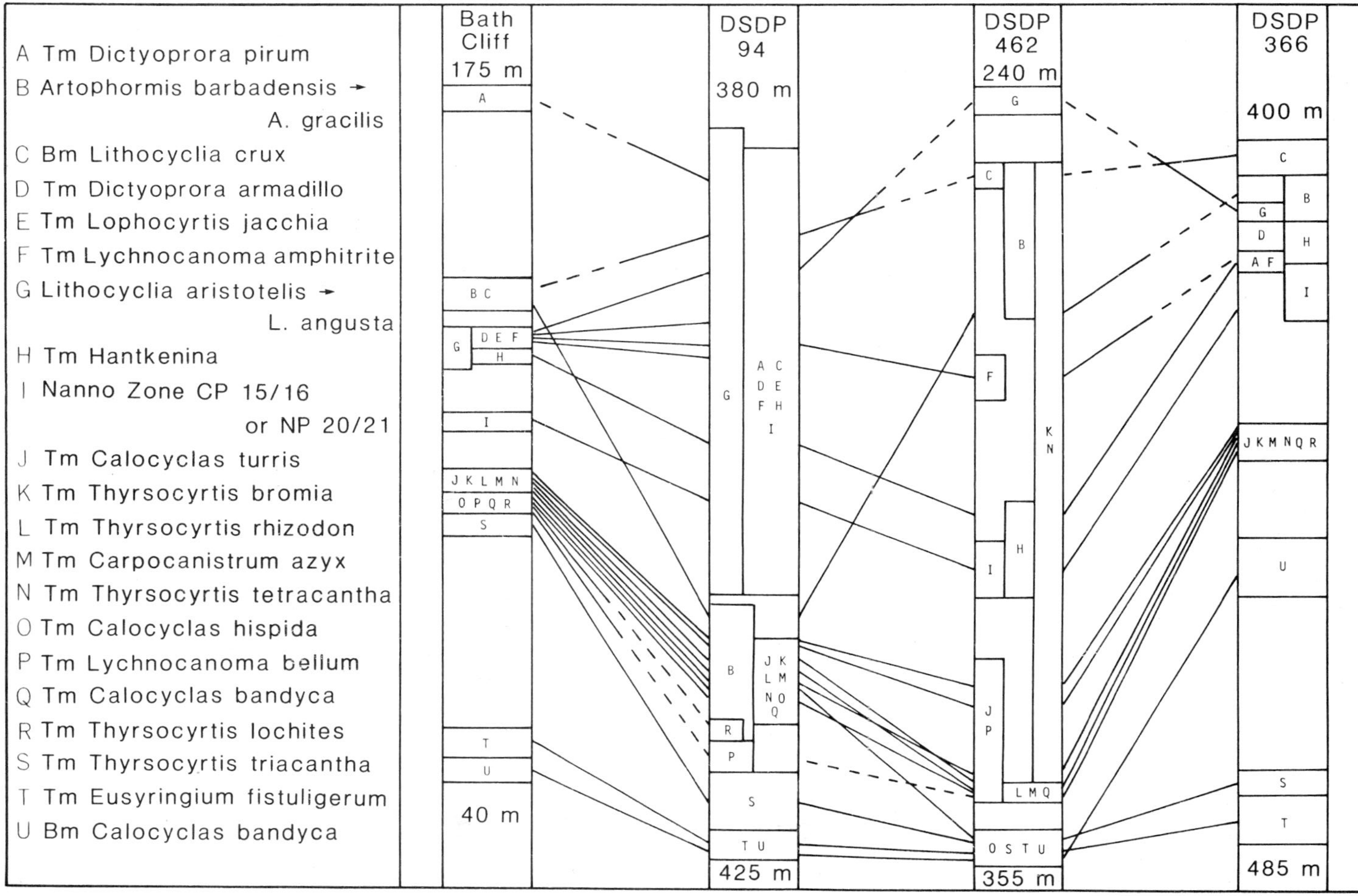
A Tm Dictyoprora pirum
B Artophormis barbadensis → A. gracilis
C Bm Lithocyclia crux
D Tm Dictyoprora armadillo
E Tm Lophocyrtis jacchia
F Tm Lychnocanoma amphitrite
G Lithocyclia aristotelis → L. angusta
H Tm Hantkenina
I Nanno Zone CP 15/16 or NP 20/21
J Tm Calocyclas turris
K Tm Thyrsocyrtis bromia
L Tm Thyrsocyrtis rhizodon
M Tm Carpocanistrum azyx
N Tm Thyrsocyrtis tetracantha
O Tm Calocyclas hispida
P Tm Lychnocanoma bellum
Q Tm Calocyclas bandyca
R Tm Thyrsocyrtis lochites
S Tm Thyrsocyrtis triacantha
T Tm Eusyringium fistuligerum
U Bm Calocyclas bandyca
Bath Cliff
175 m
40 m
DSDP 94
380 m
425 m
DSDP 462
240 m
355 m
DSDP 366
400 m
485 m

PERCENTAGES OF TAXA IN DSDP ASSEMBLAGES

TAXA / DSDP Sample (interval in cm)	Calocyclas bandyca	Eusyringium fistuligerum	Thyrsocyrtis triacantha	Thyrsocyrtis lochites	Lychnocanoma bellum	Calocyclas hispida	Thyrsocyrtis tetracantha	Carpocanistrum azyx	Thyrsocyrtis rhizodon	Thyrsocyrtis bromia	Calocyclas turris	Lithocyclia aristotelis	Lithocyclia angusta	Lychnocanoma amphitrite	Lophocyrtis jacchia	Dictyoprora pirum	Lithocyclia crux	Artophormis barbadensis	Artophormis gracilis	Dictyoprora armadillo
	1	2	3	4	5	6	7	8	9	10	11	12	13	14	15	16	17	18	19	20
366-5-2,47-50													0.03				0.1		2.1	
6-2,50-53													0.04				0.4	-	2.8	
7-3,50-53																	0.02	-	1.77	
8-3,50-53													0.01				0.28	-	1.77	
8-CC																	0.2	-	2.32	
9-1,125-129												0.005	0.04	-	-	-	-	-	0.28	-
9-4,50-53									-	-	-	-	0.006	-	-	-	-	0.006	-	-
9-CC												0.5	-	-	-	-	-	0.03	-	-
10-1,41-44												0.88	-	-	-	-	-	0.11	-	0.16
10-2,66-69	-	-	-	-	-	-	-	-	-	-	-	0.87	-	0.001	-	0.13	-	-	-	0.15
10-3,61-64	-	-	-	-	-	-	-	-	-	-	-	1.5	-	-	-	0.3	-	-	-	0.3
10-5,100-103	-	-	-	-	-	-	-	-	-	-	-	1.25	-	0.002	-	0.18	-	0.04	0.1	0.25
11-1,73-78	-	-	-	-	-	-	-	-	-	-	-	0.5	-	0.01	-	0.06	-	0.02	-	0.36
12-1,66-71	-	-	-	-	-	-	-	-	-	-	-	3.0	-	-	-	-	-	-	-	-
12-2,100-103	0.08	-	-	0.2	-	0.02	0.04	0.01	-	2.0	0.05	1.5	-	-	-	-	-	0.02	-	0.08
12-CC	0.08	-	-	0.04	-	-	0.17	0.25	-	0.4	0.04	1.7	-	-	-	-	-	0.04	-	-
13-1,144-147	0.07	-	-	1.0	-	-	0.22	0.20	-	0.4	0.1	3.5	-	-	-	-	-	0.13	-	0.24
13-2,108-111	0.003	-	0.08	0.2	-		0.24	0.24	-	0.5	0.2	1.8	-	-						
13-CC	0.005	-	0.02				0.02		-	1.03	0.03	5								
14-2,54-57	-	-	0.01	0.03	-	-	0.04		-	0.01	-	5								
14-3,109-111	-	-	0.08	0.24			0.5			0.05		5			0.5					
14-CC	-	-	-				5				0.5	5								
15-1,18-21							5			5	0.5	5								
15-2,45-50		-	0.009				5			0.05	0.5	5								
15-CC		-					0.5		-	-	0.5	5								
16-2,100-103	-	-	0.05		-	-	0.6	-	-	-	-	5			0.5					
16-3,18-21		-	5				5		-	-	0.5	5								
16-CC	-	-	0.45		-	-	0.23	-	-	-	0.5	5								
17-CC		0.4	1.5				-		-	-	0.5	0.5								

	1	2	3	4	5	6	7	8	9	10	11	12	13	14	15	16	17	18	19	20
462-25-6,123-125												0.002	0.04		-	0.04	-	-	0.9	
25-CC												0.003	0.02	-	-	-	0.004	-	0.3	
26-1,147-149											-	0.008	0.01	-	0.003	0.09	0.002	-	0.28	-
26-3,46-48						-	0.01				0.01	0.01	0.01	-	0.002	0.02	0.01	-	0.3	-
26-4,33-35											-	*0.05*	-	-	-		-			
27-1,116-118												*0.5*	-		0.2	0.02	-	-	0.04	0.02
27-3,82-83			0.02	0.02			0.05		0.02		-	0.02	0.06	-	0.03	0.06	0.01	0.01	0.6	0.03
28-6,99-101														-	-			-	0.43	
29-5,70-72	-	0.24	0.05	0.05	-	0.1	0.24	0.05	0.1	0.05	-	0.1	-	-	0.24	-	-	0.05	-	0.05
30-3,14-16														-						
31-2,23-25												0.1	-	0.03	0.05			0.05	0.001	0.125
32-4,60-62										0.01		0.21		0.05	0.28	0.14		0.14		0.12
32-CC							0.01							0.07	0.26			0.38	0.02	0.23
33-2,24-26				-	-									0.25						
33-CC				-	-		-							0.05						0.09
35-1,15-17	-	0.003	-	-	-	*0.01*	0.03	*0.01*	*0.05*	*0.03*	0.008	*0.5*		0.33	0.7	*0.04*				
35-CC	-	0.04	0.04	0.04	-	0.01	0.3	-	0.01	0.04	0.01	0.74		0.07	0.5	0.04		0.3		0.07
36-1,77-79	0.003	-	0.16	0.16	0.003	0.01	0.85	-	0.01	0.3	0.2	0.5	-	0.03	0.4	0.01		0.2	-	0.05
36-2,45-47	1.5	-	-	0.01	-	-	1.22	0.5	1.5	1.5	1.5	0.08		0.01	0.3	0.01		-		0.01
36-2,103-105	1.0	-	-	0.03	0.03	-	3.0	0.5	3.0	3.0	1.0	*0.1*		0.05	0.05	0.05		-		0.05
36-CC	1.9	-	-	0.03	0.02	-	*5*	*0.4*	*3*	*2.5*	*1*	*0.05*		*0.01*	*0.05*	-		-		*0.02*
38-3,32-34	-	0.5	6.0	-	-	0.9	-	-	*2.5*	-	-	-		-	-	-		-		-
38-4,69-71	-	0.7	4.8	-	-	0.4	-	-	*3*	-		-	-	-	-	-	-	-	-	-
94-12-4,42-44													*0.02*				*0.02*			
12-CC												-	*0.02*	-	-	-	*0.05*	-	*1*	-
13-2,42-44												-	*0.02*	-	-	-	*0.01*	-	*2*	-
13-3,42-44										-		-	0.05	-	-	-	0.04	-	2.0	-
13-CC	-	-	-	-	-	-	-	-	-	-	-	-	-	-	-	-	0.10	-	0.5	-
14-2,42-44	-	-	-	-	-	-	-	-	-	-	-	0.6	-	0.2	0.3	0.2	-	-	0.03	0.35
14-3,42-44	-	-	-	-	-	-	-	-	-	-	-	0.1	-	0.6	0.2	0.02	-	-	0.14	0.16
14-4,45-47	-	-	-	-	-	-	-	-	-	-	-	*2*	-	*0.5*	*0.3*	*0.2*	-	-	-	*0.1*
14-CC	-	-	-	-	-	-	-	-	-	-	-	*1*	-	*0.5*	*0.2*	*0.02*	-	-	-	*0.02*
15-4,70-72	1.0	-	-	-	-	0.2	1.6	2.5	2.4	0.25	2.6	0.9				-		-	-	0.2
16-1,43-45	1.5	-	-	0.06	-	0.2	5.3	0.8	5.0	1.3	2.0	0.8		*0.02*	*0.01*	0.2		0.03	-	0.2
16-2,42-44	0.5	-	-	0.01	0.008	*0.5*	*2.3*	*0.5*	1.7	0.7	1.3	0.25		0.02	*0.01*	0.25				0.1
16-CC	0.75	-	*1*	-	*0.005*	*0.2*	*2*	*0.5*	*2*	*0.5*	6.0	0.3		0.05		0.01			-	0.13
17-1,36-38	-	0.08	*1*	-	0.02	*0.5*	-	-	*2*	-	-	-		-		-		-	-	0.01
17-2,42-44	-	0.2	*1*	-		*1*	-	-	*2*	-	-	-		-		-				
17-3,43-45	-	0.2	*1*	*0.2*		*1*	-	-		-	-	-		-		-				
17-4,43-45	-	0.5	*1*	*0.2*	-	*1*	-	-		-	-	-		-		*0.05*		-		-
17-CC			*1*	*0.2*	*0.001*	*1*										-				-
18-1,43-45			*1*	*0.2*	*0.05*											-				-

DISTRIBUTION OF OSTRACODA AT THE EO - OLIGOCENE BOUNDARY IN DEEP (BARBADOS) AND SHALLOW MARINE ENVIRONMENT (GULF OF MEXICO)

W. A. van den Bold
Louisiana State University
Baton Rouge

INTRODUCTION

The distribution of ostracodes in relation to the Eo-Oligocene boundary was studied in deep marine deposits (Oceanic Formation, Barbados, and Chapapote, Horcones and Palma Real Formations, Mexico) and in shallow marine deposits (Tantoyuca, Huasteca and Meson Formations, Mexico, and Pachuta, Shubuta, Red Bluff, and Marianna Formations, U. S. Gulf Coast). The Ostracoda of this age were tabulated and described by Deboo (1969), Hazel et al. (1980), Howe (1977), Howe and Howe (1973, 1975), Howe and Chambers (1935), Howe and Law (1936) and Huff (1970) in Alabama, Mississippi and Louisiana; in Mexico by Bold (1978), Carreño (1980, 1984), Laurencich (1969) and Palacio-Fest (1982); in Barbados especially by Steineck et al. (1984). Furthermore I have used my studies on Trinidad (Bold, 1960), Cuba (Bold, 1946, 1973), Haiti (Bold, 1981) and unpublished information, as well as Steineck's (1981) work on Jamaica.

SHALLOW MARINE ENVIRONMENT

According to the ostracode distribution this is a continental shelf environment, probably between 30 and 100 m water depth. Among the 52 species present close to the Eo-Oligocene boundary in Alabama and Mississippi (Table 1), there are 32 species crossing this boundary (as established on the basis of planktonic foraminifera). Several Eocene species disappear somewhat below the boundary, and some Oligocene species appear a little above it. However, some of the latter species are known from Eocene deposits in neighboring areas, e.g. Trachyleberidea blanpiedi, Jugosocythereis vicksburgensis, and Phacorhabdotus varians (See Tables 2 and 3). Only Eucythere shubutaensis, Trachyleberis montgomeryensis, Occultocythereis broussardi, and above all Haplocytheridea montgomeryensis disappear at the boundary; Buntonia huneri appears here for the first time. In general we see, that, in the sublittoral environment of the U.S. Gulf Coast, there are only minor phylogenetic changes near the Eo-Oligocene boundary and that the majority of them does not take place at the boundary itself. Nevertheless, there is the total extinction of Haplocytheridea montgomeryensis and the almost total disappearance of the genus Haplocytheridea, which dominates this environment in the Eocene. In Mexico (Table 2) the faunal development is very similar. There is a greater percentage of species that do not cross the boundary, but there are at least 10 that do. This is probably the result of a small hiatus between the Tantoyuca and Horcones formations. Note also the absence of H. montgomeryensis in the Upper Eocene.

DEEP MARINE ENVIRONMENT

The fauna of the continental slope is dominated by species of the subfamily Krithinae, and here we see differences between the ranges of species reported from different parts of the Caribbean

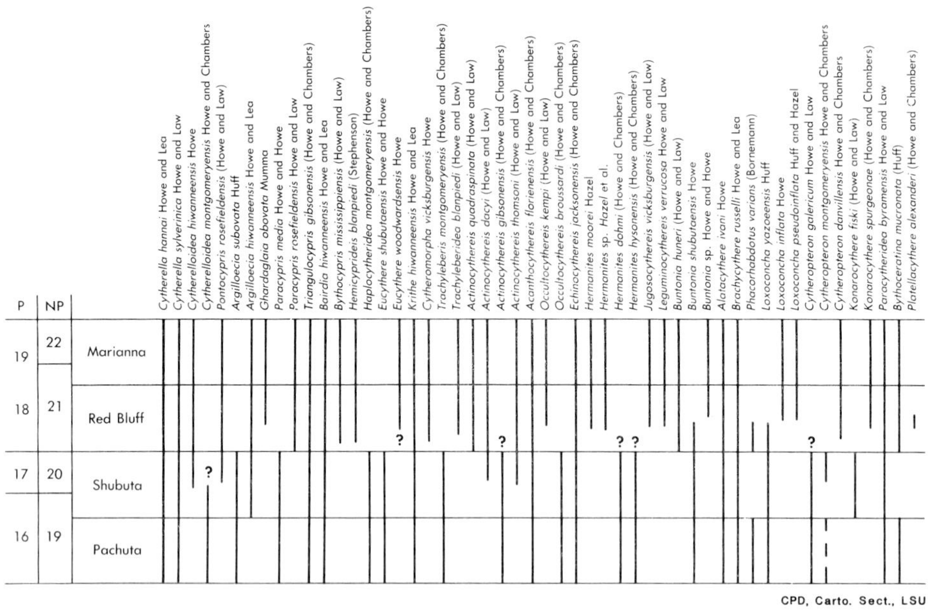

Table 1: Distribution of Ostracoda in the Pachuta, Shubbuta, Red Bluff and Marianna Formations of Alabama and Mississippi. Shallow water environment.

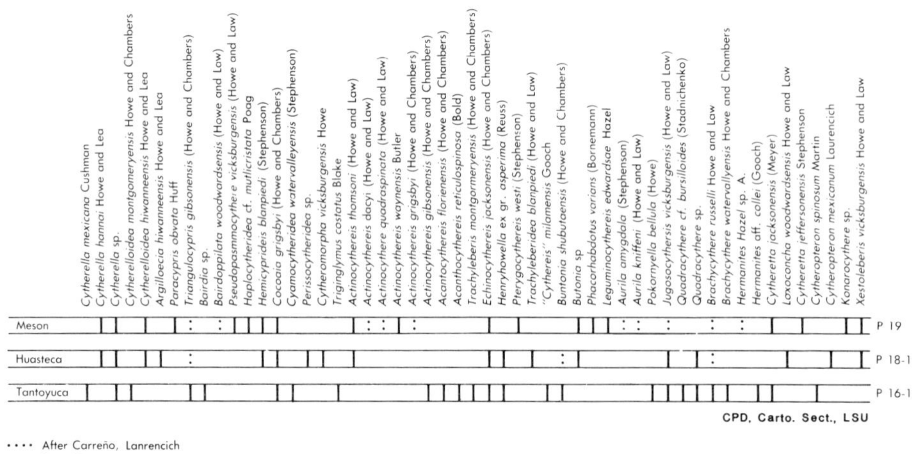

Table 2: Distribution of Ostracoda in the Tantoyuca, Huasteca and Meson Formations of the Tampico-Misantla Basin, Mexico. Shallow water environment.

and Mexico. The total range of the species is as follows: Krithe cancuenensis: Paleocene - end of Eocene; K. prolixa: from Upper Eocene upward; K. guatemalensis: Paleocene - Upper Eocene; K. cubensis: Upper Eocene - Lower Miocene; K. crassicaudata: Middle Eocene; K. morkhoveni: from Oligocene upward; K. saundersi: Upper Eocene - Upper Oligocene; K. hiwanneensis: Upper Eocene - Lower Miocene; K. vandenboldi: from Upper Eocene upward; K. trinidadensis: from Oligocene upward; Parakrithe elongata: Middle Eocene - Upper Oligocene; P. vermunti: from Middle Eocene upward; Messinella ovata: Upper Eocene - Lower Oligocene; M. guanajayensis: Lower Oligocene upward (but reported by Steineck et al. 1984, in the Eocene of Barbados). The ranges of the species of these three genera cannot be used for independant zonation until all authors have reached an agreement on their exact ranges. In the case of species of Krithe the difference is probably caused by a disagreement about the basis of specification; for me, this is the position of normal and marginal pores in the anterior part of the carapce, independant of the shape of the vestibule (Bold, 1966, 1968). K. cancuenensis seems to disappear at the Eo-Oliocene boundary. K. morkhoveni and K. trinidadensis are reported immediately above the boundary in Barbados, but are not found until the Upper Oligocene in other areas of the Caribbean. Agrenocythere antiquata is reported in the Lower Oligocene in Mexico. A. hazelae is not reported till the Upper Oligocene.

CONCLUSIONS

In the Gulf Coast the shallow marine environment persists from the Eocene into the Oligocene and the fauna does not show any change at the generic level with the only exception of the near extinction of Haplocytheridea and its replacement by the related genus Hemicyprideis. Moreover most of the appearances and disappearances of species take place a little above or below the boundary (as established by the distribution of planktonic foraminifera). In Barbados the environment remained deep (continental slope) during Eocene and Oligocene. Faunas are less diverse than in the sublittoral environment and here also most changes in fauna take place above or below the boundary, where only K. cancuenensis disappears. In the deep environment of Mexico we find very much the same, with M. ovata disappearing at the Eo-Oligocene boundary. In other areas of the Caribbean the study of ostracodes is hampered by 1) absence of Lower Oligocene beds; 2) presence of massive limestones in the Upper Eocene. In Trinidad the presence of a wildflysch facies has the effect of producing intense reworking near the Eo-Oligocene boundary.

BIBLIOGRAPHIE

Bold, W. A. van den, 1946. Contribution to the study of Ostracoda with special reference to the Tertiary and Cretaceous microfauna of the Caribbean Region. Diss. Rijksuniversiteit te Utrecht, Debussy, Amsterdam, p. 1-167.

---, 1960. Eocene and Oligocene Ostracoda of Trinidad. Micropaleontology, vol. 6, no. 2, p. 145-196.

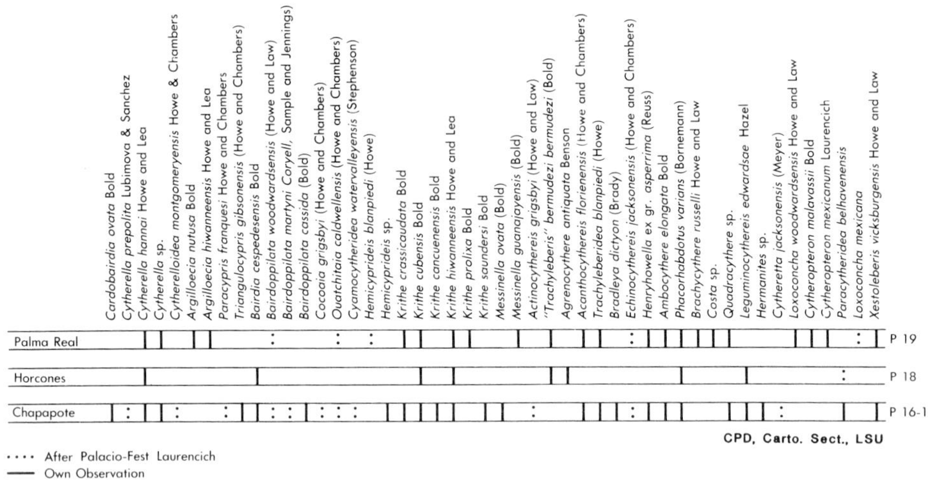

Table 3: Distribution of Ostracoda in the Chapapote, Horcones and Palma Real Formations, of the Tampico-Misantla Basin, Mexico. Deep water environment.

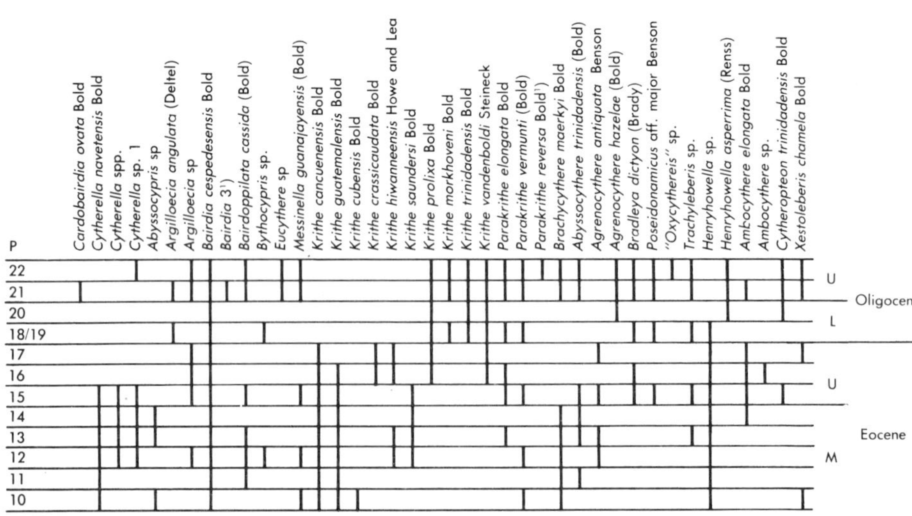

Table 4: Distribution of Ostracoda in the Oceanic Formation, Barbados. Deep water environment.

---, 1966. Répartition de certains ostracodes dans le Tertiaire des Caraïbes. 3rd Sess. IUGS, Comm. Mediterranean Neogene Stratigraphy, p. 134-139.
---, 1968. Note on the genus Krithe. Ostracodologist, 12, p.10-11.
---, 1973. Distribution of Ostracoda in the Oligocene and Lower and Middle Miocene of Cuba. Carib. Journ. Sci., vol. 13, p. 145-159.
---, 1978. Distribution of Tertiary and Quaternary Ostracoda in Central America and Mexico. Univ. Nac. Aut. Mexico, Inst. Geol. vol. 101, p. 114-137.
---, 1981. Distribution of Ostracoda in the Neogene of Central Haïti. Bulls. Amer. Paleont., vol. 79, no. 312, p. 1-136.
---, 1983. Shallow-marine biostratigraphic zonation in the Caribbean Post-Eocene. Proc. 8th Int. Symp. Ostracoda, p. 400-416.
Carreño, A. L., 1980. Los Ostrácodos de edad Oligo-Mioceno de la subprovincia Tampico-Misantla, area centro-oriental de México, Pt. 1: Oligoceno, Formación Mesón. Univ. Nac. Aut. México, Inst. Geol., Revista, vol. 4, no. 1, p. 37-45.
---, 1982. Formación Mesón. 8th Inst. Symp. Ostracoda, Libreto Guia de la excursion post-symposium: Ostrácoda de México, Planicie costera del Golfo, p. 53-57.
---, 1984. Los ostrácodos de edad Oligo-Mioceno de la subprovincia Tampico-Misantla, area centro-oriental de México, Pt. 3: Formación Alazán. Mem. 3 Congr. Latinamer. de Paleont., p. 480-487.
Deboo, P. B., 1965. Biostratigraphic correlation of the type Shubuta member of the Yazoo Clay and Red Bluff Clay with their equivalents in southwestern Alabama. Geol. Surv. Alabama, Bull. vol. 80, p. 1-84.
Hazel, J. E., Mumma, M. H., and Huff W. J., 1980. Ostracode biostratigraphy of the Lower Oligocene (Vicksburgian) of Mississippi and Alabama. Trans. Gulf Coast of Geol. Soc., vol. 30, p. 361-380.
Howe, H. J., 1977. A review of the Jacksonian-Vicksburgian boundary in the East-Central Gulf Coast by means of Ostracoda. Ibid., vol. 27, p. 291-298.
Howe, H. J. and Howe, R. S., 1975. Central Gulf Coast Jacksonian ostracodes in the H. V. Howe Collection. Ibid., vol. 25, p. 282-290.
Howe, H. V. and Chambers, 1935. Louisiana Jackson Eocene Ostracoda. Louisiana Dept. Cons., Geol. Bull., vol. 5, p. 1-48.
Howe, H. V. and Law, J., 1936. Louisiana Vicksburg Oligocene Ostracoda. Ibid., vol. 7, p. 1-78.
Howe, R. C. and Howe, H. J., 1973. Ostracodes from the Shubuta Clay (Tertiary) of Mississippi. Journ. Paleont., vol. 47, no. 4, p. 629-656.
Huff, W. J., 1970. The Jackson Eocene Ostracoda of Mississippi. Miss. Geol., Econ. & Topogr. Surv., Bull., vol. 114, p. 1-289.
Laurencich, L. A., 1969. Some Ostracoda from the Alazan Formation of Mexico. Micropal., vol. 15, no. 4, p. 493-503.
Steineck, P. L., 1981. Upper Eocene to Middle Miocene ostracode faunas and paleooceanography of the north coastal belt, Jamaica, West Indies. Mar. Micropal., vol. 6, p. 339-366.
Steineck, P. L., Breen, M., Nevins, N. and O'Hara, P., 1984. Middle Eocene and Oligocene deep-sea Ostracoda from the Oceanic Formation, Barbados. Journ. Paleont., vol. 58, no. 6, p. 1463-1496.

OSTRACODS AT THE EOCENE-OLIGOCENE BOUNDARY IN THE AQUITAINE BASIN. STRATIGRAPHY, PHYLOGENY, PALAEOENVIRONMENTS

O. Ducasse and J.P. Peypouquet

Université de Bordeaux I, Département de Géologie et Océanographie, Avenue des Facultés - 33405 Talence Cédex (France)

ABSTRACT

Eocene/Oligocene boundary ostracods in the Aquitaine Basin may serve as good regional stratigraphic markers of the margino-littoral environment. At this time there occurred renewal of microfauna and structural readjustment within the populational settlements. Paleoenvironmental conditions were different between the upper Eocene and beginning of the lower Oligocene from the littoral to the bathyal domains.

1. INTRODUCTION

The ostracod fauna at the Eocene-Oligocene boundary in the Aquitaine Basin has been the subject of intricate studies for deep water as well as shallow (littoral) water. The principal studies being those of Deltel (1961, 1962), Ducasse (1969, 1974, 1975, 1981, 1983), Ducasse and Moyes (1971), Ducasse and Peypouquet (1978, 1979), Ducasse and Rousselle (1978, 1979a, b), Keij (1955), Peypouquet (1979), Peypouquet and al. (1980, 1981).

2. MAIN STRATIGRAPHIC RESULTS

2.1. Margino-littoral environment

In the middle part of the Medoc (Fig. 1), Ducasse, Lété, Rousselle (in press) have carried out an inventory of the ostracod fauna and a complementary study of populations for some species.

The authors have shown that the faunal associations follow one another through time and permit a separation of 5 biozones characteristic of established stratigraphic horizons belonging to the early, middle and late Upper Eocene, and early Oligocene.

An evolutionary scheme has been proposed for each of the species after a study of populations. It is based on their structure, their abundance and changes through time in relation to the environment (Fig. 2).

Those examples of phylogeny demonstrate a rapide evolution of "equilibrating" pattern (Ducasse et al., 1983). They reflect a clear evolving polarity and the different established adaptive stages represent good local stratigraphic markers and excellent indicators of palaeohydrological evolutions.

The Eocene-Oligocene boundary appears then as an important break easily noticeable through the ostracod fauna. It is translated by an important qualitative renewal of the ostracod fauna (Ducasse and al., 1985) and by a major structural readjustment within populations (Fig. 2).

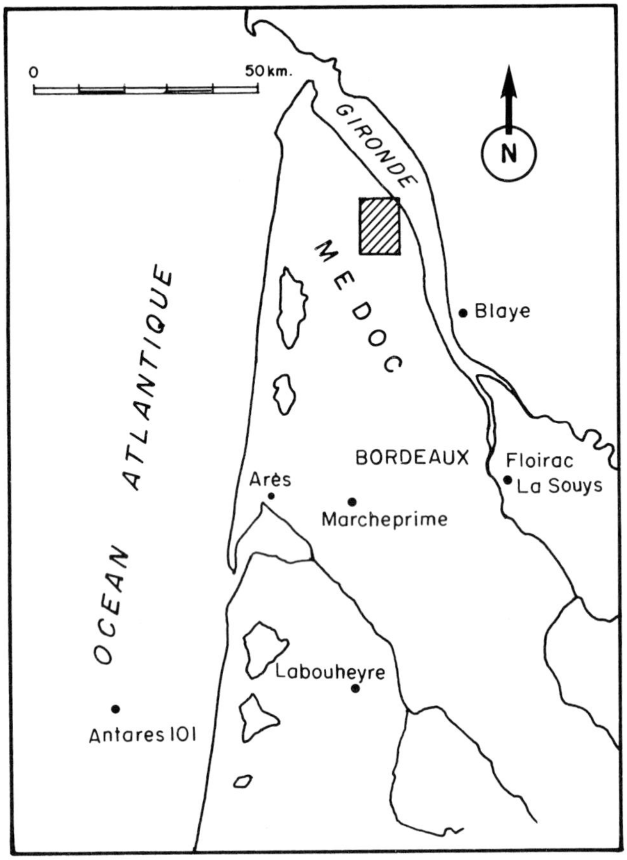

Fig. 1 - Location of the studied zone in the Northern Aquitaine Basin

 outcrops of the Medoc

One can see :

- the decrease or the disappearance of numerous Eocene species, in particular : *Pokornyella ventricosa*, *P. moyesi*, *Cytheretta vulgaris*, *Leguminocythereis div. sp.*, *Hirsutocythere horrescens*, *Cyamocytheridea aff. heizelensis*, *Schizocythere appendiculata*, *S. tessellata*, *Echinocythereis septentrionalis*, *Monsmirabilia triebeli*, *Paracytheridea grignonensis*, *Cytherelloidea dameriacensis* ;
- the development of Oligocene species such as : *Cytheretta sagri*, *Quadracythere macropora*, *Cyamocytheridea inflata*, *Pokornyella calix* and *P. limbata* ;
- the weakening of polymorphism within species and the relay of morphs

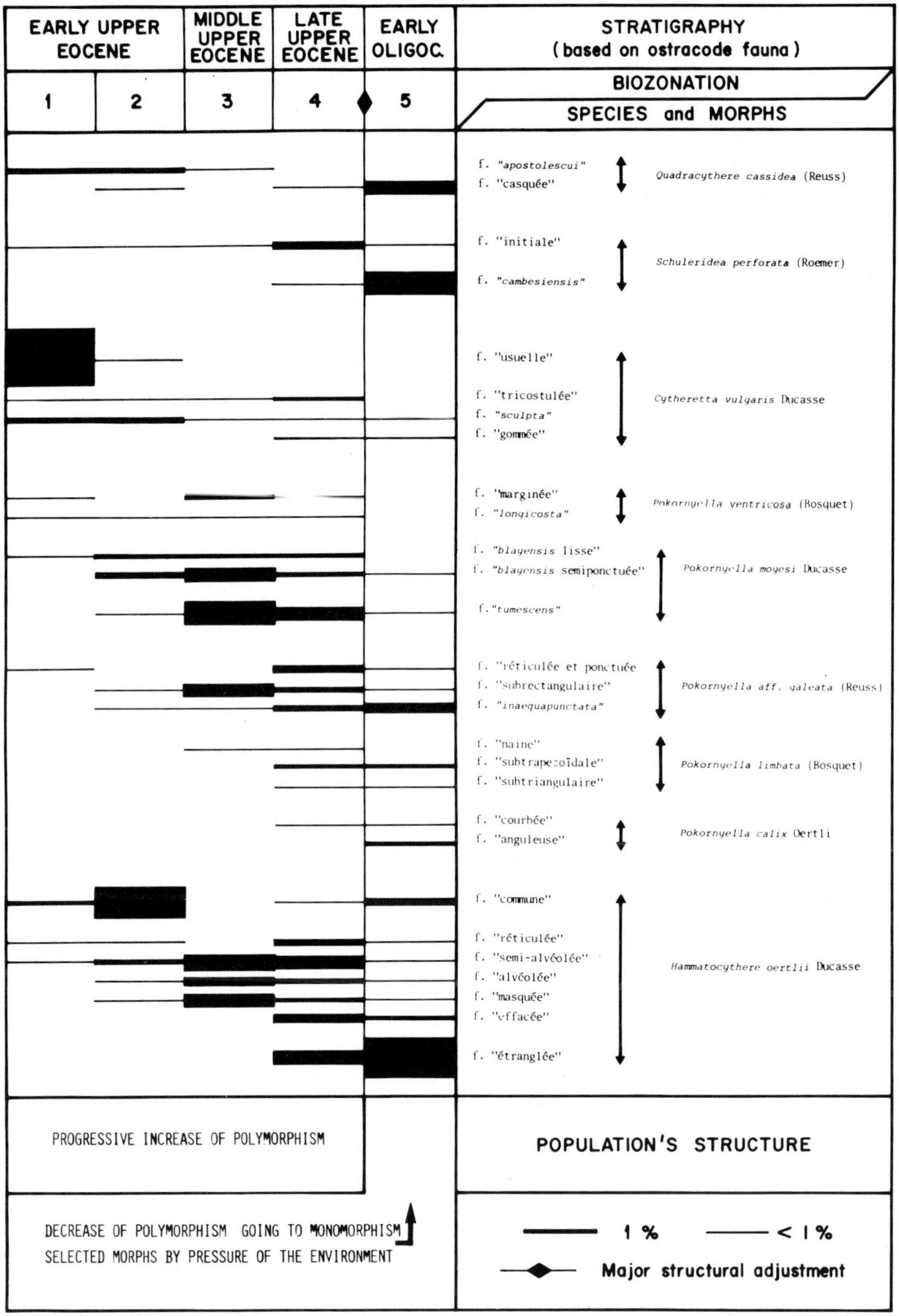

Fig. 2 - Population structures of some species at the Eocene-Oligocene boundary in the Medoc.

within *Schuleridea perforata*, *Quadracythere cassidea*, *Hammatocythere oertlii*. The new morphs, due to selection effect caused by the environment, show a strong demographic inflation ;
- the trend towards "degradation" of shells for some species such as *Hammatocythere oertlii*, *Cytheretta vulgaris*, *Quadracythere cassidea*.

2.2. Deep environment

This type of environment has been recognized in the Aquitaine Palaeogene for numerous sites localised in the central part and the southern portion of the Aquitaine Basin as well as in the Gascogne Gulf.

The ostracode fauna there is homogeneous, dull and consisting essentially of *Cytherella consueta*, *C. transversa*, *Krithe div. sp.*, *Bairdia crebra*, *Protoargilloecia angulata*, *Henryhowella asperrima*, *Pontocyprella aturica*. *Costa tricostata*, *Cardobairdia fulgens*, *Bairdia cymbula*, *Trachyleberidea prestwichiana*, *Agrenocythere ordinata*, *Cytheropteron div. sp.* can be added to the previously mentioned species. They present strong tethysian affinities.

Populations studies done on species belonging to the genera *Cytherella* and *Protoargilloecia* in the cores taken offshore Antarès 101 (Ducasse, 1981, 1983) have shown that they were maintaining a nearly constant structure through time. We have there an evolutive stage covering a long period of time which cannot provide adequate stratigraphic markers and, even a major cut, such as the Eocene-Oligocene boundary, is difficult to determine.

On the other hand, in the deepest environment in the Gascogne Gulf (Ducasse and Peypouquet, 1978, 1979), and in the Rockall Region, a significant change appears in the fauna at the end of the upper Eocene (Rockall) or at the beginning of the Lower Oligocene (Gascogne Gulf). It is characterized by "psychrospheric" forms (Benson, 1975) with essentially the genera *Poseidonamicus*, *Bradleya*, *Agrenocythere*.

3. EVOLUTION OF THE HYDROLOGICAL ENVIRONMENT AT THE EOCENE/OLIGOCENE BOUNDARY IN NORTHERN AQUITAINE AND THE NORTH ATLANTIC.

The evolution of the hydrological environment at the Eocene/Oligocene boundary, based on ostracods, has been documented in the publications of Peypouquet (1977, 1979), Peypouquet and al. (1980, 1981, 1982), Ducasse and Peypouquet (1978, 1979).

Various environmental types have been considered (Peypouquet and al., 1982) in an attempt to distinguish the changes and evolutions of the environment which present a local or regional aspect compared to those which result from a more general phenomenon.

We summarize here the main results obtained during the preceding years. We will show especially that the behaviour of the ostracod fauna and the structural changes of the populations are bound to very precise constraints of the environment and that we can now discern with more precision the impact of the physiographical, climatical and especially hydrological changes on the evolution of the ostracod microfaunas.

3.1. In the lagoon-marine bay environment

The first important point worth mentioning is the appearance in Northern Aquitaine of a more or less closed marine bay environment from the Upper Eocene. This is caused by a regression at the end of the Middle Eocene and by the last orogenic movements of the Pyrenees.

In the Medoc, the ostracod microfauna first of all testifies of a marine environment, typical of a bay which is well sheltered (biozone 1).

The continental influences with a fluvial supply becomes very significant afterwards. One can see the development of forms such as *Cyamocytheridea* and *Hammatocythere* forme "commune" (biozone 2, Fig. 2).

The biozone 3 (Fig. 2), shows that the marine environment is intensified during the occurrence of the last Eocene transgression phase and it undergoes the effects of intense evaporation, and of a ceasure of continental supply. From this, hypersaline conditions result : S $^{\circ}/_{\circ\circ}$ > 38 to 40 $^{\circ}/_{\circ\circ}$ and a more important bioprecipitation of calcium carbonate on the ostracod shells (all *Hammatocythere* species are "agraded").

With biozone 4, and especially biozone 5, the environments show mesohaline affinities coupled with a temperature decrease. Those changes cause a qualitative renewal of the ostracod fauna and a "degraded" type on the shells of *Hammatocythere* (Peypouquet et al., 1980, 1981), of *Cytheretta* and *Quadracythere* (Ducasse and al., 1985, in press).

These phenomena occur with a general decrease of the marine level which is shown by emersions in the most oriental zones of the bay and in the shoals (Gayet, 1980).

3.2. On the continental platform

The evolution of environments is controlled by the progradation of the Aquitaine continental margin. This phenomenon is du to the proximity of important deltaïc environment (Peypouquet, 1979) to the east and which supply a considerable amount of detritic and organic material.

In the area such as that of Labouheyre (Peypouquet and al., 1982), there is a progressive change from a epibathyal environment at the end of the Middle Eocene, to a high energy littoral one at the end of the Lower Oligocene. When detritic supplies are more important, phenomena associated to a halocline and a pycnocline appear and they require the development of an obvious zone with oxygen minimum layer $2 < O_2 < 3$ ml/l at the edge of the continental plateau. This is precisely the case at the Eocene/Oligocene boundary.

3.3. In the deep environment

The influence of detritic supplies on the continental slope is still very important far away from the shelf , especially at the Eocene/ Oligocene boundary.

Independently from this phenomenon, the deep water ostracods show, without any doubt, that two oceanic water masses were one on top of the other in the Gascogne Gulf. A tethysian intermediate water mass, relatively warm and not affected by the Eocene/Oligocene climatic change, lies on top of a deep bottom water mass which is largely renewed at the beginning of the Lower Oligocene. This is due to the setting of younger and colder waters probably originating from the Antarctic Realm, which brings a series of so-called "psychrospheric" ostracods ; see Benson (1975) : *Bradleya*, *Agrenocythere*, *Poseidonamicus*.

To summarize results in the Aquitaine Basin, the Eocene/Oligocene boundary is seen differently depending on the palaeogeography of the studied sites.

- In the margino-littoral environment, there is a major hydrological change (T° decrease, Salinity decrease), following an important climatic change (increased rate of humidity and temperature decrease).

- In the open marine environment the continual shoaling of the continental margin interferes with the climatic change and we need more data to reach a final conclusion.

- In the deep epi-mesobathyal environment, there is no noticeable hydrological change ; the observed faunal perturbations are brought by a continental terrigenous supply.

- In the meso-infrabathyal environment, the arrival of young deep and cold water originating from high latitudes considerably modifies the microfauna, either at the end of the Upper Eocene or at the very beginning of the Lower Oligocene.

4. GENERAL CONCLUSION

This study, covering the Eocene/Oligocene boundary, shows that the influence of the environment has a very important effect on the qualitative and quantitative behaviour of the ostracod fauna.

At that time, when the environment is unstable (margino-littoral or very deep realm), the selective pressure caused by variations of environmental conditions will stimulate faunal reactions (appearance - disappearance of species, structural readjustements within populations, monomorphism or polymorphism, phenomena of agradation - degradation).

The Eocene/Oligocene boundary is then reflected through an important renewal of the ostracod fauna and can be easily distinguished.

On the other hand, when the physico-chemical parameters of the environment are more stable (epi-mesobathyal realm), the same faunal type persists through time and the Eocene/Oligocene boundary can hardly be placed.

In consequence, and this is true for all the benthos, the studies limited to stratotypes appear insufficient for an interpretation of the stratigraphic boundary, because they only give a punctual, and often localised version of the phenomenon. In order to know the different ways through which the boundary can be represented, it is necessary to enlarge the field of investigations and follow the modifications of the environments and the fauna over an area as large as possible.

REFERENCES

Benson, R.H., 1975. The origin of the Psychrosphere as recorded in changes of deep sea ostracode assemblage. Lethaia, 8, 69-83.

Benson, R.H. et Peypouquet, J.P., 1983. The upper and mid-bathyal cenozoic ostracode faunas of the Rio Grande Rise found on leg 72, Deep Sea Drilling Project. Barker, P.F., Carlson, R.L., Johnson, D.A. et al., Initial Report of the DSDP, vol. LXXII, Washington, p. 805-818, 5 pl.

Deltel, B., 1961. Les ostracodes du Paléogène moyen et supérieur d'Aquitaine méridionale. Thèse 3e cycle, Bordeaux, n° 95 (ronéotypée), 215 p., 19 pl., 24 tabl., 2 cartes.

Deltel, B., 1962. Les ostracodes du Paléogène d'Aquitaine méridionale. Mémoires BRGM (1964), Colloque sur le Paléogène, Bordeaux, 1962, n° 28, t. 2, p. 1041-1048.

Ducasse, O., 1969. Etude micropaléontologique (ostracodes) de l'Eocène nord-aquitain. Interprétation biostratigraphique et paléogéographique. Thèse de Doctorat Sciences Naturelles, Bordeaux, n° 240 (ronéotypée), 381 p., 57 tabl., 20 pl.

Ducasse, O., 1974. Quelques remarques sur la faune d'ostracodes des faciès profonds du Tertiaire aquitain. Bull. Inst. Géol. Bassin d'Aquitaine, Bordeaux 16, p. 127-135, 2 tabl., 3 cartes.

Ducasse, O., 1975. Les associations fauniques d'ostracodes de l'Eocène moyen et supérieur dans le Sud du Bassin d'Aquitaine. Distribution schématique et valeur paléoécologique. Bull. Inst. Géol. Bassin d'Aquitaine, Bordeaux 17, p. 17-26, 4 fig.

Ducasse, O., 1981. Etude populationniste du genre *Cytherella* (ostracodes) dans les faciès bathyaux du Paléogène aquitain. Intérêt dans la reconstitution des paléoenvironnements. Bull. Inst. Géol. Bassin d'Aquitaine, Bordeaux, n° 30, p. 161-185.

Ducasse, O., 1983. Etude de populations du genre *Protoargilloecia* (ostracodes) dans les faciès bathyaux du Paléogène aquitain : deuxième test effectué en domaine profond. Comparaison avec le genre *Cytherella*. Géobios, n° 16, fasc. 3, p. 273-283, Lyon, Juin 1983.

Ducasse, O., Lété, C. et Rousselle, L., 1985. Contribution à l'étude paléontologique d'une crise paléogène : populations d'ostracodes à la limite Eocène/Oligocène dans le Médoc (Gironde). A paraître.

Ducasse, O. et Moyes, J., 1971. Intérêt des ostracodes dans une esquisse paléogéographique du Tertiaire nord-aquitain. Symp. "Paléoécologie Ostracodes", Pau, 1970. Bull. Centre Rech. Pau, SNPA 1971, 5 suppl., p. 489-514, 11 fig., 6 tabl.

Ducasse, O. et Peypouquet, J.P., 1978. Les ostracodes et l'évolution paléogéographique et paléohydrologique des marges continentales de l'Atlantique nord-oriental pendant le Cénozoïque. Bull. Soc. Géol. France, 1978, (7), t. XX, n° 4, p. 441-452.

Ducasse, O. et Peypouquet, J.P., 1979. Cenozoic ostracoda, their importance for bathymetry, hydrology and biogeography. IPOD Leg 48. In Montadert, L. et Roberts, K.J. eds. Initial Reports of Deep Sea Drilling Project, 48, p. 343-363 (U.S.Government Printing Office).

Ducasse, O. et Rousselle, L., 1978. *Hammatocythere oertlii* (Ducasse) (ostracodes) : espèce polymorphe de l'Eocène du Blayais. Bull. Inst. Géol. Bassin d'Aquitaine, Bordeaux, 24, p. 3-35.

Ducasse, O. et Rousselle, L., 1979a. Les *Hammatocythere* (ostracodes) de l'Oligocène aquitain. Bull. Inst. Géol. Bassin d'Aquitaine, Bordeaux, 25, p. 221-255.

Ducasse, O. et Rousselle, L., 1979b. L'approche populationniste chez les *Hammatocythere* du Paléogène aquitain. Une manière de suivre leurs "réponses" aux fluctuations du milieu. Incidences taxinomiques et biostratigraphiques. VIIe International Symposium on ostracodes, Belgrade, 1979, p. 185-190.

Ducasse, O., Rousselle, L. et Peypouquet, J.P., 1983. Processes of evolution in marginal-coastal and bathyal ostracods, Paleogene of Aquitaine, France. Applications of ostracode (R.F. Maddocks, ed.), Proceedings of the Eight International Symposium on ostracode, July 1982, Univ. Houston, Geosc. 1983, p. 605-611.

Gayet, J., 1980. L'ensemble des environnements oligocènes nord-aquitains : un modèle de plateforme marine stable à sédimentation carbonatée. Thèse Doctorat d'Etat, Université de Bordeaux I, n° 676, 2 tomes.

Peypouquet, J.P., 1977. Les ostracodes et la connaissance des paléomilieux profonds. Application au Cénozoïque de l'Atlantique nord-oriental. Thèse Doctorat d'Etat, 1977, n° 552, Université de Bordeaux I, 443 p.

Peypouquet, J.P., 1979. Ostracodes et paléoenvironnements. Méthodologie et application aux domaines profonds du Cénozoïque. Bull. BRGM, 1979 (2e série), section IV, p. 3-79, 34 fig.

Peypouquet, J.P., Ducasse O., Gayet, J. et Pratviel, L., 1980. "Agradation et dégradation" des tests d'ostracodes. Intérêt pour la connaissance de l'évolution paléohydrologique des domaines margino-littoraux carbonatés. Actes Réunion "Cristallisation - Déformation - Dissolution des carbonates", Bordeaux, 17-18 Nov. 1980, 13 p.

Peypouquet, J.P., Ducasse, O. et Rousselle, L., 1981. Morphogenesis and environment. Theoretical and practical aspect from *Hammatocythere* : Paleogene ostracoda of the Aquitaine basin. Inter. Symp. Concept Meth. Paleo., Barcelone, p. 173-187.

Peypouquet, J.P., Ducasse, O. et Gayet, J., 1982. Les ostracodes et la paléogéographie, la paléohydrologie et la paléoclimatologie lors de la crise "Eocène/Oligocène" dans l'Atlantique nord-oriental. Paleontology, essential of historical geology/paleontologia come Scienza geostorica. I Internat. Meeting on "Paleontology, Essential of Historical Geology", Venise, Juin 1981, p. 97-121.

CALCAREOUS NANNOFOSSIL EVENTS AT THE EOCENE/OLIGOCENE BOUNDARY

K. PERCH-NIELSEN, Geol. Inst. ETH-Z, CH-8092 Zürich; with contributions from L.M. BYBELL, J. BACKMAN, M. MADILE, S. MONECHI, C. MUELLER, A. NAGYMAROSY, E. STEURBAUT

ABSTRACT

The following sequence of LO (last occurrences) is usually observed around the Eocene/Oligocene boundary: *Cribrocentrum reticulatum, Discoaster barbadiensis, D. saipanensis* and *Ericsonia formosa*. Other useful events which need further investigation are the LO of *Calcidiscus protoannula* (near the LO of *C. reticulatum* ?), the LO of *Bramletteius serraculoides*, and the abundance peaks of *Isthmolithus recurvus* and the *Ericsonia subdisticha*-group. The LO of *I. recurvus* has been reported from before the LO of *E. formosa* to after the LO of *Reticulofenestra umbilica* which has been found to overlap with *Sphenolithus distentus* making the subdivision of the Lower/Middle Oligocene difficult.

Calcareous nannofossils disappear one by one around the E/O boundary. There is no indication of a mass mortality or mass extinction.

INTRODUCTION

It was clear from the investigations published before the establishment of IGCP 174 that the Eocene/Oligocene boundary is not marked by a drastic change in the calcareous nannofossil assemblage (Beckmann et.al.1981). The sequence of FO (first occurrences) and LO of the few marker species a few million years before and after that boundary were quite well known (Fig. 1).

CP		Okada & Bukry (1980)	Martini (1971)	NP	EVENTS: ┘ FO ┐ LO
18		*Sphenolithus distentus*		23	┘ *S. distentus*
17		*Sphenolithus predistentus*			
16	c	*Reticulofenestra hillae*	*Helicosphaera reticulata*	22	┐ *R. umbilica, R. hillae*
	b	*Ericsonia formosa*			┐ *E. formosa*
	a	*Ericsonia subdisticha*		21	┐ Acme *E. subdisticha*
15	b	*Isthmolithus recurvus*	*Discoaster barbadiensis*	20	┐ *D. barbadiensis, D. saipanens. C. reticulatum*
				19	┘ *I. recurvus*
	a	*Chiasmolithus oamaruensis*		18	┐ *C. grandis* ┘ *C. oamaruensis*

Fig. 1. Calcareous nannofossil zonations and events around the E/O boundary which is usually drawn between NP 20 and NP 21 (CP 15/16) or within NP 21.

Our tasks then were

1) assist the study of selected sections which may be chosen as boundary stratotype section (Barbados, Hungary, Italy, Spain, USA Gulf Coast)

2) establish first hand correlations between calcareous nannofossil events

and paleomagnetic records

3) keep looking for FO and LO of other species than the classic markers around the E/O boundary as possible substitute markers in sections where the classic markers are very rare or absent

4) gather quantitative or semiquantitative data about the abundance of marker species around the E/O boundary

*Sections were sampled by IGCP excursions and studied by one or several participants. Their results have been or will be published in detail. In this contribution we give an overview of our observations in the various regions and a not complete summary over published DSDP and other reports.

USA GULF COAST*

Bybell & Poore (1983) published a composite range chart for 6 localities in Alabama and Mississippi showing the consecutive LO of *Chiasmolithus titus*, *Pemma papillatum*, *Pedinocyclus larvalis*, *Calcidiscus protoannula*, *Sphenolithus pseudoradians*, *S. tribulosus*, *Isthmolithus recurvus*, *Bramletteius serraculoides* and *Coronocyclus serratus* before the LO of *Reticulofenestra umbilica* and *Ericsonia formosa* and the FO of *Sphenolithus distentus* but after the here contemperaneous LO of *Discoaster barbadiensis*, *D. saipanensis* and *Cribrocentrum reticulatum*. The assemblages of these shelf environment sections are richer than those of the open sea environment both in the Upper Eocene and in the Lower Oligocene. As in Barbados, the range of *E. formosa* was found to overlap with that of *S. distentus*, a constellation which causes problems with the zonal assignment of the higher parts of the Oligocene. See also Bybell (1982) and Siesser (1983).

BARBADOS*

The sequence of events in the Oceanic Formation of the Bath Cliff Section is presented in Saunders et al. (1984). Very detailed sampling around the NP 20/21 boundary revealed the LO of *C. reticulatum* below the LO of *D. barbadiensis* and *D. saipanensis*. The latter two species are still few to common just above the LO of *C. reticulatum* but become very rare towards their here contemperaneous LO. *S. distentus* overlaps with *E. formosa* and *R. umbilica*. Most assemblages of this section are heavily affected by dissolution, the sediments having been deposited close to the CCD. *I. recurvus* and *S. pseudoradians* occur too sporadically to be used for the subdivision of the Upper Eocene, whereas the LO of *Chiasmolithus grandis* is a useful datum level. The NP 20/21 boundary occurs several 100 000 years after the deposition of a tektite horizon but slightly before the LO of *Turborotalia cerroazulensis cunialensis*, the LO of *Hantkenina sp.* and the $\delta^{18}O$ shift to more positive values.

ATLANTIC OCEAN

The E/O boundary was missed at many DSDP Sites and cored in a few. The best investigated Sites are 522/523 on Walvis Ridge in the South Atlantic (Percival, 1984; Backman, in press). The correlation of the LO with the paleomagnetic record is shown in Fig. 2 together with the semiquantitative data on the disapearance of the discshaped discoasters and the stable isotope data. As in Barbados (Saunders et al., 1984), the $\delta^{18}O$ shift occured in NP 21, shortly after the beginning of the normal polarity interval of Chron 13.

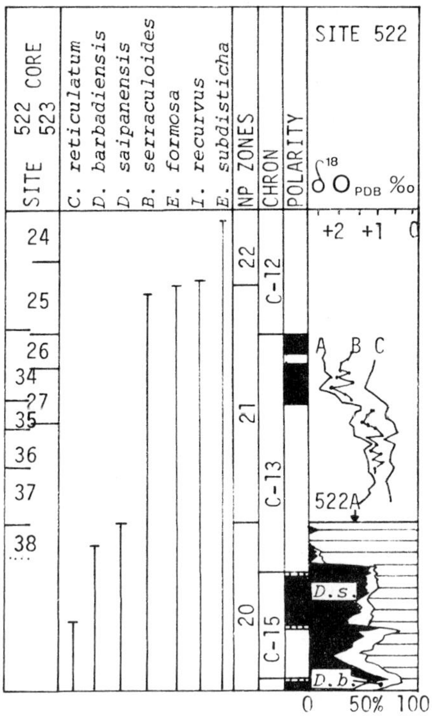

Fig. 2. Summary of events, paleomagnetic record, stable isotopes (A=*Stilostomella* spp., B=*Catapsydrax dissimilis*, C=*Globigerina venezuelana* and cumulative % of discoasters. Compiled from Backman, in press and Oberhänsli & Toumarkine, in press, for DSDP Sites 522 and 523 on Walvis Ridge, South Atlantic

On the Falkland Plateau (DSDP Sites 511, 513, 330), *C. reticulatum* disappears about 3 m below the LO of *D. saipanensis*. A barren interval separates this part of the section from NP 21 with common *I. recurvus* and few *E. formosa*. Within NP 22 Wise (1983) noted the LO of *Chiasmolithus oamaruensis* followed by the LO of *I. recurvus* and *Ericsonia subdisticha s.ampl.* (his *Clausicoccus fenestratus)* at Site 511. At Site 513A the LO of *E. formosa* is followed by the LO of *I. recurvus, E. subdisticha* and *R. umbilica*, while *C. oamaruensis* is intermittently present also higher up.

No details are extractable about the distribution and abundance of marker species at DSDP Sites 366 and 369A in the Western Central Atlantic where the NP 20/21 boundary is represented in Cores 10 and 32 respectively.

At DSDP Site 549 in the Bay of Biscay the LO of *C. reticulatum* was found slightly below the here contemporaneous LO of *D. saipanensis* and *D. barbadiensis* which occurred about 9 m below the LO of *G.c. cunialensis, G.c. cocoaensis* and *Hantkenina* sp. (Snyder, Müller & Miller, 1984). The $\delta^{18}O$ shift was found another 2 m higher up by these authors. At Site 401, the NP 20/21 boundary lies between Cores 2 and 3. At other Sites part of the Upper Eocene is missing.

At high northern latitudes, discshaped discoasters are absent. At DSDP Site 336 on the Iceland-Faroe Ridge the LO of *C. reticulatum* is used to define the NP 20/21 boundary. Here and at Site 349 on the Jan-Mayen Ridge, *I. recurvus*, as a species typical of the Late Eocene and Early Oligocene, is present over about 150m of sediments which are characterised by low diversity of calcareous nannofossils and barren intervals. *E. formosa* is very rare or absent in Eocene samples at these Sites and at Site 112 in the Labrador Sea and Sites 116 and 406 at Rockall. In the latter two areas discshaped discoasters and *C. reticulatum* are present (Müller, 1979; Perch-Nielsen, 1972).

NORTHERN EUROPE

In Germany, assemblages assignable to NP 19/20 with *D. saipanensis, D. barbadiensis, I. recurvus* and *C. reticulatum* are present in the Gehlberg Formation at Helmstedt and NP 21 and 22 were reported from the Latdorf Formation at Doberg/Piepenhagen and in the Silberberg Formation at Helmstedt (type Latdorfian; Martini & Ritzkowski, 1968; Roth, 1970; Haq, 1971; Martini, 1971).

In Denmark, the E/O boundary has been studied in the cores of the Viborg boring. Rich assemblages with *D. saipanensis, I. recurvus* and *C. reticulatum* (NP 19/20) in the Sövind Marls are overlain by the poorer assemblages of the Viborg Clay which contain no discshaped discoasters but reworked Eocene and Cretaceous coccoliths. The E/O bd was usually set at the bd between the two formations (Dinesen, 1972; Thiede et al., 1980). The NP 20/21 bd could, however, be set within the Sövind Marls above the LO of *D. saipanensis* and *C. reticulatum* in sample 137 of Thiede et al. (1980). The higher occurrences of *C. reticulatum* in samples 109 and 110, which both include considerable reworking from Middle and Upper Eocene, could be considered reworked. This interpretation is supported by the results of Dinesen (1972) who reported the LO of *Globigerapsis* cf. *G. index* from sample 139. *G. index* is known to disappear near the LO of the discshaped discoasters and *C. reticulatum* in many E/O boundary sections.

In Belgium, NP 20 with *C. reticulatum, D. saipanensis* and *S. pseudoradians* but without *I. recurvus* was found in the type area of the Bassevelde Sand Member. The stratotype section of the overlying Grimmertingen Sand Member can be assigned to NP 21 and includes *I. recurvus* and *E. formosa* but no *C. reticulatum* or *D. saipanensis*. The LO of *I. recurvus* was found above the LO of *R. umbilica* (Steurbaut, in press).

Aubry (1983) found NP 19/20 with *I. recurvus, D. saipanensis* and *C. reticulatum* in the Brockenhurst Beds at Whitecliff Bay, UK. No assemblages of NP 21 age were reported by Aubry (1983) from southern England or France. Also Lezaud (1967) described no assemblages with *I. recurvus* but without *D. saipanensis*.

ALPS AND CARPATHIANS

Charollais et al. (1980) have shown that the *Globigerina* Marls in the French Alps are of Early Oligocene age. Discshaped discoasters are absent and *I. recurvus*, *E. formosa* and *R. umbilica* occur only sporadically. Herb (personal comm. 1985) knows of no good section from the Eocene *Globigerina* Marls to the Oligocene shales in Switzerland.

Aubry (in Van Couvering et al., 1981) reported NP 19/20 and (questioned) NP 21 in the *Globigerina* Marls of southern Poland, from where also Radomski (1968) described rich assemblages including discshaped discoasters, *I. recurvus*, *C. reticulatum* and *Corannulus germanicus*. The overlying Menilitic Clay is barren or includes only a very impoverished assemblage without discshaped discoasters.

HUNGARY*

The E/O boundary lies within the Buda Marl Formation which was deposited at depths of 100-200 m and contains only few calcareous nannofossils. Reworking of Eocene forms poses problems for the zonation. Fig. 3 showing results from the Kiscell 1 borehole stands for other boreholes across the boundary. For details see Báldi et al. (1984) and Nagymarosy (1983). Note the increase in abundance of *L. minutus* and *Z. bijugatus* (and *I. recurvus* in other holes) at the NP 20/21 bd. The distribution of *C. oamaruensis*, *C. grandis*, *C. reticulatum* and *S. pseudoradians* is too sporadic to furnish any base for a zonal subdivision of the Upper Eocene. The LO of the Acme of *E. subdisticha* is used in some sections as a substitute for the LO of *E. formosa* for the NP 21/22 boundary.

The presence of Pteropods (*P*) and *Cardium lipoldi* (*Cl*) is indicated in Fig. 3, too.

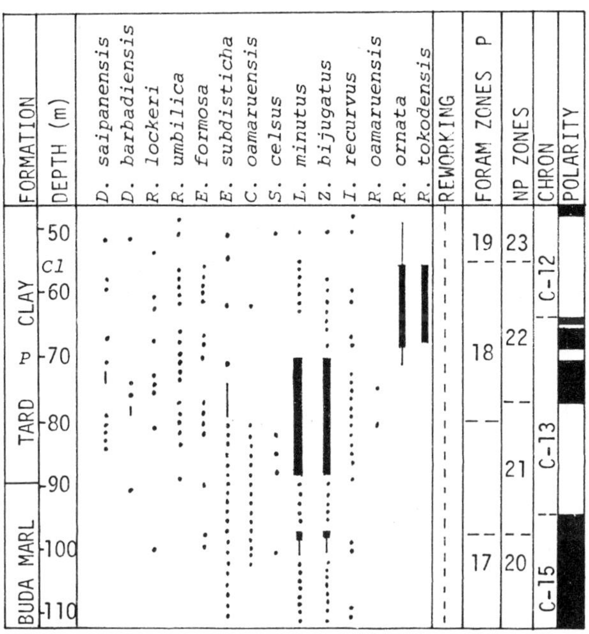

Fig. 3. Summary of ranges, paleomagnetic record and zonations in Kiscell (KL)-1 borehole in Budapest after Nagymarosy (personal comm., 1985). For a different interpretation see DISCUSSION

ITALY *

At Priabona, the type section of the Late Eocene Priabonian has been studied recently by Jossen (1982) and Verhallen & Romein (1983). The latter suggested a combined NP 20/21 Zone from the FO of *Sphenolithus predistentus* to the LO of *E. formosa*. The calcareous nannofossils are poorly preserved.

Several sections were visited by the IGCP group and studied from the Umbrian Apennines (Lowrie et al., 1982; Monechi & Thierstein, in press; Madile & Monechi, in press). The first authors presented first hand correlations of planktic fossils with the paleomagnetic record. Improved data are presented in Fig. 4.

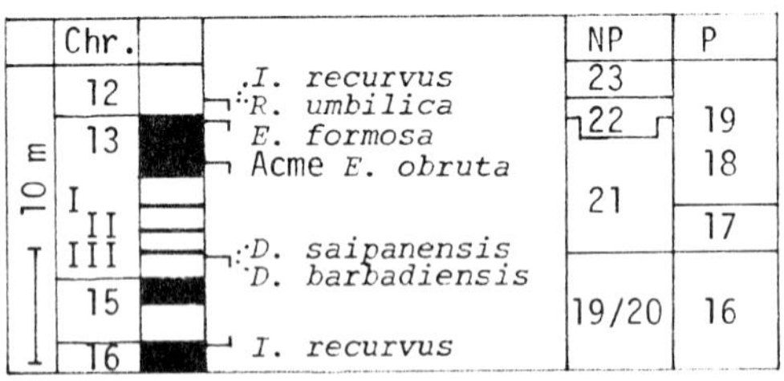

Fig. 4. Summary of events, paleomagnetic record and zonations in the Contessa Highway section after Madile & Monechi (in press in Nochi et al., this volume)

Late Eocene and Early Oligocene assemblages are very similar. A slight increase in *I. recurvus* was noted just above the LO of *D. barbadiensis* and *D. saipanensis*. *E. formosa* decreases around the LO of the discshaped discoasters. *E. obruta* (*E. subdisticha* or *E. fenestrata* of other authors) is rare in the Upper Eocene and becomes common in the upper part of NP 21. *C. reticulatum* is rare and only discontinuously present in NP 19/20.

SPAIN*

Two sections in S Spain, NE of Granada, were investigated (Molina, this vol.). At Molino de Cobo, reworking poses serious problems to the recognition of any boundaries based on LO. At Fuente Caldera (Fig. 5), the NP 20/21 boundary is suggested after the drastic decrease of the discshaped discoasters, where also an increase in *I.recurvus* could be observed, and where *G. index* has its LO. As in Italy, *E. obruta* is common in the upper part of NP 21. *C. protoannula* and *C. reticulatum* decrease before the discshaped discoasters. See also Madile & Molina, this volume.

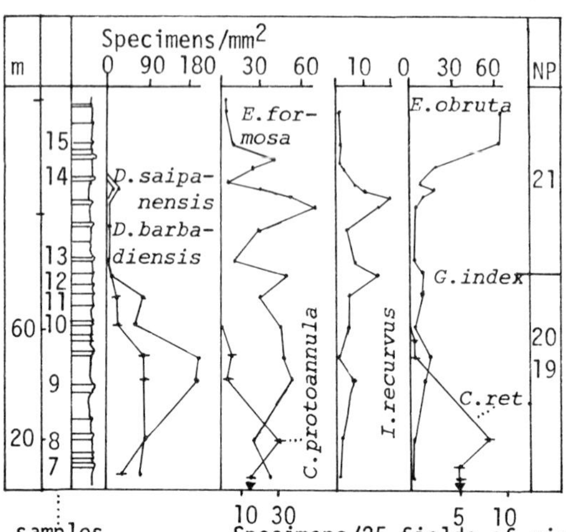

Fig. 5. Semiquantitative data of the distribution of 7 species at Fuente Caldera. After Monechi (personal comm., 1985) & P-N, unpub.

CRIMEA AND CAUCASUS

Zones NP 20 and NP 21 were found in the Almian Stage on the Crimea (locality F 3 of the Field Guide of the XII Europ. Micropal. Coll., 1971). *D. saipanensis* and *D. barbadiensis* are present in the lowermost 2 samples taken from the "*G. index* and large *Globigerina* Zone", which contain no typical *C. reticulatum*. The overlying samples from the *Bolivina* Zone include only very rare *E. formosa*, few to common *E. obruta* and few *I. recurvus* among other species, assigning them to NP 21.

Rich assemblages assignable to NP 19/20 were described from the NW Caucasus by Hay et al. (1966) and include *I. recuvus, D. saipanensis* and *C. germanicus*, a species also found just below and above the NP 20/21 boundary in the Crimea and other sections.

INDIAN OCEAN

The NP 20/21 boundary was cored at several DSDP Sites, but only little detailed information is available presently. At Site 253 on Ninetyeast Ridge, the NP 20/21 boundary lies in the uppermost part of Core 253-13 or between this core and 253-12CC, in an interval of about 2 m from where no samples were studied. Rare *D. saipanensis* are accompanied by common *I. recurvus* at the top of NP 20. At Site 267 at 59^0 southern latitude S of SE Indian Ridge, several samples with common *I. recurvus*, very rare *E. formosa* and common *Chiasmoltihsu sp.* could, at this high latitude, be assigned NP 19/20 or NP 21. They were recovered just above basement. The NP 20/21 boundary was recovered at Sites 214, 216 and 217 in the Eastern Indian Ocean. *D. saipanensis* disappears with or just above *D. barbadiensis* and *C. reticulatum*. *R. umbilica* is very rare or absent in most samples around the E/O boundary. The LO of *E. formosa* coincides with the LO of *B. serraculoides* at Site 214, is about 1 m above it at Site 216 and about 9 m above it at Site 217, where the LO of *E. formosa* occurs about 1.5 m below the LO of *D. saipanensis* (Gartner, 1974). Neither *I. recurvus* nor *Chiasmolithus* were found. At Sites 219, 220?, 223 and 224 in the Arabian Sea, Boudreaux (1974) observed the LO of *D. saipanensis, D. barbadiensis* together with the LO of *E. formosa* and/or *Cyclococcolithus lusitanicus*. Since the two species are usually considered synonymous, one of them may be *C. reticulatum*. *I. recurvus* was only found at Site 219 (*I.sp.cf.I. recurvus*). At Site 282 west of Tasmania, the LO of *C. reticulatum* is followed by a sharp increase in *I. recurvus* and *E. obruta*. Only single specimens of *D. saipanensis* were recovered above the LO of *C. reticulatum*.

PACIFIC OCEAN

At Site 445 in the Philippine Sea, common *C. reticulatum* disappear about 1.5

m below common *D. saipanensis* and rare *D. barbadiensis* also disappear. *S. distentus* overlaps with *E. formosa* and *R. umbilica* which are both considered reworked by Okada (1980). The LO of *B. serraculoides* occurs 3 m above the LO of *E. formosa* (considered reworked), if one considers the occasional higher occurrences as reworked. *R. umbilica* disappears before *E. formosa* and *I. recurvus* is rare throughout its range. No *Chiasmolithus* were found. Renewed study of Site 277 on Campbell Plateau revealed very rare *C. reticulatum* in Core 277-21CC. The LO of *D. saipanensis* in Sample 277-21-3,110cm is followed at the base of Core 277-20 by an increase in abundance of *I. recuvus*, which is rare to very rare in NP 19/20 below. *E. obruta* becomes more common in Core 277-19, where the $\delta^{18}O$ shift was found by Shackleton & Kennett (1975). *C. oamaruensis* is common in most samples of this interval while *E. formosa* has not been found.

DISCUSSION

Preparing this overview has again revealed that the sequence of events around the E/O boundary is usually, but not always, the same. It has also shown that none of the events is associated with volcanic activities, extraterrestrial impact(s) or $\delta^{18}O$ shift. Reworking and environmental limits to the distribution of the marker species still pose problems to the zoning of this interval. Comparing correlations of calcareous nannofossil events with paleomagnetic records in "complete sections" (Figs 2,4,5) should help us to better understand the distribution of FO and LO in time and space while still leaving room for individual interpretations:

	LO of *E. formosa*	LO of *I. recurvus*
Site 522	Chron 12R	Chron 12R
Italy	13N upper part	12R
Hungary	13N basal part (according to Nagymarosy's interpretation, that higher occurrences are due to reworking)	12N or above
	12R (if high occurrences are considered in place)	

REFERENCES

The references can be found in the INA Newsletters 1-7 (1979-85) and/or in Perch-Nielsen (1985, in Bolli, H.M., Saunders, J.B. & Perch-Nielsen: Plankton Startigraphy. Cambridge University Press: 329-554).

INFORMATION FROM DIATOM ANALYSIS CONCERNING THE EOCENE - OLIGOCENE BOUNDARY

J. FENNER

Diatom analysis of Eocene and Oligocene DSDP Sites in the Atlantic and Pacific Oceans showed basic differences in the response of planktic diatom assemblages in the high southern and in the low latitudes to the tectonic, oceanographic and climatic changes near the Eocene - Oligocene boundary. These differences can be demonstrated using the results from DSDP Site 366, on the Sierra Leone Rise in the eastern low latitude Atlantic, and DSDP Site 511, on the Falkland Plateau in the southwestern Atlantic Ocean.

At DSDP Site 366 (Fig.1) a ca. 90 m long, continuously cored sediment sequence through the Late Eocene and Early Oligocene was recovered, in which both siliceous and calcareous microfossil groups could be used throughout to provide stratigraphic framework (Čepek et al.,1978),- and in which no hiatus was detected.

Starting in the G.centralis Zone the abundance of diatoms per g sediment decreases in the Early Oligocene about one order of magnitude. This decrease does not seem to be due to dissolution, but rather due to a decrease in productivity as the preservation of diatoms remains good and the relative abundance of the more dissolution resistant radiolaria and sponge spicules does not increase. The decrease in abundance of diatoms is even more severe if one considers that the sediment accumulation rates decrease while the calcium carbonate content increases.

The interval in the latest Eocene and earliest Oligocene, for which strongly fluctuating $\delta^{18}O$ values are measured (Vergnaud-Grazzini and Rabussier-Lointier, 1980), is characterized by a maximum in diatom diversity (H calculated using the Shannon-Wiener equation), and an increase in the occurrence and abundance of freshwater diatoms and phytoliths. (All samples analyzed for diatom abundance were also analyzed for freshwater diatoms and phytoliths.) Both of these component groups are of terrestrial origin and must have been transported by wind to the Sierra Leone Rise, which is topographically isolated from the continental margin by a trough. Thus the increase in these terrestrial biosiliceous components represents an increase in wind intensity and probably in aridity in low southern latitude Africa during this interval.

The increase in diatom diversity is due to a stronger lateral input of neritic species (e.g. Chaetoceros resting spores) and high southern latitude species (e.g. Pyxilla reticulata), as well as due to a greater abundance of species characteristic for oceanic fronts (e.g. Rhizosolenia hebetata group).

These compositional changes in the diatom assemblage indicate stronger surface currents during this interval.

The oxygen isotope curve seems to be inversely correlated with the abundance of Cestodiscus spp., a group which becomes dominant in the Early Oligocene together with species of the genus Coscinodiscus (e.g. C. excavatus). On the other hand a large number of species characteristic for the Late Eocene decrease strongly in abundance near or at the Eocene - Oligocene boundary, but become extinct only much later: during the Oligocene. This turnover in the composition of the diatom assemblages is very distinct and was found in all low latitude sites in the Atlantic and Pacific Oceans (Corliss et al., 1984).

In the ca. 180 m long Late Eocene - Early Oligocene sediment sequence of the high southern latitude DSDP Site 511 (Figs. 2,4) such a turnover in the diatom assemblage composition is not present. Only few "Eocene" species disappear near the Eocene - Oligocene boundary. And these species were warm water loving and of subordinate prevalence already in the Late Eocene (e.g. Skeletonema barbadense). Few species have their evolutionary first occurrence in the Early Oligocene. Neither these nor those existing already in the Late Eocene became dominant in the Early Oligocene. The abundance and diversity of planktic diatoms at this site, which is characterized by high accumulation rates, do not show a special trend, but fluctuate. The three maxima of abundance and diversity of diatoms: one in the latest Eocene, two in the Early Oligocene, are correlated with an increased abundance of neritic species.

It seems that the changes in abundance, diversity, and assemblage composition of the planktic diatoms in the Late Eocene to Early Oligocene are not so much controlled by surface water temperature, which changes least in the low latitudes and strongest in the high southern latitudes, but rather by the intensity of the surface water currents and the availability of nutrients. It is probable that nutrients were not a limiting factor in the high southern latitudes but that they were in the low latitudes, where regression, increased aridity, and stratification of the water column all worked towards restricting the input of nutrients into the surface water of the oceans.

REFERENCES

Only those references are given here, which are not published in the respective Initial Reports of the Deep Sea Drilling Project.

Corliss, B.H., Aubry, M.-P., Berggren, W.A., Fenner, J.M., Keigwin, L.D., Jr., and Keller, G., 1984. The Eocene/Oligocene boundary event in the deep sea.- Science, 226: 806-810.

Vergnaud-Grazzini, C. and Rabussier-Lointier, D., 1980. Compositions isotopiques de l'oxygène et du carbone des foraminifères tertiaires en Atlantique équatorial (Site 366 du DSDP).- Rev. Géol. Dynam. et de Géogr. Phys., 22: 63-74.

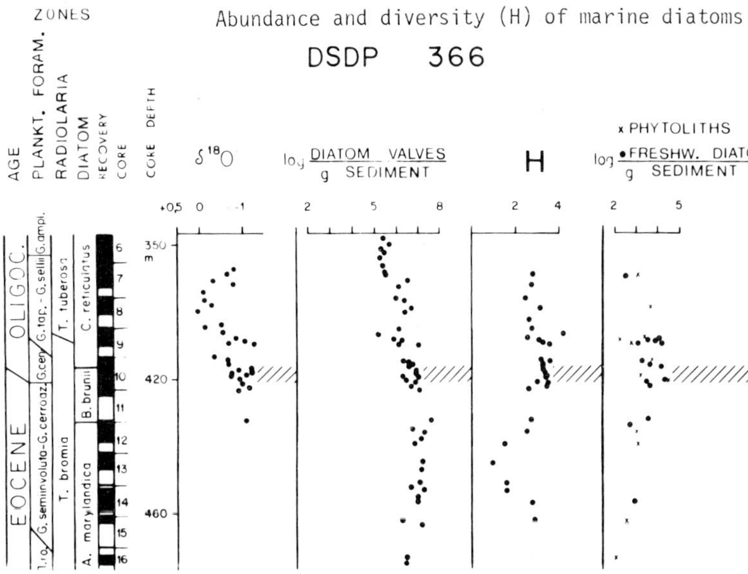

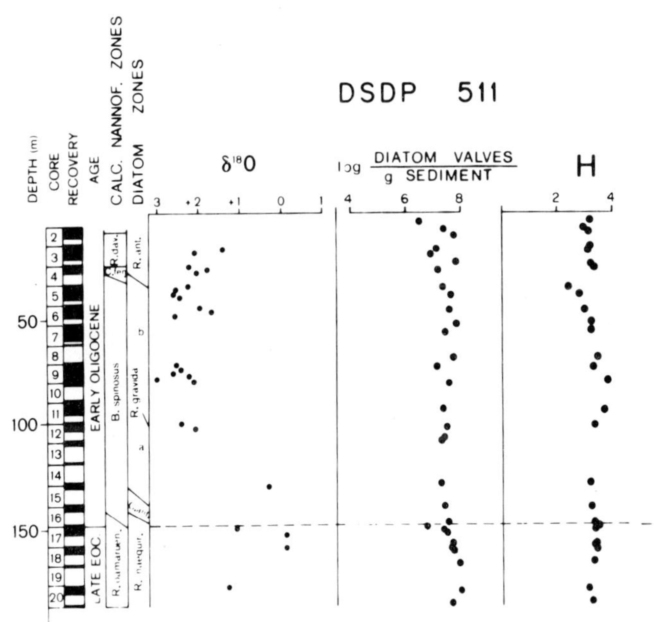

Fig. 1. DSDP Site 366. $\delta^{18}O$ of the total carbonate fraction (Vergnaud-Grazzini and Rabussier-Lointier, 1980). Plankt. Foram. Zones (Krasheninnikov and Pflaumann, 1978), Radiolaria Zones (Johnson, 1978), Diatom Zones (Fenner, 1984) The Eocene - Oligocene boundary is placed at the last occurrence of D. barbadiensis (Lancelot et al., 1978).

Fig. 2. DSDP Site 511. $\delta^{18}O$ of the planktic foraminifera (Muza et al., 1983). Calcar. Nannof. Zones (Wise, 1983), Diatom Zones (Fenner, 1984). The Eocene-Oligocene boundary is placed at the last occurrence of D. saipanensis (Wise, 1983).

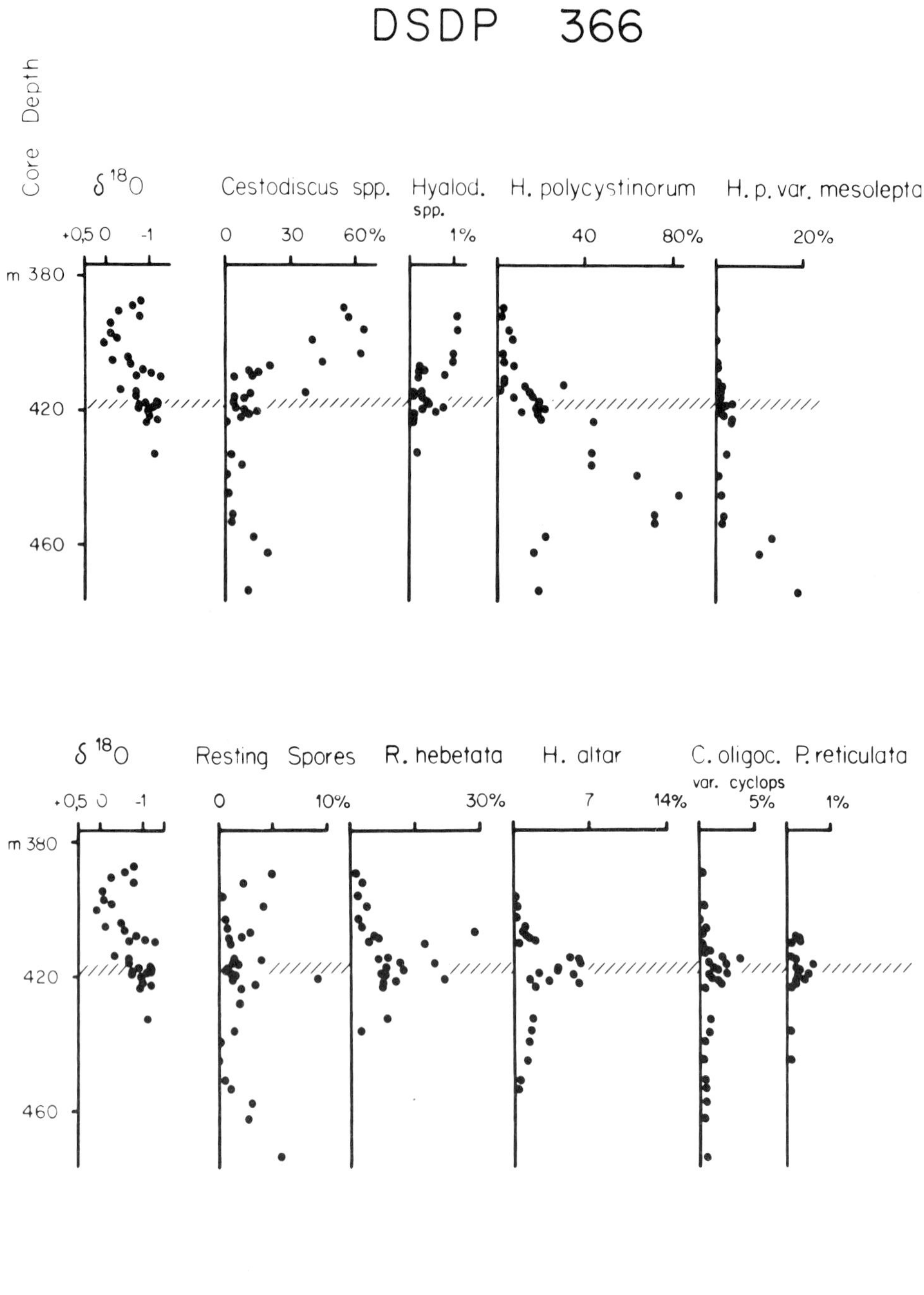

Fig. 3. Percent abundance of diatom species and species groups of the total diatom assemblage through the Late Eocene and Early Oligocene of DSDP Site 366, correlated with the oxygen isotope curve.

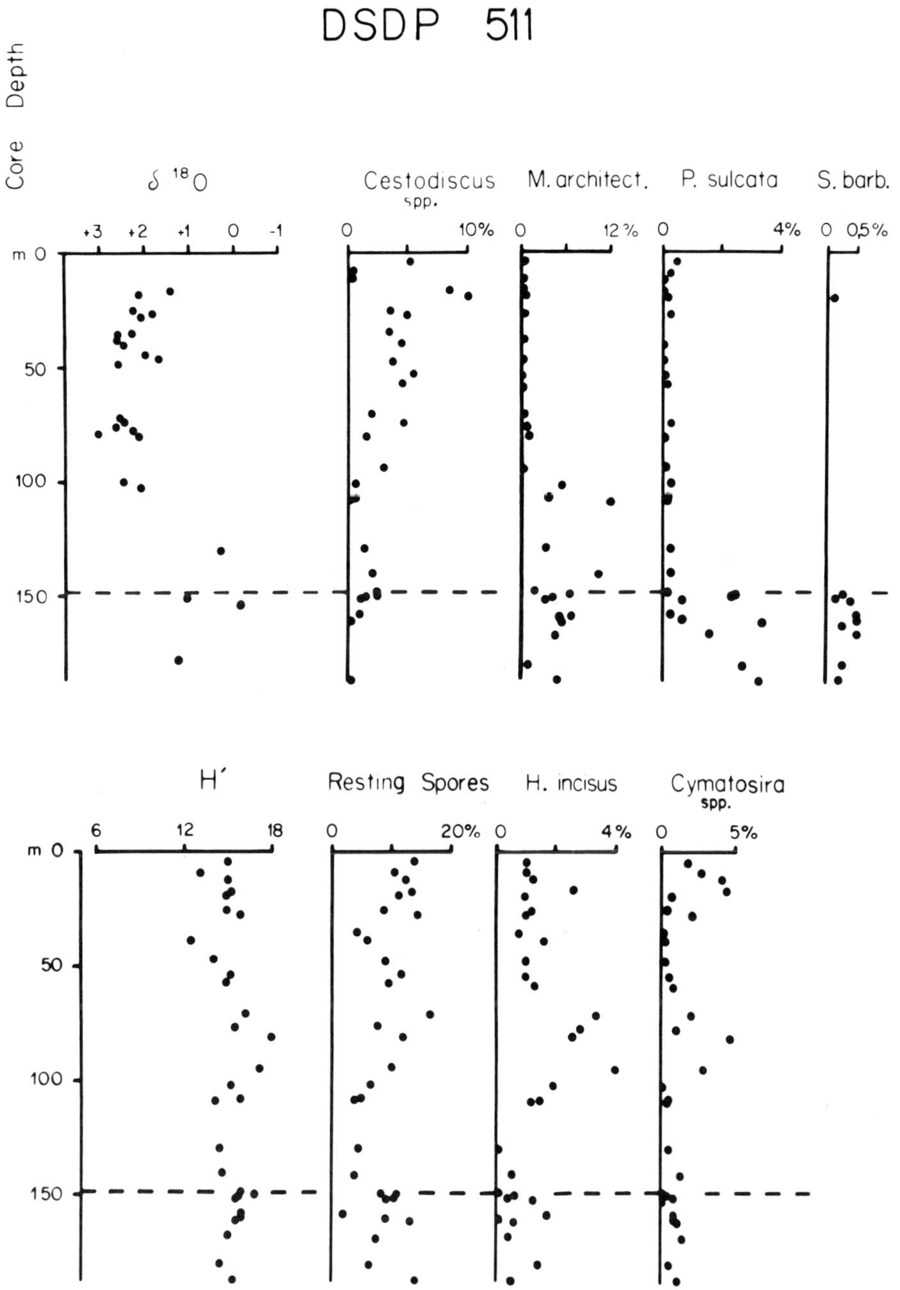

Fig. 4. Percent abundance of diatom species and species groups of the total diatom assemblage through the Late Eocene and Early Oligocene of DSDP Site 511, correlated with the oxygen isotope curve and the species equitability (H′).

EVOLUTION OF THE MICROFLORA AND DINOCYSTS AT THE EOCENE-OLIGOCENE BOUNDARY IN WESTERN EUROPE

JJ. CHATEAUNEUF

.1 INTRODUCTION

In the Anglo-Paris Basin and more generally in Western Europe the Eocene-Oligocene boundary is better marked by the evolution of the microflora than by that of the dinocysts. This is due to the fact that the sediments deposited in these regions towards the end of the Eocene and in the earliest Oligocene have a marked tendency to be continental or lagoonal, and have yielded associations essentially of spores and pollens. The dinocyst populations, generally poor or even monospecific, seldom allow precise comparison with the much more complete ones that have been described from oceanic sections. It is therefore necessary, near this boundary, to compare discontinuous sedimentary associations extending, in terms of nannoplancton, from zone NP 19-20 to zone NP 23.

.2 THE EVOLUTION OF POLLENS AND SPORES AT THE EOCENE-OLIGOCENE BOUNDARY

This boundary in the Paris Basin is marked by a deterioration of climate (Cavelier et al., 1981), and the changes in the vegetation that occur at this stage are more aptly termed regression or degradation than evolution. The commonly rich thermophile floras that occupied the borders of the basin during the Eocene were replaced towards the end of the Ludian (= Priabonian) by a coniferous and herbaceous flora, indicating a drier, colder climate. This can be seen for example in the Pantin Marls and the Romainville Green clay, formations that lie on each side of the boundary. After the deposition of the Orgemont Pebble Beds, immediately above the Green Clay an improvement in the climate (probably in terms of rainfall) enabled a few of the Eocene thermophile taxa to return, though most had permanently disappeared from the Paris Basin. The actual boundary is thus marked essentially by the extinction of genera or species (Chateauneuf, 1980), and is located at the top of the *Tricolporopollenites raguhnensis* zone and the base of the *Pityosporites 329* (a form similar to the modern genus *Cathaya*).

Boehlensipollis hohli, the classic marker of the Stampian, which appears at this level more or less throughout Europe(Sittler et Schuler, 1975) is known in the Paris Basin only from the Orgemont Pebble Beds onwards.

In the Armorican Massif, the lower sapropels of the Rennes Basin have yielded a few examples of *Boehlensipollis hohli*, while the Striatelle Marl of the Landean Basin and the *Mohria* Clay of the Rennes Basin contain *Tricolporopollenites raguhnensis*. These microfloras unfortunately do not come from the same localities and have not been seen in superposition (Ollivier - Pierre, 1980). The Eocene-Oligocene boundary can thus be placed somewhere between the deposition of the Striatelle Marl and that of the lower sapropels. The position of the St Jacut-du-Mené Clay and the Guilers Sands (regarded as intermediate) remain uncertain in the absence of either of the above markers.

On a wider scale, a palyno-stratigraphic synthesis of the **Paleogene basins of Western Europe** has been made in the context of IGCP Project 124 (Tertiary Basins of North Western Europe). It is concerned in particular with the distribution of Paleogene microflora in France, Southern England, Belgium, Germany and the North Sea. The only important event that has been demonstrated at the Eocene-Oligocene boundary was the appearance of the two taxa *Boehlensipollis hohli* and *Slowakipollis hippophaeoides*. At the same time other taxa such as *Dicolpopollis Kockeli* and *Scabratricolporites cylindricus* appeared more locally (Roche and Schuler, 1980). More recently, in the South of the Rhine graben (the Mulhouse potash basin), Schuler (1982) has shown an important change in the microflora in the middle saliferous zone, where *Boehlensipollis hohli* appears immediately after a notable development of conifers at the top of the lower saliferous zone, a change that is entirety comparable with that occuring at the end of the Ludian in the Paris Basin. The Eocene-Oligocene Boundary is similary marked by the appearance of *Boehlensipollis hohli*, in the external Alpine massifs, immediately above the Priabonian Nummulitic limestone, at the base of the Foraminiferal Marl, or in an interval of calcareous sandstone containing large arenaceous foraminifera between these two formations.

.3 THE EOCENE-OLIGOCENE BOUNDARY IN DINOFLAGELLATE TERMS

As was stated in the introduction, the evolution of dinocysts (or cysts of dinoflagellates) at the level of the Eocene-Oligocene boundary is poorly known in Europe because of the numerous gaps during this period.

Thus, **in the Paris Basin** this boundary cannot be defined on the basis of dinocysts as no association is known in the upper Ludian. The middle Ludian *Kisselovia clathrata angulosa* association is replaced at the base of the Green Clay (basal Stampian of the Sannoisian facies) by the *Phthanoperidinium amoenum Phthanoperidinium flebilis* association, immediately preceding the classic European early Oligocene *Wetzeliella gochtii* zone(Chateauneuf, 1980).

In Southern England (Hampshire and the Isle of Wight) the boundary is similary marked by emergences, with littoral and lacustrine facies. in the late Eocene, the dinocyst associations of the Brembridge Marls (Oyster beds) like those of the basal Oligocene of the Lower Hamstead beds (Nematura beds), are atypical and commonly monospecific (Liengjarern et al., 1980). Thus, *Glaphyrocysta microfenestrata* is the only species represented at certain levels in the Oyster Beds. However the *Phthanoperidinium amoenum* and *flebilis* zone recurs above these formations in the Early Oligocene (upper Hamstead Beds), preceding the *Wetzeliella gochtii* zone.

The Eocene-Oligocene boundary **in Belgium** lies between the Neerrepen Sands and the Henis Clay (these formations have been defined in quarries). Dinokysts are abundant in the Neerrepen Sands , where a diversified association belongs to the *Kisselovia clathrata angulosa* zone distinguished in the Ludian of the Paris Basin. The Henis Clay contains a few cyst forms that precede the appearance of *Wetzeliella gochtii* at the level of the Boom Clay, but which do not stricIty belong to the *Phthanoperidinium amoenum* zone defined in the Paris Basin.

In Northern Germany and in the Rhine graben the formations close to the boundary are very poor in dinocysts. The Latdorfian Sands of Lehrte and Helmstedt, which have yielded a few examples of *Kisselovia clathrata* and *Rhombodinium perforatum* can be equated with the Ludian of the Paris basin. The Rupelton, which overlies these sands already contain a *Chiropteridium partispinatum* and *C. lobospinosum* association known from the late Early Stampian of the Paris Basin (Fontainebleau Sands).

In the Alpine domain, the base of the Oligocene is poorly characterised in respect of dinocysts. The Priabonian marls of Scaffarels, Granella,Priabona and Brendola however (Nannoplancton zone NP 19-20 and base NP 21) contain the succession *Rhombodinium perforatum* - *Kisselovia clathrata*.

In the context of the various studies undertaken by European and North American researchers for **IGCP Project n° 124** (stratigraphy of the Tertiary basins of Northwestern Europe), a synthesis of the stratigraphic distribution of dinocysts has been made which also incorporates the results of research in the North Sea and the northern part of the Atlantic Ocean. The Paleogene is divided into 15 dinocyst zones, and the Eocene-Oligocene boundary is marked by the disappearance of the species *Phthanoperidinium echinatum, Areosphaeridium diktyoplokus, Rhombodinium perforatum*, and by the appearance of *Phthanoperidinium amoenum* and *Wetzeliella gochtii*.

This zonation is nevertheless somewhat artificial insofar as it attempts, on the basis of a calibration against the Nannoplancton zones, to integrate the results of zonations in middle and northern latitudes which have different key species. Thus the zonations established by Williams (1975) in the east Canadian offshore deposits, by Costa and Downie (1979) at Rockall, by Manum (1976) in the Tertiary of Norway and Greenland contain only very few Wetzeliellaceae and apparently none of the known markers of the Ludian-Priabonian and

Stampian in the Anglo-Belgian and Paris Basins. The two species *Rhombodinium perforatum* and *Wetzeliella gochtii* on the other hand seem to have a fairly wide extension in middle latitudes. They have both been found in the Carpathians (Van Couvering et al., 1981), and the latter has been found in the Thracian Basin of Turkey (Ediger, 1981).

.4 CONCLUSIONS

There is no universal scale nor widely distributed markers by which the Eocene-Oligocene boundary in Western Europe can be characterized on a basis of spores, pollens or dinocysts. There was nevertheless an important climatic change at this time, clearly marked in the evolution of the microflora by the disappearance of a large number of thermophile taxa. In addition, the Species *Boehlensipollis hohli* appears more or less throughout Europe at the base of the Oligocene. The scenario is very similar as far as the dinocysts are concerned, with the common disappearance of a variety of species at the top of the Eocene, and the appearance of *Wetzeliella gochtii* at the base of the Oligocene. A clear impression is given that climatic modifications at the Eocene - Oligocene boundary played regulating role in levelling the microfloral associations of the Oligocene relative to the much more diversified ones of the Eocene.

REFERENCES

Bujak J.P., Caro Y., Chateauneuf J.J., Costa L.I., Eaton G., Hansen J.M., Heilman-Clausen C., Ioakim C., Manum S.B. (in press). A standard Dinoflagellate zonal notation for the Tertiary of NW Europe. IGCP 124 Report.

Cavelier C., Chateauneuf J.J., Pomerol C., Rabussier D., Renard M., Vergnaud-Grazzini C. (1981). The geological events at the Eocene/Oligocene boundary. Palaeogeography, palaeoclimatology, palaeocology. vol.36, n° 3/4, 223-248.

Charollais J., Hochuli P.A., Oertli H.J., Perch-Nielsen K;, Toumarkine M., Rögl F., Pairis J.L. (1980). Les Marnes à Foraminifères et les Schistes à Meletta des chaînes subalpines septentrionales (Hte-Savoie - France). (The "Foraminiferal" Marls and "Meletta" shales of the southern Sub-alpine chains, Haute-Savoie, France). Eclogae Helveticae, vol. 73/1, 9-69.

Chateauneuf J.J. (1980). Palynostratigraphie et paléoclimatologie de l'Eocène supérieur de l'Oligocène du Bassin de Paris. (Palynostratigraphy and paleoclimatology of the Late Eocene and Oligocene of the Paris Basin). Mémoire B.R.G.M. n° 116, 357 p.

Costa L.I. and Downie C. (1979). Cenozoic dinocyst stratigraphy of sites 403 to 406 (Rockall plateau) IPOD leg 48. Init. Rep. DSDP 48, 513-529.

Costa L.I. and Muller C. (1978). Correlation of Cenozoic dinoflagellate and Nannoplankton zones from the NE Atlantic and NW Europe. Newl. stratigr., 7 (2), 65-72.

Ediger V.S. (1981). Some Tertiary dinoflagellate cysts from Thrace basin, Turkey, and their usage in paleoenvironmental analysis. Revista Espanola de Micropaleontologia, vol. XIII, n° 1, 89-103.

Liengjarern M., Costa L.I., Downie C. (1980). Dinoflagellate cysts from the Upper Eocene-Lower Oligocene of the Isle of Wight - Paleontology vol. 23, part. 3, 475-499

Manum S.B. (1976). Dinocysts in Tertiary Norwegian Sea sediments (DSDP-leg. 38) with observations on palynomorphs and palynodebris in relation to environment - Init. report DSDP 38 : 897-919.

Ollivier-Pierre M.F. (1980). Etude palynologique (spores et pollens) de gisements paléogènes du Massif Armoricain. Stratigraphie et paléogéographie. Mémoires de la Société géologique et minéralogique de Bretagne, n° 25 : 1-239

Roche E. and **Schuler M.** (1980). Etude palynologique du "complexe de Kallo". (Palynologic study of the "Kallo complex"). Ministry of Economic Affairs. Prof. paper 1980/8 n° 178, 13 p.

Schuler M. (1982). Pollens et spores des séries salifères (Eocène/Oligocène) du bassin potassique de Mulhouse (France). (Pollens and spores of the Saliferous deposits (Eocene/Oligocene) of the Mulhouse potash basin, France). Relations between microflora and evaporitic deposits. Sci. Geol. Bull., 36, 4, 255-265.

Sittler C. and **Schuler M.** (1975). Extension stratigraphique, répartition géographique et écologique de deux genres polliniques paléogènes observés en Europe occidentale : Aglaoreidia et Boehlensipollis. (Stratigraphic extension, geographic distribution and ecology of two genera of Paleogene pollens occurring in western Europe : Algoreidia and Boehlensipollis). Bull. Soc. bot. France, 122, 231-245.

Williams G.L. (1978). (Palynological biostratigraphy). DSDP sites 367 and 370. Init. Rep. DSDP. 41, 783-815.

Williams G.L. and **Bujak J.P.** (1977). Cenozoic palynostratigraphy of offshore eastern Canada. Am. Ass. strat. Palyn. Contr. Series 5A, 14-47.

Van Couvering J.A., Aubry M.P.,Bergeren W.A., Bujak J.P., Naesser C.W., Wieser T. (1981). The terminal Eocene Event and the polish connection. Palaeogeo., Palaeoclim., Palaeoecol., 36, 321-362.

CHAROPHYTA AT THE EOCENE-OLIGOCENE BOUNDARY IN WESTERN EUROPE

J. RIVELINE

Lab. de Géol. des Bassins Sédimentaires, UA 319, Université Pierre et Marie Curie, 4, Place Jussieu, 75230 Paris Cedex 05.

ABSTRACT

The detailed study of different localities in the Paris Basin and South Hampshire shows that the Charophyta evolution during the Upper Eocene and the Lower Oligocene does not display any important break but, on the contrary, gradual change.

Charophyta argue for setting the position of the Eocene-Oligocene boundary between the Bembridge Beds and the Hamstead Beds in South Hampshire or between The Marnes supragypseuses and l'Argile Verte de Romainville in the Paris Basin. That point of view is built on :

- the persistence in the Bembridge Beds and in the Marnes supragypseuses of some Eocene species;
- the disappearance in the Hamstead Beds and in the Argile Verte de Romainville ... of nearly all the Eocene species and the appearance in those formations of many forms which will be abundant in younger formations.

Among the different Western European basins, those of Paris and South Hampshire are the most interesting for the study of floristic variations at the Eocene-Oligocene boundary. It was in those regions that Castel (1968) and Grambast (1972) specified the position of that boundary based on Charophyta and placed it at the bottom of the Bembridge Beds (Hampshire) correlated with the Première Masse de Gypse (Paris Basin). They observed a specific flora in the Bembridge Beds (type-formation of the Bembridge Charophyta zone, the first one attributed to the Oligocene). This flora shows evidence of considerable renewal, compared with that observed in the Lower Headon Beds (Hampshire) or in the Marnes de Verzenay (Paris Basin). The Lower Headon Beds and the Marnes de Verzenay are the type-formations of the Verzenay Charophyta zone, considered by Grambast as the last one which can be established in the Eocene. Besides, the Bembridge Beds flora persists up to the Lower Hamstead Beds (Hampshire), the Bande Blanche (Paris Basin), the Marne à Chara of Hoogbutsel (Belgium) or the Ronzon Level (Velay, France) without any change.

In 1977 Feist-Castel again took up investigations in the English formations. She found that the floristic break between the Verzenay and the Bembridge associations is not so strongly marked. She indicated a floristic change in the Upper Headon Beds, marked by the disappearance of the greatest part of the Lower Headon Beds species and the appearance of Sphaerochara subglobosa (Groves) Horn of Rant. Also, she located the Eocene-Oligocene boundary at the base of the Middle Headon Beds. That point of view is built on the distribution of Harrisichara vasiformis-tuberculata. This species is a transition form between Harrisichara vasiformis (Reid and Groves) Gramb, known from the Verzenay Charophyta zone, and Harrisichara tuberculata (Lyell) Gramb which appears in the Bembridge Charophyta zone.

We have recently again undertaken investigations on the English and French formations. Our research has indicated that : (Riveline 1984 a and b) :

a) The change between the Verzenay and the Bembridge flora is more gradual that Feist-Castel thought. The Lower Headon Beds species does not disappear, as a whole, in the Upper Headon Beds. We observed progressive extinctions from the Middle Headon Beds up to the top of the Bembridge Beds. If the extension of Gyrogona tuberosa Gramb seems limited to the Lower Headon Beds, Chara antennata Gramb persists up to the Upper Headon Beds, Psilochara bitruncata (Reid and Groves) Feist-Castel and Psilochara repanda Gramb has still been observed in the Osborne Beds. All those disappearances are compensated by many appearances, such as Sphaerochara subglobosa at the top of the Lower Headon Beds, Rhabdochara altilis Feist-Castel and Psilochara aff conspicua Gramb in the Upper Headon Beds or Harrisichara tuberculata and Rhabdochara stockmansi Gramb in the Bembridge Beds.

b) It is possible to divide the Bembridge Charophyta zone (characterized by the presence of Harrisichara tuberculata and Rhabdochara stockmansi) into two sub-zones :

. A lower sub-zone established on the Charophyta observed in Southern Paris Basin limestones, the Marnes supragypseuses and the Bembridge Beds. This sub-zone is distinguished by :

- some particular species such as Stephanochara vectensis (Groves) Gramb, Nitellopsis (T) wonnacotti Gramb.
- some species known from the Verzenay Charophyta zone or even before.
- precursory forms of some species which will be abundant in younger formations, as Chara aff media Gramb or Sphaerochara aff hirmeri (Rasky) Mädler.

. An upper sub-zone established on Charophyta observed in the Argile Verte de Romainville, the Calcaire de Brie and the Hamstead Beds. This sub-zone is distinguished by :

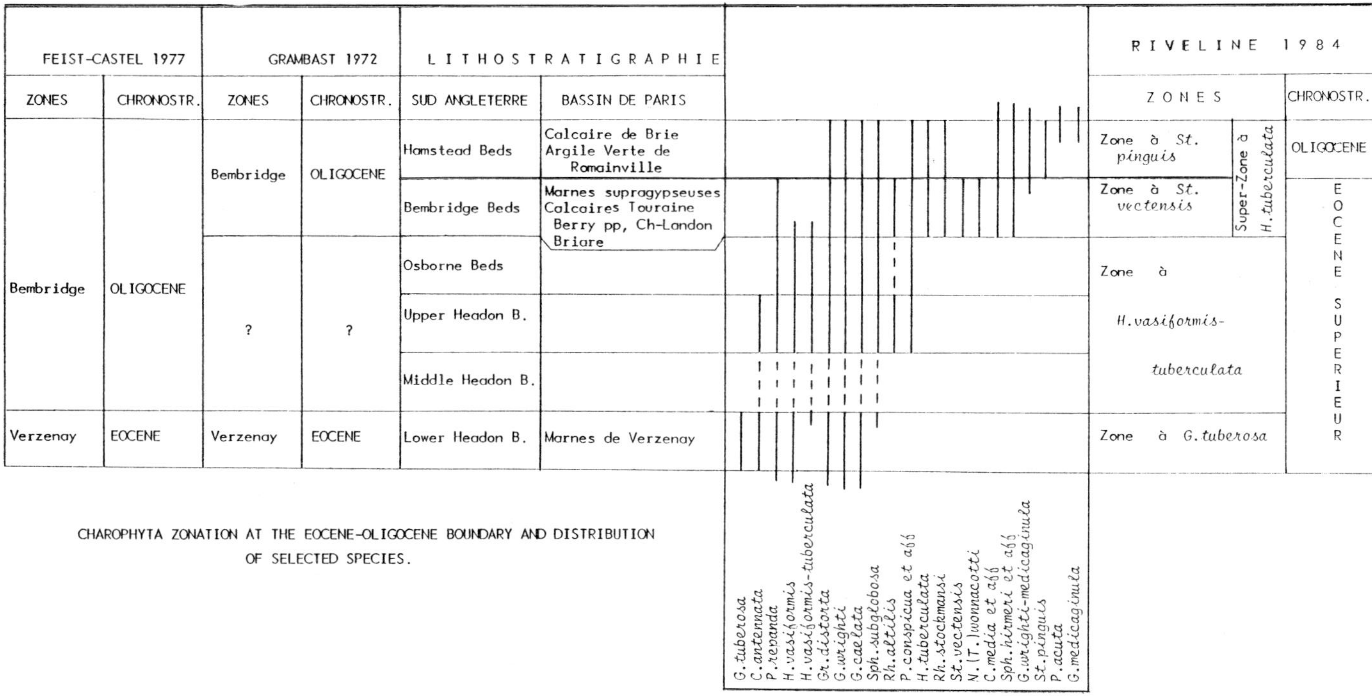

CHAROPHYTA ZONATION AT THE EOCENE-OLIGOCENE BOUNDARY AND DISTRIBUTION OF SELECTED SPECIES.

- the disappearance of nearly all the Eocene species, with the possible exception of Grovesichara distorta (Reid and Groves) Horn of Rant.
- some particular species, such as Stephanochara pinguis Gramb.
- the appearance, at the top, of some forms considered as characteristic of the Stampian Fontainebleau Charophyta zone, such as Gyrogona medicaginula lamk or Psilochara acuta Gramb.

The slight renewal observed between the Verzenay and the Bembridge Charophyta zone is not a conclusive clue to the position of the Eocene-Oligocene boundary at the base of the Bembridge Beds. The presence of some species that appeared during the Eocene in the Lower Bembridge sub-zone argues for setting that lower sub-zone in the Eocene. The absence of Eocene species and the abundance of some forms which will be developed during the Stampian argues for placing the upper sub-zone in the Oligocene.

Charophyta evolution during the Upper Eocene and the Lower Oligocene does not show any important break but, on the contrary, gradual change. Among the 47 species inventoried during the Upper Eocene, 31 do not survive through the Eocene-Oligocene boundary, as identified with the "Grande coupure de H.G. Stehlin". The species which disappear with the Oligocene are those that appeared during the Middle Eocene (Bartonian) or during the Late Upper Eocene. Only the forms having appeared during the Early Middle Ludian, in the same way as the Molluscs and the Planktonic foraminifera (Cavelier 1979), survive through the Eocene-Oligocene boundary.

REFERENCES

Castel, M., 1968. Zones de Charophytes pour l'Oligocène d'Europe occidentale. C.R. Som. Soc. Géol. Fr., fasc 4 pp. 121-122.

Cavelier, C., 1979. La limite Eocène-Oligocène en Europe occidentale. Sciences géologiques. Université Louis Pasteur. Strasbourg n° 54, 280 pp.

Feist-Castel, 1977. Evolution of the Charophyte floras in the Upper Eocene and Lower Oligocene of the Isle of Wight. Palaeontological Association, part 1, pp. 143-157, pls 21-22.

Grambast, L., 1972. Principes de l'utilisation stratigraphique des Charophytes. Application au Paléogène d'Europe occidentale. Mem. BRGM n° 77, pp. 319-328.

Riveline, J., 1984a. Les gisements à Charophytes du Cénozoïque (Danien à Burdigalien) : lithostratigraphie, biostratigraphie, chronostratigraphie. Bull. Inf. Géol. Bas. Paris, vol. hs n° 4, 523 pp.

Riveline, J., (sous presse). Les Charophytes du Cénozoïque (Danien à Burdigalien) d'Europe occidentale. Implications stratigraphiques.

MAMMALIAN EVIDENCE CONCERNING THE EOCENE-OLIGOCENE TRANSITION IN EUROPE, NORTH AMERICA AND ASIA

by Donald E. RUSSELL and Heinz TOBIEN

INTRODUCTION

The multiple problems that are contained in the Terminal Eocene Event was envisaged as worthy of a pluridisciplinary project in 1980 when the "event" was presented at the 26th International Geological Congress in Paris. The Grande Coupure, a concept dating from 1909, can be considered as something of an ancestor to the TEE. As has been noted many times, the phenomenon of the Grande Coupure was recognized in Europe by the great Swiss paleontologist H.G. Stehlin who observed a striking change of familial character in the artiodactyls and perissodactyls in passing from Eocene faunas to those of the Oligocene. In this review the Grande Coupure is taken, in a conventional sense, to represent the Eocene-Oligocene boundary, following Cavelier, 1979.

DISCUSSION

Perissodactyls, from an early dominance among Eocene European ungulates, diminish in importance and variety toward the end of that epoch; the numerous palaeotheres that characterize the middle and late Eocene essentially die out before the Oligocene, with only a few stragglers surviving after the Grande Coupure. Rhinocerotoids typify Oligocene faunas and the newcomers are presumed to have originated in Asia.

The picture is more complex as concerns artiodactyls. Late middle to late Eocene artiodactyls were widely diversified and eminently successful in Europe. While 4 families became extinct (Choeropotamidae, Mixtotheriidae, Xiphodontidae, and Haplobunodontidae), 5 persisted into the Oligocene with only few representatives (Dichobunidae, Cebochoeridae, Dacrytheriidae, Anoplotheriidae and Cainotheriidae). Amphimerycids, gelocids and anthracotheres, however, are late Eocene immigrants and indicate that the invasion began well before the Grande Coupure. New Oligocene groups (from Asia) are the entelodonts, suids, leptomerycids and tragulids. The change is radical but not of a catastrophic nature.

The evolution of rodents has been particularly well studied in Europe with the result that their history in this area is fairly well established (Hartenberger, 1973 ; Vianey-Liaud, 1979), as well as the time and routes of probable migration (Vianey-Liaud, in press). With respect to the latter, a certain degree of precision is possible in some cases ; for example, the first immigrants from Asia (in the early Oligocene) are documented -reasonably enough- earlier in southern

Germany than in the western part of Europe (Schmidt-Kittler and Vianey-Liaud, 1975).

Ischyromyoids, after an early hegemony, disappear in the late Eocene ; glirids retain a relatively minor role from the middle Eocene on through the Eocene-Oligocene transition, where a few genera are found both before and after the Grande Coupure. Much more important are the Theridomyidae. From a single genus present in the early middle Eocene they progress to attain a maximum of numbers and diversity toward the end of that epoch. Interestingly, 9 of the 14 genera present in the latter also occur in the early Oligocene, with the addition of two new forms. Based only on the above evidence it would appear that relatively little change took place among the rodents during the transition, but the general character of the Oligocene fauna is fundamentally changed by the arrival (from a North America origin) of the aplodontids, sciurids and eomyids, and (from Asia) of the cricetids and castorids. As was demonstrated by the artiodactyls, the Oligocene is characterized by the arrival of new groups superimposed on those that survived the Eocene-Oligocene transition. A major difference, however, is that the new rodents appeared after the Grande Coupure whereas some of the artiodactyl immigrants are already present in Europe before.

Also contributing to the Oligocene fauna are the lagomorphs, of which the first are present in small numbers after the Grande Coupure (Lopez and Thaler, 1974).

Creodonts are diversified in the middle Eocene; reduction of the number of taxa occurs during the late Eocene but the Asian immigrants Pterodon and Hyaenodon are present throughout the latter period with Hyaenodon also extending through the Oligocene (Lange-Badré, 1979). At that time creodonts form only a minor part of the assemblage.

The history of the carnivores is nearly the reverse ; represented in the Eocene by a few genera of miacids, they virtually explode at the beginning of the Oligocene with the appearance of felids, viverrids, mustelids, procyonids, ursids, and many amphicyonids. Miacids have disappeared. The evolutionary stages leading to this explosion remain, in large part, still to be explained.

Conservative didelphid marsupials survive the Grande Coupure with apparently little reaction to events at that time (Crochet, 1980). In contrast, out of the fairly numerous taxa of late Eocene primates only one produced an Oligocene descendant and none invaded the apparently hostile European environment.

To conclude, it must be admitted that, despite some faunal overlapping, the reality of the Grande Coupure in Europe is undeniable ; faunas immediately preceding it are readily distinguishable from those immediately following. We have here evidence of a major event that seriously disrupted the gradual prolongation of Eocene mammalian families into the Oligocene and imposed upon Europe a host of elements that originated on other continents. However, as shown above, the transition was not entirely brutal. Moreover, the time interval of maximum change in faunal constituants is not always easy to determine with precision. But even with these restrictions it is possible to say that the Grande Coupure represents the point of maximum effect heralding the end of the Terminal Eocene Event.

In North America the mammalian faunas of the late Eocene - early Oligocene transition are those characterizing the Uintan, Duchesnean and Chadronian land-mammal ages. Much discussion has taken place, however, concerning which one or how much of each age is referable to either the Eocene or the Oligocene. A consensus does not yet exist. Probably it could be said that most North American mammalian paleontologists regard the major faunal "break" as occurring in the middle of the Duchesnean ; this rupture would represent the Eocene - Oligocene limit. Unfortunately, not much is known of the mammals of that time and an element of imprecision is for the moment inevitable. Difficulty in correlating with the classic terrestrial faunas of Europe and with the world-wide marine biozones accounts for an additional large amount of uncertainty. Radiometric dating still remains rather unstable.

Since the "key" land-mammal age is the Duchesnean we will examine its problems in some detail. No other "age" in North America is so weakly substantiated (fossils are extremely rare in the Duchesne River Formation) and few have been so contested, both with regards to its validity and to its attribution to either the Eocene or the Oligocene. The fauna of the lower part of the formation is quite similar to that of the upper part of the Uinta Formation and is even considered of Uintan age by some specialists (Tedford, 1970) ; most of the Duchesnean fossils have been collected in this lower part. The meager fauna of the upper part of the Duchesne River Formation is very close in nature to that of the Chadronian and is distinctly different from that of the Uintan. It has been recently reviewed by Wilson (1984). Emry (1981) has suggested that the Chadronian could begin with this upper Duchesnean fauna. If the opinions of such workers as Tedford and Emry were followed the Duchesnean would not exist. Others, however, retain the term but consider that the Eocene extends into the base of the Chadronian (Berggren et al., 1978 ; McKenna, 1980). The familiar quandry of concepts and definitions is all too present here.

As noted by many workers (see Lillegraven, 1979, for a general review of the question) the late Eocene and early Oligocene of North America was marked by increased rates of extinctions and faunal replacements in many groups of organisms. Also at this time occurred the strongest intercontinental migration since that of the Wasatchian.

In the late (but not terminal) Eocene Uintan "age" an explosive taxonomic radiation of artiodactyls and rodents was already underway. In fact, the artiodactyls had by this time replaced perissodactyls as the dominant ungulates on the continent. Both the bunodont and the selenodont artiodactyls expanded in diversity during the Uintan ; whether or not some or all of the selenodont groups (the Hypertragulidae, Agriochoeridae and Oromerycidae) were immigrants is still debated (Golz, 1976 ; Webb and Taylor, 1980). Camelids also appear, and dichobunids become extinct. Except for the latter, all the groups continued into the Orellan ; entelodonts appeared in the Duchesnean and three new families (tayassuids, anthracotheriids and merycocoidodontids) in the Chadronian, completing the artiodactyl element of the Oligocene fauna.

According to Wilson (1980) Uintan rodents are simple descendants of Bridgerian ones. Most of the change in character of North American rodent groups occurred between the Uintan and the Orellan "ages". The largest part of this turnover took place within the Duchesnean and the first part of the Chadronian, but considerable ambiguity remains with respect to the place of origin of many families.

Lagomorphs make a timid appearance in the Uintan and expand into the Oligocene, but several archaic orders (Taeniodonta, Dinocerata (uintatheres) and Condylarthra (hyopsodontids and mesonychids) became extinct with the close of Uintan time.

From an abundance in the Uintan, primates decrease markedly in the Duchesnean but nevertheless survive into the Chadronian.

Creodonts decline even before the Uintan and with the Duchesnean are represented only by the probable Asian immigrants Pterodon and Hyaenodon ; the latter, however, continued to thrive throughout the Oligocene. Miacids somewhere underwent an evolutionary development that was responsible for the appearance in the Chadronian of the modern families of carnivores: Canidae, Felidae, Viverridae and Ursidae, as well as the extinct Nimravidae and Amphicyonidae. Details of this evolution are particularly obscure as no miacids have as yet been found in the Duchesnean ; the family is well represented in the Uintan and is known by a last survivor in the Chadronian. Radinsky (1982) has suggested that the appearance of the modern carnivore groups might be subsequent to, and not responsible for, the extinction of older forms. While this is probable the fossil record does not substantiate it.

Perissodactyls also underwent a profound change during this period, but continue to constitute an important part of the fauna, despite the increase in diversity of the artiodactyls. Isectolophids and chalicotheres disappear with the Uintan, but most of the other families continue at least through the Chadronian, when the brontotheres disappear. At that time an influx of rhinocerotids assumes a certain dominance, which persists into the Orellan.

In summary, a noteworthy change in the constitution of late Eocene - early Oligocene faunas is well documented in North America. Artiodactyls exhibit the change on a family level after the Duchesnean, as did rodents, carnivores and perissodactyls. The evidence, then, indicates the presence of an important event affecting the mammalian biota, during the Duchesnean, that produced a Chadronian fauna significantly different from that of the Uintan. Although precision in intercontinental correlation is still lacking, if the TEE was a world-wide phenomenon, one is tempted to conclude that it coincided with the Duchesnean "age" in North America.

Information from Asia is less abundant than it is from Europe or North America, but considerable progress has been made in recent years. Russell and Zhai (in press) have compiled a comprehensive review of Asian Paleogene localities and the mammalian taxa they have yielded.

Perissodactyls in the Paleogene of Asia display an abundance, in both numbers and taxa, that is unequaled elsewhere. This is particularly true of the middle Eocene ; in the late Eocene they diminish somewhat but still retain a diversity that is superior to that of the artiodactyls. Among the dominant forms, brontotheres maintain an important place in the faunas from the middle Eocene through the early Oligocene ; paralleling their evolution in North America, they suddenly become extinct before the middle Oligocene. However, they show evidence of a greater change between the middle and late Eocene than there was between the late Eocene and the early Oligocene. Other perissodactyls (Lophialetidae, Chalicotheriidae, Helaletidae and Deperetellidae) also diminish in numbers more or less dramatically in passing from the middle to the late Eocene, decreasing in any case more than at the Eocene-Oligocene limit. Hyracodontidae and Amynodontidae, however, display a reversal of this trend and increase gradually through the middle Oligocene. The amynodonts seem to have found conditions at the Eocene-Oligocene transition particularly well suited to their needs ; they increase from 8 genera in the late Eocene to 11 in the early Oligocene, of which 5 are new but nearly all of the rest are survivors from the late Eocene ; with the middle Oligocene they undergo a virtual extinction. What is especially intriguing about their history is that many of them are purportedly semi-aquatic in habits, which contrasts with the generally cold and dry climate proposed for the Terminal Eocene Event. A single (and different) genus of rhinocerotid occurs in the early and in the middle Oligocene.

Among the artiodactyls, helohyids and leptomerycids undergo a decrease in diversity in passing from the late Eocene to the early Oligocene. Entelodonts are present already in the middle Eocene ; Eoentelodon persists into the early Oligocene where it is joined by the typical Oligocene form, Entelodon. Anthracotheres are the dominant artiodactyl during this period in Asia, with a maximum of genera in the late Eocene. A strong reduction is recorded at the Eocene-Oligocene transition when 4 new families make their appearance.

Modern groups of rodents (cricetids and dipodids) are first seen in the late Eocene and pass (with generic change) into the early Oligocene where they are joined by the cylindrodonts, immigrants from North America. The period of greatest modification, as concerns the rodent population, occurred in the middle

Oligocene with the arrival of 4 new families and the flowering of the ctenodactylids.

Leporid lagomorphs show a reduction in crossing the Eocene-Oligocene boundary, although this is achieved by a single genus. Ochotonids appear in the early Oligocene and diversify greatly during the middle Oligocene.

Primates are little known in the Paleogene of Asia but a scenario similar to that cited in Europe and North America apparently prevailed : extinction occurred at Eocene-Oligocene boundary.

Typifying the creodonts, Hyaenodon occurs as early as the middle Eocene and, with apparent equinimity, traversed all periods of possible crisis up into the middle Oligocene. As a family, however, hyaenodontids decrease slightly during this time.

True carnivores are represented in the middle Eocene by miacids, canids and felids ; they are accompanied in the late Eocene by viverrids and ursids and all four persist into the middle Oligocene, when mustelids are added. There is no appreciable evidence of a faunal break in this sequence. Times of principal change are the late Eocene (with 2 new families) and the middle Oligocene (with one new family and expansion of the canids and felids).

With regard to other orders of mammals, mesonychid condylarths show a major reduction between the middle and late Eocene and die out in the early Oligocene ; the same is true for the coryphodontid pantodonts. Taeniodonts become extinct during the late Eocene.

Taken together, the data from Asia do not indicate the same importance of faunal rupture at the Eocene-Oligocene limit as do those from Europe or North America. In fact, the shifts in composition that appear both between the middle and late Eocene and between the early and middle Oligocene are also of significant amplitude. And late Eocene genera that extend across the boundary into the early Oligocene are numerous. Very few families become extinct at Eocene-Oligocene limit although primates and taeniodonts do among the orders. Viewed from the aspect of faunal renewal, the early Oligocene was nevertheless noteworthy for the appearance of many new families.

CONCLUSIONS

In the charts prepared by Savage and Russell (1983) on faunal turnover a particularly high number of genera both appeared and disappeared in the time that elapsed between the beginning of the Uintan and the end of the Chadronian in North America ; that period is unique in these respects for the entire Tertiary on that continent.

In Europe, a peak of new genera occurred between the late Eocene and the early Oligocene, with the trend continuing to a lesser degree into the middle Oligocene. Although relatively high, extinction rates remain much below those for new arrivals and actually decrease from the late Eocene to the middle Oligocene. The European Grande Coupure was clearly more concerned with the appearance of new taxa than with the disappearance of those already in place.

Based on data taken from the review of the Paleogene of Asia cited above, it is apparent that we are faced with a series of faunal breaks and not merely a single analogue of the Grande Coupure. Summing up for Asia, the maximum change in faunal composition on a generic level occurred between the middle and late Eocene ; in distant second place was the passage from the Eocene to the Oligocene, closely followed by the break between the early and middle Oligocene. On the ba-

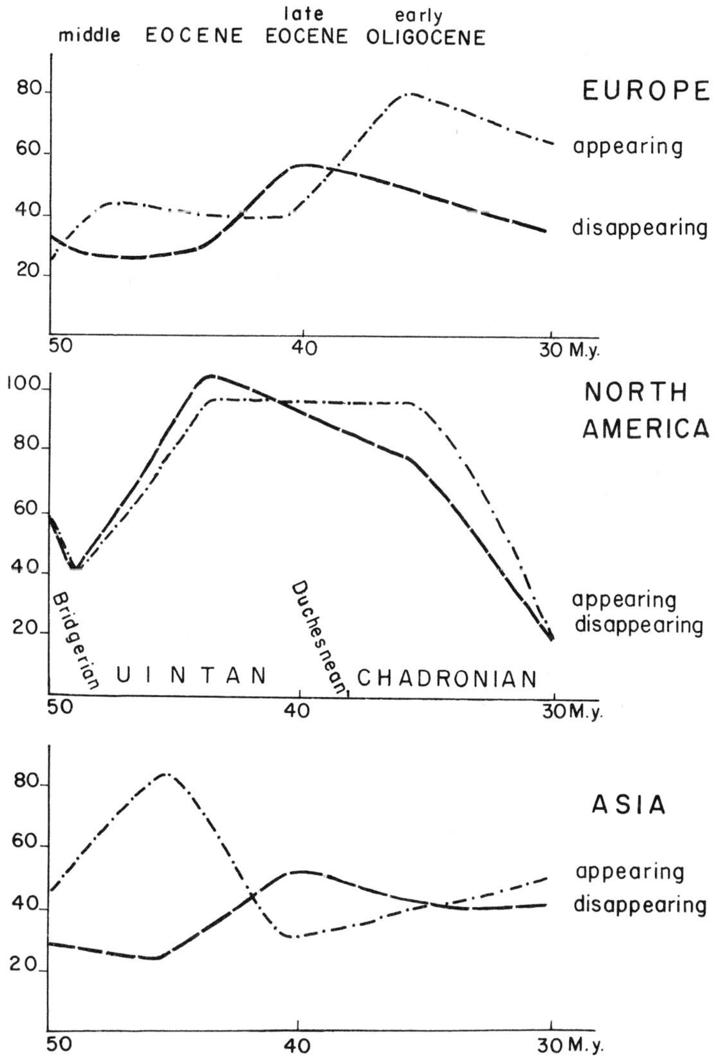

Fig. 1. Faunal turnover curves showing the number of genera (at left) plotted against time in million years. Modified from Savage and Russell (1983) and Russell and Zhai (in press).

sis of family change, however, the middle-late Eocene break and that of the late Eocene-early Oligocene both show approximately the same degree of appearance and disappearance.

In conclusion, mammalian evolution indicates that in North America particularly high rates of change were maintained over a period of at least 10 million years ; the Eocene-Oligocene boundary was situated near the middle of this time. In Europe a peak of extinctions occurred in the late Eocene (as it did in Asia) ; after a low point in the late Eocene (also as in Asia) a maximum of new genera appeared in Europe with the early Oligocene.

What the above statements point out is the presence on a world-wide scale of a time of great change, but it underlines the fact that this was not a brief period of intense biogeographic activity. In all three continents studied it had a duration of from 10 to 15 million years. The Terminal Eocene Event was not punctual in its effects on mammals.

REFERENCES

Berggren, W.A., McKenna, M.C., Hardenbol, J. and Obradovich, D., 1978. Revised Paleogene Polarity Time Scale. Journ. Geol., 86: 67-81.

Cavelier, C., 1979. La limite Eocène-Oligocène en Europe occidentale. Publ. Univ. Louis Pasteur Strasbourg, Sci. géol., mém. 50: 280 pp.

Crochet, J.-Y., 1980. Les Marsupiaux du Tertiaire d'Europe. Fondation Singer-Polignac Ed.: 279 pp.

Emry, R.J., 1981. Additions to the mammalian fauna of the type Duchesnean, with comments on the status of the Duchesnean "age". Journ. Paleont., 55 (3):563-570

Golz, D.J., 1976. Eocene Artiodactyla of Southern California. Nat. Hist. Mus. Los Angeles County Sci. Bull., 26: 85 pp.

Hartenberger, J.-L., 1973. Les Rongeurs de l'Eocène d'Europe. Leur évolution dans leur cadre biogéographique. Bull. Mus. natn. Hist. nat., Paris, 3ème sér.,174, Sci. Terre (28): 105-113.

Lange-Badré, B., 1979. Les Créodontes (Mammalia) d'Europe occidentale de l'Eocène supérieur à l'Oligocène supérieur. Mém. Mus. natn. Hist. nat., Paris, n.s., sect. C (42): 249 pp.

Lillegraven, J.A., 1979. A Biogeographical Problem Involving Comparisons of later Eocene Terrestrial Vertebrate Faunas of Western North America. In : Historical Biogeography, Plate Tectonics, and the Changing Environment, Gray, J. and Boucot, A. Eds, Oregon State University Press: 333-347.

Lopez, N. and Thaler, L., 1974. Sur le plus ancien Lagomorphe européen et la Grande Coupure oligocène de Stehlin. Palaeovertebrata, 6 (3-4): 243-251.

McKenna, M.C., 1980. Late Cretaceous and early Tertiary vertebrate paleontological reconnaissance, Togwotee Pass area, northwestern Wyoming. In : Aspects of Vertebrate History : Essays in Honor of Edwin Harris Colbert, Louis, L. Ed. Mus. of Northern Arizona Press, Flagstaff: 321-343.

Radinsky, L.B., 1982. Evolution of skull shape in carnivores. 3. The origins and early radiation of the modern carnivore families. Paleobiology, 8 (3): 177-195.

Russell, D.E. and Zhai R.-j., in press. Mammals and Stratigraphy: The Paleogene of Asia. Palaeovertebrata, mém. extraord.

Savage, D.E. and Russell, D.E., 1983. Mammalian Paleofaunas of the World. Addison-Wesley Publ. Co., Reading, Massachusetts: 544 pp.

Stehlin, H.G., 1909. Remarques sur les faunules de Mammifères des couches éocènes et oligocènes du Bassin de Paris. Bull. Soc. géol. Fr., sér. 4, 9: 488-520.

Tedford, R.H., 1970. Principles and practices of mammalian geochronology in North America. Proc. Amer. Paleont. Conv., part F: 666-703.

Vianey-Liaud, M., 1979. Evolution des Rongeurs à l'Oligocène en Europe occidentale Palaeontolographica, 166, A: 136-236.

Vianey-Liaud, M., in press. Possible evolutionary relationships among Eocene and lower Oligocene Rodents of Asia, Europe and North America. Rodent Phylogeny Symposium, Paris 1984.

Webb, S.D. and Taylor, B.E., 1980. The phylogeny of hornless ruminants and a description of the cranium of Archaeomeryx. Bull. Amer. Mus. Nat. Hist., 167 (3): 117-158.

Wilson, J.A., 1984. Vertebrate faunas 49 to 36 million years ago and additions to the species of Leptoreodon (Mammalia: Artiodactyla) found in Texas. Journ. Vert. Paleont., 4 (2): 199-207.

Wilson, R.W., 1980. The stratigraphic sequence of North American rodent faunas. Palaeovertebrata, Montpellier, mém. jubil. R. Lavocat: 273-283.

THE AMPHIBIANS AND REPTILES AT THE EOCENE-OLIGOCENE TRANSITION IN WESTERN EUROPE: AN OUTLINE OF THE FAUNAL ALTERATIONS.

by Jean- Claude RAGE

Laboratoire de paléontologie des Vertébrés, Université Paris VI, 4 place Jussieu, 75230 Paris cedex 05, France.

INTRODUCTION

In Vertebrate paleontology, when stratigraphy of the Cenozoic is concerned, attention is turned chiefly to mammals and the other groups are generally neglected. However, amphibians and reptiles can provide interesting informations. Unfortunately, thorough studies of Cenozoic amphibians and reptiles have been undertaken only in Europe and North America. Moreover, with regard to Eocene and Oligocene times, these faunas are well known only from France and to a lesser extent from Belgium and England. Specific works from those areas show that the terminal Eocene events strongly affected them. From North America, it is only known that extinctions of snakes occurred at the end of the Eocene (Holman, 1979). Therefore, only an outline of the amphibian and reptile history at the Eocene-Oligocene boundary in France, Belgium and England is given here.

THE LATE EOCENE FAUNAS

During the late Eocene (i.e., the Ludian) extinctions affected both amphibians and reptiles. On the whole, extinctions that concerned these groups were not very numerous (Rage, 1984) whereas mammals have been markedly troubled by such a phenomenon (Hartenberger, 1973). On the other hand, during the later part of late Eocene, the number of amphibians and reptiles increased. This might be an artifact of fossilization but, even if fossilization took a part it probably cannot account for all of this increase. However that may be, the late Ludian amphibians and reptiles, chiefly the latter, were rather diversified and flourishing. This is somewhat surprising because a general cooling, initiated by the beginning of the Ludian, more or less gradually increased until the marked deterioration of the climate at the end of this stage (Pomerol, 1985). Thus, apparently, the more the cooling increased, the more amphibians and reptiles were numerous, which is all the more astonishing that this fauna is, at least partly, of a tropical type. Anyhow, whatever the causes of this increasing may be, about forty taxa are known from the uppermost Ludian.

THE EOCENE-OLIGOCENE TRANSITION

From the lowermost Oligocene, less than ten taxa have been recorded. They are either Eocene forms that have survived beyond the Eocene-Oligocene boundary or asiatic immigrants. Thus, more than three fourths of the uppermost Eocene fauna died out by the Eocene-Oligocene transition. This phenomenon, called the "Grande Coupure", affected numerous land Vertebrates; it represents the major break in the amphibian and reptile Cenozoic history. It should be noted that, apart from the terrestrial faunas, an aquatic (including marine biotas) family of snakes, the Palaeopheidae, known since the late Cretaceous and widely distributed, disappeared during the late Eocene.

THE OLIGOCENE FAUNA

The very poor fauna of the lowermost Oligocene was followed by a richer but still rather poor assemblage. Besides, the amphibians and reptiles from the Oligocene are never well diversified. The settlement of Western Europe during the Oligocene is apparently the result of arrivals of oriental immigrants that occupied this area which was poorly populated after the Eocene-Oligocene boundary.

CONCLUSIONS

The Eocene-Oligocene boundary corresponds to a faunal turnover, the "Grande Coupure", that has been prominent for amphibians and reptiles in France, Belgium and England. However, tortoises appear to have not been affected by this event. The phase that preceded the Eocene-Oligocene boundary is characterized by both extinctions and diversification of faunas. Extinctions may be the consequence of the deterioration of climatic conditions. Concerning the crocodiles, the reduction of their population might be only an aggravation of a process previously initiated (Buffetaut, 1982). The diversification of the fauna is rather surprising and remains unexplained. Few Eocene forms survived beyond the Eocene-Oligocene boundary. During the early Oligocene, some oriental immigrants settled in Western Europe and their arrival probably contributed to the extinction of the Eocene autochthonous forms. Therefore, the "Grande Coupure" appears as the result of several causes. Although it is not simple, it is a stratigraphically well marked event that corresponds to the Eocene-Oligocene boundary in Western Europe.

REFERENCES

Buffetaut, E., 1982. Les crocodiliens et la coupure Eocène-Oligocène. 9º Réun. ann. Sc. Terre, p. 101.

Hartenberger, J.L., 1973. Les Rongeurs de l'Eocène d'Europe. Leur évolution dans dans leur cadre biogéographique. Bull. Mus. natn. Hist. nat., 174: 223-248.

Holman, J.A., 1979. A review of North American Tertiary snakes. Publ. Mus. Michigan State Univ., Paleont. ser., 1: 203-260.

Pomerol, C., 1985. La transition Eocène-Oligocène: est-elle un phénomène progressif ou brutal? Bull. Soc. géol. France, 1: 263-267.

Rage, J.C., 1984. La "Grande Coupure" éocène-oligocène et les herpétofaunes (Amphibiens et Reptiles): problèmes du synchronisme des évènements paléobiogéographiques. Bull. Soc. géol. France, 26: 1251-1257.

ISOTOPIC EVENTS AT THE EOCENE/OLIGOCENE TRANSITION. A REVIEW

C. VERGNAUD GRAZZINI[1] and H. OBERHAENSLI[2]

[1]Département de Géologie dynamique, 4 Place Jussieu, 75230 Paris cedex 05 (France)

[2]Geologisches Institut Eidgenössische Technische Hochschule, Zürich, CH 8092 Zürich (Switzerland)

ABSTRACT

Vergnaud Grazzini, C. and Oberhänsli, H., 1985. Isotopic events at the Eocene/Oligocene transition. A review.

Oxygen and carbon stable isotope records at the Eocene-Oligocene transition, documented in land based sections and deep sea drilling cores in the Atlantic, and Pacific Oceans (Bath Cliff Section, Core E-67-128, DSDP Sites 119, 401, 366, 522, 529, 511, 516, 292, 277) are discussed and compared.

The oxygen isotope record displays a rapid shift towards increasing values, at the transition between Zones P 17 to P 18 (= within Zone NP 21), in the early Oligocene time. Recent radiometric datings suggest an age around 35 Ma for this ^{18}O event. A duration of about 10^5 years is postulated by many authors. The exact amplitude of the $\delta^{18}O$ increase is not easy to establish because of different sampling densities at the studied sites, varying fractionation factors of the different species with time, diagenetic alterations and possible changes in the water masses which occupied the sites at that time. For benthic foraminifers, this $\delta^{18}O$ increase ranges between +0.8‰ and +1.2‰. For planktonic foraminifers the $\delta^{18}O$ increase ranges from +0.8‰ in high latitudes to +0.6-0.4‰ in low latitudes, suggesting an increase in the latitudinal temperature gradient from Eocene to Oligocene time. Because no convincing evidence from faunistic or sedimentological studies appears to support the hypothesis of a significant build up of ice caps at that precise time, we favour the interpretation that a temperature change in mid and high latitudes was the cause of the $\delta^{18}O$ increase.

Carbon isotope values correspond to the maximal peak values of the second major cycles of the Tertiary ^{13}C record. Changes in the volume or the $^{13}C/^{12}C$ ratio of some terrestrial or oceanic reservoir may explain this maximum, while changes in the rate of production of Atlantic and Pacific deep waters may explain very localized opposite trends in the ^{13}C records, at some of the sites.

INTRODUCTION

Thanks to the deep sea drillings revealing long continuous sedimentary records, the investigations that have been carried over the last 15 years have permitted to reconstruct the climate of the Tertiary with considerable accuracy. Among the tools which are useful in such reconstructions, oxygen and carbon stable isotopes studies of shells of foraminifera or calcareous nannofossils provide informations on changes in ice volume and temperature, global carbon reservoirs and abyssal circulations.

Oxygen isotopic records of Tertiary calcareous shells display a general trend towards higher values with decreasing ages. This general trend is strengthened by substantial shifts towards higher $\delta^{18}O$ values over relatively short time intervals. In high latitudes these steps are recorded by both planktonic and benthic organisms. In low latitudes they may be recorded only by benthic organisms.

Carbon isotopic records of Tertiary calcareous shells display cyclic variations. Three major cycles can be distinguished with maximal values of $\delta^{13}C$ corresponding to : (1) the late Paleocene (biozones P5-P6), (2) the Eocene-Oligocene boundary (biozones P17-P18) and (3) the middle Miocene (biozones N9-N11). Secondary cycles and very localized shifts (such as the 6.2 Ma shift) are superimposed on the major cycles. These changes, when recorded by both planktonic and benthic organisms reflect global changes in the ^{13}C budget. Furthermore, comparisons between benthic ^{13}C records from different ocean basins permit to isolate global changes from abyssal circulation changes and synoptic offsets in $\delta^{13}C$ values between basins can be used to infer abyssal circulation (Kroopnick, 1974).

An important step in the oxygen isotopic record of the Tertiary is located just above the last occurrence of *Hantkenina* and the base of the planktonic foraminiferal zone *Cassigerinella chipolensis/Pseudohastigerina micra*, within Zone NP 21.

In the last 15 years, a certain number of studies have been published which focused mainly on the identification of this ^{18}O event used as a stratigraphic marker for the Eocene-Oligocene boundary. Following the spadeworks of Savin and Douglas in the equatorial Pacific (Douglas and Savin, 1971, 1973, 1975 ; Savin, 1977 ; Savin et al., 1975) and those of Shackleton and Kennett in the Southern Ocean (Shackleton and Kennett, 1975 ; Kennett and Shackleton, 1976) a number of results have been published ; they concern deep Atlantic sites as well as land outcrops : DSDP sites 116, 119, 400-401, 366 or 398 for the North Atlantic and DSDP site 357 in the South Atlantic. Bulk carbonates from land sections such as the Cachaou section near Biarritz in the Gulf of Biscay (Rabussier Lointier, 1980) or marine molluscan shells (Buchardt, 1978) from the North Sea have been analysed in order to calibrate the amplitude of the ^{18}O change at the Eocene-Oligocene transition. All these investigations, which preceeded the use of the hydraulic piston core system, prior to DSDP leg 68, correspond to a first phase and are reported in Table 1. This first phase is characterized by the fact that :

(1) The non-continuous and incomplete recovery of the sedimentary sequences caused the imprecision in the localization of the oxygen-isotopic shifts which might appear non isochronous (as, for instance, in the Pacific) and, in some studies, led to an artificially increased positive shift in isotopic ratios due to the widely spaced sampling. Thus, the beginning of the shift was dated to start at some sites as early as the late Middle Eocene.

(2) Stable isotope analyses were performed on bulk carbonates or mixed species of benthic or planktonic foraminifers.

(3) The most attention was paid to the ^{18}O event and very few comments concern the ^{13}C event.

TABLE 1
DSDP sites and calcareous material used for the Eocene-Oligocene ^{18}O event identification.

DSDP Sites	Location and depth (m) Present depth (..) Extrapolated depth.. *	Coordinates	Biostratigraphic interval	Species analysed	$\Delta\delta^{18}O$ ‰
PACIFIC					
305 (1)	Shatsky Rise (2903)	32°00.1'N/ 157°51.0'E	P10-P18	-----> Calc. nannofossils	+ 1.30
167 (1)	Central Equatorial (3176) 3363	7°04'N/ 167°49.5'W	P18-P19	-----> Calc. nannofossils	+ 0.90
(2)			P14-P17	-----> Planktonic foram.	+ 1.08
(3)			P16-P18	-----> Benthic foram.	+ 1.00
171 (2)	North (2283)	19°07.8'N/ 169°27.6'W	P12-P18	-----> Planktonic foram.	+ 1.31
				-----> Benthic Foram.	+ 1.17
44 (2)	Central North (2478)	19°18'N/ 169°09'W	P16-P19	-----> Planktonic foram.	+ 1.50
				-----> Benthic foram.	+ 1.25
277 (4)	Campbell Plateau (1214)	52°13.43'S/ 166°11.48'E	Eocene-Oligocene	-----> Planktonic foram.	+ 2.00
				-----> Benthic foram.	+ 2.00
ATLANTIC					
116 (5)	North Rockall Plateau (1151)	57°29.76'N/ 15°55.46'W	P16-P17	-----> Calc. nannofossils	+ 1.06
398 (6)	North West Portugal	40°57.6'N/ 10°43.1'W	P13-P16	-----> Planktonic foram.	+ 2.00
			P15-P16	-----> Benthic foram.	+ 2.00
366 (5)	Sierra Leone Rise (2953) 2601	05°40.7'N/ 19°51.1'W	P16-P18	-----> Calc. nannofossils	+ 1.15
119/401 (7)	Gulf of Biscay (4447-2395) 3700	45°02.3'N/ 07°58.8'W	P16-P17	-----> Calc. nannofossils	+ 1.50

(1) Douglas and Savin (1975) ; (2) Savin (1977) ; (3) Savin et al. (1975) ; (4) Kennett and Shackleton (1976) ; (5) Cavelier et al. (1981) ; (6) Vergnaud Grazzini (1979) ; (7) Miller and Curry (1982).
* Extrapolated depths, following the determinations of Keigwin and Corliss (in press).

Only beginning with DSDP leg 68, the use of hydraulic piston coring initiates a second phase in the investigations dealing with isotopic changes at the Eocene-Oligocene boundary.

The objectives of this study are to establish the timing and the amplitude of Eocene-Oligocene stable isotopic changes which emerged from all data recently published and to present some paleo-environmental hypothesis concerning such changes (Sites plotted in Fig. 1).

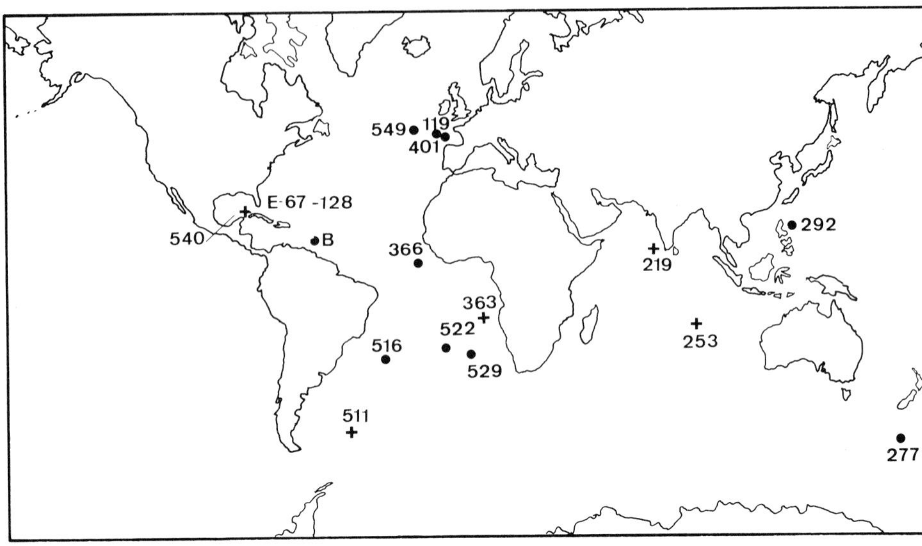

Fig. 1. Location map of the sites discussed. (•) : sites plotted in Fig. 2 to 4; (+) : sites mentionned in the discussion ; (B) Bath Cliff Section, Barbados W.I.

THE LATELY PUBLISHED ISOTOPIC DATA

The last five years several high resolution studies on the Eocene-Oligocene events have been published. These isotopic analyses are mostly based on monospecific samples of foraminifers. In a few cases magneto-stratigraphy permits a precise control for the single isotopic events. Sites which have been used recently are listed in Table 2. Isotopic results of the most complete records are plotted in Fig. 2 and 3. By now the isotope event at the Eocene/Oligocene boundary is best documented in the Atlantic Ocean.

In the following discussion, we will report the extent of the shift as the difference between average post-shift $\delta^{18}O$ values of the Early Oligocene and average latest Eocene $\delta^{18}O$ values.

In studies which deal with isotopic data of different species in the Eocene-Oligocene interval, specific fractionation may further be a source of uncertainties. Systematic differences in $\delta^{18}O$ and $\delta^{13}C$ values have been registered for both planktonic and benthic foraminifers (e.g. Douglas and Savin, 1978 ; Shackleton and Vincent, 1978 ; Belanger et al., 1981 ; Woodruff et al., 1980 ; Berger et al., 1981 ; Graham et al., 1981). Most workers consider observed $\delta^{18}O$ values of planktonic foraminifers as equilibrium values. The isotopic ranking from a

TABLE 2

Location of the sites where new isotopic data have been obtained for the Eocene/Oligocene boundary.

Leg-Hole-Section	Location Present depth(..) Extrapolated depth.. at the Eocene/Oligocene time (12)	Coordinates Paleolatitudes ( ) (12)	Species analysed	^{18}O increase at the Eocene/Oligocene transition
277 (1)	Campbell Plateau Southern Ocean (1214)	52°13'43"S/ 166°11.48'E (59°S)	*Oridorsalis* *Globigerina angiporoides*	+ 0.80 + 0.80
(12)	"	"	*Globocassidulina subglobosa* *Chiloguembelina spp.*	+ 0.85 + 0.50
292 (1) (12)	West Tropical Philippine Sea (2943) <u>1068</u>	15°49.11'N/ 124°39.05'E (6°N)	*Oridorsalis* *G. ampliapertura*	+ 1.00 + 0.40
363 (2)	Southeastern Atlantic Walvis Ridge (2248) <u>2216</u>	19°38.75'S/ 9°2.8'E (30°S)	*C. ungerianus* *G. ampliapertura*	+ 0.45 + 0.50
511 (3)	Southwest Atlantic Maurice Ewing Bank (2589)	51°00.28'S/ 46°58.30'W (59°S)	Mixed benthic foraminifers Mixed planktonic species	+ 2.50 + 1.30
516F (11)	Rio Grande Rise (1317) <u>800</u>	30°25'S/ 35°15'E	*Cibicides*	≅ + 1.00
522 (4) (5)	South Atlantic (4441) <u>2500-3000</u> <u>2693</u>	26°06.8'S/ 5°07.8'W (35°S)	*Stilostomella* *Catapsydrax dissimilis* *G. venezuelana*	+ 1.30 + 1.20 + 0.50
523 (5)	South Atlantic (4572) <u>2300</u>	28°33.13'S/ 02°15.08'W	*O. umbonatus* *Globigerinatheka mexicana*	+ 1.10
529 (6)	South Atlantic Walvis Ridge (3035) <u>2362</u>	28°55.83'S/ 02°46.08'E (40°S)	Various benthic species Various planktonic species	+ 0.60
540 (7)	Gulf of Mexico (2926)	23°49.73'N/ 84°22.25'W (20°N)	*Cibicidoides spp.*	0.50 to 0.80
Eureka 67-128	Gulf of Mexico (1494)	(25°N)	*Bulimina alazanensis* *G. ampliapertura, Chiloguembelina spp.*	+ 0.75 + 0.05
549 (8) (9)	Gulf of Biscay (2538) <u>2386</u>	49°05.23'N/ 13°05.89'W	*Planulina*	+ 0.80
Bath Cliff Section (10)	Barbados-West Indies- Land Outcrop <u>2500</u>		*O. umbonatus* *C. dissimilis*	+ 1.22 + 1.00

(1) Keigwin (1980) ; (2) Corliss et al. (1984) ; (3) Muza et al. (1983) ; (4) Poore and Matthews (1984) ; (5) Oberhänsli et al. (1984) ; (6) Shackleton et al. (1984) ; (7) Belanger and Matthews (1984) ; (8) Miller et al. (1984) ; (9) Snyder et al. (1984) ; (10) Saunders et al. (1984) ; (11) Vergnaud Grazzini and Saliege (1985) ; (12) Keigwin and Corliss (in press).

single sample is mostly explained by the fact that the different taxa live and form their test at different depth levels of the surface water column. Our discussion is based on isotopic values of the following planktonic foraminifers (Table 2) : *Chiloguembelina* spp., *Globigerina ampliapertura*, *Globigerina venezuelana*, *Globigerina angiporoides* and *Catapsydrax dissimilis*. Based on the model of depth ranking, *G. ampliapertura* and *G. venezuelana*, with light values (Poore and Matthews, 1984), indicate shallow water conditions. *Chiloguembelina* spp. are probably shallow to intermediate dwellers. *G. angiporoides* and *C. dissimilis* report on intermediate and deep surface water conditions, respectively.

In this work, isotopic data of the benthic foraminifers *Oridorsalis umbonatus*, *Cibicidoides* sp. and *Bulimina* sp. are interpreted. Apparently many of the benthic species are out of isotopic equilibrium. *Oridorsalis* and *Cibicidoides* are consistently offset from equilibrium values by 1°/₀₀.($\delta^{13}C$) and 0.5°/₀₀ ($\delta^{18}O$) respectively, whereas *Bulimina* secrets its test probably close to equilibrium conditions (Shackleton et al., 1984).

Isotopic signals at North Atlantic sites

The most recently published results are based on studies of DSDP Site 549 from the Bay of Biscay and the Bath Cliff Section in Barbados, as well as of DSDP Site 540 and the core Eureka-67-128, both located in the Gulf of Mexico.

A rapid increase in $\delta^{18}O$ is observed at Site 549 (Miller et al., 1984) and

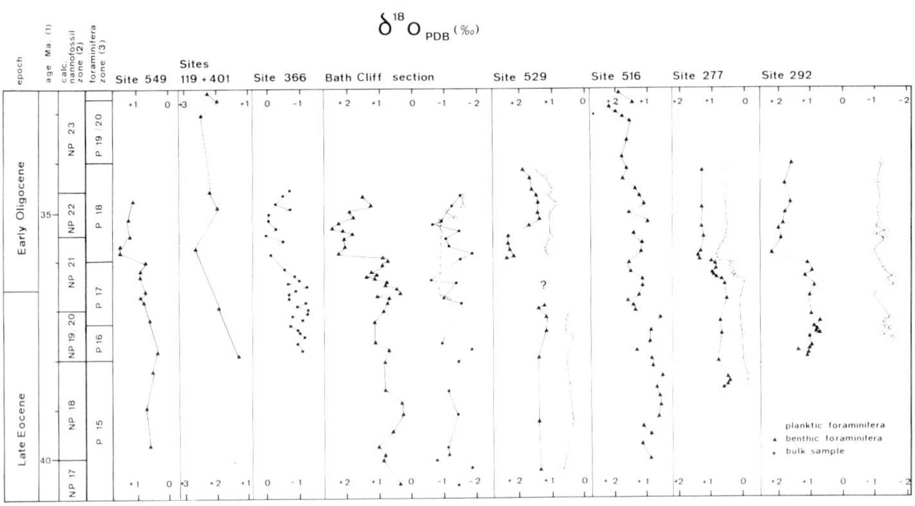

Fig. 2. $\delta^{18}O$ plot of selected Atlantic and Pacific sites (Eocene-Oligocene transition).
DSDP Site 549 : monospecific samples of *Cibicidoides* sp. (Miller et al., 1984) ; DSDP Sites 119+401 : *Cibicidoides* sp. (Miller and Fairbanks, 1983) ; Bath Cliff Section : *Oridorsalis umbonatus*, *G. venezuelana*, calcareous nannofossils (Saunders et al., 1984) ; DSDP Site 529 : *Bulimina* spp., *Globigerinatheka* spp. (Lower Eocene), *Catapsydrax* spp. (Early Oligocene) (Shackleton et al., 1984) ; DSDP Site 516 : *Cibicides* spp. (Vergnaud Grazzini and Saliege, 1985) ; DSDP Site 277 : *Oridorsalis* sp., *G. angiporoides* (Keigwin, 1980) ; DSDP Site 292 : *Oridorsalis* sp., *G. ampliapertura* (Keigwin, 1980). (1) Berggren et al. (1985) ; (2) Martini (1971) ; (3) Blow (1969) and Berggren and Van Couvering (1974).

at Bath Cliff Section (Saunders et al., 1984). It culminates in Zone P 18 (Fig. 2). In the benthic record the amplitude of the shift is 0.8°/oo at Site 549 (*Cibicidoides* sp.) and 1.22°/oo at the Bath Cliff Section (*Oridorsalis umbonatus*). At Site 540 (*Cibicides* sp., Belanger and Matthews, 1984) and core E-67-128 (*Bulimina alazanensis*, Keigwin and Corliss, in press) the shift is 0.5°/oo and 1.0°/oo respectively. $\delta^{18}O$ values of the deep dwelling planktonic species (*Catapsydrax dissimilis*) increase by about 1°/oo across the P17-P18 transition at the Bath Cliff Section. However the shallow dwellers (*Globigerina venezuelana*, *G. ampliapertura*, *Chiloguembelina cubensis* (?)) and mixed nannofossils from the Bath Cliff Section and core E-67-128 do not display any significative trend across P17-P18 zones boundary (upper part of Zone NP 21). At the equatorial Site 366, however, $\delta^{18}O$ values of bulk carbonate increase by more than 1°/oo through P17-P18 transition. Diagenetic alterations may probably account at least for part of the shift.

$\delta^{13}C$ data at Sites 540 and 549 do not show any comparable trends. The benthics, at Site 549, reach maximum values in the lower part of Zone P 18 (Fig. 3) while at Site 540, the maximal $\delta^{13}C$ values are located at the Eocene/Oligocene

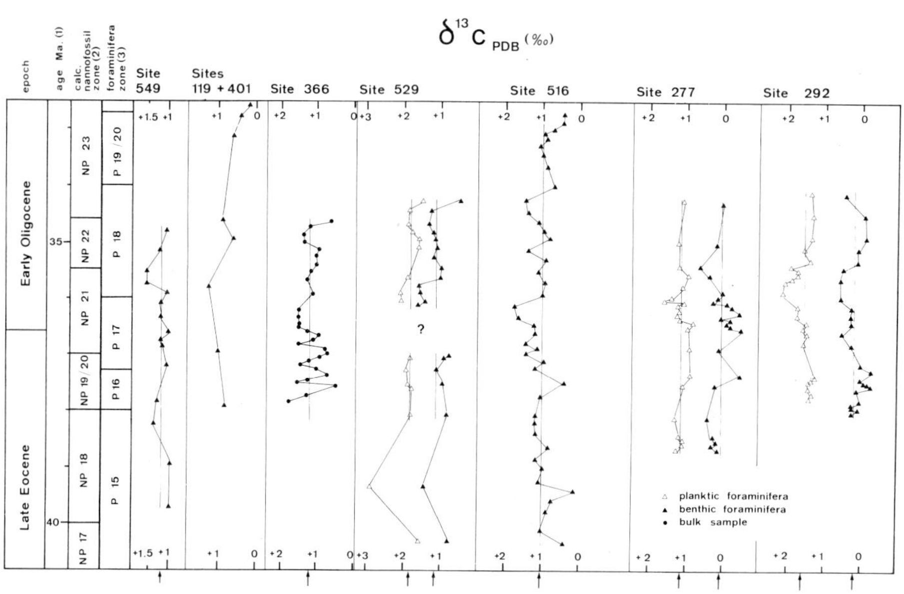

Fig. 3. $\delta^{13}C$ plot of selected Atlantic and Pacific sites (Eocene-Oligocene transition) ; the lines indicate average $\delta^{13}C$ compositions of benthic and planktonic records. For further details, see legend to Fig. 2.

boundary. After what, $\delta^{13}C$ values decrease at both sites. The $\delta^{13}C$ record of the bulk carbonate at Site 366 does not show any trend. It appears that the ^{13}C record at the Eocene-Oligocene transition is located at the maximal ^{13}C enrichment of the second major cycles of the Tertiary.

Isotopic signals at South Atlantic Sites

Oxygen and carbon isotopic records from the Walvis Ridge (DSDP Sites 363 and 529), the Angola basin (Site 522), the Rio Grande Rise (Site 516) and the Falkland Plateau (Site 511) are discussed (Fig. 2 and 3).

In the low resolution isotopic record of Site 363 (Corliss et al., 1984), an increase of about 0.5°/₀₀ in $\delta^{18}O$ values of *G. ampliapertura* and *Chiloguembelina* spp. is recorded, while the benthic foraminifer *Cibicides ungerianus* records only a 0.45°/₀₀ change. These increases are recorded at the transition between Zones *Turborotalia cerroazulensis* and *P. micra* (Keigwin and Corliss, in press).

At Site 529 (Shackleton et al., 1984), oxygen isotopic analyses of *Bulimina* spp., *Globigerinatheka* spp., *Catapsydrax* spp. result in a 1°/₀₀ increase located between the upper part of Zone P 17 and the lower part of Zone P 18. The shift observed in the surface water is most probably enhanced because *Globigerinatheka* spp. (measured in Eocene sediments) and *Catapsydrax* spp. (Oligocene record) had different depth habitats. A hiatus at the transition between P 17 and P 18 separates most probably minimal from maximal values (Fig. 2 and 3).

At Site 522 (Fig. 4) a high resolution record has been generated for the time interval of Zones NP 20 and NP 21 (Oberhänsli et al., 1984). The benthic foraminiferal species *Stilostomella* spp. and the planktonic species *G. venezuelana* and *C. dissimilis* were analysed. In the uppermost part of Zone NP 20, the $\delta^{18}O$ values of benthic and planktonic foraminifers start to increase steplike. The maximum increase is recorded slightly beyond the Eocene-Oligocene boundary. The overall increase is around 0.5°/₀₀ for the shallower planktonic species and over 1.0°/₀₀ for the benthic species.

At Site 516 (Vergnaud Grazzini and Saliege, 1985), the benthic foraminiferal species *Cibicides* spp. has been analysed (Fig. 2 and 3). A steplike increase towards higher $\delta^{18}O$ values starts as early as Zone P 14. The first maximal values are obtained in Zone P 18. The average increase of $\delta^{18}O$ values, between Zones P 17 and P 18 corresponds to approximately 1.0°/₀₀.

At DSDP Site 511, isotopic analyses were performed on mixed species assemblages (Muza et al., 1983). An increase in $\delta^{18}O$ values by approximately 2°/₀₀, however, is also recorded between the LAD of *Discoaster saipanensis* and the uppermost part of foraminiferal zone *Globigerina brevis*. The amplitude of this shift is considerably higher than expected. It is probably distorted because analyses have been performed on mixed assemblages.

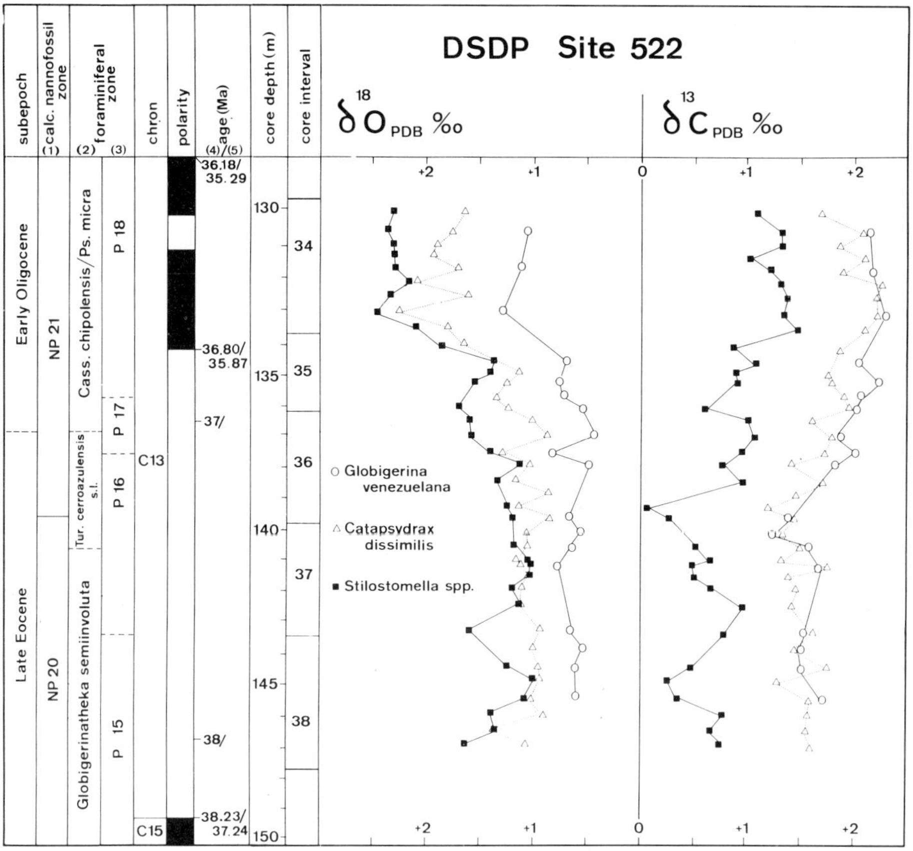

Fig. 4. $\delta^{18}O$ and $\delta^{13}C$ plot of the South Atlantic DSDP Site 522 (Oberhänsli and Toumarkine, 1985). (1) Martini (1971) ; (2) Bolli et al. (1985) ; (3) Blow (1969) and Berggren and Van Couvering (1974) ; (4) Hsü et al. (1984) ; (5) Berggren et al. (1985).

At Sites 516 and 529 (Fig. 3) and 363 (Keigwin and Corliss, in press) maximum $\delta^{13}C$ values for benthic and planktonic foraminifers are reached at the transition of Zones P 17 to P 18. At Site 522 (Fig. 4) the $\delta^{13}C$ record for surface and bottom water inhabitants increases steadily throughout Zone P 17 and reaches maximum values within Zone P 18. During this change, the vertical $\delta^{13}C$ gradient remains stable.

Isotopic results from Pacific Ocean sites

At Site 277 from the southern Ocean, high resolution isotopic study was published by Keigwin (1980) using monospecific samples of foraminifers. Both the planktonic species *G. angiporoides* and the benthic species *Oridorsalis* sp. display a similar $\delta^{18}O$ increase (around 0.8‰) at the transition from Zones P 17 and P 18 (Fig. 2). More recent measurements (Keigwin and Corliss, in press), however, indicate an increase of about 0.5°/₀₀ in the $\delta^{18}O$ values of the planktonic *Chiloguembelina* spp., while the benthic species *Globocassidulina subglobosa* records an increase of the $\delta^{18}O$ values by 0.85°/₀₀ (Keigwin and Corliss, in press).

At the tropical Site 292 from the Philippine Sea, the benthic foraminifer *Oridorsalis* sp. displays a $\delta^{18}O$ increase of about 1°/₀₀ while the planktonic *G. ampliapertura* reveals an increase of less than 0.4°/₀₀ from Zones P 17 to P 18.

Coevally with the $\delta^{18}O$ shift, the $\delta^{13}C$ values of benthic and planktonic foraminifers become higher (Fig. 3). During this change the vertical ^{13}C gradient does not vary significantly.

THE TIMING AND THE AMPLITUDE OF THE ISOTOPIC EVENTS AT THE EOCENE-OLIGOCENE TRANSITION

The timing of the oxygen isotopic event

Biostratigraphical assignements for the sites discussed are available from studies performed by Miller (Site 549, Miller et al., 1984), Müller (Site 366, in Rabussier Lointier, 1980), Boersma (Site 529, Boersma, 1984), Pujol (Site 516, Pujol, 1983), Jenkins (1975) and Edwards and Perch-Nielsen (Site 277, Edwards and Perch-Nielsen, 1975), Ujiie (1975) and Ellis (Site 292, Ellis, 1975), Toumarkine and Percival (Site 522, Oberhänsli and Toumarkine, 1985 ; Percival, 1984) and Perch-Nielsen and Saunders (Bath Cliff Section, Saunders et al., 1984).

For our purposes (Fig. 2-4) we adapted the different zonation schemes used for planktonic foraminifers and calcareous nannofossils to the zonal scheme of Blow (1969), Berggren and Van Couvering (1974 : P-Zones) and Martini (1971 : NP-Zones). In the present study we place the Eocene-Oligocene boundary tentatively at the level of the last specimens of *Turborotalia cerroazulensis*. In relation to the calcareous nannofossils the epoch boundary is located slightly above the NP 20/NP 21 Zone boundary.

Generally there is a good agreement in the timing of the ^{18}O event in records with a sufficiently high resolution. The Eocene-Oligocene ^{18}O event corresponds in fact to a rapid shift towards higher values marking a step in a global increasing trend in $\delta^{18}O$ values. This globally recorded trend of increasing values starts as early as the late Early Eocene (Shackleton et al., 1984). A duration of about 10^5 years for the ^{18}O event at the Eocene-Oligocene transition has been postulated by many authors (e.g. Kennett and Shackleton, 1976).

The Eocene-Oligocene boundary event is located at the transition between Zones P 17 to P 18 (= within NP 21). This corresponds in fact to the Early Oligocene time. Also at DSDP 529, where an important hiatus is noted within the upper part of Zone P 17, the $\delta^{18}O$ values recorded at the base of Zone P 18 are by 0.8°/₀₀ higher for planktonics and by 1.0°/₀₀ higher for benthics than the $\delta^{18}O$ values recorded below the hiatus, suggesting that the shift occurred somewhere between Zone P 17 and the base of Zone P 18.

The best precision, however, for the timing of the event is provided by the high resolution record obtained at DSDP Site 522 (Fig. 4 ; Oberhänsli and Toumarkine, 1985). At this site, the magnetostratigraphic record indicates that the shift is exactly located in magnetochrone 13, within the interval of C 13 N2 to C 13 R2.

The time scale of the radiometric age used in Fig. 2 and 3 is that of Berggren et al. (1985) . In Fig. 4, two alternative time scales are given (Hsü et al., 1984 and Berggren et al., 1985). Recent data (Montanari et al., this volume) suggest that the radiometric age of the Eocene-Oligocene boundary might be younger than indicated on Fig. 2-4 and could be approximately 35.7 Ma with a 2 standard-deviation uncertainty of 0.4 Ma. From the other data from the same authors, we may also deduce that top of Anomaly 13 N is dated around 35 Ma. It follows that the $\delta^{18}O$ event should be bracketted by these two dates : 35 Ma and 35.7 Ma.

The amplitude of the oxygen-isotope shift

The extent of the increase in $\delta^{18}O$ values recorded by benthic foraminifers ranges between +0.8°/₀₀ (DSDP Site 549) and +1.22°/₀₀ (Bath Cliff Section). At DSDP Site 516, a steplike increase of 1°/₀₀ occurs between Zones P 17 and P 18, but the shift located at the base of Zone P 18 accounts only for a 0.6°/₀₀ increase. There does not appear to be any relation between the extent of the $\delta^{18}O$ shift and the paleodepths of the various sites ; the extents of the $\delta^{18}O$ shifts however may probably not be straightforward compared because of different sampling densities at the studied sites and possibly varying fractionation factors of the different species with time.

Concerning the planktonic foraminifers, it may be observed, as a first approximation, that the amplitude of the ^{18}O shift decreases when passing from high latitude ($\Delta\delta^{18}O \cong +0.8$°/₀₀ at DSDP Site 277, 52° South) to mid and low latitudes (0.6°/₀₀ at DSDP Site 522, 26° South, and 0.4°/₀₀ at DSDP Site 292, 15° North). At the Bath Cliff Section, which is equally a low latitude site, the ^{18}O record of the shallow dwelling *G. venezuelana* shows no shift at all.

The carbon isotopic event

Carbon isotope records are not easy to compare in such a limited time span because they make part of a more global cycle. In the discussed sites, maximal values are located in the second major cycle of the Tertiary ^{13}C record, displaying a maximum within Zones P 17-P 18.

The varying positions of the ^{13}C maximal peak (upper part or lower part of Zone NP 21) result in negative correlations between some of the records, such as those of Sites 277 and 516 or 529.

At some sites, the ^{13}C gradient between surface and bottom water inhabitants remains more or less constant (DSDP Sites 292 and 522). At other sites this vertical gradient may change from the late Eocene to the early Oligocene (DSDP Site 277) or fluctuates (DSDP Sites 522 and 529), suggesting that local factors may have additionally influenced the ^{13}C pattern in surface and bottom waters.

THE MEANING OF STABLE ISOTOPE EVENTS AT THE EOCENE-OLIGOCENE TRANSITION

The oxygen isotope shift

The temperature effect. The most widely accepted interpretation of the rapid ^{18}O enrichment is that it reflects a major cooling of antarctic surface water (and thus of deep bottom water worldwide).

Shackleton and Kennett (1975) have interpreted the dramatic cooling as the crossing of some critical threshold not yet identified but in some way related to the gradual isolation of Antarctica. Although no continental ice sheet had yet formed, these authors argued that the planktonic isotopic paleotemperatures drop implied sea-level glaciation and that benthic paleotemperatures indicated the onset of deep water circulation dominated by the formation of cold bottom waters around the Antarctic continent. For other authors, however, this temperature drop is not related to any sea ice formation (Savin, 1977). Kvasov and Verbitsky (1981) stated that deep circulation south of Tasmania began 38 Ma ago and caused the abrupt cooling. However they cite no supporting evidence for this assumption.

The scenario proposed by Thierstein and Berger (1978) is based on the assumption that the Arctic Ocean became hyposaline due to a transient closing of the Labrador passage sometimes during the Eocene. The introduction of a low salinity surface water layer on the world oceans would temporarily obstruct the heat exchange between deep and shallow water layers. Thus the moderating influence of the ocean on climate was reduced. As a consequence seasonality increases. Temperature in low latitude surface water rises due to the stable stratification of the watermasses. Transport of humidity to high latitude areas may have been enhanced. According to Thierstein and Berger (1978) the terminal Eocene injection event could have been a trigger for covering Antarctica with sufficient snow to move the polar front out of the shelf which would have been crucial for the formation of cold bottom water.

An alternative interpretation of the ^{18}O record is given by the model of continental glaciation (Matthews and Poore, 1980). Forming continental ice caps, the sea water is depleted in ^{16}O. As a consequence the $\delta^{18}O$ values of shells increase during these intervals. Despite the lack of independent evidences for such ice formation on Antarctica during the Eocene-Oligocene transition, this hypothesis cannot be dismissed. In favour of this model are the observations reported by Corliss (1981). At DSDP Site 277, benthic foraminiferal assemblages do not display any major change at the Eocene-Oligocene boundary and thus would not justify a cooling of 3°C or even more. Corliss concluded that either the foraminifera possess wide environmental tolerance, or the drop in temperature was less than 3°C. In conclusion part of the isotopic shift could be due to an accumulation of continental ice. This was also stated by Snyder et al. (1984).

As mentionned above, Kvasov and Verbitsky (1981) stated that deep circulation between Antarctica and Australia began at the Eocene-Oligocene boundary at approximately 38 Ma ago. According to these authors the resultant cooling on the Antarctic continent may initiate the growth of major ice fields on several mountain massifs in East Antarctica in about 100,000 years, which would be consistent with the duration of the isotopic event (Kennett and Shackleton, 1976).

Kennett et al. (1975) and Weissel et al. (1977) however believe that this circulation did not start until 30-25 Ma ago.

More recently Keigwin and Corliss (in press) revised a series of middle Eocene to early Oligocene stable isotope data from Atlantic, Pacific and Indian ocean locations. They interpreted a 0.45°/₀₀ increase in $\delta^{18}O$ values of surface dwelling foraminifers, recorded at all latitudes including those in the tropics, as an evidence for ice accumulation on the continents, during the earliest Oligocene. These authors also evaluated the average increase in $\delta^{18}O$ values of benthic foraminifers, as 0.6°/₀₀. So that, with a sea water compositionnal effect of 0.4°/₀₀ the deep sea cooled in the early Oligocene time by as much as 2°C at some locations and 1°C overall. Furthermore, they did not note any change in the planetary temperature gradient from Eocene to Oligocene time.

In our opinion a glacial effect cannot be entirely excluded and might be as high as 0.4°/₀₀. This value represents the minimal increase in $\delta^{18}O$ values of shallow dwelling planktonics recorded at some of the low latitude sites. At the Bath Cliff Section, however, the surface-water dwelling *G. venezuelana* does not show any positive oxygen shift within the questionable time interval, although we may not exclude yet the scenario that low latitudes experienced a temperature increase when high latitudes suffered a cooling and ice caps formed on Antarctica.

By now we do not have any independent evidence from faunistic or sedimentological studies which would support a significant build up of ice caps within Zone NP 21. Therefore we favor the interpretation that the oxygen-shift recorded within the upper part of Zone NP 21 is based mainly on a temperature change in high and mid latitudes. As a consequence the latitudinal temperature gradient may have increased considerably from Eocene and Oligocene time.

The carbon isotope event

The maximum in $\delta^{13}C$ values of planktonic and benthic foraminifers which is recorded within Zone NP 21 makes part of major global changes.

Changes in the $\delta^{13}C$ of foraminiferal shells are usually taken to reflect changes in the isotopic composition of the dissolved carbon in the water surrounding the shell.

Global changes in the $\delta^{13}C$ values of the oceanic CO_2 may result from a change in the size of other carbon reservoirs such as the atmosphere, the biosphere (for rapid changes) or carbonate sediments (for changes lasting about 1.8×10^5 yrs), or riverine bicarbonate, or changes in the ratio of carbon buried as $CaCO_3$ to carbon buried as organic carbon. Global changes may also result from the changes in the $\delta^{13}C$ values of some carbon reservoirs.

Differences between the $\delta^{13}C$ values of the deep CO_2 of different oceanic basins are related to the apparent oxygen utilization (A.O.U.) and indicate differences in the aging of water masses and, thus, in abyssal circulation.

Various models have recently been published which explain a 0.7°/₀₀ increase in the $\delta^{13}C$ values recorded in holocene sediments. Shackleton et al. (1984) relate it to an increase in atmospheric pCO_2 during deglaciation ; evidences for atmospheric pCO_2 changes were recently made available through the study of polar ice cores. Other models - the phosphate extraction model (Broecker, 1982), the coral-reef model (Berger, 1982 ; Berger and Killingley, 1982), the denitrification model, the C/P model, the deglacial forest build-up (Shackleton, 1977)- have been discussed by Berger and Keir (1984). In all these models, the glaciation also appears to be associated with a lower atmospheric pCO_2 and a lower ^{13}C oceanic content.

The three major cycles of the Tertiary, however, correspond to larger changes in $\delta^{13}C$ values indicating either that larger volume changes occurred in some of the carbon reservoirs or in the ratio of input/output carbon from reservoirs with highly different $\delta^{13}C$ values - as, for instance, terrestrial organic matter (-20°/₀₀) versus marine organic matter (-25°/₀₀). Very recently, Vincent and Berger (in press) demonstrated that the mid-Miocene oxygen isotope step occurred within a substantial $\delta^{13}C$ excursion toward heavier values, which they call "the Monterey carbon isotope excursion". This large excursion represents the last of the major ^{13}C cycles of the Tertiary. It signifies excess extraction of organic matter above the normal long-term conditions of dynamic equilibrium. Extraction of organic matter in Miocene is mainly the fact of increasing equatorial upwelling and coastal upwelling in the Pacific, at the same time. The resulting increases in carbon deposition might have exerted a considerable "pull" in the ocean-atmosphere reservoir : this pull might in turn have lowered the atmospheric pCO_2 and then the global temperature.

Although one could observe that the positive ^{13}C excursion which culminates around the Eocene-Oligocene boundary is likely followed by a certain amount of ice storage ($\delta^{18}O$ increase during Zone P 21), we do not have any further evidence for a higher organic production at that time. The collapse of a carbon reservoir with a high $^{13}C/^{12}C$ ratio or the build-up of a reservoir with a low $^{13}C/^{12}C$ ratio may have occurred at that time.

The opposite trend observed between the ^{13}C records of Site 277 and other sites, within biozone NP 21, however, might have been the result of a change in the rate of production of Pacific and Atlantic bottom waters or reflect a supply of ^{13}C-enriched bottom waters in Atlantic and North Pacific areas at that time.

CONCLUSIONS

The review of the most complete stable isotope records at the Eocene-Oligocene transition documented in land based sections as well as in deep sea drillings of the Atlantic and Pacific Oceans, leads to the following conclusions :

1) The oxygen-isotope event at the Eocene-Oligocene transition corresponds, in fact, to a rapid shift towards higher $\delta^{18}O$ values marking a step in a global increasing trend. The ^{18}O event is located at the transition between Zones P 17 to P 18 (=within Zone NP 21) ; this corresponds, in fact, to the early Oligocene time. The best timing of the ^{18}O event, however, is provided by the high resolution record obtained at DSDP Site 522, where magnetostratigraphy indicates that the shift is exactly located in magnetochrone 13 within the interval of C 13 N2 to C 13 R2. Recent radiometric datings lead to propose an age around 35 Ma for this event. A duration of about 10^5 years has been postulated by many authors.

2) The amplitudes of the $\delta^{18}O$ shift cannot be straightforward compared because of different sampling densities at the studied sites and possibly varying fractionation factors of the different species with time ; furthermore, at some sites, diagenetic alterations may probably account for part of the shift.

For benthic foraminifers, the ^{18}O shift ranges between 0.8‰ and 1.22‰. For planktonic foraminifers, it appears, as a first approximation, that this amplitude decreases when passing from high latitudes ($\Delta^{18}O$ = +0.8‰) to mid and low latitudes (+0.6 to +0.4‰).

3) Although a maximal glacial effect of about 0.4‰ couldn't be entirely excluded, no independent evidence from faunistic or sedimentological studies would support the hypothesis of a significant build-up of ice caps within the time interval corresponding to Zone NP 21. Therefore, the present authors favour

the interpretation that the oxygen isotopic shift recorded in the upper part of Zone NP 21 is based mainly on a temperature change in high and mid latitudes.

4) $\delta^{13}C$ values display a maximum within Zones P 17-P 18, which corresponds to the peak-values of the second major cycles of the Tertiary ^{13}C record. Local factors may have additionally influenced the ^{13}C pattern in surface and deep waters.

5) Maximal $\delta^{13}C$ values may have resulted either from the collapse of a carbon reservoir with a high $^{13}C/^{12}C$ ratio or the build-up of a reservoir with a low $^{13}C/^{12}C$ ratio. The opposite trends observed between the North Pacific, the Atlantic and the South Pacific ^{13}C records at the Eocene-Oligocene transition may indicate a change in the rate of production of Pacific and Atlantic bottom waters ; they may also reflect a supply of ^{13}C enriched bottom waters in Atlantic and North Pacific areas, at that time.

A better understanding of the isotopic events recorded at the Eocene-Oligocene transition requires that, in the future, continuous sedimentary sequences be recovered in "key areas" such as high, low and mid latitudes in all oceans, at different longitudes ; it also necessitates the magnetostratigraphic control of the sequences and a denser sampling to obtain a high resolution biostratigraphic and isotopic record. In this article, we have tried to point to the facts which limit the paleoclimatic interpretations and prevent from any global reconstruction in the paleocirculations.

REFERENCES

Belanger, P.E., Curry, W.B. and Matthews, R.K., 1981. Core-top evaluation of benthic foraminiferal isotopic ratios for paleo-oceanographic interpretations. Palaeogeogr.,Palaeoclimat.,Palaeoecol., 33: 205-220.

Belanger, P.E. and Matthews, R.K., 1984. The foraminiferal isotope record across the Eocene/Oligocene boundary at Deep Sea Drilling Project Site 540. In : R.T. Buffler, W. Schlanger et al., Init. Repts. DSDP, 77, Washington, D.C. (U.S. Gov't. Printing Office) : 589-592.

Berger, W.H., 1982a. Increase of carbon dioxide in the atmosphere during deglaciation : the coral reef hypothesis. Naturwissenschaften, 69: 87-88.

Berger, W.H., 1982b. Deglacial CO_2 buildup : constraints on the coral-reef model. Palaeogeogr.,Palaeoclimat.,Palaeoecol., 40: 235-253.

Berger, W.H. and Keir, R.S., 1984. Glacial-Holocene changes in atmospheric CO_2 and the deep sea record. In : J.E. Hansen and T. Takahashi (Editors), Climatic Processes and Climate Sensitivity, Geophysical Monograph, 29, Maurice Ewing vol. 5, Washington D.C. : 337-351.

Berger, W.H. and Killingley, J.S., 1982. The Vorthington effect and the origin of the Younger Dryas. Journ. Mar. Res., 40 (suppl.) : 27-38.

Berger, W.H., Bé, A.W.H. and Vincent, E. (Editors), 1981. Oxygen and carbon isotopes in Foraminifera. Palaeogeogr.,Palaeoclimat.,Palaeoecol., 33: 1-277.

Berggren, W.A. and Van Couvering, J.A., 1974. The late Neogene. Biostratigraphy, geochronology and palaeoclimatology of the last 15 millions years in marine and continental sequences. Palaeogeogr.,Palaeoclimat.,Palaeoecol., 16, 1-2, 216 pp.

Berggren, W.A., Kent, D. and Flynn, J.J., 1985. Paleogene geochronology and chrono-stratigraphy. In : N.J. Snelling (Editor), Geochronology and Geologic Timescale, Geol. Soc. Lond. Special Paper.

Blow, W.H., 1969. Late Middle Eocene to Recent planktonic foraminiferal biostratigraphy. Proc. 1st Intern. Conf. on Planktonic Microfossils, Geneva 1967, Leiden, E.J. Brill : 199-422.

Boersma, A., 1984. Cretaceous-Tertiary foraminifers from the South Eastern Atlantic, Walvis Ridge Area, Deep Sea Drilling Project Leg 74. In : T.C. Moore Jr., P.D. Rabinowitz et al., Init. Repts. DSDP, 74, Washington (U.S. Gov't Printing Office) : 501-523.

Bolli, H.M., Saunders, J.B. and Perch-Nielsen, K., 1985. Plankton stratigraphy. Cambridge University Press.

Broecker, W.S., 1982. Ocean chemistry during glacial time. Geochim. Cosmochim. Acta, 46: 1689-1705.

Buchardt, B., 1978. Oxygen isotope paleotemperatures from the Tertiary period in the North Sea area. Nature, 275: 121-123.

Cavelier, C., Chateauneuf, J.J., Pomerol, C., Rabussier, D., Renard, M. and Vergnaud Grazzini, C., 1981. The geological events at the Eocene-Oligocene boundary. Palaeogeogr.,Palaeoclimat.,Palaeoecol., 36: 223-248.

Corliss, B.H., 1981. Deep sea benthonic foraminiferal turnover near the Eocene/Oligocene boundary. Mar. Micropal., 6: 367-384.

Corliss, B.H., 1985. Microhabitats of benthic foraminifera within deep-sea sediments. Nature, 314: 435-438.

Corliss, B.H., Aubry, M.P., Berggren, W.A., Fenner, J.M., Keigwin, L.D. Jr. and Keller, G., 1984. The Eocene/Oligocene boundary event in the Deep-Sea. Science, 226: 806-810.

Curry, D. and Odin, G.S., 1982. Dating of the Paleogene. In : G.S. Odin (Editor) Numeral Dating in Stratigraphy, J. Willey and Sons Ltd., 34: 607-630.

Douglas, R.G. and Savin, S.M., 1971. Isotopic analyses of planktonic foraminifera from the Cenozoic of the North Pacific, Leg 6. In : A.G. Fischer et al., Init. Repts. DSDP, 6, Washington (U.S. Gov't Printing Office) : 1123-1127.

Douglas, R.G. and Savin, S.M., 1973. Oxygen and carbon isotope analyses of Cretaceous and Tertiary foraminifera from the North Pacific. In : E.L. Winterer, J.I. Ewing et al., Init. Repts. DSDP, 17, Washington (U.S. Gov't Printing Office) : 591-605.

Douglas, R.G. and Savin, S.M., 1975. Oxygen and carbon isotope analyses of Tertiary and Cretaceous nannofossils from Shatsky Rise and other sites in the North Pacific Ocean. In : R.L. Larson, R. Moberly et al., Init. Repts. DSDP, 32, Washington (U.S. Gov't Printing Office) : 509-520.

Douglas, R.G. and Savin, S.M., 1978. Oxygen isotopic evidence for the depth stratification of Tertiary and Cretaceous planktonic foraminifera. Mar. Micropal., 3: 175-196.

Edwards, A.R. and Perch-Nielsen, K., 1975. Calcareous nannofossils from southern southwest Pacific, Deep Sea Drilling Project, Leg 29. In : J.P. Kennett, R.E. Houtz et al., Init. Repts. DSDP, 29, Washington (U.S. Gov't Printing Office) : 469-539.

Ellis, C.H., 1975. Calcareous nannofossil biostratigraphy, Leg 31 DSDP. In : D.E. Karig, J.C. Ingle Jr. et al., Init. Repts. DSDP, 31, Washington (U.S. Gov't Printing Office) : 655-676.

Graham, D.W., Corliss, B.H., Bender, M.L. and Keigwin, L.D. Jr., 1981. Carbon and oxygen isotopic disequilibria of recent deep sea benthic foraminifera. Mar. Micropal., 6 : 483-497.

Hsü, K.J., Percival, S.F. Jr., Wright, R.C. and Petersen, N.P., 1984. Numerical ages of magnetostratigraphically calibrated biostratigraphic zones. In : K.J. Hsü, J.L. La Brecque et al., Init. Repts. DSDP, 73, Washington (U.S. Gov't Printing Office) : 623-635.

Jenkins, D.J., 1975. Cenozoic planktonic foraminiferal biostratigraphy of the southwestern Pacific and Tasman Sea, DSDP Leg 29. In : J.P. Kennett, R.E. Houtz et al., Init. Repts. DSDP, 29, Washington (U.S. Gov't Printing Office): 449-467.

Keigwin, L.D. Jr., 1980. Paleoceanographic changes in the Pacific at the Eocene/ Oligocene boundary. Nature, 287: 722-725.
Keigwin, L.D. Jr. and Corliss, B.H., in press. Stable isotopes in Eocene/ Oligocene foraminifera.
Kennett, J.P. and Shackleton, N.J., 1976. Oxygen isotopic evidence for the development of the psychrosphere 38 Myrs ago. Nature, 260: 513-515.
Kennett, J.P., Houtz, R.E., Andrews, J.E., Edwards, A.R., Gostin, V.A., Hajos, M., Hampton, M.A., Jenkins, D.G., Margolis, S.V., Ovenshine, A.T. and Perch-Nielsen, K., 1975. Cenozoic paleoceanography in the Southwest Pacific Ocean, Antarctic glaciation and the development of the circum Antarctic current. In : J.P. Kennett, R.E. Houtz et al., Init. Repts. DSDP, 29, Washington (U.S. Gov't Printing Office) : 1155-1169.
Kroopnick, P., 1974. Correlations between ^{13}C and CO_2 in surface waters and atmospheric CO_2. Earth and Planet. Sc. Lett., 22: 397-403.
Kvasov, D.D. and Verbitsky, M.Y., 1981. Causes of Antarctic glaciation in the Cenozoic. Quat. Res., 15: 1-17.
Martini, E., 1971. Standard Tertiary and Quaternary calcareous nannoplankton zonation. In : A. Farinacci (Editor), Plankt. Conf. Second. Proc. Rome, Tecnoscienza, 2 : 739-785.
Matthews, R.K. and Poore, R.Z., 1980. Tertiary $\delta^{18}O$ record and glacio-eustatic sea level fluctuations. Geology, 8: 501-504.
Miller, K.G. and Curry, W.B., 1982. Eocene to Oligocene benthic foraminiferal isotopic records in the Bay of Biscay. Nature, 296: 347-350.
Miller, K.G. and Fairbanks, R.G., 1983. Evidence for Oligocene-Middle Miocene abyssal circulation changes in the western North Atlantic. Nature, 306, 5940: 250-253.
Miller, K.G., Curry, W.B. and Ostermann, D.R., 1984. Late Paleogene (Eocene to Oligocene) benthic foraminiferal paleoceanography of the Goban Spur region, DSDP Leg 80. In : P.C. de Graciansky, C.W. Poag et al., Init. Repts. DSDP, 80, Washington (U.S. Gov't Printing Office).
Muza, J.P., Williams, D.F. and Sherwood, W.W. Jr., 1983. Paleogene oxygen isotope record for deep sea drilling sites 511 and 512 subantarctic South Atlantic Ocean : paleotemperatures, paleoceanographic changes and the Eocene/ Oligocene boundary event. In : W.J. Ludwig, V.A. Krasheninnikov et al., Init. Repts. DSDP, 71, Washington (U.S. Gov't Printing Office) : 409-422.
Oberhänsli, H., in press. Latest Cretaceous-Early Neogene oxygen and carbon isotopic history at DSDP sites in the Indian Ocean. Mar. Micropal., Special issue on stable isotopes.
Oberhänsli, H., Mckenzie, J., Toumarkine, M. and Weissert, H., 1984. A paleoclimatic and paleoceanographic record of the Paleogene in the central South Atlantic (Leg 73, Sites 522, 523 and 524). In : K.J. Hsü, J.L. La Brecque et al., Init. Repts. DSDP, 73, Washington (U.S. Gov't Printing Office) : 737-747.
Oberhänsli, H. and Toumarkine, M., 1985. The Paleogene oxygen and carbon isotope history of Sites 522, 523 and 524 from the central South Atlantic. In : K.J. Hsü and H. Weissert (Editors), Symposium on South Atlantic Paleooceanography, Cambridge University Press.
Percival, S.F., 1984. Late Cretaceous to Pleistocene calcareous nannofossils from the South Atlantic, Deep Sea Drilling Project Leg 73. In : K.J. Hsü, J.L. La Brecque et al., Init. Repts. DSDP, 73, Washington (U.S. Gov't Printing Office) : 391-424.
Poore, R.Z., 1984. Middle Eocene through Quaternary planktonic foraminifers from the southern Angola Basin : Deep Sea Drilling Project Leg 73. In : K.J. Hsü, J.L. La Brecque et al., Init. Repts. DSDP, 73, Washington (U.S. Gov't Printing Office) : 429-448.
Poore, R.Z. and Matthews, R.K., 1984a. Oxygen isotope ranking of Late Eocene and Oligocene planktonic foraminifers : implications for Oligocene sea surface temperatures and global ice volume. Mar. Micropal., 9: 111-134.
Poore, R.Z. and Matthews, R.K., 1984b. Late Eocene-Oligocene oxygen and carbon isotope records from South Atlantic Ocean, Deep Sea Drilling Project Site

522. In : K.J. Hsü, J.L. La Brecque et al., Init. Repts. DSDP, 73, Washington (U.S. Gov't Printing Office) : 725-735.

Pujol, C., 1983. Cenozoic planktonic foraminiferal biostratigraphy of south western Atlantic. In : P. Baker, D. Johnson et al., Init. Repts. DSDP, 72, Washington (U.S. Gov't Printing Office) : 623-673.

Rabussier Lointier, D., 1980. Variations de composition isotopique de l'oxygène et du carbone en milieu marin et coupures stratigraphiques du Cénozoïque. Thèse de 3ème cycle, Paris, 182 pp.

Saunders, J.B., Bernoulli, D., Müller-Merz, E., Oberhänsli, H., Perch-Nielsen, K., Riedel, W.F., Sanfilippo, A. and Torrini, R. Jr., 1984. Stratigraphy of the late middle Eocene to early Oligocene in the Bath Cliff Section, Barbados, West Indies. Micropaleontology, 30, 4: 390-425.

Savin, S.M., 1977. The history of the earth's surface temperature during the past 100 millions years. Ann. Rev. Earth Planet. Sc., 5: 319-335.

Savin, S.M., Douglas, R.G. and Stheli, F.G., 1975. Tertiary marine paleotemperatures. Geol. Soc. Amer. Bull., 86: 1499-1510.

Shackleton, N.J., 1977. Carbon-13 in *Uvigerina* : tropical rainforest history and the Equatorial Pacific Carbonate dissolution cycles. In : N.R. Andersen and A. Malahoff (Editors), The fate of fossil fuel CO_2 in the oceans, Plenum, New York : 401-427.

Shackleton, N.J. and Kennett, J.P., 1975. Paleotemperature history of the Cenozoic and the initiation of Antarctic glaciation : oxygen and carbon isotope analyses in DSDP Sites 277, 279, 281. In : J.P. Kennett, R.E. Houtz et al., Init. Repts. DSDP, 29, Washington (U.S. Gov't Printing Office) : 743-756.

Shackleton, N.J. and Vincent, E., 1978. Oxygen and carbon isotope studies in recent foraminifera from the Southwest Indian Ocean. Marine Micropal., 3: 1-13.

Shackleton, N.J., Hall, M.A. and Boersma, A., 1984. Oxygen and carbon isotope data from Leg 74 foraminifers. In : T.C. Moore, P.D. Rabinowitz et al., Init. Repts. DSDP, 74, Washington (U.S. Gov't Printing Office): 599-612.

Snyder, S.W., Müller, C. and Miller, K.G., 1984. Eocene-Oligocene boundary : biostratigraphic recognition and gradual paleoceanographic changes at DSDP Site 549. Geology, 12: 112-115.

Thierstein, H.R. and Berger, W.H., 1978. Injection events in ocean history. Nature, 276: 461-466.

Ujiié, H., 1975. Planktonic foraminiferal biostratigraphy in the western Philippine Sea, Leg 31 of DSDP. In : D.E. Karig, J.C. Ingle Jr. et al., Init. Repts. DSDP, 31, Washington (U.S. Gov't Printing Office) : 677-684.

Vergnaud Grazzini, C., 1979. Cenozoic paleotemperatures at Site 398, eastern North Atlantic : diagenetic effects on carbon and oxygen isotopic signal. In : J. Sibuet, W.B. Ryan et al., Init. Repts. DSDP, 47-2, Washington (U.S. Gov't Printing Office) : 507-511.

Vergnaud Grazzini, C. and Saliège, J.F., 1985. Les événements isotopiques en milieu océanique à la transition Eocène-Oligocène dans le Pacifique et l'Atlantique. Paléocirculations profondes en Atlantique Sud. Bull. Soc. géol. Fr., (2), I, 3: 441-455.

Vincent, E. and Berger, W.H., in press. Carbon dioxide and polar cooling in the Miocene : the Monterey hypothesis. In : E.T. Sundquist and W.S. Broecker (Editors), The carbon cycle and atmospheric CO_2 : Natural variations Archean to present, Amer. Geophys. Union, Monograph series 32.

Weissel, J.K., Hayes, D.E. and Herron, E.M., 1977. Paleotectonic synthesis : the displacements between Australia, New Zealand and Antarctica since the Late Cretaceous. Marine Geol., 25: 231-277.

Woodruff, F., Savin, S.M. and Douglas, R.G., 1980. Biological fractionation of oxygen and carbon isotopes by recent foraminifera. Mar. Micropal., 5: 3-11.

GEOCHEMICAL EVENTS (TRACE ELEMENTS AND STABLE ISOTOPES) RECORDED ON BULK CARBONATES NEAR THE EOCENE-OLIGOCENE BOUNDARY. Application to the Contessa section (Gubbio, Umbria, Italia).

M. RENARD[1], F. BERTHENET[1], S. CLAUSER[1.2], and G. RICHEBOIS[1]

[1]Dept. Geol. Sed. and LA CNRS 319, Univ. P. & M. Curie, 4, place Jussieu, 75252, PARIS Cedex 05, (France).

[2]BRGM, BP 6009, 45060 , ORLEANS Cedex, (France).

The purpose of this paper is to try to characterize the Eocene-Oligocene boundary by a geochemical way or more precisely to test of this boundary is related to a single drastic phenomenon or takes place during a period of progressive change in the physical and chemical conditions of the world.

Recently, we have proposed (Renard 1984, 1985) global evolution curves of Sr and Mg contents and stable isotopes of bulk pelagic carbonates since 140 Ma. These curves are founded on the chemical analysis of more 1500 samples from various oceanic sites and pelagic land sections. Sample ages are based on the radiometric time scale of Odin (1982). We shall look how these curves illuminate the problem of Eocene-Oligocene boundary location.

I - STRONTIUM CONTENTS OF PELAGIC CARBONATES

We have shown (Renard 1984, 1985) that the variations of pelagic carbonate Sr contents was not reducible to diagenetic phenomena but reflect, for the main part, fluctuations of sea water Sr/Ca ratio.

The long term fluctuations of Sr contents are related to oceanic ridge hydrothermal activity variation which leads to a more or less important input of Ca into the ocean (Sr content of sea water remains constant but its isotopic ratio fluctuates). Thus, during hydrothermal high activity periods, Ca input into the ocean increases and Sr/Ca ratio of sea water (and consequently of pelagic carbonates) decreases. As the oceanic residence time of Ca is about 2 Ma, pelagic carbonates record these variations with a some order delay. During low activity periods, Ca input decreases and Sr/Ca ratio of sea water and pelagic carbonates increases. Short term variations are superimposed on that; they are related to alternation of transgression-regression cycles which result in the trapping of a more or less important quantity of Ca and Sr in platform carbonates (but as oceanic residence time of Sr is greater (12 Ma), the effects of its variations are more delayed than those of Ca). Thus, pelagic carbonates synthetized during transgression present higher Sr contents than those produced during regression. These short term variations were used to establish a geochemical stratigraphy (chemiostratigraphy) but in some parts of the global curve (and especially near the Eocene-Oligocene boundary) it is not always easy to separate long and short term fluctuations (fig. 1). After low values during Middle Eocene (400 ppm $\leqslant$ Sr $\leqslant$ 900 ppm), Sr contents increase during the Late Eocene to reach about 1000-1300 ppm at the Eocene-Oligocene boundary. Then values slightly decrease (900-1100 ppm) during the end of the Early Oligocene and the Late Oligocene. They increase again during the Miocene (1200-1700 ppm near 12 Ma). In contrast to the principal stratigraphic breaks of the Cenozoic (such as Maastrichtian-Danian, Paleocene-Eocene or Oligocene-Miocene boundaries), the Eocene-Oligocene boundary is not characterized by a drastic negative shift and the Middle Eocene-Late Eocene boundary seems more important with regard to Sr geochemistry.

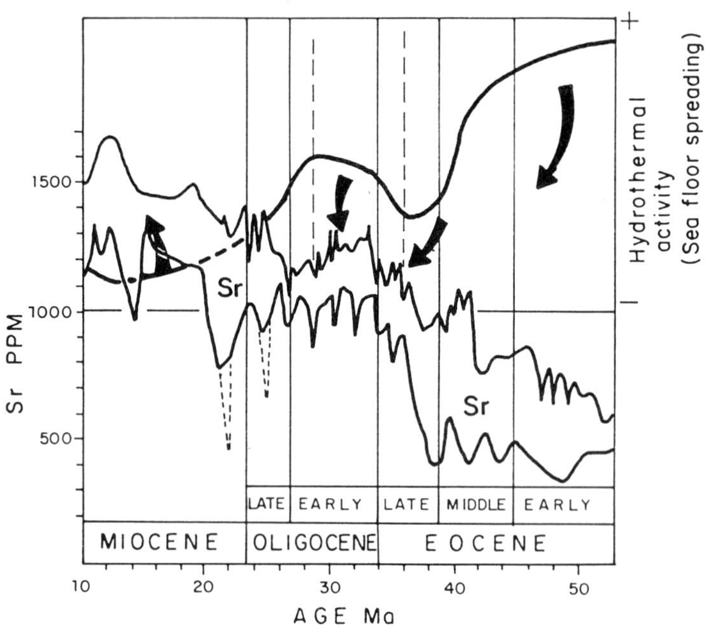

Fig. 1. Relationship between pelagic carbonate Sr content and hydrothermal activity.

If we look at the phenomena that produce such Sr variations (see above), we can admit (fig. 1) that a period of high hydrothermal activity of mid oceanic ridges (therefore, in first approximation, a period of rapid sea-floor spreading) occurs during the Early Eocene and leads to Sr-poor carbonates . To explain the increase of Sr content during the Late Eocene, we must conclude that hydrothermal activity (and sea-floor spreading) was considerably slakened at the end of Middle Eocene. Sea-floor spreading rate should increase again at the end of the Late Eocene and during the main part of the Early Oligocene to explain the decrease of Sr content during the Early Oligocene. Finally, hydrothermal activity will strongly decrease from the end of the Early Oligocene to the Late Miocene and this leads to the highest Sr contents which can be observed since the Late Miocene. In a global way, this interpretation agrees with geophysical data and in details, an accelaration of sea-floor spreading during the Late Eocene is also assumed by Montanari et al. (this volume) on the basis of radiometric ages of biotites from the Contessa section.

II - MAGNESIUM CONTENTS OF PELAGIC CARBONATES

Magnesium content evolution curve (fig. 2) is less conclusive during this period than Sr evolution curve. This is due, in part, to the sensitivity of Mg in early diagenesis and problems that exist for samples from site 390-390A which present contents higher than those of other sites (3000 against 1500 ppm). A part of this difference may be due to either contamination by Mg-rich interstitial water or the presence of traces of dolomite (not recognized by XR diffraction). Biostratigraphic problems as in the Late Eocene of site 516F also play a part and introduce more troubles than for Sr. In spite of these difficulties, we can observe that the general evolution curve of pelagic carbonate Mg content fluctuates, on a large scale, in an opposite way from the Sr curve (1200-2100 ppm during the Early-Middle Eocene (up to 3000 ppm with site 390), 900-1800 ppm during the Early Oligocene and 600-120 ppm during the Miocene).

As in the case of Sr and for the same reasons, the long-term evolution of Mg contents may not be affected by late burial diagenetic influence (arguments for this can be found in Renard 1984). These fluctuations are also controlled by the variation in hydrothermal activity on mid-ocean ridges. But this is more intricate than for Sr because hydrothermalism has a double influence : it liberates Ca and it takes up Mg from the ocean (50 % of the Mg outflow). However, as the oceanic residence times are very different (Ca = 2 Ma and Mg = 40 Ma), effects are delayed in time. An attempt of modelization can be found in Renard 1984 and 1985.

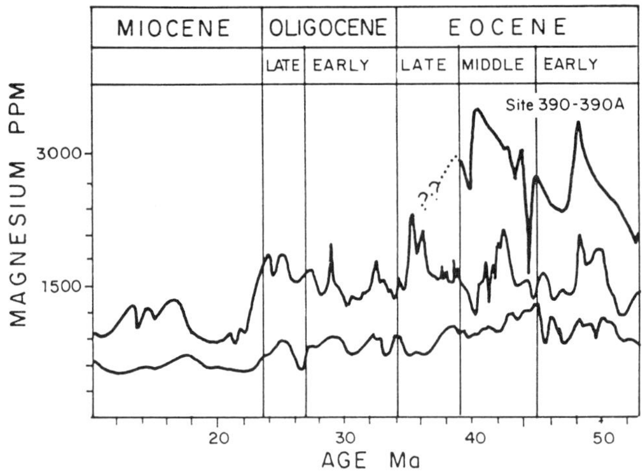

Fig. 2. Evolution with time of pelagic carbonate Mg contents.

III - OXYGEN ISOTOPIC COMPOSITION IN THE PELAGIC CARBONATES

The use of isotopic composition in carbonates as a geothermometer and a stratigraphic tool has now become classic in sedimentary geology (see Vergnaud-Grazzini et Oberhäusli, this volume). However, due to the effects of late burial diagenesis, measures on bulk carbonates are more or less obsolete. We have shown (Renard, 1984, 1985) that although diagenesis may be a major factor controlling oxygen isotopic ratio in bulk pelagic carbonates, the long-term evolution of this signal is not exclusively the result of diagenesis. Comparison of the curves obtained from isolated planktonic Foraminifera and from bulk carbonate (Killingley 1983, Renard 1984) shows that these two types of analysis are not as different as is usually acknowledged :

- (i) The global trend is the same.
- (ii) The amplitude of variation between the Late Paleocene and the Early Quaternary is comparable (-1 to + 4 °/₀₀ for our data, -2 to + 3 °/₀₀ for Shackleton and Kennett (1975) data at sites 277, 279 A and 281).
- (iii) Major isotopic shifts occur at the same time.

Diagenesis does not remove the original environmental signal from either Foraminifera or from bulk carbonate; and from the Late Jurassic there are numerous shifts in the $\delta 18_O$ curve in bulk carbonates corresponding to the record of climatic and/or paleooceanographic fluctuations by pelagic carbonates.

One of these shifts is located near the Eocene-Oligocene boundary or more precisely in the earliest Oligocene (fig. 3). After a progressive increase (of about 1,5 °/₀₀) during the Middle and the Late Eocene, an important positive shift (of about 1,5 - 1,7 °/₀₀) occurs between 34.5 to 33 Ma in site 516F (South Atlantic) and site 116 (North Atlantic). In contrast, in the Tethyan section of the Contessa, the Eocene-Oligocene boundary positive excursion is lower (of about 0,5 °/₀₀).

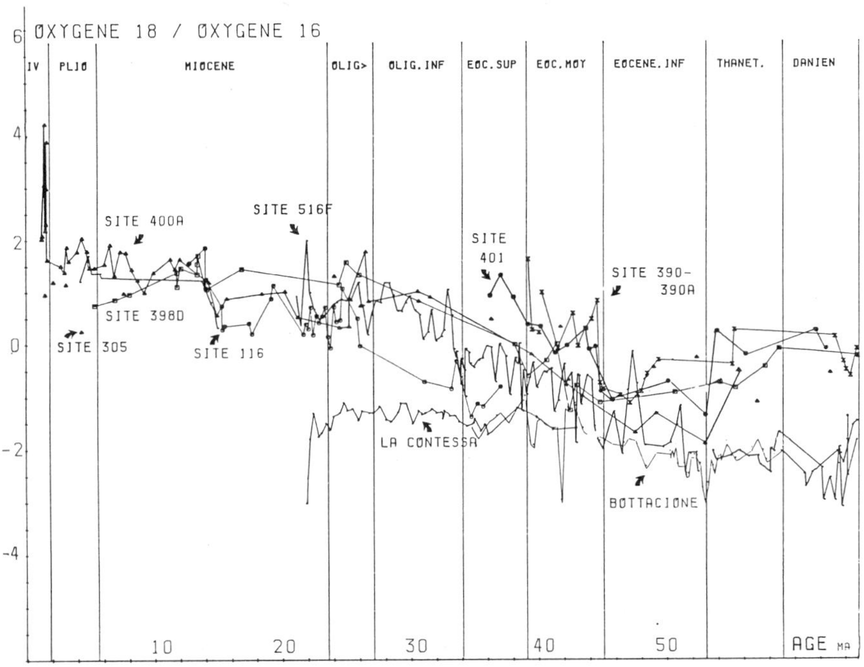

Fig. 3 Evolution with time of bulk carbonate $\delta^{18}O$ in various DSDP sites.

Details of the various hypothetic explanations of this positive shift (temperature effect, vairation of salinity, building of a large amount of ice on Antarctica ...) can be found in Vergnaud-Grazzini and Oberhäusli (this volume).

Numerous studies have attempted to explain the Eocene-Oligocene boundary $\delta^{18}O$ shift amplitude difference observed between various sites. For planktonic Foraminifera there is no relationship between the extent of the $\delta^{18}O$ shift and paleodepth of the site but, as a first approximation, we can observe that the amplitude of the $\delta^{18}O$ shift decreases when passing from high latitudes to mid and low latitude. This may be related to an increase of the latitudinal temperature grandient at that time in the world ocean (Vergnaud-

Grazzini and Oberhäusli, this volume). However, we think that world-wide paleooceanographic changes which occur at that time play an important role in the control of the $\delta^{18}O$ excursion amplitude. A comparison of the bulk carbonate $\delta^{18}O$ curve from Tethyan, South Atlantic and North Atlantic sites since the Late Jurassic (Renard 1984) is conclusive. From the Cenomanian to the Eocene (fig. 4) there is a difference of about 2 ‰ between the $\delta^{18}O$ of carbonates from Tethys-South Atlantic sites (about 0 to + 0.5 ‰) and those from the North Atlantic (and probably Pacific) sites (-2 to -2.5 ‰). This disparity does not show any correlation with diagenetic parameters (microfacies, sedimentary overload ..., Renard et al., 1982, 1983).

This difference of about 2 ‰ suggests a variation of temperature close to 10°C between Tethys-South Atlantic surface waters (warmer) and North Atlantic surface waters (colder); but the difference may not fully depend on temperature as the $\delta^{18}O$ of water may have been relatively different in the two area. However it may be, this shows that communication was very restricted between North and South Atlantic during this time; on the contrary, exchanges between Tethys and the South Atlantic was easier.

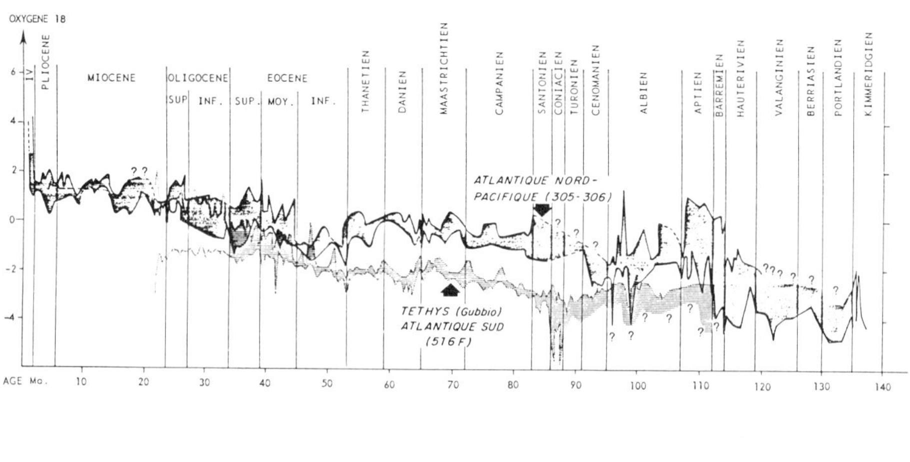

Fig. 4. Long-term evolution of $\delta^{18}O$ in Pacific-North Atlantic area and South Atlantic-Tethys area.

The $\delta^{18}O$ curves (fig. 4) can be used to describe the geodynamic evolution of the Tethys area. During the Early Cretaceous, communications with the North Atlantic seem relatively open. A first phase of isolation occurs during the Aptian and Early Albian but isotopic curves are again similar during the Late Albian. The second phase of isolation occurred during the Cenomanian and exchanges between the two oceans were very restricted during the Late Cretaceous and the Early Paleocene. During the Late Paleocene (inducing by a partial recovery of communications and by installation of a dominant north-south circulation system in Atlantic Ocean at the expense of previous east-west system) isotopic curves tend to become similar again. The exchange rate increased during the Eocene between the South Atlantic and the Tethys (Gubbio and site 516F isotopic curves increase to North Atlantic values). From the Middle Eocene and especially from the Late Eocene, the evolution of oxygen isotopic ratio curves is dissociated for site 516F (which comes closer and closer to those of North Atlantic sites) and for Gubbio (which remains stable). After a weak attempt of communication during the Early Eocene, Atlantic Tethys communications are again restricted. This reconstruction of Tethys evolution from oxygen isotopic ratio curves agrees well with geodynamic reconstructions (Biju-Duval et al. 1977, Biju-Duval 1980, Lancelot 1980, Ricou 1985, this volume).

Isotopic evolution at site 516F also shows difficulty in distinguishing, in the case of only one site study, local and global phenomena. Thus the Eocene-Oligocene boundary is isotopically well marked in this site (increase of about 1.7 °/₀₀) but a part of this comes from the paleooceanographic modifications described above. On the contrary, isotopic shift is slighter in the Contessa section (about 0.5 °/₀₀) because Tethys-Atlantic communications remain restricted.

IV - $\delta 13_C$ IN PELAGIC CARBONATES

The $\delta 13_C$ is one of the most useful paleooceanographic and stratigraphic tools. This is due to :

- (i) Its thermodependence is very weak and thus its evolution curve is, in the oceanic realm, pratically independent of burial diagenesis.
- (ii) The "vital effect" is slighter for calcareous nannoplankton than for Foraminifera.
- (iii) Its evolution curve with time shows numerous shifts which have stratigraphic significance. The most known are located at the Cretaceous Tertiary boundary, at the Paleocene-Eocene boundary and in the Late Miocene near 6.2 Ma.
- (iiii) Pelagic carbonate $\delta 13_C$ seems to be, more or less directly, a paleodepth indicator :
  - . All the negative shifts occur during regressions, all the positive ones during transgressions (Letolle & Renard 1980). Thus the global

evolution curve of $\delta^{13}C/\delta^{12}C$ (Cavelier et al.1981) is parallel to those of the sea level variations of Vail et al. (1977).
For a given period, there is a positive relationship between the mean value of $\delta^{13}C$ and the paleodepth of the site (Renard et al. 1982, Renard & Letolle 1984).

Explanation long term evolution of the $\delta^{13}C$ can be found in Renard (1984-1985) and Vergnaud-Grazzini and Oberhäusli (this volume). Three major factors play a part :

- (i) Ocean-atmosphere exchanges.
- (ii) Variations of organic carbon/carbonate carbon ratio.
- (iii) Variations of oceanic paleocirculations and deep water/ surface water exchanges.

These different mechanisms are not independent and explanation of pelagic carbonate $\delta^{13}C$ cannot be unicausal.

The pelagic $\delta^{13}C$ curve presents numerous second order variations which seem to be synchronous (fig. 5). They may be related to regressive/transgressive cycles (Cavelier et al. 1981). The modeling of Broecker (1982) may explain these relationships by variations of phosphorus trapping on the continental shelf (which lead to variations of production rate of organic matter) during transgressive and regressive periods.

In contrast to other Cenozoic breaks such as the Cretaceous/Tertiary Eocene, Oligocene-Miocene boundaries, the Eocene-Oligocene boundary does not coïncide with a major negative shift in the $\delta^{13}C$ curve (fig. 5). This boundary corresponds to a reversal of the trend of $\delta^{13}C$ curve. After a period of increasing $\delta^{13}C$ (Early Eocene (+ 0,75 ‰ to + 1,25 %) to Late Eocene (+ 1.25 ‰ to + 2.25 %)), values decrease during the Oligocene (+ 0,75‰ to + 1,50‰) near the Miocene-Oligocene boundary.

However, superposed on this long term evolution, we can observe a minor negative shift (of about 0.25 to 0.50 ‰ amplitude) located at the Eocene-Oligocene boundary.

In conclusion, the Eocene-Oligocene boundary does not correspond to a major geochemical break in the Cenozoic. However, numerous geochemical changes occur during this period and have stratigraphic significance. We shall try now to use them in a precise study of the Contessa section (Umbria, Italy).

V - APPLICATION TO THE CONTESSA SECTION

The Contessa section (Gubbio, Umbria, Italy) provides an opportunity for precise correlation between bio, magneto and chemiostratigraphy. Samples analysed correspond to whose of Lowrie et al. (1982) study (Contessa Quarry section). Details of location, sedimentology biostratigraphy and magneto-

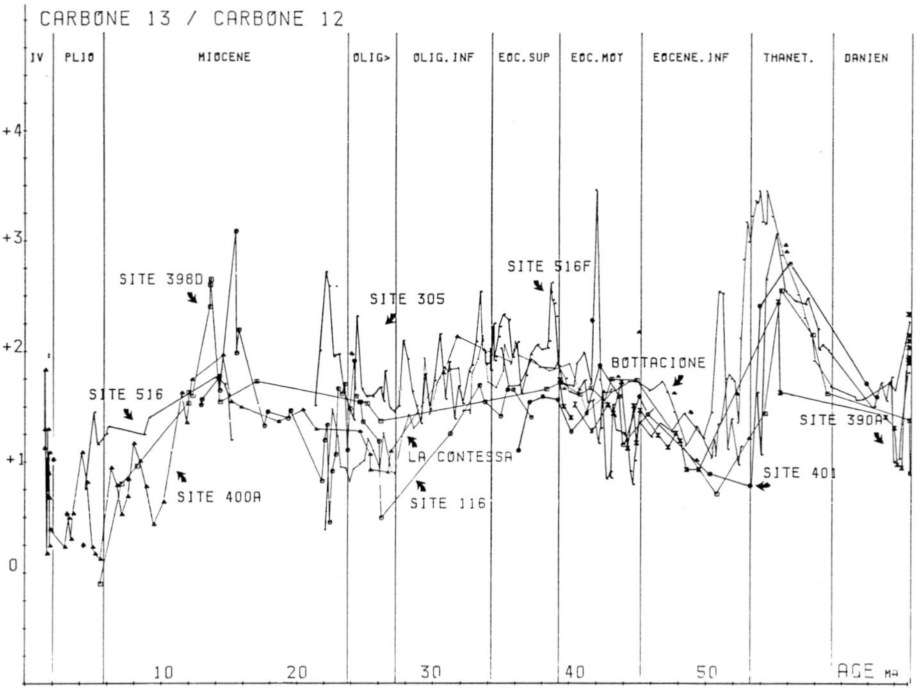

Fig. 5. $\delta^{13}C$ evolution in various DSDP sites.

stratigraphy can be found in this paper.

All analytical results can be found in Renard (1984) or are also available on request.

A) Strontium (fig. 6)

The strontium curve is very conclusive. The base of the Middle Eocene is a Sr low period ([Sr] lower than 600 ppm). The upper part of the Middle Eocene, the Late Eocene and the Earliest Oligocene correspond to an important increase from 540 ppm at 185 m to 1050 ppm at 225.05 m. The beginning of the increase of the Sr content is more or less synchronous with the CP13/CP14 nannozone boundary, the end, more or less, is with the Eocene-Oligocene boundary. This increase is broken by two negative shifts; the first occurs at the Late-Middle Eocene boundary (205 m, base of 16/17 magnetozone), the second at 215 m. During the Early and Middle Oligocene the Sr concentration slightly increases from 1050 at 225.05 m to 1200 ppm at 264.50 m. A negative shift, followed by a rapid increase, breaks this trend at 233.30 m (NP21/NP22 boundary) and an other negative shift ends this evolution at 267.10 m(950 ppm). Then, Sr contents

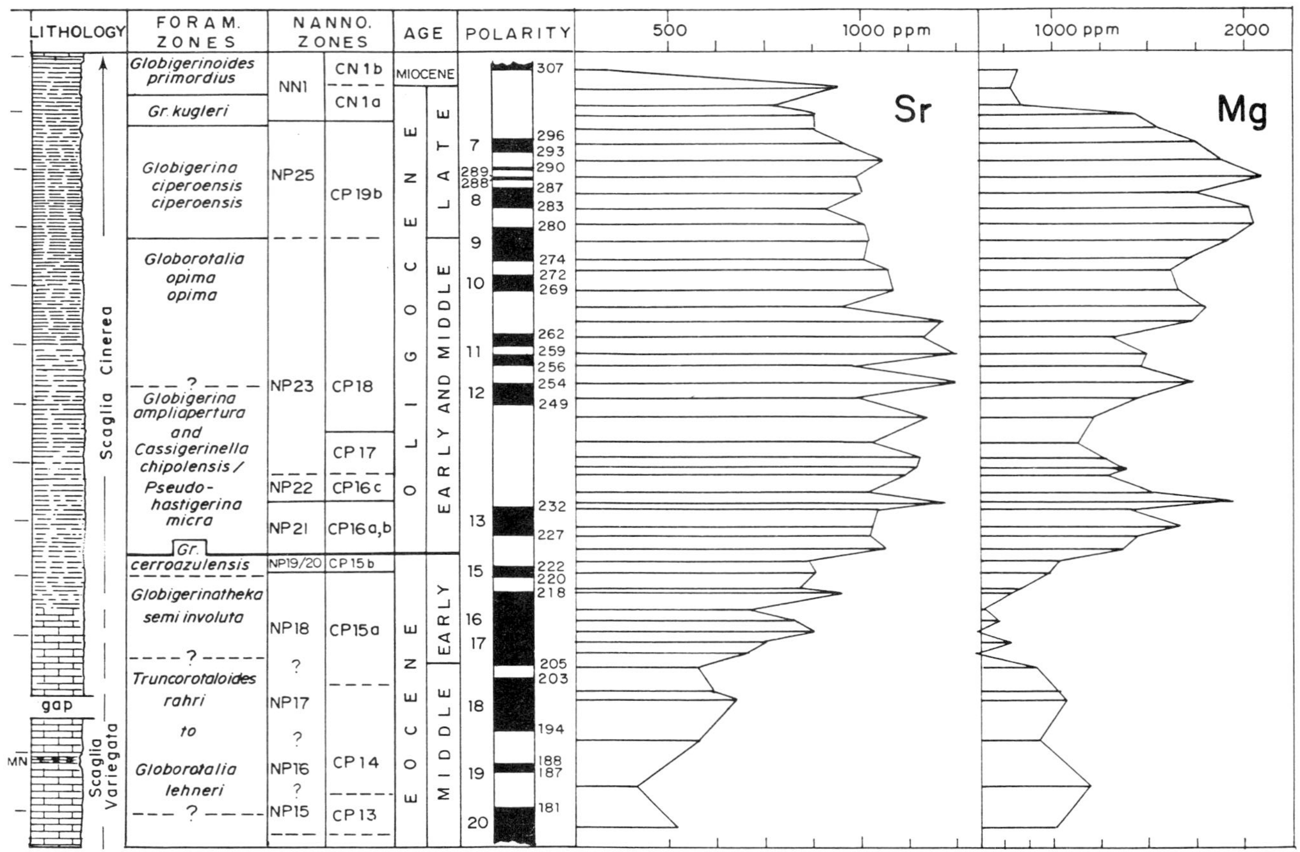

Fig. 6. Evolution of Sr and Mg contents of bulk carbonates from the Contessa Quarry section.

remain more or less constant to about 1000 ppm to 292.35 m. Afterward, Sr contents decrease, at first slowly (820 ppm at 303.25 m) then rapidly at the Oligocene-Miocene boundary (470 ppm at 307.30 m).

B) Magnesium (fig. 6)

For magnesium the Middle Eocene is also a low concentration period ([Mg] about 1000 ppm). An important negative shift coincides with the Middle-Late Eocene boundary (about 700 ppm at 207.5 m) and the Mg content remains very low during the lower part of the Late Eocene. An important geochemical event occurs at 215 m, the Sr and Mg concentrations, which fluctuate in an opposite way since the Cenomanian, begin to fluctuate in a parallel way. Mg contents rapidly increase to reach 1750 ppm at 233.90 m. The Eocene-Oligocene boundary is relatively well obvious in the Mg curve (increase of Mg content). In contrast, a negative shift occurs at the top of the NP 21 nannozone (233.30 m).

Mg concentrations decrease from 1750 ppm at 233.90 m to 1120 ppm at 245 m (near the CP17/CP18 boundary) then increase to reach 1880 ppm at 289.10 m. A negative shift occurs at 261,25 m (1260 ppm). During the Late Oligocene, Mg concentrations decrease, at first progressively (1300 ppm at 300 m), then abruptly (830 ppm at 303.25 ppm).

C) Iron (fig. 7)

It is more difficult to interpret Fe and Mn curves because their global evolution curves are not as well known as those of Mg and Sr. The major factors which play a role are (Andrianiazy and Renard 1984) :

- (i) The production rate of Mn and Fe by mid-oceanic ridges.
- (ii) The production rate of Mn and Fe by continental weathering.
- (iii) The redox conditions of the ocean.
- (iiii) The paleodepth of the site and its distance from the mid-oceanic ridge.

After a Fe low period during the Middle Eocene (350 ppm at 193 m), the concentrations greatly increase to reach 1460 ppm at 232.40 ppm in the Early Oligocene. Three negative shifts occur during this increase. The first at 215 m, the second at 218.50 m (NP18/NP19 boundary) and the third coincides with the Eocene-Oligocene boundary.

An important negative shift ends this evolution (930 ppm at 233.90 m) and corresponds to the NP 21/NP22 nannozone boundary. Then, the general trend is a decrease of concentrations to 720 ppm at 251.50 m, followed by an increase to 1350 ppm at 256.35 m. A new decrease occurs during the Middle Oligocene (810 ppm at 283.40 m). The Late Oligocene begins by an increase of Fe content to reach 1150 ppm at 300 m and the Oligocene-Miocene boundary

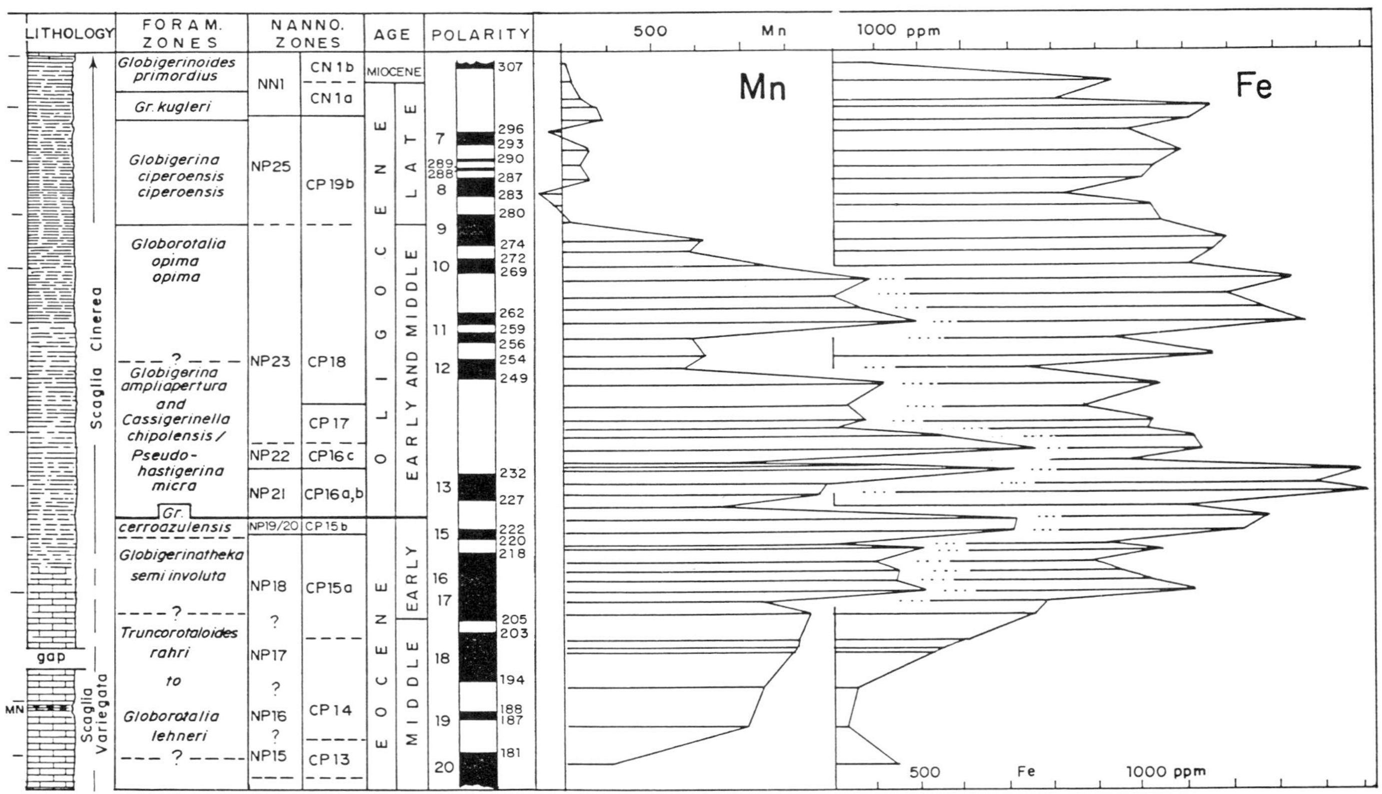

Fig. 7. Mn and Fe contents of bulk carbonates from the Contessa Quarry section.

corresponds to a rapid decrease of Fe (370 ppm at 307.30 m).

D) Manganese (fig. 7)

From the Middle Eocene to the Earliest Oligocene, there is an increase of iron concentration from 390 ppm at 178.5 m to 1350 ppm to 236.40 m. Four negative shifts break this general trend. The first corresponds to the Middle-Late Eocene boundary (860 ppm at 205 m, 740 ppm at 207.50 m), the second occurs at 218.5 m (880 ppm), the third coincides with the Eocene-Oligocene boundary (1300 ppm at 221.50 m, 640 ppm at 225.50 m) and the fourth with NP21/NP22 nannozone boundary (680 ppm at 233.9 m).

The end of the Early Oligocene corresponds to a decrease of Mn content from 1350 ppm at 236.40 m to about 600 ppm between 281.5 and 256.35 m. Then concentrations abruptly increase to 1100 ppm at 259 m and progressively decrease to the Middle-Late Miocene boundary (220 ppm at 277.75 m). The Late Oligocene is a Mn low period (Mn $<$ 400 ppm).

E) Oxygen isotopes (fig. 8)

The Middle Eocene is a period of increasing $\delta^{18}O$ (from -1,68 ‰ at 178.60 m to - 1.30 ‰ at 200 m). A negative shift seems to occur at the Middle-Late Eocene boundary, but due to a bad sampling it is difficult to evaluate its importance (-1.30 ‰ at 200 m; -1.81 ‰ at 211.80 m). From the Late Eocene to the Early Oligocene the $\delta^{18}O$ progressively increases from - 1.80 ‰ at 218 m to 1.21 ‰ at 254 m. A slight positive excursion occurs in the upper part of the NP21 nannozone and corresponds to the "so called" Eocene-Oligocene boundary $\delta^{18}O$ shift (-1.46 ‰ at 279.70 m to - 1.35 ‰ at 232.40 m, see above).

The $\delta^{18}O$ remains more or less constant to 283.40 m (- 1.20 ‰). Three negative shift occur at :

- 248.05 m : - 1.60 ‰
- 259 m : - 1.57 ‰
- 274.75 m : - 1.54 ‰ (corresponds to the Middle-Late Oligocene boundary).

During the Late Oligocene, the $\delta^{18}O$ decreases progressively from - 1.20‰ at 283.40 m to - 1.84 ‰ at 300 m. A positive shift occurs at 303.25 m (- 1.4 ‰) followed by an important decrease at the Oligocene-Miocene boundary (- 2.61 ‰ at 307.30 m).

F) Carbon isotopes (fig. 8)

The $\delta^{13}C$ increases from the Middle Eocene (+ 1.66‰ at 178.60 m) to the Early Oligocene (+ 2.14 ‰ at 227.40 m). A negative shift occurs near the Eocene/Oligocene boundary (1.44 ‰ at 223.30 m). The $\delta^{13}C$ decreases during the main part of the Oligocene from 2,14 ‰ at 227.40 m to + 0,82 ‰ at

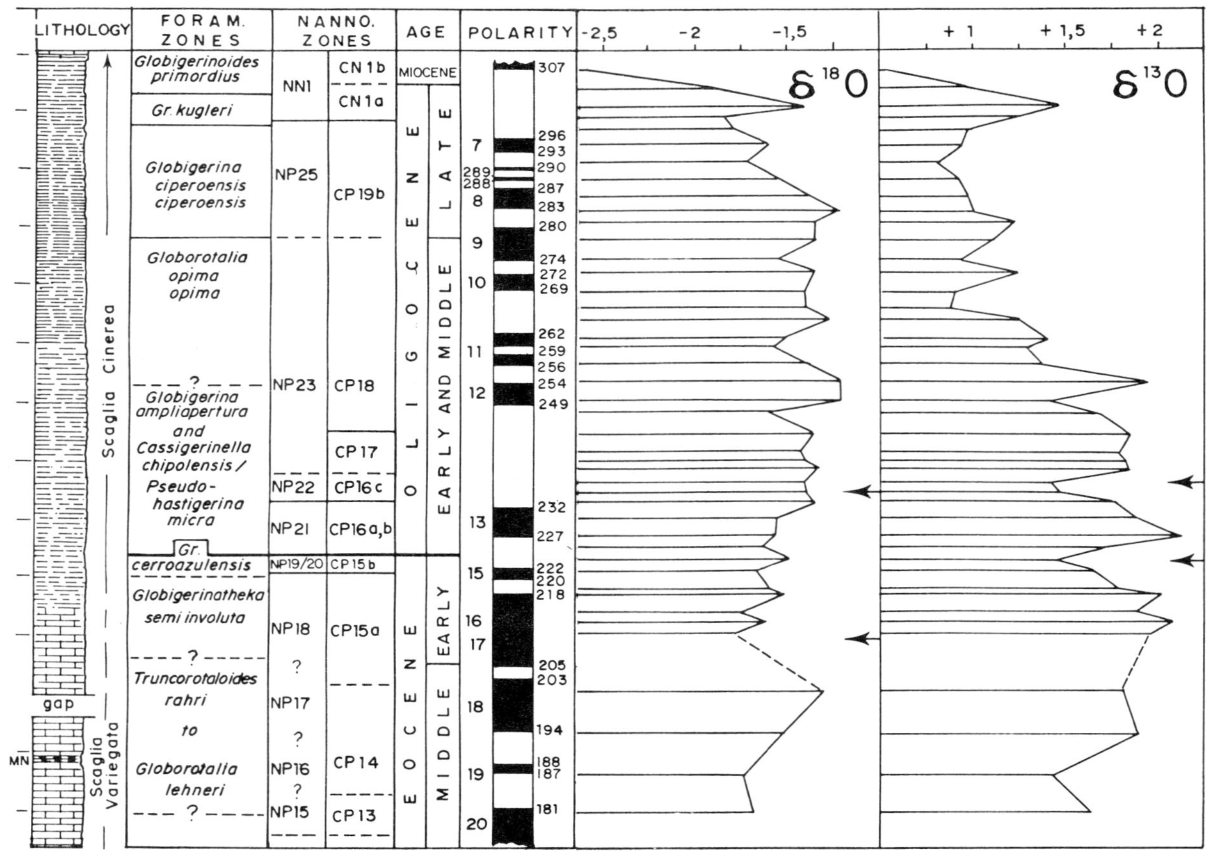

Fig. 8. $\delta^{18}O$ and $\delta^{13}C$ evolutions of bulk carbonates from the Contessa Quarry section.

292.35 m. Two negatives shifts occur at the NP 21/NP 22 nannozones boundary at 233.90 m (1.47 ‰) and at 267.10 m (0.89 ‰). During the Latest Oligocene, the isotopic ratio of carbon increases to + 1.48 ‰ at 303.25 m. The Oligocene-Miocene boundary coincides with an important negative shift (+ 0.55 ‰ at 307.30 m).

G) Conclusions

This study of Contessa section pelagic carbonate geochemistry shows that :

1) Numerous geochemical events occur during Eocene-Oligocene times.
2) Many of these events are synchronous with biostratigraphic events.
3) For several geochemical events we observe a progressive evolution during the Late Eocene-Early Oligocene times and it is difficult to put one of these geochemical events forward as a major stratigraphic break.

We have tried to estimate (fig. 9) the relative importance of each geochemical event by a semi-quantitative estimation (a star for a slight event, five stars for a strong event). The relative importance is given by the total of stars.

Seven major events occur between the Middle Eocene and the Earliest Miocene :

- The event 1 is located in the Middle Eocene at the CP13/CP14 nannozone boundary and corresponds to slight events for Sr and $\delta^{18}O$ and medium events for Fe, Mn and $\delta^{13}C$ (total = 9 stars).

- The event 2 which coincides with the Middle (Late Eocene boundary) is medium for Sr and Mn and strong for Mg and $\delta^{18}O$ (total 13 stars).

- The event 3 corresponds to the classical Eocene-Oligocene boundary. It is a strong event for Sr and $\delta^{13}C$ medium for Mg and Fe, strong for Mn (total 19 stars).

- The event 4 is the most important geochemical event that occurs during this period. It is located at the NP21/NP22 nannozone boundary and it is a strong event for Sr, Mg, Mn, Fe, $\delta^{13}C$ and medium for $\delta^{18}O$ (total 23 stars).

- The event 5 occurs during the 12 magnetozone time and is more or less synchronous with the base of the Globorotalia opima Zone and the Early-Middle Oligocene boundary. It is slight for Sr, medium for the $\delta^{18}O$, $\delta^{13}C$ and strong for Mn (total 12 stars).

- The event 6 (Latest Oligocene) is synchronous with the NP25/NN1 boundary; it is medium for Sr, Fe, $\delta^{18}O$ and strong for Mg (total 13 stars). Slightly later a strong event for $\delta^{18}O$ and $\delta^{13}C$ occurs (total 7 stars).

- The event 7 coincides with the Oligocene-Miocene boundary. It is a strong event for Fe, $\delta^{18}O$, $\delta^{13}C$ and medium for Sr (total 18 stars).

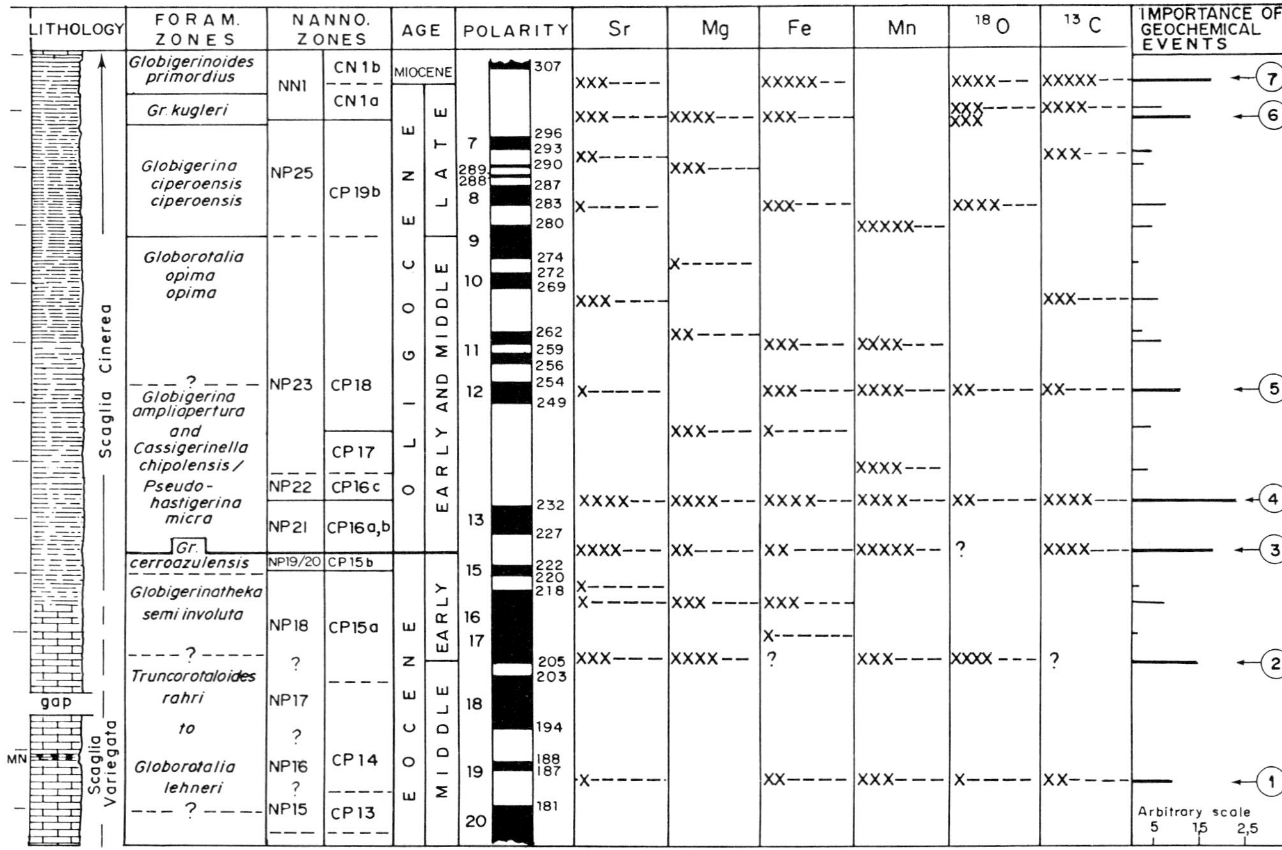

Fig. 9. Correlation between the various geochemical breaks observed in the Contessa Quarry section and the bio/magnetostratigraphic data.

At the end of this study, many remarks are imperative :

1) The classical Eocene-Oligocene boundary (event 3) does not correspond to the most important geochemical event occurring during these times. The event 4, which coincides with the NP21/NP22 nannozone boundary, is the most important geochemical break. Event 2 (Middle-Late Eocene boundary) is also an important break. In contrast to the Eocene-Oligocene boundary the Oligocene-Miocene boundary is geochemically well obvious.

2) Numerous diachronic geochemical events occur during Middle and Late Oligocene times but the Middle-Late boundary is not well marked geochemically (although it corresponds to a major break for Mn).

3) The succession of geochemical events between the Middle Eocene and the Early Oligocene illuminates problems of Eocene-Oligocene boundary location and for the choose of stratotypes :

- Event 1, more or less, corresponds to the base of the Bartonian stratotype.
- Event 2 is close to the base of the Priabonian stratotype.
- Event 3 is close to the base of the Stampian stratotype.
- Event 4 is close to the base of the Rupelian stratotype.

Thus numerous chemical events occur in the ocean during these times. In the epicontinental area, related to local conditions, sediments record these events with more or less strength. This leads to the displacements in time of various stratotypes and stratigraphic breaks.

AKNOWLEDGEMENTS

Many thanks are due to Dr Monique Toumarkine and Dr William Lowrie which provide us Contessa Quarry samples, and Dr Russel for his help in English translation.

REFERENCES

Biju-Duval B., Dercourt J. & Le Pichon X., 1977 . From the Tethys ocean to the Mediterranean sea : a plate tectonic model of the evolution of the western alpine system. Int. symp. struct. history. Mediterranean bassins. Split. Biju-Duval et Montadert edit. Technip, Paris, pp. 143-164.

Biju-Duval B., 1980. De la Téthys aux mers intra-alpines actuelles : introduction in colloque C5 - 26è C.G.I., Mémoire B.R.G.M., 115, pp. 232-236.

Broecker W.S., 1982. Ocean chemistry during glacial time. Geochim. Cosmochim., Acta v 46, pp. 1689-1705.

Cavelier C., Chateauneuf J.J., Pomerol Ch., Rabussier D., Renard M. & Vergnaud-Grazzini C., 1981. Geological Events at E/O boundary. Paleogeo., Paleoclimat., Paleoecology. , 36: 223-248.

Killingley J.S., 1983. Effect of diagenetic recristallization on $\delta^{18}O$ / $\delta^{16}O$ values of deep-sea sediments. Nature, 201: 594-597.

Lancelot Y., 1980. Birth and evolution of the "Atlantic tethys" (Central North Atlantic). In coll. C5, 26è CGI. Mémoire du B.R.G.M., n° 115, pp. 215-223.

Letolle R. & Renard M., 1980. Evolution des teneurs en $\delta^{13}C$ des carbonates pélagiques aux limites Crétacé/Tertiaire et Paléocène/Eocène. C.R. Acad. Sc. Paris, t. 290, pp 827-830.

Odin G.S., 1982. Numerical dating in stratigraphy. Odin G.S. Edit. John Wiley & Sons Ltd, 1040 pp.

Renard M., Delacotte O., Letolle R., 1982. Le strontium et les isotopes stables dans les carbonates totaux de quelques sites de l'Atlantique et de la Téthys. Bull. Soc. Géol. Fr., (7), t. XXIV, n° 3, pp. 519-534.

Renard M., Letolle R., 1983. Essai d'interprétation du rôle de la profondeur de dépôt dans la répartition des teneurs en manganèse et dans l'évolution du rapport isotopique du carbone des carbonates pélagiques : influence de l'oxygénation du milieu. C.R. Acad. Sc. Paris, t. 296, pp. 1737-1740.

Renard M., Richebois G. & Letolle R., 1983. Trace element and stable isotope geochemistry of Paleocene to Coniacian samples from hole 516F, comparison with North-Atlantic and Tethys sites. In barker P.F. et al., Init. Reports, DSDP, Washington, U.S. Gov. print. office, 72: 399-420.

Renard M., 1984. Géochimie des carbonates pélagiques. Mise en évidence des fluctuations de la composition des eaux océaniques depuis 140 Ma. Essai de chimiostratigraphie. Mém. Sc. Terre Univ. P. & M. Curie, Paris, n° 84-16, et Mém. B.R.G.M. n° 85 (1985).

Vail P.R., Mitchum R.M. & Thompson S. (1977). Seismic stratigraphy and global change of sea level : Point 4 global cycles of relative changes of sea level. Am. Assoc. Petrol. Geol. mem. 26, pp. 83-97.

GEOCHRONOLOGY OF THE EOCENE/OLIGOCENE BOUNDARY

W. A. Berggren

ABSTRACT

Current age estimates for the Eocene/Oligocene boundary range from ~33 Ma to ~38 Ma. A review is presented of the nature and methods employed in deriving these ages. Integrated studies employing the combined techniques of bio-, magneto-, and isotope chronology suggest a value between 36-37 Ma; that preferred here is 36.6 Ma.

INTRODUCTION

Current age estimates of the Eocene/Oligocene boundary range from <33 Ma (Glass and Crosbie, 1982) to ~38 Ma (Lowrie and Alvarez, 1981; Lowrie *et al.*, 1982). Intermediate values centering on 33-34 Ma (Odin, 1982; Curry and Odin, 1982) and 36-37 Ma (Prothero *et al.*, 1982, 1983; Prothero and Armentrout, 1985; Berggren *et al.*, 1985a) may also be found in the literature. The methods employed in deriving these age estimates vary considerably and it is important to understand how they have been determined in any discussion of the age of this boundary. In the following discussion the reader is referred to Figure 1.

DISCUSSION

Potassium argon dates, predominantly on glauconites, and predominantly from NW Europe (see compilation of dates in Odin, ed., 1982) have led Curry and Odin (1982) to suggest an age of 34 (+2,-1) Ma for the Eocene/Oligocene boundary (correlated with the base of the Latdorf Sands in Germany and the Grimmertingen Sands in Belgium, the Brockenhurst Beds in England, and the Marnes a Lucines of the Paris Basin). This boundary is not equivalent to that used by Berggren *et al.* (1985a) in their studies of Paleogene geochronology. They use the base of the Rupelian = Stampian as the base of the Oligocene, for which Curry and Odin (1982) suggest an age of 33 (+2, -1) Ma.

In the Gulf Coastal Plain a number of K-Ar dates (on low potassium glauconites and biotites) have yielded values between 36.7-38.5 Ma (uncorrected) for upper Eocene levels in the Jackson Formation (Ghosh, 1972) which, together with a number of similar glauconite dates on stratigraphically

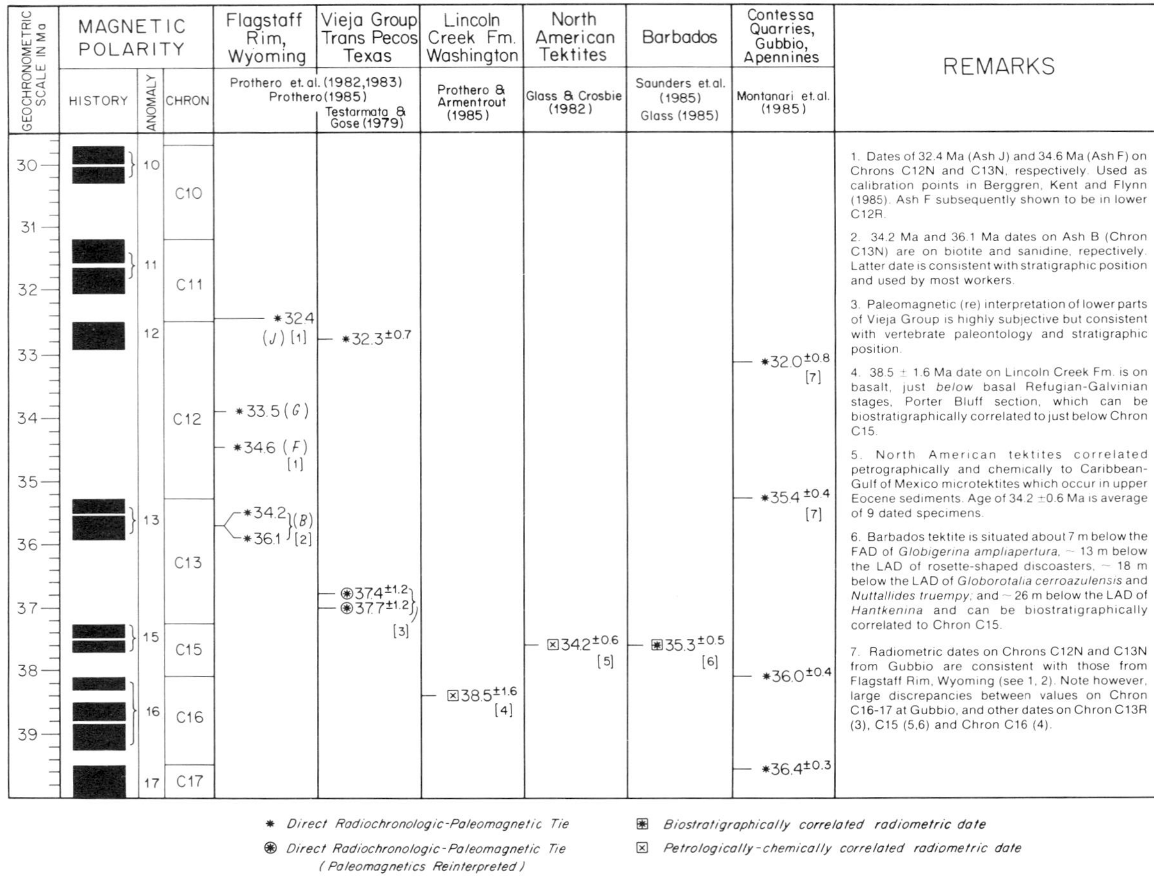

Figure 1. Biostratigraphic and/or magnetostratigraphic control of some late Eocene - early Oligocene isotopic dates.

correlative levels in NW Europe, led Hardenbol and Berggren (1978) to estimate an age of 37 Ma for the Eocene/Oligocene boundary.

The age estimated by Glass and Crosbie (1982) of 32.5±0.9 Ma for the E/O boundary is a biochronologic estimate based on sediment rate calculations for the age of the biostratigraphically correlated boundary level in several DSDP cores using an average date (based on 9 specimens) of 34.2±0.6 Ma for the North American tektite strewnfield (Fig. 1). These tektites are interpreted to be petrologically and chemically similar to the microtektites found in the upper Eocene of several DSDP cores. More recently the late Eocene tektite event has been recorded in the Bath Cliff and Gay's Cove sections of Barbados (Saunders et al., 1984; Sanfilippo et al., 1985). A K-Ar date of 35.3±0.5 Ma has been measured (B. Glass, oral communication, IGCP 174 meeting, Paris, 29 March 1985) on the Barbados microtektites. The microtektite level is situated approximately 7 m below the FAD of Globigerina ampliapertura, ~13 m below the LAD of rosette shaped discoasters, ~18 m below the LAD of Globorotalia cerroazulensis and Nuttallides truempyi and ~26 m below the LAD of Hantkenina. The large O^{18} shift to positive values, elsewhere characteristic of Chron C13N, occurs 3-4 m above the LAD of Hantkenina (Saunders et al., 1984). The microtektite level can be biostratigraphically correlated to a level within Chron C15, of late Eocene age.

A series of high temperature radiometric dates at Flagstaff Rim, Wyoming, bracketing the interval from Chron C12N to C13N (Prothero et al., 1982, 1983; Prothero, 1985) in the Vieja Group, Trans Pecos, Texas (op. cit.) provide important data for age estimates of the Eocene/Oligocene boundary (Fig. 1). Dates of 32.4 Ma and 34.6 Ma on ash beds J and F, respectively, were used as calibration points for the younger limits of Chrons C12N and C13N by Berggren et al. (1985). Subsequent work has shown (Prothero, 1985) that ash bed F (34.6 Ma) is actually situated within the lower half of Chron C12R. The dates from Flagstaff Rim suggest that the Eocene/Oligocene boundary is older than 36.1 Ma (sanidine date on lower part of Chron C13N). This is supported by the 37.4±1.2 Ma and 37.7±1.2 Ma dates on the paleomagnetically reinterpreted C13R interval of the lower part of the Vieja Group, Trans Pecos, Texas (Prothero et al., 1982, 1983; Prothero, 1985).

More recently Prothero and Armentrout (1985) have identified a magnetic polarity stratigraphy corresponding to Chrons C15R to C6C in the Lincoln Creek Formation of Washington (Fig. 1). The Narizian/Refugian benthic foraminiferal "Stage" boundary occurs within Chron C15R. A high temperature basalt date of

38.5±1.6 Ma at or just below this boundary at the Porter Bluff section can be correlated to a level just below Chron C15, suggesting an age of about 38 Ma for Chron C15R, and an age of 36-37 Ma for the Eocene/Oligocene boundary (Prothero and Armentrout, 1985).

A series of ^{40}K-^{39}Ar dates (on biotites) spanning Chrons C9N to C16-17 have been recently measured in the Contessa Quarries, Gubbio, in northern Italy (Montanari, oral communication, IGCP 174 meeting, Paris, 29 March 1985; Montanari et al., 1985). Radiometric dates on Chrons C12N (32.0±0.8 Ma) and C13N (35.4±0.4 Ma) are consistent with those from Flagstaff Rim, Wyoming (Prothero et al., 1982, 1983; Prothero, 1985). The dates of 36.0±0.4 Ma and 36.4±0.3 Ma on Chron C16N and indistinct Chrons C16-17N, respectively, are seen to contrast significantly with those of 38.5±1.6 Ma on Chron C16 correlative in the southern Olympic Peninsula, Washington (Prothero and Armentrout, 1985) and with dates of 37.4±1.2 Ma and 37.7±1.2 Ma on the lower part of the Vieja Group, Trans Pecos, Texas, reinterpreted as Chron C13R (Prothero et al., 1982, 1983; Prothero, 1985). The difference of only 1 my between K-Ar dates on Chrons C13N and C16-17N (an interval considered to span 4-5 my on sea-floor anomaly time scales) has important implications for global marine geotectonics (i.e., rapid increase in sea-floor spreading rates), but it is beyond the scope of this paper to consider that subject here. Suffice to observe that comparison of the ratio of magnetic anomaly spacings on different spreading centers in different ocean basins of the world indicates relatively constant global spreading rates during the late Paleogene and early Neogene (Klitgord, personal communication, May, 1985).

The Paleogene magnetobiochronologic scales of Lowrie and Alvarez (1981) and Lowrie et al. (1982) and that of Berggren et al. (1985a) used different radiometric calibration points and different methods of deriving the chronology itself. Lowrie and Alvarez (1981) fixed the ages of 9 levels in the late Cretaceous to Oligocene-Miocene portion of the geomagnetic reversal sequence on the basis of magnetobiostratigraphic correlations in the Gubbio section of the Umbrian Apennines, in Italy. The age of 38.0 Ma for the Eocene/Oligocene boundary was based upon an earlier estimate of 37.0 Ma (Hardenbol and Berggren, 1978). The later age estimate was simply recalculated using updated decay constants by Ness et al. (1980). Such stringent use of calibration tie-points, however, increases the possibility of introducing as an artifact apparent acceleration in sea-floor spreading as the number of calibration tie-points increases within a finite time interval. This is likely to occur because the inherent errors in the age estimates of the calibration points become more important in calculating interval spreading rates as the calibration points used in this manner become more closely spaced in time.

An alternative method employed by Berggren _et al._ (1985a,b) assumes a minimum number of changes in sea-floor spreading rates that will still satisfy the constraints of the calibration tie-points. Linear segments, each encompassing significant portions of the magnetic reversal sequence are thus identified and a chronology is determined by linear regression analysis. In the general polarity time scale of Berggren _et al._ (1985a,b) the following calibration points were used: a) 3.40 Ma - anomaly 2A or Gauss/Gilbert boundary; b) 8.87 Ma - anomaly 5N; c) 32.4 Ma - anomaly 12 (Chron C12N); d) 34.6 Ma - anomaly 13 (Chron C13N, since shown to be located in C12R); e) 49.5 Ma - anomaly 21 (Chron C21); f) 84.0 Ma - anomaly 34 (Chron C31). Three linear calibration age-apparent segments result with inflection points at anomalies 5 and 24. Ages for magnetic polarity intervals or chrons were calculated according to linear regression of these three segments. The Eocene/Oligocene boundary, located in Chron C13R, has an estimated age of 36.6 Ma on this scale. A more detailed discussion of the methods and problems of deriving a magnetobiochronologic time scale is presented in Berggren _et al._ (1985a).

It is important to remember that age estimates of geologic epoch boundaries are by now difficult to derive completely independently. This results from the fact that rocks developed elsewhere than on the ocean floor provide dates for calibration, and with correlation to the geomagnetic reversal sequence and assessment of the dates themselves are often developed within a biostratigraphic framework. This lack of independence is particularly relevant in the Neogene where age estimates of important boundaries are often already obtained in close conjunction with correlations to the geomagnetic time-scale (e.g., Ryan _et al._, 1974). Unless long lava sequences, devoid of fossils but possible to date radiometrically, are found (see for instance, McDougall _et al._, 1984) or a reliable method is developed to date oceanic crustal rocks, a certain degree of circular reasoning (or more optimistically positive feedback) is virtually inevitable. Nevertheless, there is an impelling motivation and justification for considering both data sets simultaneously because the highly developed correlation between bio- and magnetostratigraphy demand a set of ages consistent within both frameworks. Thus any change on the estimated ages within one framework automatically implies a corresponding change in the other, unless the correlations can be shown to be incorrect (for a good example of this, see the recent change to middle Miocene magnetobiochronology made by Miller et al., 1985; Berggren et al., 1985b).

CONCLUSIONS

Additional work is needed to resolve problems of the geochronology of the Eocene/Oligocene boundary. An integrated approach is required, i.e., magneto-bio-radiochronology, but the potential sources of error are multiple: precision in dating (glauconites vs. high temperature minerals; fission track vs. K-Ar, etc.), identification of magnetostratigraphic patterns, biostratigraphic accuracy and resolution. It would appear that the extreme values currently suggested (~ 33-34 Ma and ~ 38 Ma) for the Eocene/Oligocene boundary can be considered unlikely. Values between 35-37 Ma would appear to be more acceptable. That preferred here is 36.6 Ma (see Berggren et al., 1985a,c).

ACKNOWLEDGMENTS

I should like to express my thanks to Professor C. Pomerol (Paris) for the invitation and support to attend the final meeting of IGCP Project 174 in Paris (March, 1985) in order to present this review of the current status of the Eocene/Oligocene boundary geochronology. Discussions with numerous colleagues have played an important part in the brief synthesis presented here, but I would like to mention in particular M-P. Aubry (Lyon - Woods Hole), D. V. Kent (New York), K. Klitgord (Woods Hole), B. P. Glass (Newark, Delaware), D. Prothero (Galesburg, Illinois), A. Montanari (Berkeley, California). I would like to thank in particular M-P. Aubry, D. V. Kent, D. Prothero, and J.J. Flynn for their comments on an early draft of this paper.

This work has been supported by a grant from the Submarine Geology and Geophysics Division of the National Science Foundation, OCE-8417040. This is Woods Hole Oceanographic Institution Contribution No. 5971.

REFERENCES:

Berggren, W.A., Kent, D.V. and Flynn, J.J., 1985a. Paleogene geochronology and chronostratigraphy. In: Snelling, N.J. (ed.), Geochronology and the Geological Record, Geological Society of London, Special Paper.

Berggren, W.A., Kent, D.V. and Van Couvering, J.A., 1985b. Neogene geochronology and chronostratigraphy. In: Snelling, N.J. (ed.), Geochronology and the Geological Record, Geological Society of London, Special Paper.

Berggren, W.A., Kent, D.V., Flynn, J.J., and Van Couvering, J.A., 1985c. Cenozoic Geochronology. Bull. Geological Society of America, vol. 96 (in press).

Curry, D., and Odin, G.S., 1982. Dating of the Palaeogene. In: Odin, G.S. (ed.). Numerical Dating in Stratigraphy. John Wiley and Sons, Ltd., New York, p. 607-630.

Ghosh, P.K. 1972. Use of bentonites and glauconites in potassium 40/argon 40 dating in Gulf Coast stratigraphy (Ph.D. Thesis): Houston, Texas, Rice University, 136. p.

Glass, B.P. and Crosbie, J.R. 1982. Age of the Eocene/Oligocene boundary based on extrapolation from North American microtektite layer. Amer. Assoc. Petrol Geol. Bull., 66(4): 41-476.

Hardenbol, J. and Berggren, W.A. 1978. A new Paleogene numerical time scale. Am. Assoc. Petrol. Geol., Studies in Geology, no. 6: 213-234.

LaBrecque, J.L., Kent, D.V. and Cande, S.C., 1977. Revised Magnetic Polarity Time Scale for the Late Cretaceous and Cenozoic Time. Geology, 5: 330-335.

Lowrie, W. and Alvarez, W., 1981. One hundred million years of geomagnetic polarity history. Geology, 9: 392-397.

Lowrie, W., Alvarez, W., Napoleone, G., Perch-Nielsen, K., Premoli Silva, I. and Toumarkine, M. 1982. Paleogene magnetic stratigraphy in Umbrian pelagic carbonate rocks: The Contessa sections, Gubbio. Geol. Soc. Amer. Bull. 93: 414-432.

McDougall, I., Kristjansson, L., and Saemundsson, K., 1984. Magnetostratigraphy and geochronology of Northwest Iceland. Jour. Geophysical Research, 89 (B8): 7029-7060.

Miller, K.G., Aubry, M.-P., Kahn, M.J., Melillo, A.J., Kent, D.V. and Berggren, W.A., 1985. Oligocene - Miocene biostratigraphy, magnetostratigraphy, and isotope stratigraphy of the western North Atlantic. Geology, 13: 257-261.

Montanari, A., Drake, R., Bice, D.M., Alvarez, W., and Curtis, G.H., 1983. Radiometric time scale for the Upper Eocene and Oligocene based on K/Ar and Rb/Sr dating of volcanic biotites from the pelagic sequence of Gubbio, Italy. Geology (in press).

Ness, G., Levi, S. and Crouch, R., 1980. Marine magnetic anomaly timescales for the Cenozoic and Late Cretaceous: a precis, critique and synthesis. Reviews of Geophysics and Space Physics, 18(4): 753-770.

Odin, G.S. (Ed.) 1982. Numerical dating in stratigraphy. John Wiley and Sons, Chichester, 1040 p.

Prothero, D.R., 1985. Chadronian (early Oligocene) magnetostratigraphy of eastern Wyoming: implications for the age of the Eocene-Oligocene boundary. Journal of Geology.

Prothero, D.R. and Armentrout, J.M., 1985. Magnetostratigraphic correlation of the Lincoln Creek Formation, Washington: implications for the age of the Eocene/Oligocene boundary. Geology, 13: 208-211.

Prothero, D.R., Denham, C.R. and Farmer, H.G., 1983. Magnetostratigraphy of the White River Group and its implications for Oligocene geochronology. Palaeogeogr., Palaeoclimatol., Palaeooecol., 42 (1983): 151-166.

Prothero, D.R., Denham, C.R. and Farmer, H.G. 1982. Oligocene calibration of the magnetic polarity timescale. Geology, 10(11): 650-653.

Ryan, W.B.F., Cita, M.B., Rawson, M.D., Burckle, L.H. and Saito, T. 1974. A paleomagnetic assignment of Neogene stage boundaries and the development of isochronous datum planes between the Mediterranean, the Pacific and Indian Oceans in order to investigate the response of the world ocean to the Mediterranean "Salinity Crisis". Riv. Ital. Paleont., 80: 631-688.

Sanfilippo, A., Riedel, W.R., Glass, B.P. and Kyte, F.T., 1985. Late Eocene microtektites and radiolarian extinctions on Barbados. Nature, 314: 613-615.

Saunders, J.B., Bernoulli, D., Muller-Merz, E., Oberhansli, H., Perch-Nielsen, K.P., Riedel, W.R., Sanfilippo, A., and Torini, R., Jr., 1984. Stratigraphy of the late Middle Eocene to early Oligocene section, Barbados, West Indies. Micropaleontology, 30(4): 390-425.

Testarmata, M.M. and Gose, W.A. 1979. Magnetostratigraphy of the Eocene-Oligocene Vieja Group, Trans-Pecos Texas. In: "Cenozoic geology of the Trans-Pecos volcanic field of Texas", A.W. Walton and C.D. Henry (eds.). Bureau of Economic Geol. Guidebook 19, Bur. Econ. Geol. Univ. Texas, Austin, 55-66.

MAGNETIC STRATIGRAPHY OF THE EOCENE/OLIGOCENE BOUNDARY

W. LOWRIE
Institut für Geophysik, ETH-Hönggerberg, 8093 Zürich, Switzerland

INTRODUCTION

Detailed interdisciplinary magnetostratigraphic and biostratigraphic investigations of the Eocene/Oligocene boundary were first reported for the pelagic carbonate rocks of the Umbrian sequence in the Northern Apennines of Central Italy (Lowrie et al., 1982). The main features of this research were confirmed in sediment cores from sites 522 and 523 on Leg 73 of the Deep Sea Drilling Project (Tauxe et al., 1983).

RESULTS FROM THE UMBRIAN SEQUENCE

In the Umbrian sequence of pelagic carbonate rocks the Late Cretaceous to Eocene Scaglia Rossa limestone grades into the grey Oligocene Scaglia Cinerea marlstone. The varicolored transitional interval is sometimes designated the Scaglia Variegata and corresponds to the upper Eocene. The Eocene/Oligocene boundary falls in the friable grey beds near the base of the Scaglia Cinerea. The lithology of this marly formation is mechanically weak and most outcrops are strongly sheared by tectonic deformation. They are so fissile that accurate paleomagnetic sampling is difficult. However, unusually continuous, fresh outcrops exposed in active quarries on both sides of the Contessa valley near Gubbio have given good magnetostratigraphic results.

The magnetic properties of fresh Scaglia Cinerea samples are comparatively straightforward. A stable primary component of remanent magnetization can be defined by progressive alternating field (AF) or thermal demagnetization (Lowrie et al., 1982). However, the magnetic behavior of weathered samples is akin to that of Scaglia Variegata, in which a persistent secondary component carried by hematite can only be removed satisfactorily by thermal treatment. The Contessa quarry section (Lowrie et al., 1982) was evaluated using a combination of AF and thermal demagnetization. The stable remanent magnetization defines discrete magnetozones of uniform polarity (Fig. 1). The sampling interval of 30-60 cm ensured that mos magnetozones are represented by several samples each, but a few of the shorter zones in the upper Oligocene are described by only a single sample.

The biostratigraphy of the Contessa quarry section was based on planktonic foraminifera and calcareous nannofossils. Although the microfauna and microflora were abundant, they were generally poorly preserved. All Eocene and Oligocene

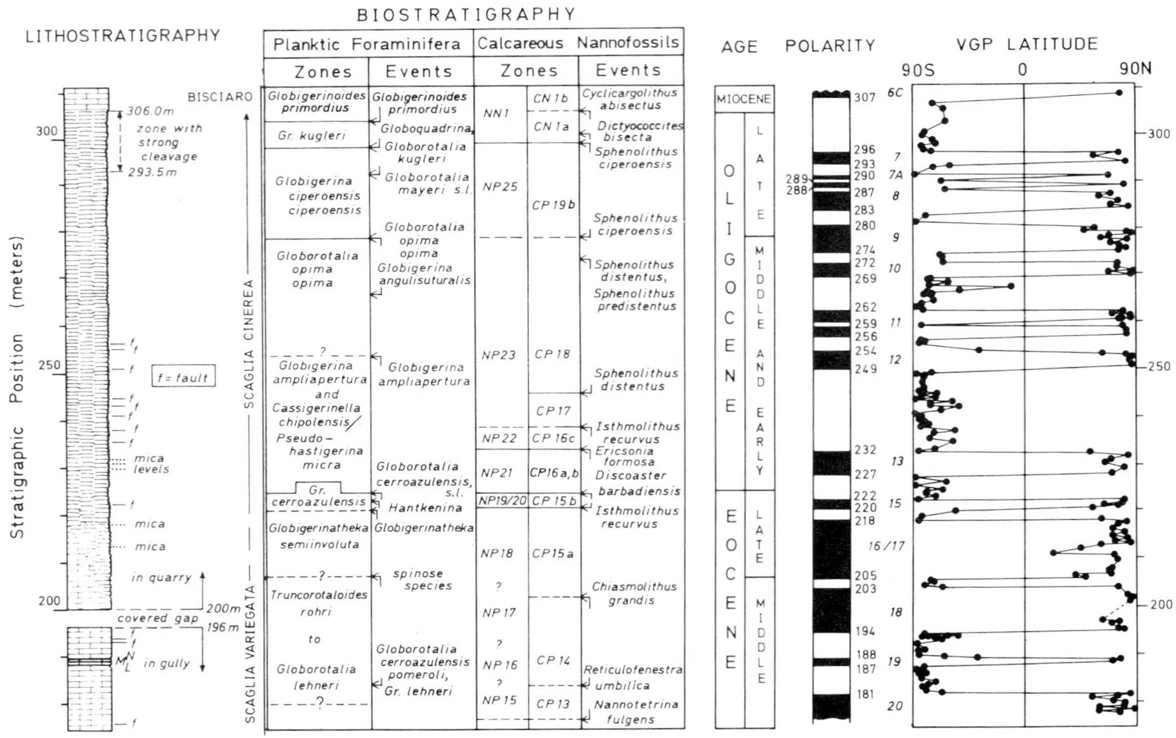

Fig. 1. Magnetostratigraphic and biostratigraphic results from the Contessa Quarry section.

foraminiferal zones were recognized but in some parts neighboring paleontological zones were not differentiated and some zonal boundaries were located only approximately. Foraminiferal preservation was poor around the Eocene/Oligocene boundary and the lower Oligocene could not be zoned. However, a nannofossil zonation was obtained in this interval. Although heavy overgrowth prevented identification of some species, the calcareous nannofossil zonation complemented the planktonic foraminiferal results and reinforced the interpreted stage boundary locations. The uppermost Eocene coccolith zones NP19/20 and the corresponding *G. cerroazulensis* foraminiferal zone were observed to be unusually thin, suggesting that a fault might have cut out part of the section (Lowrie et al., 1982). Several other small faults appear not to have disturbed the magnetic polarity sequence or biozonations greatly (Fig. 1).

The Contessa quarry section preserved an excellent record of geomagnetic polarity history. Cox (1982) suggested a practical nomenclature for identifying the magnetic polarity chrons in the sequence interpreted from oceanic magnetic anomalies. With this nomenclature the Contessa quarry section correlates with chrons 6C to 20 and the Eocene/Oligocene boundary falls within reversed polarity chron 13r, between normal chrons 13 and 15 (Fig. 1).

RESULTS FROM DSDP LEG 73

The sediment cores at DSDP sites 522 and 523 were taken with a hydraulic piston corer which causes minimal disturbance of the sediment or its magnetic properties. Magnetic cleaning of pilot samples showed good stability, with a partial overprint in some samples that could be removed by AF or thermal demagnetization. DSDP sediment recovery is often incomplete and only the inclinations of azimuthally unoriented cores can be used for paleomagnetic interpretations. The results at sites 522 and 523, based primarily on shipboard AF demagnetization, agreed well with the polarity sequence for chrons 6C to 16 at site 522 and chrons 12 to 20 at site 523 (Tauxe et al., 1983). The Eocene/Oligocene boundary, placed at the top of nannofossil zone NP20, correlated with reversed chron 13r, as in the Contessa quarry section.

COMPARISON OF BIOCHRONOLOGICAL DATUM PLANES

Oligocene first appearance datum (FAD) and last appearance datum (LAD) levels in DSDP Leg 73 holes and the Gubbio Contessa section (Fig. 2) sometimes had different correlations with the magnetic polarity time scale (Poore et al., 1982). Nannofossil discrepancies appear to be larger than those of planktonic foraminifera. For 14 datum levels the average difference in event chronology is 0.18 Ma (standard error 0.26) and is not significant statistically. However, if the 6 FAD events are considered separately, a significant difference is found. The DSDP ages average 0.45 Ma (standard error 0.39 Ma) older than the Contessa ages. The 8 LAD events differ even more significantly; the DSDP ages average 0.57 Ma (standard error 0.28 Ma) younger than the Contessa ages. These results may imply that species appeared sooner and disappeared later in the South Atlantic than in the southern Tethys. The paleontological investigators encountered difficulties with foraminiferal and nannofossil preservation in the DSDP sites and Contessa quarry section. The difference may represent the resolution of a paleontological datum level under such conditions, rather than an evolutionary difference (LaBrecque et al., 1983).

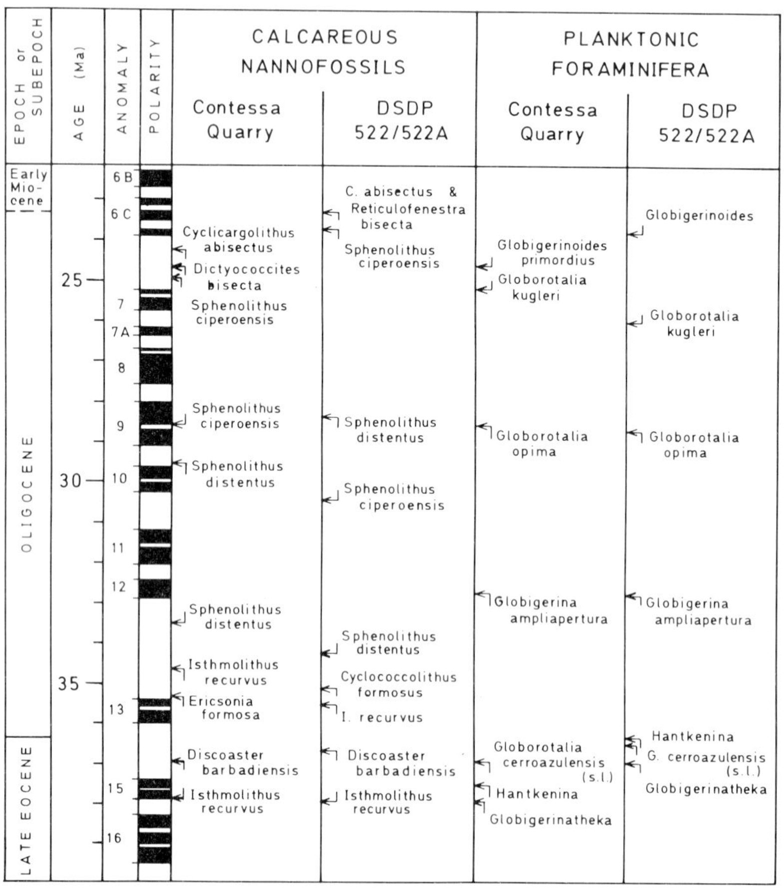

Fig. 2. Comparison of FAD and LAD levels in the Contessa Quarry and DSDP Site 522/522A with the LaBrecque et al. (1977) magnetic polarity time scale (after Poore et al., 1982).

SHORT MAGNETIC EVENTS

DSDP sites 522 and 523 show several short magnetozones for which there are no corresponding oceanic magnetic anomalies (Tauxe et al., 1983). Some zones are represented by only one sample which could be accidentally misoriented, but other short magnetozones are documented by multiple sampling. At DSDP site 522, reversed chron 13r, to which the Eocene/Oligocene boundary is correlated, contains three short positive magnetozones. The recovery of the chron 13r interval at site 523 was incomplete, but there is a single-sample positive magnetozone within the longer negative magnetozone.

The Contessa quarry section contained no magnetozones without a counterpart in the oceanic magnetic record, although the sedimentation rate and sampling interval were similar to the DSDP sites. The absence of short magnetic events may be due to a fault in the 13r interval (Fig. 1). A quarry across the valley is free of this faulting. It gives a complete Eocene/Oligocene boundary section correlated magnetostratigraphically with reversed polarity chron 13r (Nocchi et al., this volume); three short normal magnetozones are found within reversed polarity chronozone 13r in this section and the Contessa Highway section.

The existence of short magnetic "events" has important consequences for some theoretical analyses of the statistical behavior of geomagnetic reversals and for the corresponding processes in the earth's core (Lowrie and Kent, 1983). It is important that the existence of short magnetozones in magnetostratigraphic sections be verified thoroughly by multiple sampling, detailed demagnetization and rock magnetic analysis.

DISCUSSION

Magnetostratigraphic and biostratigraphic research have resulted in improved correlation of the Paleogene stage boundaries to the oceanic magnetic polarity sequence (Lowrie and Alvarez, 1981; LaBrecque et al., 1983). The Eocene/Oligocene boundary has been located relative to the oceanic magnetic polarity sequence in several studies. It lies within reversed chron 13r, between normal chrons 13 and 15, which may also contain a few very short normal polarity chrons. FAD and LAD levels in the DSDP sites and Contessa section have been correlated to the magnetic polarity time scale with an average discrepancy of around 0.5 Ma. This figure can hopefully be improved at sites with better preserved biostratigraphies and good magnetic stratigraphy.

REFERENCES

Cox, A.V., 1982. Magnetostratigraphic time scale. In: W.B. Harland, A.V. Cox, P.G. Llewellyn, C.A.G. Pickton, A.G. Smith and R. Walters, A Geologic time scale. Cambridge Univ. Press, Cambridge, 63-84.

LaBrecque, J.L., Kent, D.V., and Cande, S.C., 1977. Revised magnetic polarity time scale for Late Cretaceous and Cenozoic time. Geology, 5: 330-335.

LaBrecque, J.L., Hsu, K.J., Carman, M.F., Karpoff, A.-M., McKenzie, J.A., Percival, S.P., Petersen, N.P., Pisciotto, K.A., Schreiber, E., Tauxe, L., Tucker, P., Weissert, H.J. and Wright, R., 1983. DSDP Leg 73: Contributions to Paleogene stratigraphy in nomenclature, chronology and sedimentation rates. Palaeogeogr., Palaeoclimatol., Palaeoecol., 42: 91-125.

Lowrie, W., and Alvarez, W., 1981. One hundred million years of geomagnetic polarity history. Geology, 9: 392-397.

Lowrie, W., Alvarez, W., Napoleone, G., Perch-Nielsen, K., Premoli Silva, I., and Toumarkine, M., 1982. Paleogene magnetic stratigraphy in Umbrian pelagic carbonate rocks: the Contessa sections, Gubbio. Geol. Soc. Amer. Bull., 93: 414-432.

Lowrie, W. and Kent, D. V., 1983. Geomagnetic reversal frequency since the Late Cretaceous. Earth Planet. Sci. Lett., 62: 305-313.

Nocchi, M., Parisi, J., Monechi, S., Madile, M., Monaco, T., Napoleone, G., Orlando, M., Premoli Silva, I., Ripepe, M. and Bice, D. The Eocene/Oligocene boundary in the Umbrian pelagic sequences, Italy. This volume.

Poore, R. Z., Tauxe, L., Percival. S. F., and LaBrecque, J. L., 1982. Late Eocene/Oligocene magnetostratigraphy and biostratigraphy at South Atlantic DSDP Site 522. Geology, 10: 508-511.

Tauxe, L., Tucker, P., Petersen, N. P. and LaBrecque, J. L., 1983. The magnetostratigraphy of Leg 73 sediments. Palaeogeogr., Palaeoclimatol., Palaeoecol., 42: 675-90.

LATE EOCENE-EARLY OLIGOCENE CARBONATE SEDIMENTATION IN THE DEEP SEA

ROBERT C. THUNELL* AND BRUCE H. CORLISS*

ABSTRACT

Calcium carbonate accumulation rate patterns for the late Eocene-early Oligocene interval in 12 DSDP sites show inter- and intra-basin variations. Significant increases in calcium carbonate accumulation at the Eocene/Oligocene boundary are seen at Sites 77B, 292, 289 (Pacific Ocean), and Sites 214 and 219 (Indian Ocean), whereas a decrease is observed at Site 516F (South Atlantic). In contrast, no major changes are seen at Sites 73 and 277 (Pacific Ocean) or at Sites 363, 366 and 360 (Atlantic Ocean). Intra-oceanic variation in the accumulation records from the Pacific is observed as well. In comparison to Sites 292, 289 and 77B, Sites 73 and 277 show no change at the boundary. Instead increases are observed in Site 277 during the late Eocene and in Site 73 at the early/late Oligocene boundary.

In general, the observed changes in carbonate accumulation rates follow the patterns established for fluctuations in the position of the CCD across the Eocene/Oligocene boundary. The Pacific CCD undergoes the greatest drop at the boundary and the greatest increase in accumulation rates occurs in this ocean as well. The Indian Ocean is characterized by an intermediate change in both the depth of the CCD and carbonate accumulation rates at the Eocene/Oligocene boundary, while the South Atlantic records only a small change in accumulation rates and an intermediate change in the depth of the CCD.

The inter-ocean differences in both accumulation rate patterns and changes in the position of the CCD argue against a global causal event, such as a change in sea level, to account for the observed sedimentation changes. Instead, this variability may best be explained by changes in surface water productivity and bottom water circulation patterns. The intra-ocean variability in the Pacific can be explained by these variables as well.

*Addresses: R. Thunell, Dept. of Geology, Univ. of South Carolina, Columbia, South Carolina 29208; B. Corliss, Dept. of Geology, Duke University, Durham, North Carolina 27706

INTRODUCTION

Changes in global climatic conditions and ocean basin configuration during the Cenozoic are responsible for the evolution of the modern oceanic environment. The transformation from a "thermospheric" ocean in late Cretaceous time to the present day "psychrospheric" state (Brunn, 1957) has occurred through a series of distinct steps or events (Berggren and Hollister, 1977; Kennett, 1977; Berger et al., 1981).

One of the most dramatic of these steps occurred near the Eocene/Oligocene boundary (Corliss et al., 1984). Based on a turn-over in deep-sea ostracode assemblages, Benson (1975) suggested that the deep oceans became filled with cold bottom waters at this time. Kennett and Shackleton (1976) subsequently estimated that bottom water temperatures dropped by 4-5°C within a period of less than 100,000 years near the Eocene/Oligocene boundary. More recent oxygen isotopic studies have confirmed that a major change occurred at that time in the vertical temperature structure of the oceans (see Corliss and Keigwin, 1985 for a review). In addition, isotopic studies of planktonic foraminifera (Keigwin, 1980) suggest that latitudinal temperature gradients were enhanced around the Eocene/Oligocene boundary and this is reflected in a reorganization of plankton biogeographic provinces (Haq et al., 1977; Kennett, 1978; Sancetta, 1979; Haq, 1981) and a decrease in planktonic microfossil diversity (Berggren, 1969; Ciffelli, 1969; Lipps, 1970; Frerichs, 1971; Thunell, 1981; Corliss et al., 1984). Recent studies indicate that the marine faunal and floral response to the changing climatic and oceanographic conditions near the Eocene/Oligocene boundary was gradual (Corliss, 1981; Corliss et al., 1984; Snyder et al., 1984), rather than catastrophic (Alvarez et al., 1982; Ganapathy, 1982).

Sedimentation in the deep sea was also impacted by this global cooling around the Eocene/Oligocene boundary. The onset of cold, bottom water production at this time in high southern latitudes resulted in an invigoration of deep sea circulation which caused substantial submarine erosion and widespread hiatuses (Kennett et al., 1972; Rona, 1973; Davies et al., 1975; Moore et al., 1978). This was accompanied by a basic change in ocean chemistry which resulted in a dramatic deepening of the carbonate compensation depth (CCD) in the Pacific, with a lesser drop occurring in the other oceans (Heath, 1969; Berger and von Rad, 1972; Berger, 1973 and 1978; Van Andel and Moore, 1974; Van Andel, 1975; Van Andel et al., 1975 and 1977; Tucholke and Vogt, 1979; Thiede et al., 1980). Based on a synthesis of the results of the early legs of the Deep Sea Drilling Project (DSDP), Davies and Supko (1972) demonstrated that there was a major change in the lithology of deep sea sediments between the Eocene and Oligocene, with Oligocene sediments containing a much higher calcareous component. This is consistent with the above mentioned trend in the CCD, and the fact that much of the carbonate deposited during Eocene time was sequestered on continental shelves (Worsley and Davies, 1979; Davies and Worsley, 1981).

In the present study we attempt to quantitatively estimate late Eocene and early Oligocene carbonate accumulation rates at various DSDP locations in the Atlantic, Pacific and Indian Oceans. Our objective is to compare detailed carbonate accumulation records from each of the oceans and evaluate these records in terms of changing oceanographic conditions.

SEA LEVEL FLUCTUATIONS AND DEEP SEA SEDIMENTATION

Eustatic changes in sea level are very important in controlling sedimentation in the deep sea (Berger, 1970; Rona, 1973; Davies et al., 1977; Hay and Southam, 1977; Worsley and Davies, 1979a and 1979b; Davies and Worsley, 1981). The supply of both clastic and dissolved material to the oceans is a function of the amount of exposed continental land mass. During high stands of sea level, most of the clastic input is trapped on the shelves and very little reaches the deep sea. In contrast, during low stands of sea level, the shelf tends to be bypassed and terrigenous material is carried directly to the deep sea. Carbonate sedimentation in the oceans follows a similar basin-shelf fractionation pattern related to sea level change (Berger, 1970; Davies and Worsley, 1981; Kendall and Schlager, 1981). During times of high sea level and extensive submerged shelves, much of the calcium carbonate entering the oceans in the dissolved state is sequestered on the margin, leaving little available for accumulation in the deep sea. Conversely, the bypassing of the shelves during regressions changes the predominant site of carbonate deposition from the shelves to the deep sea.

According to Fischer and Arthur (1977), sea level is also an important factor in controlling the fluctuation between an "oligotaxic" and a "polytaxic" marine environment. "Polytaxic" episodes are times of high sea level, sluggish circulation, a shallow CCD, widespread chert deposition, and high diversity in pelagic biotas. These should also be times of decreased pelagic productivity, since the shelves are acting as a sink for the biolimiting nutrients (Berger, 1970; Fischer and Arthur, 1977; Broecker, 1982).

"Oligotaxic" episodes are characterized by lowstands of sea level, highly convective oceans, a deep CCD and low taxonomic diversity. These should also be times of increased fertility and productivity in the open ocean (Fischer and Arthur, 1977). Nutrients trapped on the shelf during sea level highstands will be eroded and carried into the open ocean resulting in higher productivity and an increase in the flux of biogenic material to the seafloor (Berger, 1970; Broecker, 1982).

DATA BASE

Twelve DSDP sites (Table 1) were selected for this study, and their approximate positions during late Eocene-early Oligocene time are indicated on the 40 million year paleogeographic reconstruction of Barron and others (1981) (Figure 1). These twelve sites were chosen on the basis of having fairly good stratigraphic continuity and core recovery. Two additional Indian Ocean sites (242 and 253) were initially considered in this study, but were later eliminated because of the presence of major unconformities at the Eocene/Oligocene boundary. For each site, a late Eocene-early Oligocene (~38 Ma) paleodepth was calculated using the technique outlined in Berger and Winterer (1974) (Table 1). We have assumed that those sites located on aseismic ridges (Sites 214, 253 and 516) have subsided at rates comparable to that for normal oceanic crust (Detrick et al., 1977).

Time series records of carbonate accumulation at each site were generated using the following equation:

$$AR_{CaCO_3} = \%CaCO_3(D-1.025\ P)\ SR$$

where AR = accumulation rate of carbonate ($gm/cm^2/10^3 yrs.$)

D = wet bulk density (gm/cm^3)

P = porosity (vol. percent)

SR = sedimentation rate ($cm/10^3 yrs.$)

Carbonate content of individual samples was determined using a gasometric method similar to that described in Jones and Kaiteris

TABLE 1. Site locations.

SITE	LAT.	LONG.	WATER DEPTH (Present Day)	PALEO-DEPTH (40ma)
A. Eastern Pacific				
1. Site 73	01°54.58'S	137°28.12'W	4387m	3250m
2. Site 77B	00°28.90'N	133°13.70'W	4291m	2560m
3. Site 167	07°04.10'N	176°49.50'W	3176m	3360m
B. Western Pacific				
1. Site 277	52°13.43'S	166°11.48'E	1232m	1232m
2. Site 289	00°29.92'S	158°30.69'E	2206m	2400m
3. Site 292	15°49.11'N	124°39.05'E	2943m	1070m
C. Indian Ocean				
1. Site 214	11°20.21'S	88°43.08'E	1671m	860m
2. Site 219	09°01.75'N	72°52.67'E	1764m	1170m
D. South Atlantic				
1. Site 360	35°50.75'S	18°05.79'E	2949m	2215m
2. Site 363	19°38.75'S	09°02.80'E	2248m	2000m
3. Site 366	05°40.70'N	19°51.10'W	2853m	2600m
4. Site 516F	30°16.59'S	35°17.11'W	1313m	800m

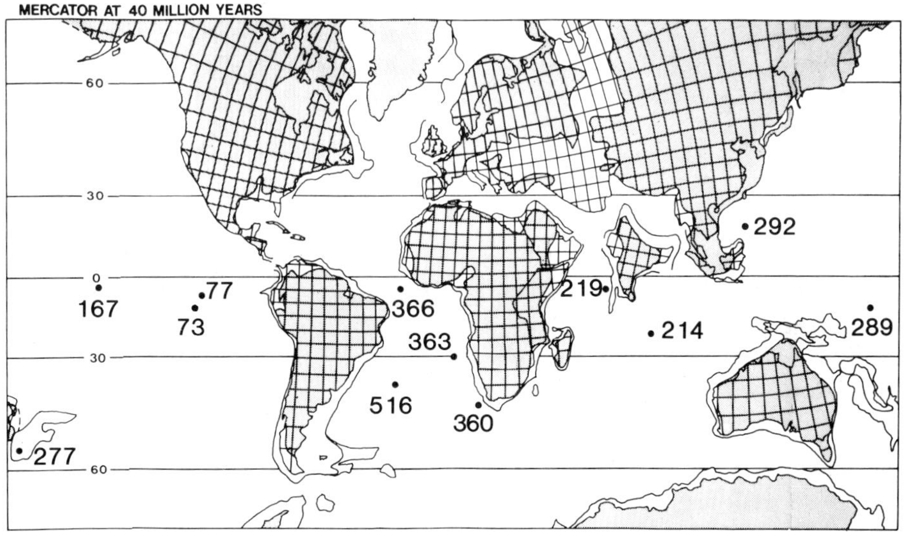

Fig. 1. Locations of DSDP sites used in this study plotted on the 40 million year paleogeographic reconstruction of Barron and others (1981).

(1983). Wet bulk density and porosity data are routinely generated for all DSDP core material and are available through the Information Handling Group at DSDP in La Jolla, California. Sedimentation rates were estimated using available biostratigraphic data from the Initial Reports and from Keller (1983), together with the timescale of Hardenbol and Berggren (1978). A more in-depth discussion of the procedure for calculating accumulation rates is available in Van Andel and others (1975).

The advantage of using carbonate accumulation rates instead of simple carbonate content data is that the accumulation rate calculation removes the effect of dilution by non-calcareous material. As a result, variation in carbonate accumulation through time can be attributed only to changes in supply (productivity) or dissolution.

RESULTS

Western Pacific

The carbonate accumulation records for the three western Pacific sites all show the same basic pattern during the late Eocene-early Oligocene (Fig. 2). Late Eocene accumulation rates are lower than those for the early Oligocene, with the shift from low to high values occurring close to the Eocene-Oligocene boundary in Sites 289 (Ontong Java Plateau) and 292 (Philippine Sea). The calculated accumulation rates for Sites 289 and 292 are also very similar in absolute value: late Eocene carbonate accumulation rates are approximately 2.5 $gm/cm^2/10^3$ years, dropping to about 2.0 $gm/cm^2/10^3$ years across the boundary, and then rising to 3.5 $gm/cm^2/10^3$ years in the early Oligocene.

At Site 277 on the Campbell Plateau, late Eocene carbonate accumulation rates are negligible below 230 m sub-bottom depth and then rise to greater than 4.5 $gm/cm^2/10^3$ years. However, the increase in carbonate accumulation at this site occurs in the late Eocene and not at the Eocene/Oligocene boundary as observed at Sites 289 and 292. Thus, although there is a basic similarity in the patterns at all three sites, the timing of the increase in carbonate accumulation appears to be diachronous between the high latitude site (277) and the lower latitude sites (289 and 292).

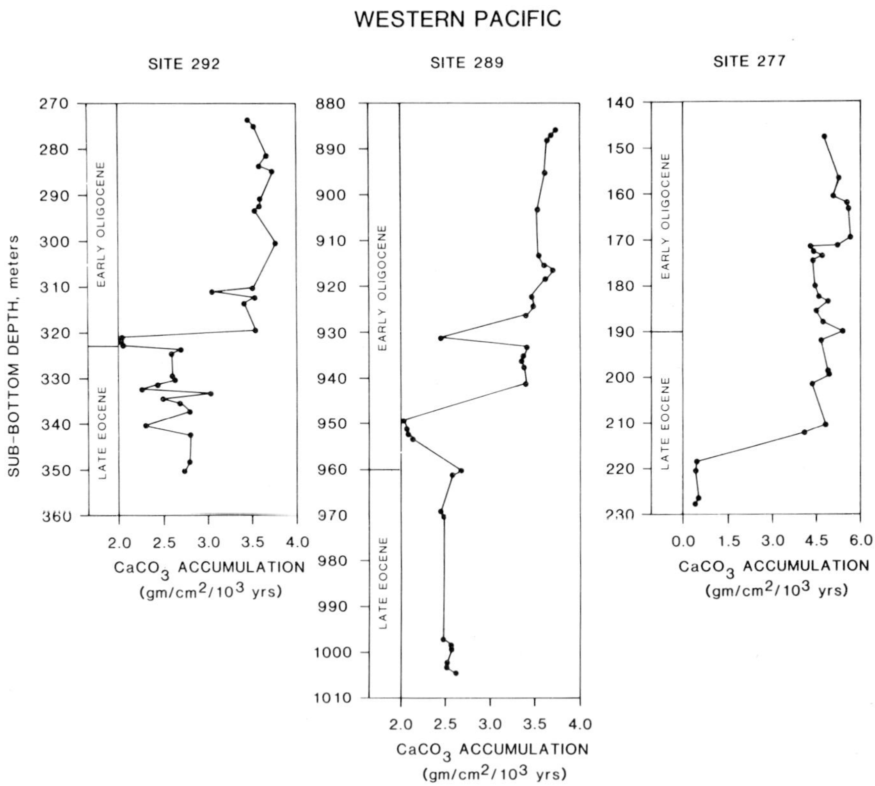

Fig. 2. Late Eocene-early Oligocene carbonate accumulation records for Western Pacific DSDP sites.

Eastern Pacific

The eastern Pacific sites (73, 77B and 167) have carbonate accumulation patterns (Fig. 3) that are variable, with the Site 77B record being similar to that observed for the western Pacific (Fig. 2). In particular, Sites 77B and 167 contain increases in carbonate accumulation in the earliest Oligocene. In contrast, there is no change in carbonate accumulation rate across the Eocene/Oligocene boundary in Site 73, but there is an abrupt increase in accumulation at the base of the late Oligocene. This event is also evident in the Site 77B record. At Site 167 an increase in carbonate accumulation occurs at the Eocene/Oligocene boundary, but the early Oligocene values are similar to those found in the middle and early Eocene part of the section. At all three sites, late Eocene carbonate accumulation rates are less than 1.0 $gm/cm^2/10^3$ years. These rates are also less than those estimated for the three western Pacific sites and may be due to slightly poorer preservation since the eastern Pacific sites examined are all deeper than the western Pacific sites (Table 1).

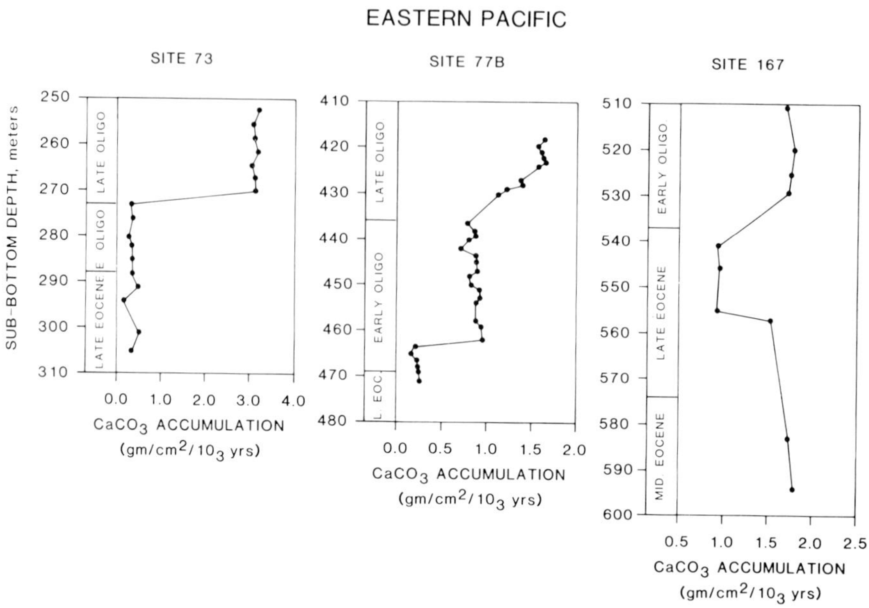

Fig. 3. Late Eocene-early Oligocene carbonate accumulation records for Eastern Pacific DSDP sites.

Indian Ocean

The Two Indian Ocean sites examined (Sites 214 and 219) display increases in carbonate accumulation near the Eocene-Oligocene boundary (Fig. 4). Both of these sites also indicate that carbonate accumulation was relatively low throughout the late Eocene-early Oligocene interval in this region (less than 1.0 $gm/cm^2/10^3$ years). The very shallow paleodepths for these sites (Table 1) would suggest that these low accumulation rates are not an artifact of dissolution.

South Atlantic

The South Atlantic carbonate accumulation records are somewhat more ambiguous with regards to change near the Eocene-Oligocene boundary than those from the Indian and Pacific Ocean (Fig. 5).

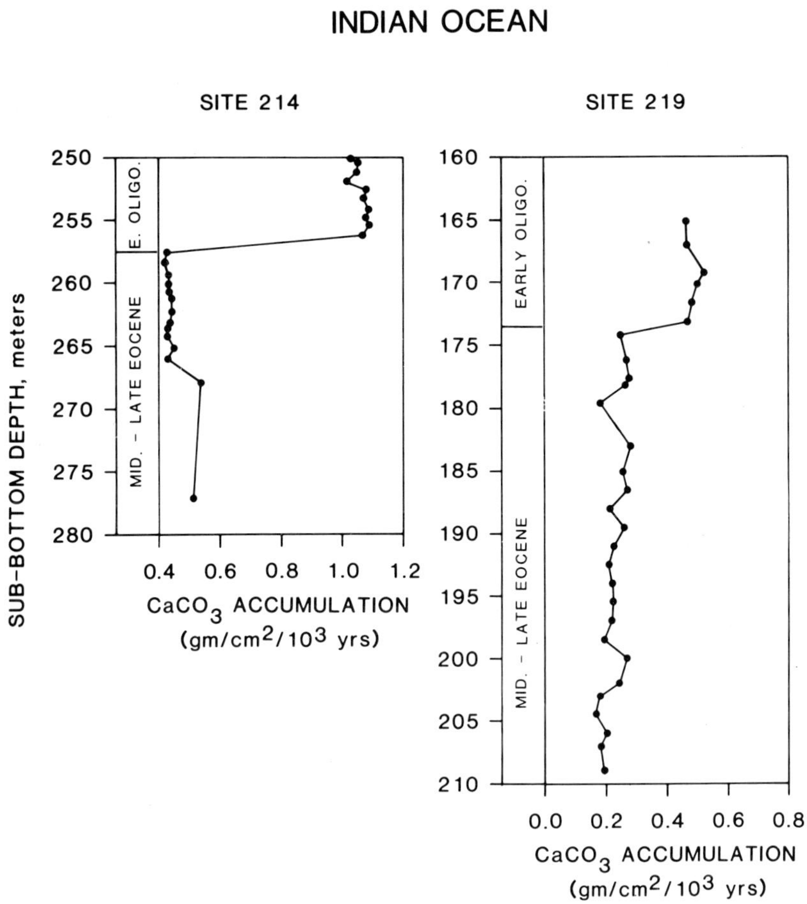

Fig. 4. Late Eocene-early Oligocene carbonate accumulation records for Indian Ocean DSDP sites.

The records for Sites 363 and 366 are quite similar, with a slight increase in carbonate accumulation occurring in the early Oligocene. At both sites carbonate accumulation tends to be relatively low, fluctuating between 0.5-1.0 $gm/cm^2 /10^3$ yrs. throughout the late Eocene-early Oligocene interval (Fig. 5).

The site 516F record shows a decrease in carbonate accumulation across the boundary, dropping from around 3.0 $gm/cm^2 /10^3$ yrs. in the late Eocene to about 2.0 $gm/cm^2/10^3$ yrs. in the early Oligocene. Finally, the record for Site 360 displays a high degree of variability in the late Eocene, and combined with a limited number of early Oligocene data points makes it difficult to evaluate any changes across the boundary.

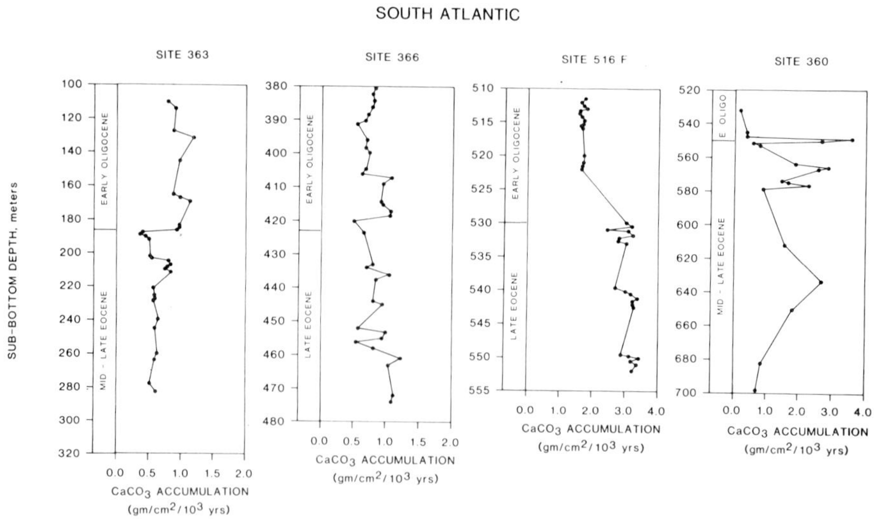

Fig. 5. Late Eocene-early Oligocene carbonate accumulation records for the South Atlantic DSDP sites.

INTER-BASIN COMPARISON

Timing of Accumulation Rate Changes

It is important to determine if the observed changes in carbonate accumulation in each of the ocean basins are isochronous or diachronous. If these late Eocene-early Oligocene accumulation rate changes were primarily controlled by sea level induced changes in basin-shelf fractionation patterns (Berger, 1970), then we would expect the major changes in accumulation to be synchronous both within and between basins. However, a comparison of the accumulation rate patterns presented here and published sea level curves (Vail et al., 1977; Vail and Hardenbol, 1979) does not support such a simple scenario.

The late Eocene-early Oligocene (Priabonian-Rupelian; approximately 40 Ma to 32 Ma according to Hardenbol and Berggren, 1977) is basically a long transgressive interval, that is not punctuated by any major eustatic events (Vail and Hardenbol, 1979). With regards to the Eocene-Oligocene boundary, Vail and others (1977) reported a small regression, while Loutit and others (1983) have suggested that sea level actually rose across the boundary. Although the sea level record during the late Eocene-early Oligocene is not clear at this time, the interpretations do not suggest a dramatic change in sea level (either up or down) which would cause major changes in sediment accumulation in the deep sea. More importantly, a comparison of the carbonate accumulation records indicates that there are distinct intra- and inter-basin differences in the timing and direction of the major changes in these records (Figs. 2-5). In the western Pacific there is a significant increase in carbonate accumulation in the earliest Oligocene at the two low latitude sites (Sites 289 and 292), while a similar change occurs earlier (late Eocene) at the higher latitude Site 277 (Fig. 2). A similar situation exists in the eastern equatorial Pacific. Site 77B records an increase in carbonate accumulation in the earliest Oligocene, while the only significant increase in accumulation at Site 73 occurs near the early-late Oligocene boundary (Figure 3). The other eastern equatorial site, Site 167, shows an increase in accumulation at the boundary, but only to values found previously in the late Eocene. Both of the Indian Ocean sites (214 and 219) examined contain a major increase in carbonate accumulation at the Eocene/Oligocene boundary (Fig. 4). In contrast, Site 516F from the South Atlantic records a decrease in accumulation across the boundary (Fig. 5). Thus, although each of the ocean basins examined reveals, to a varying extent, major changes in carbonate accumulation during the late Eocene-early Oligocene period, these changes are not necessarily synchronous or unidirectional. This suggests that the observed changes in carbonate accumulation may be the result of a number of independent or regional paleoceanographic changes, rather than a single global event which would have resulted in an isochronous response throughout all of the oceans.

Dissolution vs. Productivity

The rate at which carbonate accumulates on the seafloor is a function of the net difference between the supply rate and the recycling rate. The observed changes in late Eocene-early Oligocene carbonate accumulation rates can therefore be attributed to changes in productivity and/or dissolution, although distinguishing between these is not always straightforward. The paleodepths for many of the sites studied were well above the CCD throughout the entire late Eocene-early Oligocene interval (Table 1). This is particularly true for Sites 214, 219, 277 and 516F, which are all located on topographic highs and had very shallow paleodepths (less than 1500m). Even in the late Eocene when the CCD was very shallow, the carbonate being deposited at these sites would not have been subjected to a substantial amount of dissolution.

It has been well documented that a significant deepening of the CCD occurred near the Eocene/Oligocene boundary (Fig. 6), although the magnitude of this drop appears to have varied from ocean to ocean (Heath, 1969; van Andel and Moore, 1974; van Andel, 1975; van Andel et al., 1975 and 1977). During the late Eocene, the CCD was at a depth of 3800 m or shallower in all of the ocean basins. In the early Oligocene the CCD dropped over a 1,000 m in the Pacific and between 600 to 700 m in the Indian and Atlantic Oceans to a depth of between 4,000 and 4,500 m. This drop in the CCD had the obvious effect of greatly increasing the area of seafloor on which carbonate could accumulate without being totally dissolved. If the observed changes in carbonate accumulation during the late Eocene-early Oligocene were due

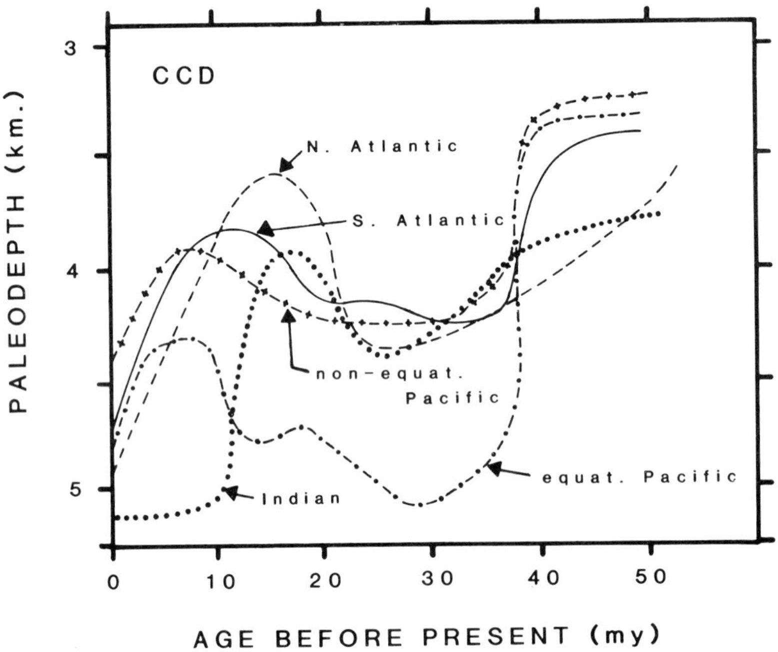

Fig. 6. Variation of the calcite compensation depth (CCD) with time for the equatorial Pacific and non-equatorial Pacific Ocean (van Andel et al., 1975), the Indian Ocean (van Andel, 1975), the South Atlantic Ocean (van Andel et al., 1977) and the North Atlantic Ocean (Berger and van Rad, 1972).

solely to this global deepening of the CCD, we would expect the changes to be in the same direction at all of the sites. Specifically, all of the records should indicate an increase in carbonate accumulation near the Eocene/Oligocene boundary. As previously mentioned, this is not the situation.

This would seem to suggest that changes in surface water productivity may have been the primary factor responsible for the observed changes in these late Eocene-early Oligocene carbonate accumulation records. Since productivity is a function of local or regional hydrographic conditions, it is not unreasonable to consider that different sectors of the ocean basins may have been subjected to very different patterns of productivity during this late Eocene-early Oligocene period. If this were the case, it becomes much easier to explain the inter- and intra-basin differences in carbonate accumulation patterns. For example, the late Eocene increase in carbonate accumulation at Site 277 in the south Pacific may have been due to a local increase in productivity that is totally independent of and unrelated to a later increase in productivity found near the Eocene/Oligocene boundary in Sites 73 and 167.

LATE EOCENE - EARLY OLIGOCENE DEPOSITIONAL ENVIRONMENT

Late Eocene - early Oligocene oceanographic conditions and associated depositional environments were the result of interplay between global climatic change and more local tectonically induced changes in basin configuration, and as such would have varied considerably from region to region. The depositional sequence found at Site 277 on the Campbell Plateau reflects the tectonic evolution of the southwest Pacific. Following initial rifting between Antarctica and Australia in the early Cenozoic (Weissel and Hayes, 1972), continued seafloor spreading gradually produced an open ocean environment in this region and terrigenous sediments were replaced by biogenic oozes (Kennett, 1977). The late Eocene increase in carbonate accumulation at Site 277 is considered to be due to an associated increase in open ocean productivity, possibly related to upwelling over this topographic high. Similar lithologic changes are found throughout the Southern Ocean, although the timing of these changes is diachronous due to different ages of formation of the various sedimentary basins (Kennett et al., 1975).

Dramatic changes in deep water circulation conditions near the Eocene/Oligocene boundary are inferred from a variety of sedimentological, paleontological and geochemical data. Changes in deep-sea ostracode assemblages (Benson, 1975) and an enrichment in benthic foraminiferal oxygen isotopes (Kennett and Shackleton, 1976) led to the suggestion that the development of the psychrosphere took place near the Eocene/Oligocene boundary in response to a decrease in Antarctic surface water temperatures. Initiation of North Atlantic deep water circulation at this time has also been suggested as an important component in the deep thermohaline circulation based on seismic stratigraphic data (Miller and Tucholke, 1983). The deep circulation event at the Eocene/Oligocene boundary was suggested to have been one component of a series of bottom water changes occurring in the middle to late Eocene (Corliss, 1981). Recent isotopic studies have corroborated this interpretation by showing that bottom water coolings in the deep ocean occurred at different times during the Eocene (Keigwin and Corliss, 1985).

Surface circulation changed markedly near the Eocene/Oligocene boundary as well. Biogeographic patterns of calcareous nannoplankton and planktonic foraminifera (Haq and Lohmann, 1976; Haq et al., 1977; Sancetta, 1979) show changes across this boundary which were interpreted as reflecting a cooling of surface waters. Detailed time-series studies of these microfossil groups show the Eocene/Oligocene boundary cooling to be part of a longer term cooling trend that began in the middle Eocene (Keller, 1983).

The micropaleontological and isotopic data indicate that surface water temperature gradients increased near the Eocene/Oligocene boundary and that surface and deep water circulation became more vigorous. The greater mixing of the oceans is reflected in the sediment record as well, with the occurrence of a large number of hiatuses near the boundary (Rona, 1973; Davies et al., 1975; Moore et al., 1978).

The increased accumulation of $CaCO_3$ can be related to an intensification of surface and deep circulation in two ways. Increased atmospheric circulation caused by enhanced latitudinal temperature gradients would result in an increase in upwelling of nutrient-rich near-surface water which would lead to higher surface water productivity. Secondly, the preservation of carbonate would be enhanced by the formation of young bottom water with relatively high amounts of oxygen and low CO_2 concentrations.

SUMMARY

A model has been developed for the global carbon cycle which suggests a close interrelationship between depth of the CCD, height of sea level and basin-shelf fractionation of carbonate sediments (Berger 1973 and 1977; Berger and Winterer, 1974; Sclater et al., 1979). According to this model, during high stands of sea level the CCD is relatively shallow and carbonate deposition is concentrated on the shelves. Alternately, during low stands of sea level the CCD is depressed and the deep oceans become the major site of carbonate deposition. This simple model cannot adequately explain the inter- and intra-ocean variability observed in this detailed study of late Eocene-early Oligocene carbonate accumulation patterns in the deep sea.

Although the CCD dropped significantly in all of the ocean basins near the Eocene/Oligocene boundary, there was no major regression associated with this drop and hence no reason for there to be a significant change in basin-shelf carbonate fractionation patterns. If the drop in CCD near the Eocene/Oligocene boundary was the primary factor affecting the accumulation of carbonate in the deep sea at this time, then we would expect all of the records to display an increase in carbonate accumulation in the early Oligocene. The data presented here shows that this was not the situation, since the timing of the major changes in these accumulation records is not synchronous from site to site.

We believe that the variability observed in these late Eocene-early Oligocene carbonate accumulation records is a reflection of global climatic change with more local paleogeographic changes. Differences in surface water productivity both in space and time appear to be the most viable explanation for the site to site variability in carbonate accumulation patterns.

ACKNOWLEDGEMENTS

The samples used in this study were provided by the Deep Sea Drilling Project through funding from the National Science Foundation.

REFERENCES

Alvarez, W., Asaro, F., Michael, H. and Alvarez, L. W., 1982. Iridium anomaly approximately synchronous with terminal Eocene extinctions. Science 216, 886-888

Barron, E. J., Harrison, C. G. A., Sloan, J. L. and Hay, W. W., 1981. Paleogeography, 180 million years ago to the present. Eclogae Geologicae Helvetiae 74, 443-470.

Benson, M. L., 1975. The origin of the psychrosphere as recorded in changes of deep-sea ostracode assemblages. Lethaia 8, 69-83.

Berger, W. H., 1970. Biogenous deep-sea sediments: fractionation by deep-sea circulation. Geol. Soc. Amer. Bull. 81, 1385-1402.

Berger, W. H., 1973. Cenozoic sedimentation in the eastern tropical Pacific. Geol. Soc. Amer. Bull. 84, 1941-1954.

Berger W. H., 1977. Carbon dioxide excursions and the deep-sea record: aspects of the problem. In: Anderson, N. R. and Malahoff, A., eds., Fate of Fossil Fuel CO , Plenum Pub. Co., 505-542.

Berger, W. H. and von Rad, U., 1972. Cretaceous and Cenozoic sediments from the Atlantic Ocean. In: Hayes, D. E., Pimm, A. C. and others, Initial Reports of the Deep Sea Drilling Project, vol. 14 (U. S. Government Printing Office) Washington, DC., 787-954.

Berger, W. H., and Winterer, E. L., 1974. Plate stratigraphy and the fluctuating carbonate line. In: Pelagic Sediments on Land and Under the Sea, eds., Hsu, K. J. and Jenkyns, H. C., Blackwell Scientific Publications Ltd., England, 11-48.

Berger, W. H., Vincent, E. and Thierstein, H., 1981. The deep-sea record: major steps in Cenozoic ocean evolution. S.E.P.M. Spec. Pub. 32, 489-504.

Berggren, W. A., 1969. Rates of evolution in some Cenozoic planktonic foraminifera. Micropaleontology 15, 351-365.

Berggren, W. A. and Hollister, C. D., 1977. Plate tectonics and paleocirculation-commotion in the ocean. Tectonophysics 38, 11-48.

Broecker, W. S., 1982. Glacial to interglacial changes in ocean chemistry. Progress in Oceanography 11, 151-197.

Brunn, A. F., 1957. Deep sea and abyssal depths. Geol. Soc. Amer. Mem. 67, 641-672.

Cifelli, R., 1969. Radiation of Cenozoic planktonic foraminifera. Syst. Zool. 18, 154-168.

Corliss, B. H., 1981. Deep-sea benthic foraminiferal turnover near the Eocene/Oligocene boundary. Marine Micropaleontology 6, 367-384.

Corliss, B. H., Aubry, M. P., Berggren, W. A., Fenner, J. M., Keigwin, L. D. and Keller, G., 1984. The Eocene/Oligocene boundary event in the deep sea. Science 226, 806-810.

Davies, T. A. and Supko, P. R., 1973. Oceanic sediments and their diagenesis: some examples from deep sea drilling. Jour. Sed. Petrology 43, 381-390.

Davies, T. A. and others, 1975. Unconformities in the sediments of the Indian Ocean. Nature 253, 15-19.

Davies, T. A. and Worsley, T. R., 1981. Paleoenvironmental implications of oceanic carbonate sedimentation rates. Soc. Econ. Pal. Min. Spec. Pub. 32, 169-180.

Davies, T. A., Hay, W. W., Southam, J. R. and Worsley, T. R., 1979. Estimates of Cenozoic oceanic sedimentation rates. Science 197, 53-55.

Detrick, R. S., Sclater, J. G. and Thiede, J., 1977. The subsidence of aseismic ridges. Earth Planet. Sci. Lett. 34, 185-196.
Fisher, A. L. and Arthur, M. A., 1977. Secular variations in the pelagic realm. In: Deep Water Carbonate Environments, eds., Cook, H. E. and Enos, P., S. E. P. M. Spec. Pub. 25, 19-50.
Ganapathy, R., 1982. Evidence for a major meteorite impact on the earth 34 million years ago: implication for Eocene extinctions. Science 216, 885-886.
Haq, B. U., 1981. Paleogene paleoceanography: early Cenozoic oceans revisited. Oceanologica Acta, Sp., 71-82.
Haq, B. U. and Lohmann, G. P., 1976. Early Cenozoic calcareous nannoplankton biogeography of the Atlantic Ocean. Mar. Micropaleo. 1, 119-194.
Haq, B. U., Premoli-Silva, I. and Lohmann, G. P., 1977. Calcareous plankton paleobiogeographic evidence for major climatic fluctuations in the early Cenozoic Atlantic Ocean. Jour. Geophy. Res. 82, 3861-3876.
Hardenbol, J. and Berggren, W. A., 1978. A new Paleogene numerical time scale. AAPG Studies in Geology No. 6, 213-234.
Hay, W. W. and Southam, J.R., 1979. Modulation of marine sedimentation by continental shelves, In: Anderson, N. and Malahoff, A., eds., The Fate of Fossil Fuel CO in the Oceans, Plenum Pub. Corp., 569-604.
Heath, G. R., 1969. Carbonate sedimentation in the abyssal equatorial Pacific during the past 50 million years. Geol. Soc. Amer. Bull. 80, 689-694.
Jones, G. A. and Kaiteris, P., 1983. A vacuum-gasometric technique for rapid and precise analysis of calcium carbonate in sediments. Jour. Sed. Petrology 53, 655-660.
Keigwin, L. D., 1980. Paleoceanographic change in the Pacific at the Eocene/Oligocene boundary. Nature 287, 722-725.
Keigwin, L. D. and Corliss, B. H., 1985. Stable isotopes in Eocene/Oligocene foraminifera (in prep.).
Keller, G., 1983. Biochronology and paleoclimatic implications of middle Eocene to Oligocene planktonic foraminiferal faunas. Mar. Micropaleo. 7, 463-486.
Kendall, C. G. and Schlager, W., 1981. Carbonates and relative changes in sea level. Marine Geology 44, 181-212.
Kennett, J. P., 1977. Cenozoic evolution of Antarctic glaciation, the circum-Antarctic Current, and their impact on global paleoceanography. Jour. Geophys. Res. 82, 3843-3860.
Kennett, J. P., 1978. The development of planktonic biogeography in the Southern Ocean during the Cenozoic. Mar. Micropaleo. 3, 301-345.
Kennett, J. P. and Shackleton, N. J., 1976. Oxygen isotopic evidence for the development of the psychrosphere 38 million years ago. Nature 360, 513-515.
Kennett, J. P. and others, 1972. Australian-Antarctic continental drift, paleocirculation changes, and Oligocene deep-sea erosion. Nature 239, 51-55.
Kennett, J. P., Houtz, R. E. and others, 1975. Cenozoic paleoceanography in the southwest Pacific Ocean, Antarctic glaciation, and the development of the circum-Antarctic current. Initial Reports of the Deep Sea Drilling Project, vol. 29, U. S. Govt. Printing Office (Washington, DC) 1155-1169.
Lipps, J. H., 1970. Plankton Evolution. Evolution, 24, 1-22.

Loutit, T., Baum, G. R. and Wright, R. C., 1983. Eocene/Oligocene sea level changes as reflected in Alabama outcrop secions. Amer. Assoc. Pet. Geol. Annual Convention (abstract) 118.

Miller, K. G. and Tucholke, B. E., 1983. Development of Cenozoic abyssal circulation south of the Greenland-Scotland Ridge. In: Bott, Sozov, Talwani and Thiede, eds., Structure and Development of the Greenland-Scotland Ridge, Plenum Publishing Co.

Moore, T. C., van Andel, T. H., Sancetta, C. and Pisias, N. G., 1978. Cenozoic hiatuses in pelagic sediments. Micropaleontology 24, 113-138.

Rona, P. A., 1973. Worldwide unconformities in marine sediments related to eustatic changes of sea level. Nature 244, 25-26.

Sancetta, C., 1979. Paleogene Pacific microfossils and paleoceanography. Mar. Micropaleo. 4, 363-398.

Sclater, J. G., Boyle, E. and Edmond, J. M., 1979. A quantitative analysis of some factors affecting carbonate sedimentation in the oceans. In: Talwani, M., Hay, W. and Ryan, W. B. F., eds., Deep Sea Drilling Results in the Atlantic Ocean: Continental Margins and Paleoenvironment, Amer. Geophys. Union, 235-248.

Snyder, S. W., Muller, C. and Miller, K., 1984. Eocene/Oligocene boundary: biostratigraphic recognition and gradual paleoceanographic charge at DSDP Site 549. Geology 12, 112-115.

Thiede, J., Agdestein, T. and Strand, J. E., 1980. Depth distribution of calcareous sediments in the Mesozoic and Cenozoic Atlantic Ocean. Earth Planet. Sci. Lett. 47, 416-422.

Thunell, R. C., 1981. Cenozoic paleotemperature changes and planktonic foraminiferal speciation. Nature 289, 670-672.

Tucholke, B. E. and Vogt, P. R., 1979. Western North Atlantic: sedimentary evolution and aspects of tectonic history, In: Tucholke, B. E., Vogt, P. R. et al. Initial Reports of the Deep Sea Drilling Project. Leg 43 (U.S. Government Printing Office) Washington, DC.

Vail, P. R. and Harbenbol, J., 1979. Sea-level changes during the Tertiary. Oceanus 22, 71-79.

CLAY MINERALOGY AT THE EOCENE/OLIGOCENE BOUNDARY.

Hervé CHAMLEY
Sédimentologie et Géochimie, Université de Lille I, 59655 Villeneuve d'Ascq, France.

1 INTRODUCTION

Clay mineralogical investigations have been carried out in many Paleogene series dominated by terrigenous influences. Eocene-Oligocene series have been peculiarly investigated in the Atlantic and Western Pacific Oceans, by X-ray diffraction, electromicroscopy and atomic absorption analyses. Studies concern also some european and north-american land sections. As in the whole Cenozoic columns, the clay assemblages chiefly reflect changes in the continental paleoenvironment, and only local volcanogenic influences and diagenetic recristallizations occur (e.g. Chamley, 1981 ; Desprairies, 1981 ; Robert, 1982 ; Holtzapffel et al., 1985). Our purpose is to summarize the main data and interpretations available, by considering the clay successions at three different time scales.

2 CENOZOIC

Most Cenozoic series of the world ocean show a step-by-step increase of illite, chlorite and associated quartz, feldspars and heavy minerals, and a concomitant decrease of smectite, fibrous clays and often kaolinite (e.g. Chamley, 1979, 1981). This general change reflects, in a tectonically-stable context, the decrease of the rock-derived supply and the increase of the soil-derived supply to the ocean. The cause is the Cenozoic world-wide cooling, favoring the physical alteration processes, instead of the chemical ones, and permitting better latitudinal water exchanges. The phenomenon starts close to the Eocene-Oligocene passage, and the clay assemblage behaviour parallels those of many other climate-induced sedimentary markers (e.g. foraminifera, palynomorphs, stable isotopes ; see this vol.).

3 EOCENE AND OLIGOCENE

In a general way, the mineralogical trend recorded for the whole Cenozoic is confirmed when focusing the investigations on the Eocene-Oligocene period. Numerous data are provided by

Robert (1982) in the South Atlantic. For instance Site 366 DSDP, on the Sierra Leone Rise, reveals a very progressive increase of illite, chlorite, irregular mixed-layers and feldspars, starting in the Late Eocene (Fig. 1). At the same time the non-clay early diagenetic minerals disappear (opal, clinoptilolite), while kaolinite and iron oxides tend to develop, suggesting progressive more humid conditions in West Africa (reworking of upstream and well-drained soils). Similar data are recorded on the eastern flank of the Mid-Atlantic ridge (Karpoff, 1984). On the Walvis Ridge (Site 528) the same trend occurs, but chlorite appears in the Middle Eocene and kaolinite in the Late Eocene, while palygorskite amount irregularly decreases from the Early Eocene upwards. All these data, which point to cooler climate, and probably to more regular rainfall conditions, can be precisely expressed by the smectite/illite ratio, whose decrease roughly parallels the oxygen-isotope signal (e.g. sites 14 and 19 DSDP, South of Brazil Basin ; in Robert, 1982).

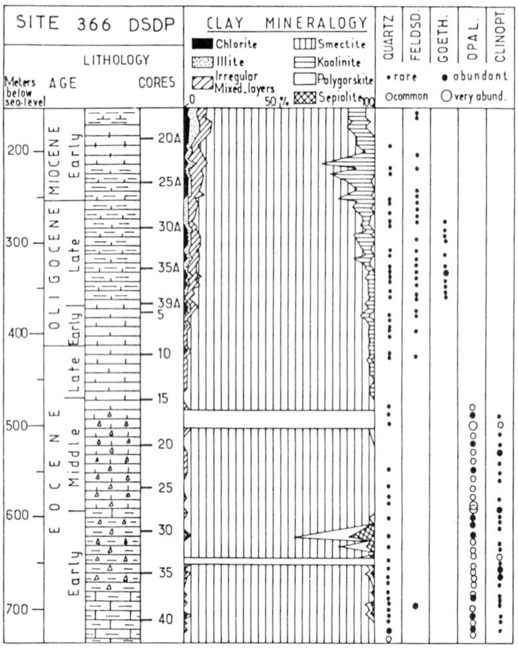

Fig. 1. Eocene and Oligocene clay mineralogy at Site 366, Sierra Leone Rise (from Robert, 1982).

Peculiar situations occur sometimes. At Site 20 DSDP (NE of Rio Grande Rise ; Robert, 1982), a very rapid increase of the illite group corresponds to a hiatus in the sedimentation during the Late Eocene-lowermost Oligocene, which results from increased bottom circulation and not from a strong climatic change. The importance of sedimentary gaps in the North-East Atlantic between the Early Eocene and the Middle Miocene is also hydrodynamically controlled and prevents the obtention of a suitable climatic record (Latouche and Maillet, 1980). At the Angola Basin/Walvis Ridge junction (Site 363), chlorite appears in the Early Eocene, while illite and palygorskite abundance does not significantly change until the uppermost Eocene, what probably results from local petrographic, eustatic and tectonic conditions (Robert, 1982). On Lord Howe Rise, in the South-West Pacific, the fast increase and then decrease of the illite group at the Eocene-Oligocene transition correlates with extensive tectonic activity in New Zealand, hidding the climatic effects (Robert et al., in press). On Goban Spur off the Armorican margin the Late Eocene locally corresponds to condensed volcanic-sedimentary sections, allowing the development of authigenic zeolites, badly-crystallized smectites, Fe-Mn oxides and nodules ; only few mineralogical and geochemical characters still reflect the climatic cooling (Karpoff, 1984).

4 EOCENE-OLIGOCENE TRANSITION

Different sections have been studied in Rumania, Slovakia, Sicily, Barbabos, Alabama (USA), at the precise passage between Eocene and Oligocene (sampling Ch. Pomerol). Clay mineral assemblages generally show changes of a small amplitude, with different trends. For instance the Brebi section in Rumania, which corresponds to a carbonate platform environment, is characterized by a temporary appearance of chlorite in traces, the appearance of kaolinite and low increase of illite (see smectite/illite and smectite/kaolinite ratios), and the increase of quartz and feldspars (Fig. 2). These changes parallel those recorded at a less-detailed scale (see above, § 3), and suggest climatic conditions getting cooler and a bit more humid (increase of rock-derived and well-drained soil-derived minerals).

The Runcu section is also located in Rumania, and corresponds to flysch facies. The Eocene-Oligocene transition shows opposite clay mineralogical changes as in Brebi. Illite and kaolinite abundances decrease compared to smectite, and certainly reflect changes in erosion processes due to a tectonically instable environment (Fig. 3). Slovakia samples show random and fairly important fluctuations at the biostratigraphic boundary while Sicily, Barbados and Alabama sections reveal no significant or only very small changes. Thus the clay mineralogy at the Eocene-Oligocene boundary seems to depend rather on local conditions than on global ones.

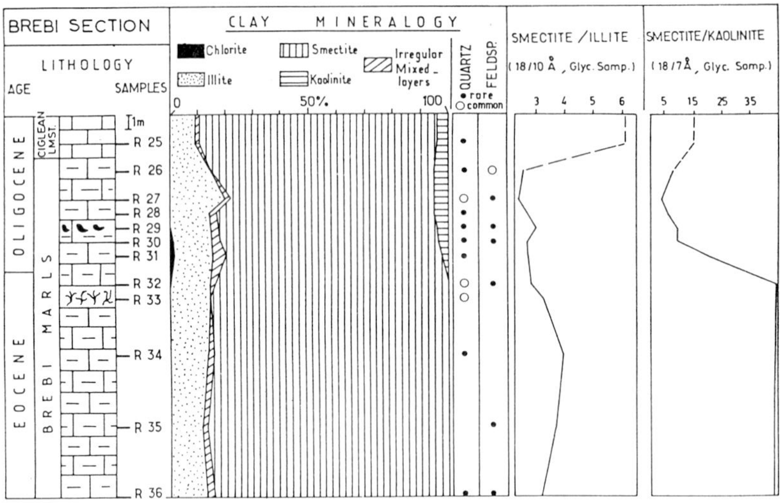

Fig. 2. Brebi section, Rumania - Clay mineralogy at the Eocene-Oligocene boundary.

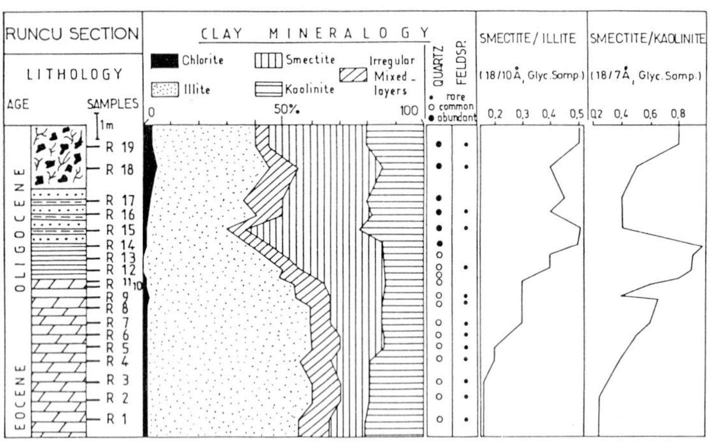

Fig. 3. Runcu section, Rumania. Clay mineralogy at the Eocene-Oligocene boundary.

5 CONCLUSION

The clay mineral successions recorded at the Eocene-Oligocene transition in various sea and land sections express a world cooling, inserted in the general Cenozoic glacial history, and caused by a general increase in physical alteration processes on land-masses. This cooling, moderately marked when compared to some ones occurring later on (Miocene, Pliocene), seems to be associated with a more regular annual humidity and an increase of the North-South marine circulation in the ocean. The climatic change may be hidden by regional pecularities, such as tectonic activity, volcanism or active diagenesis.

The climatic change is gradual, and only the sections marked by a sedimentary gap give the impression of a strong break. The phenomenon usually starts in the latest Middle Eocene-earliest Late Eocene. At the Eocene-Oligocene biostratigraphic boundary properly, the mineralogical evolution can be parallel, opposite or unrelated to the global one, according to the sections considered. The more detailed is the scale, the more local are the paleoenvironmental influences and significance. From a clay mineralogical point of view, the Eocene-Oligocene boundary is a transition rather than a sharp limit.

ACKNOWLEDGMENTS

I strongly acknowledge C. Pomerol for conceding the mineralogical study of materials he sampled in various continental sections, and A.-M. Karpoff and C. Robert for very useful data and scientific exchanges.

REFERENCES

Chamley, H., 1979. North Atlantic clay sedimentation and paleoenvironment since the late Jurassic. In : M. Talwani, W. Hay and W. B. F. Ryan (Editors), Deep Drill. Res. Atlantic Ocean : Continental Margins and Paleoenvironment. M. Ewing Ser., AGU, 3, pp. 342-361.

Chamley, H., 1981. Long-term trends in clay deposition in the ocean. Oceanol. Acta, sp. nb, pp. 105-110.

Desprairies, A., 1981. Authigenic minerals in the volcanogenic sediments, Leg 60. Init. Repts Deep Sea Drill. Proj. (U.S. Governm. Print. Off.), pp. 455-466.

Holtzapffel, T., Bonnot-Courtois, C., Chamley, H., and Clauer, N., 1985. Héritage et diagenèse des smectites du domaine sédimentaire nord-atlantique (Crétacé, Paléogène). Bull. Soc. géol. France, (8), 1, pp. 25-33.

Karpoff, A.-M., 1984. Miocene red clays of the South Atlantic : dissolution facies of calcareous oozes at Deep Sea Drilling Project Sites 519 to 523, Leg 73. Init. Repts Deep Sea Drill. Proj. (U.S. Governm. Print. Off.), 73, pp. 515-535.

Karpoff, A.-M., Bourbon, M., Ancel, B. and Graciansky P. C. de, 1984. Diagenetic polymetallic crusts at Sites 550 and 548 of Leg 50, Northeastern Atlantic Ocean. Init. Repts Deep Sea Drill. Proj. (U.S. Governm. Print. Off.) 80, pp. 823-844.

Latouche, C. and Maillet, N., 1985. Minéraux argileux et évolution des environnements sédimentaires du domaine atlantique nord-oriental durant le Paléogène. Bull. Soc. géol. France, (7), 22, pp. 757-761.

Robert, C., 1982. Modalité de la sédimentation argileuse en relation avec l'histoire géologique de l'Atlantique Sud. Thèse, Sci. nat., Aix-Marseille 2, 169 pp.

Robert, C., Stein, R. and Acquaviva, M. (in press). Cenozoic history of the New-Zealand sector of the Southwest Pacific, DSDP Leg 90. Init. Repts Deep Sea Drill. Proj. (U.S. Governm. Print. Off.), 90.

EVOLUTION OF THE TETHYAN SEAWAYS AND IMPLICATIONS FOR THE OCEANIC CIRCULATION AROUND THE EOCENE-OLIGOCENE BOUNDARY.

by L.E. RICOU, B. MERCIER de LEPINAY and J. MARCOUX

Laboratoire de Géologie Structurale et UA 215 CNRS, 4 pl. Jussieu, T. 26, 75230 PARIS Cedex 05 (FRANCE).

Abstract.

The geometric evolution of the Tethyan basins during the Eocene-Oligocene period is reviewed in order to check how far their closure could have modified the circulation of oceanic water and thus participated to the Eocene/Oligocene events. At that time the Himalayan Tethys is already closed but the corresponding seaway is simply shifted South of India. A possible major change in the Caribbean Tethys is poorly documented. The critical area is the Mediterranean Tethys where fast-going collision between Africa and Eurasia is likely to have put an end to the circumterrestrial low-latitude circulation which existed since the breakup of the Pangea.

This seaway between the Indian and Atlantic oceans was not completely cut off before the Miocene but suffered a drastic narrowing during the Eocene which reduced the westward flux of warm water to a non-significant amount. Moreover it is suggested that, due to intense evaporation in the Mediterranean Sea, an intermediate stage took place during which water from both oceans flowed toward the Mediterranean.

INTRODUCTION.

The study of the Eocene/Oligocene events, either progressive or catastrophic (Pomerol, 1985) cannot be separated from the study of the reorganisation of the continental and oceanic areas which occured simultaneously. As pointed out by Shackleton and Kennett (1975) and Haq (1981) the important climatic change which occured at the same time must be linked with a major modification of the oceanic currents.

During the Cretaceous, the thermal gradient was low and relatively warm water reached high latitudes. The breakup of the Pangea had created an East-West seaway which encircled the Earth along the low latitudes, the Tethyan ocean. During the Cenozoic, this seaway became interrupted between India and Eurasia, Africa and Eurasia, South America and North America, while the Atlantic ocean opened northward and a circum-antarctic circulation was established. The circulation of warm water at the low latitudes is now restricted to separated basins. This present-day organisation of the oceanic currents, which is accompagnied by formation of ice caps and migration of cold water away from the high latitudes appeared around the Eocene-Oligocene boundary (Shackleton and Kennett, 1975; and Haq, 1981).

It is clear that the disappearance of the Tethyan seaway played a major role in the process but this disappearance was a multi-step phenomenon which began long before the Eocene-Oligocene boundary and was completed only a long time after.

It is thus necessary to check which step could have played the major role and to review the geometrical history of the three segments of the seaway which have been closed successively, i.e. the Himalayan, the Mediterranean and the Caribbean parts of the Tethys.

THE HIMALAYAN TETHYS.

The closure of this seaway is clearly older than the Eocene-Oligocene events. The paleomagnetic data favor a collision since the late Paleocene (Patriat and Achache, 1984). The speed of relative motion between India and Eurasia was drastically reduced around anomaly 23 (52 Ma) and this is taken as evidence for a complete collision (Besse et al., 1984). Moreover, the stratigraphic data show that marine deposits ceased at the end of Early Eocene (Van Haver et al., 1984).

However, this closure of the Himalayan seaway was compensated by the coeval widening of the Indian ocean. An east-west seaway at the low latitude still persisted. It had simply shifted South of India. It is thus unlikely that this Himalayan collision could have destroyed the Tethyan type of water exchange which could still persist south of India and reach the seaway between Africa and Eurasia.

THE MEDITERRANEAN TETHYS.

From the western Mediterranea to Iran, the whole area is affected during the Eocene-Oligocene epoch by strong compressive tectonics, while the speed of relative motion between Africa and Eurasia neatly slows down, the overall situation indicating a general collision (Ricou et al., 1985).

More precisely, the Oligocene mean speed of relative motion is reduced by a factor of 2 to 3 with respect to the Eocene mean speed; moreover the motion became strongly oblique and the normal component of the relative speed has been reduced by a factor of 5 to 10 (Savostin et al., 1985). It means that the major narrowing of the seaway occured during the Eocene.

However, a marine connection allowing faunal exchange with the Indian ocean persisted till the early Miocene (Adams et al., 1983) while the connection between the Mediterranean Sea and the Atlantic ocean was interrupted only during the Messinian. A set of paleogeographic maps illustrating the alpine evolution of the Mediterranean Tethys (Dercourt et al., 1985) allows a more precise approach.

At the beginning of the Cenozoic (fig.1) three major groups of basins are fully interconnected: the triangular remnant of the Triassic Tethys still widely open eastward, the south-european marginal basins (the future miocene Paratethys) and the Pre-Mediterranean basins. An oblique barrier extends from the Alps to the Arabian promontory but it is largely submerged and the submersion will increase during the Eocene transgressions. Being very long (more than 3000 km) this submerged barrier leaves enough space for a significant water flow over it. The system communicates with the Atlantic ocean through a deep and 400 km wide passage established upon an old (Jurassic) oceanic crust.

The major part of the system is situated in the trade-wind latitudes and a westward flow toward the Atlantic is likely (Berggreen and Hollister, 1974; Arthur and Jenkins, 1981; Pomerol, 1985). The Maestrichtian to Eocene phosphate deposits of NW Africa appear linked through upwelling to this Mediterranean westward current (Arthur and Jenkins, 1981). The same remark can apply to the coeval phosphates of Lybia and Syria, all located along the upwelling flank (i.e. left in the Northern Hemisphere) of the wind-driven current.

The Eocene rapid narrowing led to considerable changes in the central and eastern parts of the system. At the Eocene-Oligocene boundary (fig. 2) the submerged oblique barrier has evolved into a

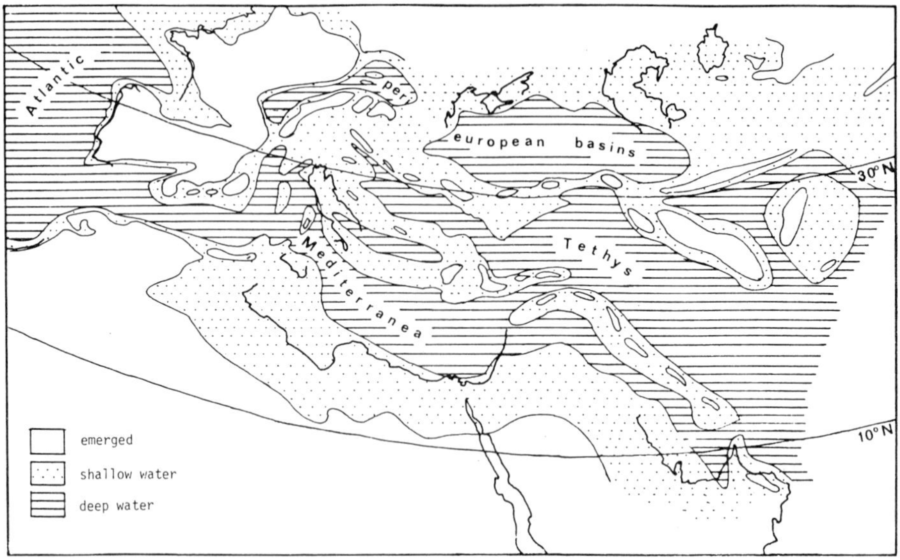

Fig. 1 Lowermost Tertiary (after Dercourt et al. 1985)

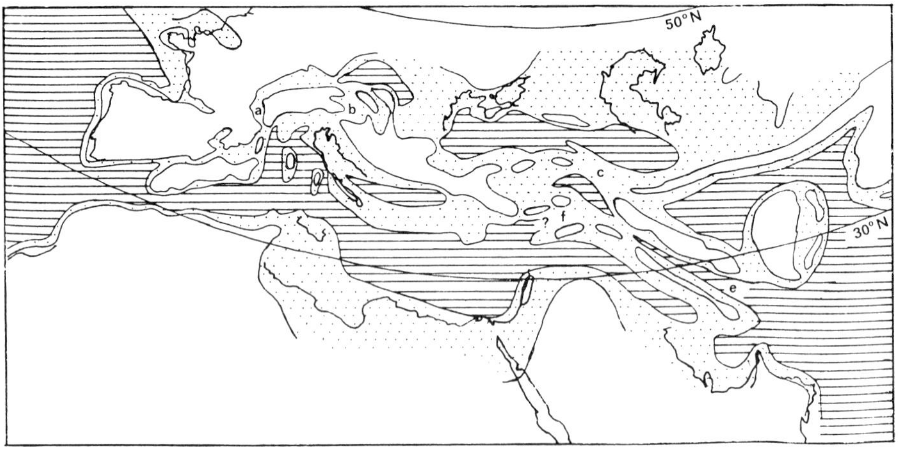

Fig. 2 Eocene - Oligocene (ibid.)

mountain chain. The Paratethys is almost isolated, the remaining outward links beeing: a) in the French Alps, along a filling-up flysch basin; b) across the Eastern Alps, through the shallow and discontinuous slovenian seaway; c) south of the Black sea and Caspian sea, across an arc of volcanic isles. Data on the basin itself led Baldi (1984) to date back its first isolation at the very beginning of the Oligocene. The Mediterranean basins are still linked with the Atlantic ocean but their links with the Indian ocean are now restricted to the north arabian passage.

This passage has been described by Bizon et al. (1972) as two paralllel seaways separated by a longitudinal high showing frequent emersions. The southern seaway was established upon the Arabian Platform. At the beginning of the Tertiary it showed an axial low (low-energy mudstone deposits) bordered by two shoal zones (high-energy limestones). Submitted to progressive regression, its two shoals merged together on the Qatar transverse at the beginning of the Oligocene and the marine connection disappeared at the beginning of the Miocene. The history of the northern seaway is more difficult to decipher as large parts of it have disappeared below Neogene thrusts. Along its whole length Eocene marine deposits and transgressive lower Miocene limestones followed by a poorly dated flysch are known. In between, a pelagic Oligocene is documented in its iranian part (e on fig. 2), but no marine Oligocene is known from its turkish part (f on fig. 2). From these data we deduce that most probably any Oligocene marine exchange between the Mediterranea and the Indian ocean had to flow over the shoals of the southern seaway. The water flow was certainly drastically lower during the Oligocene than during the Eocene. These data furnish only a general frame and we cannot point out the precise moment (if any) when the reduction of the Indian-Atlantic circulation through the Mediterranean Tethys led to a renewal of the global oceanic circulation. Probably any eustatic drop of the sea level would have precipitated the event. Moreover, this reduction probably led to a deficit of water in the Mediterranean basins and thus to pre-closure stage during which the Tethyan westward circulation would have been replaced by a double input from both Atlantic and Indian oceans. Actually, these basins were situated at the latitude of the present Sahara (20°-30°N) and we must assume a strong evaporation while the input from the Paratethys basins and the Indian ocean was strongly reduced. The exchange with the Atlantic ocean would have then foreshadow the present situation notwithstanding the persisting marine links with the Indian ocean.

THE CARIBBEAN TETHYS.

The Caribbean seaway between the Pacific and Atlantic oceans lasted till the Pliocene (Keigwin, 1978). For the Eocene-Oligocene epoch, all the cinematic reconstructions (e.g. Burke et al., 1984) show that the total width between South and North Americas was reduced by a factor of 2. This only reduction could not lead to a significant drop of water exchange. What is probably more relevant to our problem is the middle-late Eocene birth of the volcanic outer arc of Lesser Antilles which reached a noticeable development during the Early Oligocene (Martin-Kaye, 1969) . It probably constituted a barrier, at least for deep water circulation. However, our knowledge is insufficient to discard the hypothesis of eventual breaks through this barrier.

CONCLUSION.

Accepting after Haq (1981) that the Eocene/Oligocene boundary event is linked with a reorganisation of the water masses in the world ocean, this review shows in turn how the tectonic evolution of the Tethyan basin played an important role in this reorganisation.

As a matter of fact, there are two major events during the alpine history of the Tethys for what concerns the oceanic circulation: the breakup of Pangea which allowed a Tethyan circulation encircling the globe at the low latitudes; the closure of the Tethyan seaways which put an end to this circulation.

The eocene closure of the Himalayan Tethys could not play this role as it was compensated by the widening of the Indian Ocean south of India. There are some indications that the Carribean Tethys could have been dammed at the end of the Eocene by a transverse volcanic arc, but the efficiency of this barrier remains questionable.

The Tethyan circulation has been cut off in the Mediterranean Tethys between Africa and Eurasia and we can propose the following scenario: 1) during the Eocene an active circulation from the Indian ocean towards the Atlantic ocean through the Mediterranean basins participated through upwelling to the important phosphate deposits; 2) while shoals progressively invaded the Eastern passage, a double input towards the strongly evaporating Mediterranean basin took place; 3) the marine links through the eastern passage disappeared during the Lower Miocene. The transition between 1) and 2) marks the death of the Tethyan circulation. The geometrical analysis of the system cannot by itself tell us precisely when this event took place. It can be bracketted between the Middle Eocene and the Early Oligocene as the compared situations for these two epochs are drastically different and as the fastest narrowing evolution of the systemm occurred during the Eocene.

REFERENCES.

Adams, C.G., Gentry A.W. and Whybrow, P.J., 1983. Dating the terminal Tethyan event. In:J.E. MeulenKamp (Ed.)- Reconstruction of Marine paleoenvironments. Utrecht Micropal. Bull. 30: 273-298.

Arthur, M.A. and Jenkyns, H.C., 1981. Phosphorites and paleoceanography. Oceanol. Acta, 1981, Proceedings 26 th. International Congress, Geology of oceans symposium, Paris, July 7-17 1980: 83-90.

Baldi, T., 1984. The terminal Eocene and Early Oligocene events in Hungary and the separation of an anoxic, cold Paratethys. Eclogae geol. Helv., v.30,N°1: 1-27.

Bizon, G., Bizon, J.J. and Ricou, L.E., 1972. Etude stratigraphique et paléogéographique des formations tertiaires de la région de Neyriz (Fars interne. Zagros iranien). Rev. Inst. Fr. Pétr., 27 (3): 369-405.

Berggren, W.A. and Hollister, C.D., 1974. Paleogeography, Paleobiogeographiy and the history of Circulation in the Atlantic Ocean. Soc. Econ. Paleont. Mineral. Special Publ., n° 20: 126-186.

Besse, J., Courtillot V., Pozzi, J.P., Westphal, M. and Zhou, Y.X., 1984. Paleomagnetic estimates of crustal shortening in the Himalayan thrusts and Zangbo suture. Nature, v. 311, n° 5987: 621-626.

Burke, K., Cooper, C., Dewey, J.F., Mann, P., Pindell, J.L., 1984. Caribbean tectonics and relative plate motions. Geol. Soc. of Amer. Memoir n° 162: 31-63.

Dercourt, J., Zonenshain, L.P., Ricou, L.E., Kazmin V.G., Le Pichon, X., Knipper, A.L., Grandjacquet, C., Sborshchikov,I.M., Boulin, J., Sorokhtin, O., Geyssant, J., Lepvrier, C., Biju-Duval, B., Sibuet J.C., Savostin, L.A., Westphal, M. and Lauer, J.P., 1985. Présentation de 9 cartes paléogéographiques au 1/20.000.000 e s'étendant de l'Atlantique au Pamir pour la période du Lias à l'Actuel. Bull.Soc. géol. Fr., (8), t. I, n° 5: 637-652

Haq, B.U., 1981. Paleogene paleoceanography: Early Cenozoic oceans revisited. Oceanol. Acta, 1981, Proceedings 26 th International Congress, Geology of oceans symposium, Paris, July 7-17 1980: 71-82.

Keigwin, Jr., 1978. Pliocene closing of the Isthmus of Panama, based on biostratigraphic evidence from nearby Pacific Ocean and Caribbean Sea cores. Geology, v. 6/10: 630-634.

Martin-Kaye, P.H.A., 1969. A summary of the Geology of the Lesser Antilles. Overseas Geol. and Mineral Res. 10 (2): 172-206.

Patriat, P. and Achache J., 1984. India-Eurasia collision chronology has implications for crustal shortening and driving mechanism of plates. Nature, v. 311, n° 5987: 615-621.

Pomerol, Ch., 1985. La transition Eocene-Oligocene est-elle un phénomène progressif ou brutal? Bull. Soc. géol. Fr., (8), t. I: 263-267.

Ricou, L.E., Zonenshain L.P., Dercourt, J., Kazmin, G.V., Le Pichon, X., Knipper, A.L., Grandjacquet, C., Sborshchikov, I., Geyssant, J., Lepvrier, C., Pechersky, D.M., Boulin, J., Sibuet, J.-C., Savostin, L.A., Sorokhtin, O., Westphal, M., Bazhenov, M., Lauer,J.-P., and Biju-Duval, B., 1985. Méthodes pour l'établissement de neuf cartes paléogéographiques de l'Atlantique au Pamir depuis le Lias. Bull. Soc. géol. Fr., (8), t. I, n° 5 : 625-635.

Savostin, L.A., Sibuet, J.C., Zonenshain, L.P., Le Pichon, X. and Roulet, M.J., 1985. Klimatic evolution of Tethys Belt from the Atlantic ocean to Pamir since Triassic. Tectonophysics, sous presse.

Shackleton, N.J., Kennett, J.P., 1975. Paleotemperature history of the Cenozoic and the initiation of Antartic glaciation: oxygene and carbon isotope analysis in DSDP Sites 277, 279 and 281. In "Initial Reports Deep Sea Drilling Project", v. 29, ed. by J.P. Kennett,R.E. Houtz et al., Washington, US Gov. Print. Off.: 743-755.

Van-Haver, Th., Bassoullet, J.P., Blondeau, A. and Mascle, G., 1984. Les séries détritiques du Bassin de l'Indus au Ladakh: Nouvelles données stratigraphiques et structurales. Riv. it. Paleont. Strat., v. 90, n°1: 87-102.

LATE EOCENE MICROTEKTITES AND CLINOPYROXENE-BEARING SPHERULES

B. P. GLASS

Geology Department, University of Delaware, Newark, DE 19716

ABSTRACT

Late Eocene microtektites, belonging to the North American tektite strewn field, have been found in the Gulf of Mexico, Caribbean Sea, on Barbados, and in the western Atlantic. The North American microtektite layer does not appear to be associated with any major extinctions or climatic changes; however, a slightly older layer ($\sim$10,000 yrs. older) of clinopyroxene-bearing spherules appear to be associated with an iridium anomaly and the extinction of several radiolarian species. The clinopyroxene-bearing spherules have been found in the eastern Indian Ocean and equatorial Pacific, as well as, the Gulf of Mexico and Caribbean Sea. The North American tektites and microtektites have K-Ar, Ar-40/Ar-39, and fission-track ages of 34 to 35 m.y. Extrapolation from the microtektite layer to the overlying Eocene/Oligocene boundary indicates an age of 33-34 m.y. for the Eocene/Oligocene boundary.

INTRODUCTION

Microtektites are microscopic (<1 mm dia.) tektites (Glass, 1967; Glass et al., 1979). Tektites are glass bodies similar in appearance and composition to volcanic glass (obsidian), but can be distinguished from volcanic glass by their low water content, high FeO/Fe_2O_3 ratio, lack of microliths, and presence of lechatelierite (SiO_2 glass) (Chao, 1963). Most investigators believe that tektites are terrestrial surficial deposits melted and splashed out by an impact event (e.g., Taylor, 1973; King, 1977); however, some authors (e.g., O'Keefe, 1976) support a lunar volcanic origin.

Tektites are found scattered over regions of the Earth's surface called strewn fields. Tektites found in Texas and Georgia have similar ages and compositions and are believed to belong to a single strewn field--the North American. The North American tektites are not found *in situ*, but field studies suggest they weathered out of the late Eocene Jackson Formation (Barnes, 1951; King, 1962). K-Ar, ^{40}Ar-^{39}Ar, and fission-track dating indicate that the North American tektites were formed $\sim$ 34 m.y. (e.g., Zähringer, 1963; Storzer and Wagner, 1971; Bottomley et al., 1977).

North American microtektites have been found in late Eocene deep-sea deposits. A slightly older layer of clinopyroxene-bearing spherules also occurs in deep-sea deposits, but there are no published data that support the hypothesis of multiple late Eocene tektite events.

NORTH AMERICAN MICROTEKTITES

Geographic Distribution and Mass

Late Eocene microtektites have been found in the Gulf of Mexico, Caribbean Sea, western equatorial Atlantic, and on Barbados (Donnelly and Chao, 1973; Glass et al., 1973; Glass and Zwart, 1979; Saunders et al., 1984; Sanfilippo et al., 1985) (Table 1). The total mass of late Eocene microtektites is estimated to be $\sim 10^9$ metric tons (Glass and Zwart, 1979).

TABLE 1
Late Eocene microtektite and clinopyroxene-bearing spherule occurrences.

Site*	Location	Depth**	Reference ±	Comments
		North American Microtektites		
DSDP 94	Gulf of Mexico	14-1, 109 cm	1	reworked
DSDP 94	Gulf of Mexico	15-3, 110 cm	2, 3	
DSDP 149	Caribbean Sea	31-1, top	2, 3	
DSDP 543	western Atlantic	27?	4	no published data
E67-128	Gulf of Mexico	1581.9-1587.8 m	1, 5	rare, scattered
RC9-58	Caribbean Sea	254 cm	6, 7	
Barbados	western Atlantic	-------	4, 8	
		Clinopyroxene-bearing Spherules		
DSDP 65	equatorial Pacific	15	5	rare, scattered
DSDP 69A	equatorial Pacific	9-5, 30 cm	3	
DSDP 70A	equatorial Pacific	27-3, 30 cm	3	rare
DSDP 94	Gulf of Mexico	15-3 or 4	5, 9	layer ill-defined
DSDP 149	Caribbean Sea	31-1, 5 cm	9	
DSDP 161A	equatorial Pacific	10-5 & 6	3	rare, scattered
DSDP 162	equatorial Pacific	4	5	scattered
DSDP 166	equatorial Pacific	12-6, 45 cm	3	
DSDP 167	equatorial Pacific	28-1, 35 cm	3	
DSDP 216	eastern Indian Ocean	16-2, 5 cm	3	
DSDP 292	western Pacific	36	1, 3	reworked
DSDP 292	western Pacific	38-2, 80 cm	1, 3	peak at 35 cm according to (1)
RC9-8	Caribbean Sea	280 cm	7	

* DSDP = Deep Sea Drilling Project Sites; E67-128 = Shell Oil Company core; RC9-58 = Lamont-Doherty core

** Depth of Late Eocene microtektite or clinopyroxene-bearing spherule layer. Depth for DSDP sites indicated by core-section, depth (cm)

± 1) Keller et al. (1983), 2) Glass and Zwart (1979), 3) Glass and Crosbie (1982), 4) Sanfilippo et al. (1985), 5) This paper, 6) Glass et al. (1973), 7) Glass et al. (1982), 8) Saunders et al. (1984), 9) Zwart (1977)

Identification

The late Eocene microtektites found in the Gulf of Mexico, Caribbean Sea, and on Barbados have major oxide compositions similar to each other and to the North American tektites found in Texas and Georgia (Donnelly and Chao, 1973; Glass et al., 1973; Glass and Zwart, 1979; Sanfilippo et al., 1985). Ngo et al. (in press) found that the Barbados microtektites (and tektite fragments) have Sm and Nd isotopic ratios indistinguishable from North American tektites. Furthermore, the Barbados tektite fragments have a ^{40}Ar-^{39}Ar age that is, within statistical error, the same as the K-Ar and ^{40}Ar-^{39}Ar age of the North American tektites (York and Hall, personal communication, 1985). Thus, based on the above data, it appears that the late Eocene microtektites belong to the North American tektite strewn field.

Implications for Age of the Eocene/Oligocene Boundary

Based on data published in the Initial Reports of the Deep Sea Drilling Project, the sediments containing the North American microtektite layer occur in the *Thyrsocyrtis bromia* radiolarian Zone. Although calcareous microfossils are absent or poorly preserved at most of the North American microtektite-bearing sites, it appears that the North American microtektite layer occurs within the *Discoaster barbadiensis* calcareous nannofossil Zone or at the base of the *Isthmolithus recurvus* or CP15b calcareous nannofossil Zone (Glass and Crosbie, 1982; Keller et al, 1983). It also appears to be at the top of the P15 or *Globigerinatheka semiinvoluta* planktic foraminifera Zone (Glass and Crosbie, 1982; Keller et al., 1983); however, at Bath Cliff, Barbados, it appears to be within the *Turborotalia cerroazulensis* Zone rather than the *G. semiinvoluta* Zone (Saunders et al., 1984). Thus, it is clear from the biostratigraphy that the North American microtektite layer occurs in sediments of late Eocene age.

One of the most widely accepted Paleogene time scales is that by Hardenbol and Berggren (1978). According to these authors the Eocene/Oligocene boundary has an age of 37 m.y. Therefore, since the North American microtektite layer occurs in late Eocene deposits, it should be older than 37 m.y.; but North American tektites are well-dated at 34-35 m.y. based on the K-Ar, ^{40}Ar-^{39}Ar, and fission-track methods (e.g., Zähringer, 1963; Storzer and Wagner, 1971; Bottomley et al., 1977).

Glass and Crosbie (1982) estimated an age of 32.5 ± 0.9 m.y. for the Eocene/Oligocene boundary by extrapolation from the North American microtektite layer using sediment accumulation rates. More recently, tektite fragments recovered from the North American microtektite layer at Bath Cliff, Barbados, have been dated at 35.5 ± 0.4 m.y. using ^{40}Ar-^{39}Ar (York and Hall, personal communication, 1985). At Bath Cliff the North Americn microtektite layer is $\sim$26 m below the Eocene/Oligocene boundary (Saunders et al., 1984). According to Saunders et al. (1984), the late Eocene section at Bath Cliff has a sediment accumulation rate of 27 m/m.y. This would indicate an age of $\sim 34.5 \pm 0.4$ m.y. for the Eocene/Oligocene boundary. This is older than the age of

32.5 m.y. proposed by Glass and Crosbie (1982), but still significantly younger than the age of 37 m.y. proposed by Hardenbol and Berggren (1978).

MULTIPLE LATE EOCENE MICROTEKTITES HORIZONS?

Keller et al. (1983) have suggested that there were multiple late Eocene tektite events based on their discovery of multiple "microtektite" horizons. However, Keller et al. (1983) did not present the data to support such a conclusion (Glass, 1984) nor did Keller et al. distinguish between microtektites and clinopyroxene-bearing (cpx) spherules. A reexamination of all the "microtektite-bearing" sites reported by Keller et al. (1983) (except for St. Stephen quarry, Alabama) indicates that there is only one microtektite layer and it is part of the North American tektite strewn field. Later occurrences, or horizons, appear to be due to reworking from the North American microtektite layer or the cpx spherule layer (Glass, 1984; Glass et al., in preparation).

CLINOPYROXENE-BEARING SPHERULES (CPX SPHERULES)

Geographic Occurrence and Mass

Glassy spherules containing clinopyroxene have been found closely associated with the North American microtektites at two sites in the Caribbean Sea (RC9-58 and DSDP 149) and at one site in the Gulf of Mexico (DSDP 94) (John and Glass, 1974; Glass and Zwart, 1979) (Table 1). At these two DSDP sites the microtektites and cpx spherules are intermixed, but in core RC9-58 from the Caribbean Sea, the cpx spherules occur in a separate layer with a peak abundance $\sim$25 cm below the peak abundance of the microtektites (Glass et al., 1982). Cpx spherules have also been found in several equatorial Pacific sites and in an eastern equatorial Indian Ocean site. The cpx spherule layer thus appears to extend at least half-way around the earth; and Glass et al. (in preparation) estimate a total mass of $\sim 10^9$ metric tons for the spherules in the cpx layer.

Description and Origin

The cpx spherules are generally opaque and smaller than the North American microtektites. Some of the spherules in the cpx layer have a missing phase (or phases). In some cases only a glassy groundmass remains. Some of the spherules appear to be completely glassy. However, most of the glassy spherules in the cpx layer have compositions similar to the cpx spherules. The spherules in the cpx layer can generally be distinguished from the North American microtektites by their higher MgO and CaO and lower TiO_2 and Al_2O_3 contents for a given SiO_2 content (Glass et al., in preparation).

Because of the presence of crystalline phases in the cpx spherules, they should not be referred to as microtektites, which by definition are essentially devoid of crystalline material. However, their widespread distribution and unusual composition indicate that they, like the microtektites, were probably formed by an impact event. The differences in crystallinity between the cpx spherules and microtektites is probably due to differences in cooling history and/or composition.

Associated Iridium Anomaly

An iridium anomaly is associated with the cpx spherule layer (Ganapathy, 1982; Asaro et al., 1982; Alvarez et al., 1982; Glass et al., 1982). The associated iridium anomaly provides further support for the impact origin of these spherules.

Associated Radiolarian Extinctions

The extinction of several species of Radiolaria (e.g., Thrysocyrtis bromia, T. tetracantha, T. rhizodon, Calocyclas turris) is closely associated with the North American microtektite/cpx spherule layer in the Caribbean and Gulf of Mexico sites (Maurrasse and Glass, 1976; Glass and Zwart, 1977; Glass and Zwart, 1979). These same species became extinct near the cpx spherule layer in the Pacific and Indian Ocean sites (Glass and Crosbie, 1982). Detailed work at Barbados shows that the extinctions occur below the North American microtektite layer (Sanfilippo, 1985). No cpx spherules have been found at the Barbados sites, but an iridium anomaly is associated with the radiolarian extinctions. This suggests that the extinctions are associated with the cpx spherule layer, rather than with the North American microtektite layer as previously thought.

Age

The cpx spherule layer is closely associated with the North American microtektite layer in the Caribbean and Gulf of Mexico and therefore occurs in the same biostratigraphic interval. In core RC9-58 from the Caribbean Sea, the cpx layer is $\sim$ 25 cm below the North American microtektite layer; but the sediment accumulation rate for that core is not known, and, therefore, the age difference cannot be determined. The cpx spherules are not found on Barbados, but an iridium anomaly indicates where the cpx layer should have been. Their absence in the Barbados sections is probably due to destruction by groundwater solution. At Bath Cliff the and iridium anomaly occurs $\sim$ 27 cm below the peak abundance of North American microtektites. The sediment accumulation rate for the late Eocene section at Bath Cliff is estimated to be $\sim$ 27.1 m/m.y. (Saunders et al., 1984). Thus the age difference between the North American microtektite layer and the iridium anomaly and, therefore, the cpx layer is $\sim$ 10,000 yrs. [Sanfilippo et al. (1985) used a sediment accumulation rate of 20 m/m.y. and obtained an age difference of 13,500 m.y.]

DISCUSSION AND CONCLUSION

At the present time there is evidence for only one late Eocene microtektite layer and it belongs to the North American tektite strewn field. Later occurrences in the late Eocene appear to be due to reworking. Several authors have suggested (e.g., O'Keefe, 1980; Glass, 1982; Ganapathy, 1982; Alvarez, 1982) that the North American tektite event might have been associated with mass extinctions and/or climatic changes at the end of the Eocene; however, at the present time there is no evidence in the deep-sea record for extinctions or climatic changes associated with the North American microtektite layer.

Clinopyroxene-bearing spherules (cpx spherules) are found closely associated with the North American microtektites in the Caribbean and Gulf of Mexico; however, in core RC9-58 from the Caribbean, they occur in a layer about 25 cm below the North American microtektite layer. Cpx spherules are also found across the equatorial Pacific and into the eastern Indian Ocean. The cpx spherule layer is associated with the extinction of several species of Radiolaria and with an iridium anomaly. The cpx spherule layer has not been found at any of the Barbados sites; however, based on an iridium anomaly it appears that the cpx spherule layer should have been $\sim$ 27 cm below the peak abundance of North American microtektites at Bath Cliff. This indicates that the cpx spherule event preceded the North American tektite event by about 10,000 yrs. There is no evidence for a climatic change at this time, but no detailed studies have been made across either the cpx or the North American microtektite layer.

REFERENCES

Alvarez, W., Asaro, F., Michel, H. V. and Alvarez, L. W., 1982. Iridium anomaly approximately synchronous with terminal Eocene extinctions. Science, 216, 886-888.

Asaro, F., Alvarez, L. W., Alvarez, W. and Michel, H. V., 1982. Geochemical anomalies near the Eocene-Oligocene and Permian-Triassic boundaries. Geol. Soc. Am. Spec. Papers, 190, 517-528.

Barnes, V. E., 1951. New tektite areas in Texas. Bull. Geol. Soc. Am., 62, 1422 (abstract).

Bottomley, R. J., York, D. and Grieve, R. A. F., 1977. Possible source craters for the North American tektites - a geochronological investigation. Am. Geophys. Union Trans., 50, 309 (abstract).

Chao, E. C. T., 1963. In: J. A. O'Keefe (Editor), Tektites. University of Chicago Press, Chicago, pp. 51-94.

Donnelly, T. W. and Chao, E. C. T., 1973. Microtektites of late Eocene age from the eastern Caribbean Sea. Initial Rep. Deep Sea Drill. Proj., 15, 1031-1037.

Ganapathy, R., 1982. Evidence for a major meteorite impact on the Earth 34 million years ago: Implication for Eocene extinctions. Science, 216, 885-886.

Glass, B., 1967. Microtektites in deep-sea sediments. Nature, 214, 372-374.

Glass, B. P., 1982. Possible correlation between tektite events and climatic changes? Geol. Soc. Am. Spec. Papers, 190, 251-256.

Glass, B. P., 1984. Multiple microtektite horizons in upper Eocene marine sediments? Science, 224, 309.
Glass, B. P. and Crosbie, J. R., 1982. Age of Eocene/Oligocene boundary based on extrapolation from North American microtektite layer. Bull. Am. Assoc. Petrol. Geologists, 66, 471-476.
Glass, B. P. and Zwart, M. J., 1977. North American microtektites, radiolarian extinctions and the age of the Eocene/Oligocene boundary. In: F. M. Swain (Editor), Stratigraphic Micropaleontology of Atlantic Basin and Borderlands. Amsterdam, Elsevier, pp. 553-568.
Glass, B. P. and Zwart, M. J., 1979. North American microtektites in Deep Sea Drilling Project cores from the Caribbean Sea and Gulf of Mexico. Bull. Geol. Soc. Am., 90, 595-602.
Glass, B. P., DuBois, D. L. and Ganapathy, R., 1982. Relationship between an iridium anomaly and the North American microtektite layer in core RC9-58 from the Caribbean Sea. J. Geophys. Res., 87, 425-428.
Glass, B. P., Swincki, M. B. and Zwart, P. A., 1979. Australasian, Ivory Coast, and North American tektite strewn fields: size, mass, and correlation with geomagnetic reversals and other earth events. Proc. Lunar Planet. Sci. Conf. 10th, 2535-2545.
Glass, B. P., Burns, C. A., Crosbie, J. R. and DuBois, D. L., (in preparation). Submitted to Lunar Planet. Sci. Conf. 16th.
Glass, B. P., Baker, R. N., Storzer, D. and Wagner, G. A., 1973. North American microtektites from the Caribbean Sea and Gulf of Mexico. Earth Planet Sci Lett., 19, 184-192.
Hardenbol, J. and Berggren, W. A., 1978. A new Paleogene numerical time scale. AAPG Studies in Geology, 6, 213-234.
John, C. and Glass, B. P., 1974. Clinopyroxene-bearing glass spherules associated with North American microtektites. Geology, 2, 599-602.
Keller, G., D'Hondt, S. and Vallier, T. L., 1983. Multiple microtektite horizons in upper Eocene marine sediments: no evidence for mass extinctions. Science, 221, 150-152.
King, E. A., 1962. Field investigation of Georgia tektites and description of new specimens. Georgia Mineral Newsletter, 15, 84-89.
King, E. A., 1977. The origin of tektites: A brief review. Am. Scientist, 64, p. 212-218.
Maurrasse, F. and Glass, B. P., 1976. Radiolarian stratigraphy and North American microtektites in Caribbean core RC9-58: Implications concerning the age of the Eocene-Oligocene boundary. Proc. 7th Caribbean Geol. Conf., 205-212.
Ngo, H. H., Wasserburg, G. J. and Glass, B. P., in press. Nd and Sr isotopic composition of tektite material from Barbados and their relationship to North American tektites. Geochim. Cosmochim. Acta.
O'Keefe, J. A., 1976. Tektites and Their Origin. Elsevier, Amsterdam, 254 pp.
O'Keefe, J. A., 1980. The terminal Eocene event: formation of a ring system around the earth? Nature, 285, 301-311.
Sanfilippo, A., Riedel, W. R., Glass, B. P. and Kyte, F. T., in press. Late Eocene microtektites and radiolarian extinctions on Barbados, Nature, 314, 613-615.
Saunders, J. B., Bernoulli, D., Müller-Merz, E., Oberhansli, H., Perch-Neilsen, K., Riedel, W. R., Sanfilippo, A. and Torrini, Jr., R., 1984. Stratigraphy of the late Middle Eocene to Early Oligocene in the Bath Cliff section, Barbados, West Indies, Micropaleontology, 30, 390-425.
Storzer, D. and Wagner, G. A., 1971. Fission-track ages of North American tektites. Earth Planet. Sci. Lett., 10, 434-444.
Taylor, R., 1973. Tektites: A post-Apollo view. Earth Sci. Rev. 9, 101-123.
Zähringer, J., 1963. K-Ar measurements of tektites. In: Radioactive Dating: Proc. Symposium Athens, Nov. 19-23, 1962. International Atomic Energy Agency, Vienna, pp. 389-305.
Zwart, M. J., 1977. North American microtektites From Deep Sea Drilling Project cores, M. S. Thesis, 74 pp., University of Delaware, Newark.

LATE EOCENE IMPACT EVENTS AND STEPWISE MASS EXTINCTIONS

GERTA KELLER

Department of Geological and Geophysical Sciences, Princeton University, Princeton, NJ 08544

ABSTRACT

Late Eocene extinctions are neither catastrophic nor mass extinctions, but occur stepwise in a sequence of four steps over a period of 3.4 million years. Closely associated with two of the stepwise extinction events are three microtektite horizons; one in the upper part of *Globigerapsis semiinvoluta* Zone at 38.2 Ma, and two closely spaced layers in the lower part of *Globorotalia cerroazulensis* Zone at 37.3 and 37.2 Ma. Species extinctions and relative species abundance declines are closely associated with microtektite layers and suggest, but do not prove, a cause-effect relationship between impact events and some of the stepwise mass extinctions.

INTRODUCTION

Paleontological research has made it increasingly clear that both evolution and climatic changes are characterized by long periods of stability separated by brief episodes of rapid faunal turnover and climatic oscillations. The late Eocene represents such an episode but, contrary to repeated claims of mass extinctions at the Eocene/Oligocene boundary this faunal turnover occurs in a series of discrete steps over a period of about 3 million years beginning in the late Eocene and culminating in the early Oligocene. Each step is characterized by successive replacement of tropical marine faunas and floras by cooler subtropical and temperate elements as observed by Haq and Lohmann (1976), Kennett (1977, 1978), Keller (1983a,b) and Corliss and Keigwin (in press).

What makes the late Eocene different from other faunal turnovers, however, is the close association of at least three impact events with this stepwise faunal turnover. What effect did these impact events have on planktonic marine faunas? Is there a cause-effect relationship between impact events and faunal turnovers? Are climatic changes triggered by impact events?

Not all workers agree, however, that there is evidence of multiple impact events in the form of microtektites in late Eocene sediments. Glass and coworkers (this volume, in press) believe that only one microtektite layer exists in late Eocene sediments. Their argument rests partly on the definition of microtektites and partly on disagreement of stratigraphic position and the geochemical nature of microtektites.

The first part of this report will address the controversy of multiple late Eocene impact events. Due to space limitations, however, the geochemical and stratigraphic data are presented elsewhere (Keller et al, in prep.; Keller, in preparation). The second part will discuss the nature of stepwise extinctions both in terms of species extinctions and faunal abundance changes and explore questions of cause and effect between stepwise extinctions and impact events.

MULTIPLE IMPACT EVENTS: THE CONTROVERSY

The recent discovery of at least three impact events in deep-sea sediments in the form of microtektites by Keller, D'Hondt and Vallier (1983) as well as anomalously high iridium concentrations by Ganapathy (1982), Alvarez and others (1982) and Keller and others (in preparation), has again raised the spectre of catastrophic extinctions near the Eocene/Oligocene boundary. The closely spaced microtektite horizons suggest that the earth was in the path of a comet shower lasting about 1 to 2 million years (Hut et al, in preparation). Three impact craters with late Eocene K/Ar ages are known: Popigai, Siberia, USSR 39±9 Ma; Mistastin, Labrador, Canada 38±4 Ma; Wanapitei, Ontario, Canada 37±2 Ma (see Figure 1, Keller, this volume).

Glass and co-workers (1982, in press, this volume) believe, however, that only one late Eocene microtektite layer is present stratigraphically near the extinction of four radiolarian species at the Subzone *Calocyclas bandyca*/ *Cryptopora ornata* (b/c) boundary of *Thyrsocyrtis bromia* Zone and in the lower part of foraminiferal Zone P16. This microtektite layer corresponds to the uppermost of three layers recognized by Keller and co-workers (1983, in preparation). However, Glass and co-workers (1982) have also differentiated a second layer of clinopyroxene (cpx) bearing spherules about 25cm below the uppermost microtektite layer in the Caribbean Core RC9-58, which they ascribed to differential settings of the upper layer. They now consider this cpx-bearing spherule layer to represent a second impact events (Glass et al, in press, this volume). Because a larger percentage of these spherules contain clinopyroxene crystals (microlites) which are not part of the original definition of microtektites, they label them "cpx-bearing spherules" instead of microtektites. Nonetheless, they agree that cpx-bearing spherules, like microtektites, are impact derived and not a product of terrestrial volcanism; hence, with respect to multiple "microtektite layers" as evidence of multiple impact events, we are dealing first with a problem of semantics.

Tektite or microtektite glass differs from terrestrial volcanic glasses (obsidians) in its lower water content and other volatiles and its generally amorphous structure. Under the microscope obsidians have abundant microlites or microscopic crystals whereas tektites have "essentially none" (O'Keefe, 1976, p.1). We have found, however, that spherules with microlites are commonly found among spherules without microlites as for instance at Sites 315A, 69A and RC9-58 as also observed by Glass et al, in press. Moreover, the glassy matrix of spherules with microlites appears chemically undifferentiated from microtektites without microlites of the same layer (D'Hondt, Keller, and Stallard, in preparation). Therefore, the original definition of microtektites should be expanded to include all impact derived spherules, whether or not they contain microlites. Based on this revised definition of microtektites, the cpx-bearing spherules of Glass and others (1982) are included in our second late Eocene microtektite layer and impact event.

Keller and others (1983, in preparation) have found evidence of a still earlier third microtektite layer, located stratigraphically in the upper foraminiferal Zone P15 of Site 292, Core 38-2 (75cm) (Figure 1). Scattered microtektites were originally reported by Glass and Crosbie (1982) in Site 292, Core 36-4 and 36-2 of Zone P16 and considered as part of the North American strewn field (our uppermost microtektite layer); but now they consider the Core 38-2 layer as part of the North American strewn field and the Core 36 microtektites as upward reworking. This interpretation has at least two severe and fatal problems, one stratigraphic and one geochemical.

Stratigraphically, Core 38-2 of Site 292 is in the upper part of foraminiferal Zone P15 whereas Core 36 is within the lower part of Zone P16. Moreover, the two microtektite occurrences are nearly 20 m apart; hence reworking would have to come from a nearby eroded exposure of the lower microtektite layer, but without depositing the characteristic Zone P15 fauna associated with this layer - a practical improbability. Furthermore, radiolarian stratigraphy also places the Core 36-4 microtektites as coeval with the uppermost Caribbean layer at the Subzone *C. bandyca*/*C. ornata* boundary of *T. bromia* Zone (Figure 1). Thus, neither foraminiferal nor radiolarian stratigraphy is in agreement with Glass and co-workers' assumption that Core 38-2 of Site 292 is coeval with their Caribbean microtektite layer.

Based on sediment accumulation rates, using the Berggren et al (1985) time scale, the Core 38-2 layer is 38.25 m.y. old and the scattered microtektites in Core 36-4 and 36-2 are 37.30 and 37.25 m.y. old respectively (Figure 2), similar to the ages of the upper two layers in the Caribbean and equatorial Pacific cores. Thus, stratigraphically the microtektite layer of Core 38-2 cannot be coeval with either of the two Caribbean microtektite layers, but Core 36 is coeval. The fact that only scattered microtektites were found in Core 36 could be due to distance from source of impact, reworking, or removal by bottom currents which is common at this stratigraphic interval (Keller, in preparation).

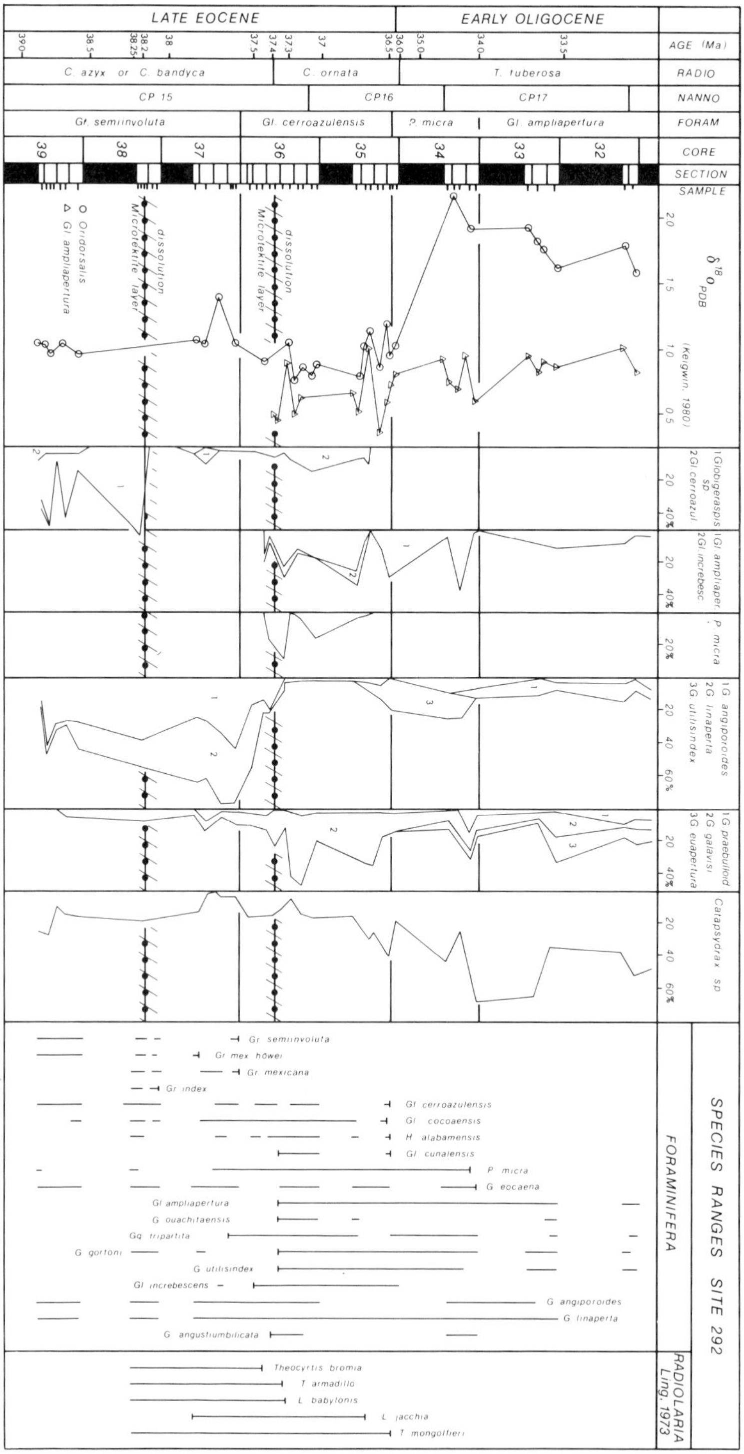
LATE EOCENE
EARLY OLIGOCENE
AGE (Ma)
RADIO
NANNO
FORAM
CORE
SECTION
SAMPLE
C. azyx or C. bandyca
C. ornata
T. tuberosa
CP 15
CP16
CP17
Gr. semiinvoluta
Gl. cerroazulensis
P. micra
Gl. ampliapertura
dissolution
Microtektite layer
O. O. dorsalis
Δ Gl. ampliapertura
(Keigwin, 1980)
1 Globigeraspis sp
2 Gl. cerroazul
1 Gl. ampliaper.
2 Gl. increbesc.
P. micra
1 G. angiporoides
2 G. linaperta
3 G. utilisindex
1 G. praebulloid
2 G. galavisi
3 G. euapertura
Catapsydrax sp
SPECIES RANGES SITE 292
FORAMINIFERA
Gr. semiinvoluta
Gr. mex. höwei
Gr. mexicana
Gr. index
Gl. cerroazulensis
Gl. cocoaensis
H. alabamensis
Gl. cunialensis
P. micra
G. eocaena
Gl. ampliapertura
G. ouachitaensis
Gq. tripartita
G. gortoni
G. utilisindex
Gl. increbescens
G. angiporoides
G. linaperta
G. angustiumbilicata
RADIOLARIA
Ling, 1973
Theocyrtis bromia
T. armadillo
L. babylonis
L. jacchia
T. mongolfieri

Fig. 1. Selected foraminifer and radiolaria species ranges, relative species abundance of dominant species and oxygen isotope data of planktonic and benthonic foraminiferal species of Site 292. Black intervals in core-section indicate no core recovery. Microtektite horizons marked by black dots and intervals of increased carbonate dissolution marked by diagonal lines.

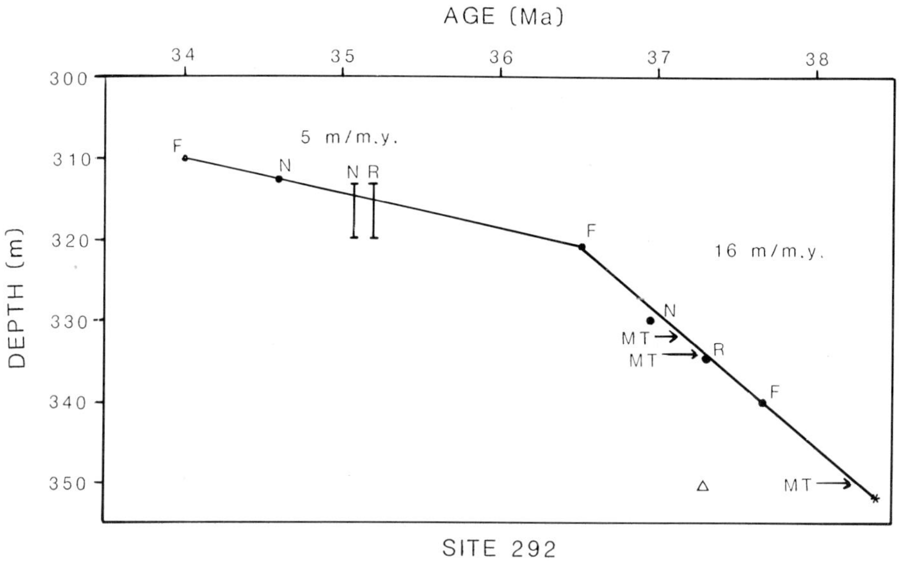

Fig. 2. Sediment accumulation curve of site 292 based on foraminiferal (F), nannoplankton (N) and radiolarian (R) datum events. Time scale after Berggren et al (1985). Asterisk marks last abundance peak of *Globigerapsis*. Location of the Core 38-2 microtektite layer (MT) and scattered microtektites of Core 36-4 and 36-2 indicated by arrows. Triangle shows that stratigraphically the Core 38-2 microtektite layer is not equivalent in age to the North American strewn field as claimed by Glass et al (this volume, in press).

Recent geochemical analysis of Site 292 microtektites as well as microtektites from many other sites clearly show Core 38-2 microtektites to be significantly higher in CaO and MgO than those of the uppermost Caribbean layer (D'Hondt, Keller and Stallard, in preparation). Thus, stratigraphic and geochemical analyses show Site 292 Core 38-2 microtektites to be both older and geochemically different from the upper and younger two microtektite layers.

In summary, the Glass-Keller controversy boils down to: (a) semantics, i.e. whether to arbitrarily define microtektites as spherules without microlites even though it can be shown that otherwise similar spherules with and without microlites occur within the same layer; and (b) the existence of a third impact event in Zone P15. We agree, however, on the presence of two closely spaced impact events in Zone P16.

STEPWISE MASS EXTINCTIONS

Contrary to repeated claims of catastrophic or mass extinctions at the Eocene/Oligocene boundary, this faunal turnover occurs over a period of 3-4 million years beginning in the late Eocene and culminating in the early Oligocene. Corliss and others (1984) have shown that during this time an average of 2 to 4 (6-12%) species become extinct per million years and an equal number of species originate among three microfossil groups (benthic and planktonic foraminifers, nannofossils). An equally low number of species extinctions and originations is observed in radiolarians (Saunders et al, 1985; Sanfilippo et al, 1985). This represents a maximum of 6-12% species extinct in marine plankton every 1 m.y. interval and can hardly be construed as mass extinction.

Close examination of the species record indicates that successive periods of extinctions occur abruptly during short stratigraphic intervals creating a stepwise extinction effect (Figure 3) and steps are separated by relatively stable or quiescent periods (Keller, 1983, in prep.). Each step is marked by extinction of 3-5 species (< 15%) usually associated with one to several originations over a few tens to a few 100,000 years and represents an accelerated faunal turnover. Such accelerated stepwise extinctions occur near the Truncorotaloides rohri/Globigerapsis semiinvoluta, Gr. semiinvoluta/Globorotalia cerroazulensis, Gl. cerroazulensis/Pseudohastigerina micra, and the Gl. ampliapertura/Gl. opima Zone boundaries (Figure 3). In addition, a stepwise extinction occurs in the late Gr. semiinvoluta Zone at 38.3 Ma (Figure 3). Although the number of species extinct at each stepwise extinction event is less than 15%, the sum total of the late Eocene stepwise extinctions over a 3.4 million year interval (40.0-36.6 Ma) results in a near complete faunal turnover with only about 20% of the species surviving into the Oligocene where they become extinct in the early Gl. opima Zone (Figure 3).

When population dynamics in terms of relative numbers of individuals per species is considered, the faunal changes at each stepwise extinction event appear more drastic. Planktonic foraminiferal populations are usually dominated by about 6 to 7 species, or about 20% of the species population, yet they comprise about 80-90% of the individual members of the population (Figure 1). Hence, 80% of the species make up only 10-20% of the individual members of the total population. Extinction of several of these rare species will have a minor impact on the total population as their combined species abundances are not likely to exceed 2-10%. Extinction of one or more of the dominant species,

---

Fig. 3. Microplankton zonations, datum events, microtektite horizons, stepwise mass extinctions and hiatuses of the late Eocene to early Oligocene. Asterisk mark last abundance peak of foraminiferal species. Time scale after Berggren et al (1985).

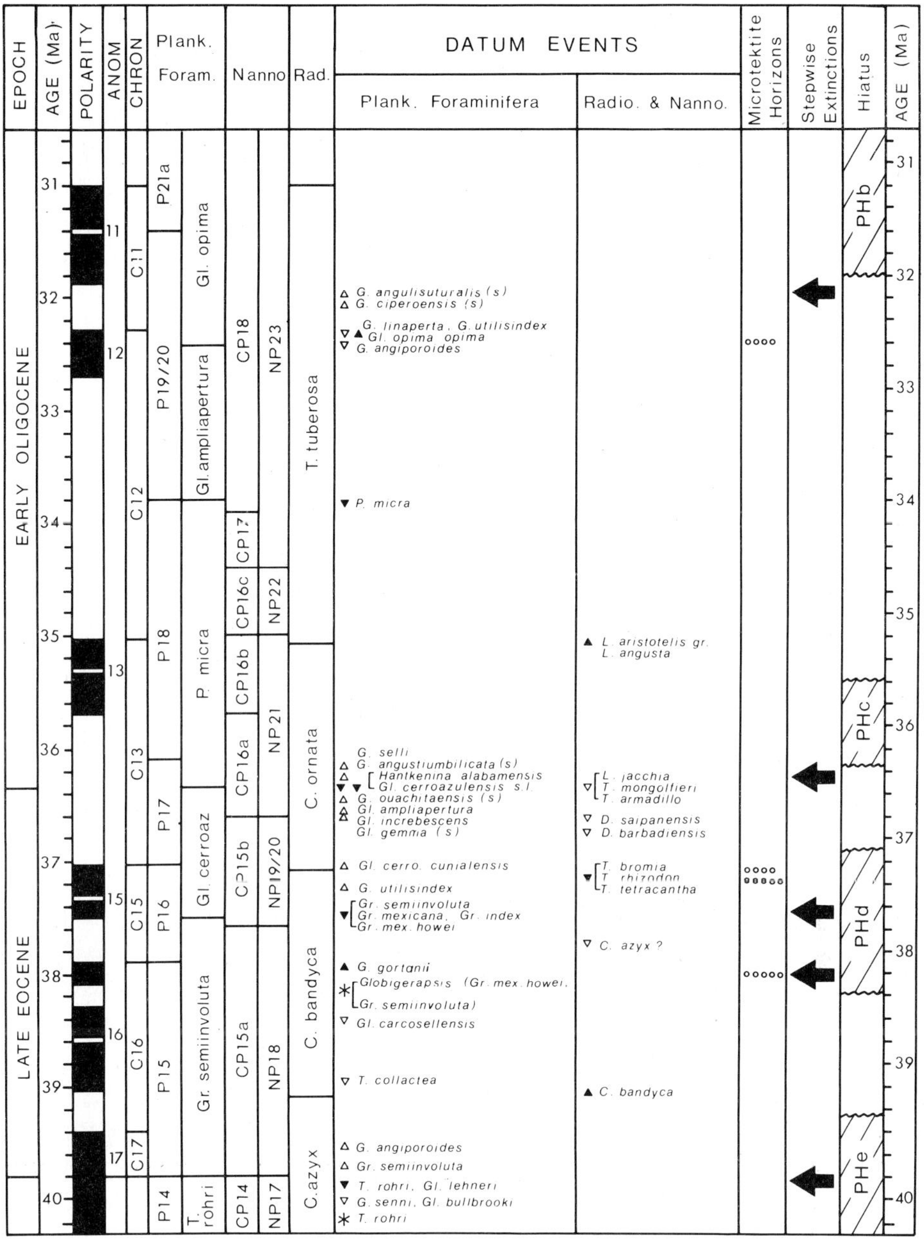
EPOCH
AGE (Ma)
POLARITY
ANOM
CHRON
Plank. Foram.
Nanno
Rad.
DATUM EVENTS
Plank. Foraminifera
Radio. & Nanno.
Microtektite Horizons
Stepwise Extinctions
Hiatus
AGE (Ma)
EARLY OLIGOCENE
LATE EOCENE
31
32
33
34
35
36
37
38
39
40
11
12
13
15
16
17
C11
C12
C13
C15
C16
C17
P21a
P19/20
P18
P17
P16
P15
P14
Gl. opima
Gl. ampliapertura
P. micra
Gl. cerroaz
Gr. semiinvoluta
T. rohri
CP18
CP17
CP16c
CP16b
CP16a
CP15b
CP15a
CP14
NP23
NP22
NP21
NP19/20
NP18
NP17
T. tuberosa
C. ornata
C. bandyca
C.azyx
G. angulisuturalis (s)
G. ciperoensis (s)
G. linaperta, G. utilisindex
Gl. opima opima
G. angiporoides
P. micra
G. selli
G. angustiumbilicata (s)
Hantkenina alabamensis
Gl. cerroazulensis s.l.
G. ouachitaensis (s)
Gl. ampliapertura
Gl. increbescens
Gl. gemma (s)
Gl. cerro. cunialensis
G. utilisindex
Gr. semiinvoluta
Gr. mexicana, Gr. index
Gr. mex. howei
G. gortanii
Globigerapsis (Gr. mex. howei,
Gr. semiinvoluta)
Gl. carcosellensis
T. collactea
G. angiporoides
Gr. semiinvoluta
T. rohri, Gl. lehneri
G. senni, Gl. bullbrooki
T. rohri
L. aristotelis gr.
L. angusta
L. jacchia
T. mongolfieri
T. armadillo
D. saipanensis
D. barbadiensis
T. bromia
T. rhizodon
T. tetracantha
C. azyx ?
C. bandyca
PHb
PHc
PHd
PHe

however, will have a major effect as they may comprise 50% or more of the total population.

Species abundance plots of Site 292 (Figure 1) illustrate that stepwise extinctions affecting major population shifts occurred near times of impact events. A cause-effect relationship is suggested, however, between the Gr. semiinvoluta (P15) Zone impact event and the terminal population decline of the Globigerapsis group. It must be noted, however, that the extinction record may be obscured by carbonate dissolution at both times of impacts (for further discussion see Keller, in preparation).

DISCUSSION

Are these periods of accelerated faunal turnovers, or stepwise mass extinctions really periodic occurring every 26 million years as Raup and Sepkoski (1984) claim? The Cenozoic record does not show a periodicity in the range of 26-32 m.y. Population changes in the magnitude of the late Eocene stepwise mass extinctions occur in the early Miocene (19-22 Ma) early late Miocene (10-12 Ma) and late Oligocene (28-32 Ma) (Keller, 1981, 1983a) and quantitative faunal analysis will undoubtedly discover further stepwise extinction events.

Can we invoke impacts as causal effects to stepwise mass extinctions? Based on analysis of 16 late Eocene marine sections the answer is a qualified no for the 2 closely spaced upper impact events, and a qualified yes for the lower impact event (Keller, in prep.). The answer is qualified because there is some ambiguity in the sections studied due to $CaCO_3$ solution at the microtektite horizons. At the upper two microtektite layers both species extinctions and species abundance declines appear to antedate the late Eocene impact events at 37.3 and 37.2 Ma. A cause-effect relationship is suggested, however, between the Zone P15 impact event and the terminal population decline of the Globigerapsis group. The record may be obscured however by solution of species. There appears to be a two fold $CaCO_3$ solution profile, one oscillating between 70-90% $CaCO_3$ affects primarily solution prone species; the second in the range of 0-40% $CaCO_3$ is associated with microtektite horizons (Keller et al, in prep.). The latter removes nearly all species and may be due to a shortlived lowered productivity at the time of the impact. Carbonate solution may be due to lower fertility and/or increased production of young corrosive bottom water during global cooling.

However, even if it could be shown that two of the late Eocene stepwise extinctions were caused by impacts, another causal mechanism would still by necessary for the stepwise extinctions at the middle/late Eocene boundary for which no evidence of an impact has been found to date.

Although it appears that impact events may have accelerated population changes already in progress, and in some cases caused the demise of species populations, the ultimate cause of the late Eocene stepwise mass extinctions must be looked for in the long range paleoclimatic, paleooceanographic and tectonic changes of the middle Eocene to early Oligocene. These changes are most closely related to the development of the circum-Antarctic circulation, subsequent cooling of Antarctica followed by production of Antarctic bottom water which ultimately led to the permanent bottom water temperature drop at the Eocene/Oligocene boundary.

REFERENCES

Alvarez, W., Asaro, F., Michel, H.V. and Alvarez, L.W., 1982. Iridium anomaly approximately synchronous with terminal Eocene extinctions. Science, 216: 886-888.

Berggren, W.A., Kent, D.V. and Flynn, J.J., 1985. Paleogene geochronology and chronostratigraphy. In: Snelling, N.J. (ed.), Geochronology and the Geological Record. Geol. Soc. London Spec. Paper.

Corliss, B.H., Aubry, M.-P., Berggren,W.A., Fenner, J.M., Keigwin, L.D. Jr. and Keller, G., 1984. The Eocene/Oligocene Boundary Event in the Deep Sea. Science, 226: 806-810.

Corliss, B.H. and Keigwin, L.D. Jr., in press. Eocene-Oligocene Paleooceanography.

Ganapathy, R., 1982. Evidence for a major meteorite impact on the earth 34 million years ago: Implications for Eocene extinctions. Science, 216: 885-886.

Glass, B.P. and Crosbie, J.R., 1982. Age of Eocene/Oligocene boundary based on extrapolation from North American Microtektite layer. Amer. Assoc. Pet. Geol. Bull., 66(4): 471-476.

Glass, B.P., DuBois, D.L. and Ganapathy, R., 1982. Relationship between an Iridium anomaly and the North American microtektite layer in Core RC9-58 from the Caribbean Sea. Proceedings of the 13th Lunar and Planetary Science Conference, Part I, J. Geophys. Res. Vol. 87, supplement, p. A425-A428, Nov. 15, 1982.

Glass, B.P., Burns, C.A., Crosbie, J.R. and DuBois, D.L., in press. Late Eocene North American Microtektites and clinopyroxene-bearing spherules. Proceedings of the 16th Lunar and Planetary Science Conference, J. Geophys. Res.

Grieve, R.A.F., 1982. The record of impact on Earth: Implications for a major Cretaceous/Tertiary impact event. In: L.T. Silver and P.H. Schultz (eds.) Geological Implications of Impacts of Large Asteroids and Comets on the Earth. Geol. Soc. Amer., Spec. Paper 190, p. 25-38.

Haq, B.U. and Lohmann, G.P., 1976. Early Cenozoic nannoplankton biogeography of the Atlantic Ocean. Mar. Micropaleontol., 1: 119-194.

Hut, P., Kauffman, E.G., Alvarez, W., Weissman, P. and Keller, G., in prep. Constraints on the duration of Comet Storms as possible causes of paleontological stepwise mass extinctions. Science.

Keigwin, L.D. Jr., 1980. Paleooceanographic change in the Pacific at the Eocene Oligocene boundary. Nature, 287(5784): 722-725.

Keller, G., 1981. Miocene biochronology and paleooceanography of the North Pacific. Marine Micropaleontology, 6: 535-551.

Keller, G., 1983a. Biochronology and paleoclimatic implications of middle Eocene to Oligocene planktic foraminiferal faunas. Marine Micropaleontology, 7: 464-486.

Keller, G., 1983b. Paleoclimatic analysis of middle Eocene through Oligocene planktic foraminiferal faunas. Paleoclimatol., Paleoecol., Paleooceanogr., 43: 73-94.

Keller, G., D'Hondt, S. and Vallier, T.L., 1983. Multiple Microtektite Horizons in upper Eocene Marine Sediments: No evidence for Mass Extinctions. Science, 221: 150-152.
Keller, G., D'Hondt, S., Keigwin, L.D., Onstott, T., Orth, C. and Shoemaker, E., in prep. Age and Geochemistry of Late Eocene Microtektite Horizons.
Keller, G., in prep. Stepwise Mass Extinctions and Impact Events: Late Eocene to Early Oligocene. Marine Micropaleontology.
Kennett, J.P., 1977. Cenozoic Evolution of Antarctic Glaciation, the Circum-Antarctic Ocean, and their Impact on Global Paleooceanography. J. Geophys. Res. 82(27): 3843-3860.
Kennett, J.P., 1978. The development of planktonic biogeography in the southern ocean during the Cenozoic. Marine Micropaleontology, 3: 301-345.
Ling, H.Y., 1973. Radiolaria: Leg 31 of the Deep Sea Drilling Project. In: Initial Reports of the Deep Sea Drilling Project, 31: 655-676.
O'Keefe, J.A., 1976. Tektites and their Origin. Elsevier Scientific Publishing Company, 254 p.
Raup, D and Sepkoski, J., 1984. Periodicity of extinctions in the Geologic Past. Proc. Natl. Acad. Sci. U.S.A., v. 81 : 801-805.
Sanfilippo, A., Riedel, W.R., Glass, B.P. and Kyte, F.T., 1985. Late Eocene microtektites and radiolarian extinctions on Barbados. Nature 314 :613/615
Saunders J.B., Bernoulli,D., Muller-Merz, E., Oberhansli, H., Perch-Nielsen, K., Riedel, W.R., Sanfilippo, A. and Torrini, R. Jr., 1985. Stratigraphy of the Late Middle Eocene to Early Oligocene in the Bath Cliff Section, Barbados, West Indies. Micropaleontology, 30 : 390-425.

ADDITIONAL LIST OF ADDRESSES

1 - AKHMETIEV M.A. : Geol. Inst., Acad. Sc., Pyzhevsky, 7, 109017 Moscow Zh - 17 (USSR).

2 - BERGGREN W.A. : Dept of Geol - W H O I, Woodshole, Ma 02543 (USA) and Dept. of Geol. Brown University, Providence. R.I. 02912 (USA).

3 - BORISOV B.A. : Geol. Inst. of the Ministry of Geol., Leningrad (USSR).

4 - CHATEAUNEUF J.J. : BRGM, BP 6009, 45060 Orleans (France).

5 - DASHZEVEG D. : Geol. Inst., Ac. Sc., Ulan Bator (Mongolia).

6 - DEVYATKIN E.V. : Geol. Inst. Acad. Sc., Pyzhevsky, 7, 109017 Moscow Zh - 17 (USSR).

7 - DIAZ de GAMERO M. : Escuela de Geologia, Universidad Apartado 47350, Caracas 1041 (Venezuela).

8 - EROFEEV V.S. : Geol. Inst. of the Kazatch SSR, Alma Ata (URSS).

9 - FENNER J. : Geol. Inst. Universitad, Olshausenstr. 40.60, D 2300 Kiel (F.R.G.).

10 - GRAMANN F. : Niedersächsisches Landesamt für Bodenforschung Stilleweg 2, D 3000 Hannover 51 (F.R.G.).

11 - GRIGORIAN S.M. : Geol. Inst., Ac. Sc. Yerevan, Armenian SSR (USSR).

12 - HAGN H. : Inst. für Paläontologie und Historische Geologie, Abt. für Mikropaläontologie, Richard Wagner Str. 10/11, D 8000 München (F.R.G.).

13 - JENKINS D.G. : Dept. of Earth Sciences. The Open University, Milton Keynes MK 7 6 AA (G.B.).

14 - KRASHENINNIKOV V.A. : Geol. Inst., Acad. Sc., Pyzhevsky, 7, 109017 Moscos Zh - 17 (USSR).

15 - MARTIROSIAN Y.A. : Geol. Inst., Ac. Sc. Yerevan, Armenian SSR (USSR).

16 - MÜLLER-MERZ E. : Natural History Museum, Basel, CH 4001 (Switzerland)

17 - PTUCHIAN A.E. : General Establish. of Geol. of the Armenian SSR, Yerevan (USSR).

18 - RIEDEL W.R. : Scripps Institution, La Jolla, Ca 92093 (USA).

19 - RITZKOWSKI S. : Inst. und Museum für Geologie und Paläontologie, Geologisches Landesamt Rheinland-Pfalz, Emmeransstr. 36, D 6500 Mains (F.R.G.).

20 - RUSSELL D.E. : Inst. de Paléontologie, Museum, 8, rue Buffon, 75005 Paris (France).

21 - SANFILIPPO A. : Scripps Institution, La Jolla, Ca 92093 (USA).

22 - SAUNDERS J.B. : Natural History Museum, Basel, CH 4001 (Switzerland).

23 - SEROVA M.Y. : Geol. Inst.,Acad. Sc., Pyzhevsky, 7, 109017 Moscow Zh - 17 (USSR).

24 - SONNE V. : Geologisches Landesamt Rheinland-Pfalz, Emmeransstr. 36
D 6500 Mainz (F.R.G.)

25 - TOBIEN H. : Inst. fur Geowissenschten, Gutenberg Univ., Postfach 3980, 6500 Mainz (F.R.G.)

26 - TSEKHOVSKY V.G. : Geol. Inst., Acad. Sc., Pyzhevsky, 7,109017,Moscow Zh - I7 (USSR).

27 - ZAPOROZHETS N.I. : Geol. Inst., Acad. Sc., Pyzhevsky,7,109017,Moscow Zh - 17 (USSR).